Analytic Perturbation Theory and Its Applications

Konstantin E. Avrachenkov
Inria Sophia Antipolis
Sophia Antipolis, France

Jerzy A. Filar
Flinders University
Adelaide, Australia

Phil G. Howlett
University of South Australia
Adelaide, Australia

Analytic Perturbation Theory and Its Applications

Society for Industrial and Applied Mathematics
Philadelphia

10 9 8 7 6 5 4 3 2 1

Library of Congress Cataloging-in-Publication Data

Avrachenkov, Konstantin, author.
Analytic perturbation theory and its applications / Konstantin E. Avrachenkov, Inria Sophia Antipolis, Sophia Antipolis, France, Jerzy A. Filar, Flinders University, Adelaide, Australia, Phil G. Howlett, University of South Australia, Adelaide, Australia.
pages cm
Includes bibliographical references and index.
ISBN 978-1-611973-13-6
1. Perturbation (Mathematics) I. Filar, Jerzy A., 1949- author. II. Howlett, P. G. (Philip G.), 1944- author. III. Title.
QA871.A97 2013
515'.392--dc23

2013033335

To our students, who, we believe,
will advance this topic far beyond
what is reported here.
Though they may not realize it,
we learned from them at least as much
as they learned from us.

Contents

Preface

We live in an era in which ever more complex phenomena (e.g., climate change dynamics, stock markets, complex logistics, and the Internet) are being described with the help of mathematical models, frequently referred to as systems. These systems typically depend on one or more parameters that are assigned nominal values based on the current understanding of the phenomena. Since, usually, these nominal values are only estimates, it is important to know how deviations from these values affect the solutions of the system and, in particular, whether for some of these parameters even small deviations from nominal values can have a big impact.

Naturally, it is crucially important to understand the underlying causes and nature of these big impacts and to do so for neighborhoods of multiparameter configurations. Unfortunately, in their most general settings, multiparameter deviations are still too complex to analyze fully, and even single-parameter deviations pose significant technical challenges. Nonetheless, the latter constitute a natural starting point, especially since in recent years much progress has been made in analyzing the asymptotic behavior of these single-parameter deviations in many special settings arising in the sciences, engineering, and economics.

Consequently, in this book we consider systems that can be disturbed, to a varying degree, by changing the value of a single perturbation parameter loosely referred to as the "perturbation." Since in most applications such a perturbation would be small but unknown, a fundamental issue that needs to be understood is the behavior of the solutions as the perturbation tends to zero. This issue is important because for many of the most interesting applications there is, roughly speaking, a discontinuity at the limit, which complicates the analysis. These are the so-called singularly perturbed problems.

Put a little more precisely, the book analyzes—in a unified way—the general linear and nonlinear systems of algebraic equations that depend on a small perturbation parameter. The perturbation is analytic; that is, left-hand sides of the perturbed equations can be expanded as a power series of the perturbation parameter. However, the solutions may have more complicated expansions such as Laurent or even Puiseux series. These series expansions form a basis for the asymptotic analysis (as the perturbation tends to zero). The analysis is then applied to a wide range of problems including Markov processes, constrained optimization, and linear operators on Hilbert and Banach spaces. The recurrent common themes in the analyses presented is the use of fundamental equations, series expansions, and the appropriate partitioning of the domain and range spaces.

We would like to gratefully acknowledge most valuable contributions from many colleagues and students including Amie Albrecht, Eitan Altman, Vladimir Ejov, Vladimir Gaitsgory, Moshe Haviv, Jean-Bernard Lasserre, Nelly Litvak, (the late) Charles Pearce, and Jago Korf. Similarly, the institutions where we have worked during the long period of writing, University of South Australia, Inria, and Flinders University, have also generously supported this effort. Finally, many of the analyses reported here were carried out as parts of Discovery and International Linkage grants from the Australian Research Council.

Konstantin E. Avrachenkov, Jerzy A. Filar, and Phil G. Howlett

Chapter 1

Introduction and Motivation

1.1 ▪ Background

In a vast majority of applications of mathematics, systems of governing equations include parameters that are assumed to have known values. Of course, in practice, these values may be known only up to a certain level of accuracy. Hence, it is essential to understand how deviations from their nominal values may affect solutions of these governing equations. Naturally, there is a desire to study the effect of all possible deviations. However, in its most general setting, this is a formidable challenge, and hence structural assumptions are usually required if strong, constructive results are to be explicitly derived.

Frequently, parameters of interest will be coefficients of a matrix. Therefore, it is natural to begin investigations by analyzing matrices with perturbed elements. Historically, there was a lot of interest in understanding how such perturbations affect key properties of the matrix. For instance, how will the eigenvalues and eigenvectors of this matrix be affected?

Perhaps the first comprehensive set of answers was supplied in the, now classical, treatise of Kato [99]. Indeed, Kato's treatment was more general and covered the analysis of linear operators as well as matrices. However, Kato [99] and a majority of other researchers have concentrated their effort on the perturbation analysis of the eigenvalue problem.

In this book we shall study a range of problems that is more general than spectral analysis. In particular, we will be interested in the behavior of solutions to perturbed linear and polynomial systems of equations, perturbed mathematical programming problems, perturbed Markov chains and Markov decision processes, and some corresponding extensions to operators in Hilbert and Banach spaces.

In the same spirit as Kato, we focus on the case of analytic perturbations. The latter have the structural form where the perturbed data specifying the problem can be expanded as a power series in terms of first, second, and higher orders of deviations multiplied by corresponding powers of an auxiliary perturbation variable. When that variable tends to zero the perturbation dissipates and the problem reduces to the original, unperturbed, problem. Nonetheless, the same need not be true of the solutions that are of most interest to the researchers studying the system. These can exhibit complex behaviors that involve discontinuities, singularities, and branching.

Indeed, since the 1960s researchers in various disciplines have studied particular manifestations of the complex behavior of solutions to many important problems.

For instance, perturbed mathematical programs were studied by Pervozvanski and Gaitsgori [126], and the study of perturbed Markov chains was, perhaps, formally initiated by Schweitzer [137]. It is this, not uncommon, complexity of the limiting behavior of solutions that stimulated the present book.

1.2 ▪ Raison d'Être and Exclusions

Imagine that the perturbed matrix mentioned in the previous section had the form

$$\tilde{A} = A + D, \tag{1.1}$$

where A is a matrix of nominal coefficient values, $\tilde{A}$ is a matrix of perturbed data, and D is the perturbation itself. There are numerous publications devoted to this subject (see, e.g., the books by Stewart and Sun [147] and Konstantinov et al. [103] and the survey by Higham [80]). However, without any further structural assumptions on D, asymptotic analysis as the norm of D tends to zero is typically only possible when the rank of the perturbed matrix $\tilde{A}$ is the same as the rank of A. Roughly speaking, this corresponds to the case of what we later define to be a *regular perturbation*. Generally, in such a case solutions of the perturbed problem tend to solutions of the original unperturbed problem.

In this book we wish to explain some of the complex asymptotic behavior of solutions such as discontinuity, singularity, and branching. Typically, this arises when the rank of the perturbed matrix $\tilde{A}$ is different from the rank of A. For instance, consider the simple system of linear equations

$$\tilde{A}x = \begin{bmatrix} 1 & 1 \\ 1+\varepsilon & 1+2\varepsilon \end{bmatrix} \begin{bmatrix} x_1 \\ x_2 \end{bmatrix} = \begin{bmatrix} 1 \\ 0 \end{bmatrix}. \tag{1.2}$$

Clearly, $\tilde{A}$ is of the form (1.1) since we can write

$$\tilde{A} = A + D = \begin{bmatrix} 1 & 1 \\ 1 & 1 \end{bmatrix} + \begin{bmatrix} 0 & 0 \\ \varepsilon & 2\varepsilon \end{bmatrix}.$$

Now, for any $\varepsilon \neq 0$, the inverse of $\tilde{A}$ exists and can be written as

$$\tilde{A}^{-1} = \frac{1}{\varepsilon}\begin{bmatrix} 1+2\varepsilon & -1 \\ -1-\varepsilon & 1 \end{bmatrix} = \frac{1}{\varepsilon}\begin{bmatrix} 1 & -1 \\ -1 & 1 \end{bmatrix} + \begin{bmatrix} 2 & 0 \\ -1 & 0 \end{bmatrix}.$$

Hence, the unique solution of (1.2) has the form of *Laurent series*

$$\tilde{x} = \frac{1}{\varepsilon}\begin{bmatrix} 1 \\ -1 \end{bmatrix} + \begin{bmatrix} 2 \\ -1 \end{bmatrix}.$$

Despite the fact that the norm of D tends to 0 as $\varepsilon \to 0$, we see that $\tilde{x}$ diverges. The singular part of the Laurent series indicates the direction along which $\tilde{x}$ diverges as $\varepsilon \to 0$.

The above example indicates that a singularity manifests itself in the series expansion of a solution. This phenomenon is common in a wide range of interesting mathematical and applied problems and lends itself to rigorous analysis if we impose the additional assumption that the perturbed matrix is of the form

$$A(\varepsilon) = A_0 + \varepsilon A_1 + \varepsilon^2 A_2 + \dots, \tag{1.3}$$

where the above power series is assumed to be convergent in some neighborhood of $\varepsilon = 0$. Hence it is natural to call this particular type of perturbation an *analytic perturbation*.

Consequently, it is also natural to consider a *singular perturbation* to be one where solutions to the perturbed problem are not analytic functions with respect to the perturbation parameter ε.

It will be seen that with the above analytic perturbation assumption, a unified treatment of both the regular and singular perturbations is possible. Indeed, the approach we propose has been inspired by Kato's systematic analysis of the perturbed spectrum problem but applied to a much wider class of problems. Thus, while Kato's motivating problem is captured by the eigenvalue equation

$$A(\varepsilon)x(\varepsilon)=\lambda(\varepsilon)x(\varepsilon), \tag{1.4}$$

our motivating problem is the asymptotic behavior of solutions to the perturbed system of equations

$$f(x,\varepsilon)=0,$$

where $f(x,\varepsilon)$ can be a system of linear or polynomial equations. In the linear case this reduces to

$$L(\varepsilon)x(\varepsilon)=c(\varepsilon).$$

In particular, if $L(\varepsilon)$ has an inverse for $\varepsilon \neq 0$, and sufficiently small, then we investigate the properties of the perturbed inverse operator $L^{-1}(\varepsilon)$ (or matrix-valued function $A^{-1}(\varepsilon)$ in the finite dimensional case). For example, we rely on the fact that $A^{-1}(\varepsilon)$ can always be expanded as a Laurent series

$$A^{-1}(\varepsilon)=\frac{1}{\varepsilon^s}B_{-s}+\cdots+\frac{1}{\varepsilon}B_{-1}+B_0+\varepsilon B_1+\dots . \tag{1.5}$$

The preceding system equation $f(x,\varepsilon)=0$ arises as a building block of solutions to many practical problems. In particular, there is an enormous number of problems that are formulated as either linear or nonlinear mathematical programs. Hence a fundamental question that arises concerns the stability (or instability) of a solution when the problem is slightly perturbed.

Perhaps surprisingly, this can be a very difficult question. Even in the simplest case of linear programming, standard Operations Research textbooks discuss only the most straightforward cases and scrupulously avoid the general issue of how to analyze the effect of a perturbation when the whole coefficient matrix is also affected.

The next example (taken from [126]) illustrates that even in the "trivial" case of linear programming the effect of a small perturbation can be "nontrivial." Consider the simple optimization problem in two variables

$$\max_{x_1,x_2} \; x_2$$

$$\text{s.t.} \quad \begin{array}{rcrcl} x_1 & + & x_2 & = & 1, \\ (1+\varepsilon)x_1 & + & (1+2\varepsilon)x_2 & = & 1+\varepsilon, \end{array}$$

$$x_1 \geq 0, \quad x_2 \geq 0.$$

It is clear that for any $\varepsilon > 0$ there is a unique (and hence optimal) feasible solution at $x_1^* = 1$, $x_2^* = 0$. However, when $\varepsilon = 0$, the two equality constraints coincide, the set of feasible solutions becomes infinite, and the maximum is attained at $\hat{x}_1 = 0$, $\hat{x}_2 = 1$.

More generally, techniques developed in this book permit us to describe the asymptotic behavior of solutions[1] to a generic, perturbed, mathematical program:

[1] The word *solution* is used in a broad sense at this stage. In some cases the solution will, indeed, be a global optimum, while in other cases it will be only a local optimum or a stationary point.

$$\begin{aligned}&\max f(x,\varepsilon)\\ \text{s.t.}\quad & g_i(x,\varepsilon)=0,\ i=1,\dots,m,\\ & h_j(x,\varepsilon)\le 0,\ j=1,\dots,p,\end{aligned} \tag{MP(ε)}$$

where $x \in \mathbb{R}^n, \varepsilon \in [0,\infty)$, and f, g_i's, h_j's are functions on $\mathbb{R}^n \times [0,\infty)$. We will be especially concerned with characterizing solutions, $x^*(\varepsilon)$, of (MP(ε)) as functions of the perturbation parameter, ε. This class of problems is closely related to the well-established topics of sensitivity or postoptimality, or parametric analysis of mathematical programs (see Bonnans and Shapiro [29]). However, our approach covers both the regularly and singularly perturbed problems and thereby resolves instances such as that illustrated in the above simple linear programming example.

Other important applications treated here include perturbed Markov chains and decision processes and their applications to Google PageRank and the Hamiltonian cycle problems.

Let us give an idea of applicability of the perturbation theory to the example of Google PageRank. PageRank is one of the principal criteria according to which Google sorts answers to a user's query. It is a centrality ranking on the directed graph of web pages and hyperlinks. Let A be an adjacency matrix of this graph. Namely, $a_{ij}=1$ if there is a hyperlink from page i to page j, and $a_{ij}=0$ otherwise. Let D be a diagonal matrix whose diagonal elements are equal to the out-degrees of the vertices. The matrix $L=D-A$ is called the graph Laplacian. If a page does not have outgoing hyperlinks, it is assumed that it points to all pages. Also, let v^T be a probability distribution vector which defines preferences of some group of users, and let ε be some regularization parameter. Then, PageRank can be defined by the following equation:

$$\pi = \varepsilon v^T[L+\varepsilon A]^{-1}D.$$

Since the graph Laplacian L has at least one zero eigenvalue, $L+\varepsilon A$ is a singular perturbation of L, and its inverse can be expressed in the form of Laurent series (1.5). This application is studied in detail in Chapter 6.

Consequently, the book is intended to bridge at least some of the gap between the theoretical perturbation analysis and areas of applications where perturbations arise naturally and cause difficulties in the interpretation of "solutions" which require rigorous and yet pragmatic resolution. To achieve this goal, the book is organized as an advanced textbook rather than a research monograph. In particular, a lot of expository material has been included to make the book as self-contained as practicable. In the next section, we outline a number of possible courses that can be taught on the basis of the material covered. Nonetheless, the book also contains sufficiently many new, or very recent, results to be of interest to researchers involved in the study of perturbed systems.

Finally, it must be acknowledged that a number of, clearly relevant, topics have been excluded so as to limit the scope of this text. These include the theories of perturbed ordinary and partial differential equations, stochastic diffusions, and perturbations of the spectrum. Most of these are well covered by several existing books such as Kato [99], Baumgärtel [22], O'Malley [125], Vasileva et al. [153], Kevorkian and Cole [102], and Verhulst [156]. Singular perturbations of Markov processes in continuous time are well covered in the book of Yin and Zhang [162]. Elementwise regular perturbations of matrices are extensively treated in the books of Stewart and Sun [147] and Konstantinov et al. [103].

Although the question of numerical computation is an extremely important aspect of perturbation analysis, we shall not undertake per se a systematic study of this topic.

We are well aware that the difference between an exact solution and a numerically computed solution is a prima facie case where perturbation theory may be used to define suitable error bounds. Nevertheless we do recommend that best practice should be used for all relevant numerical computations. This applies particularly to the numerical solution of any collection of key equations.

1.3 ▪ Organization of the Material

Since problems induced by perturbations manifest themselves in a variety of settings, some of which already led to established lines of research, the parts and chapters of this book are arranged so as to facilitate quick access to a wide range of results. The three main parts group chapters containing material related to (I) finite dimensional perturbations, (II) application of results in Part I to optimization and Markov processes, and (III) infinite dimensional perturbations.

The figure below displays some of the logical connections among various chapters. The solid arrows in the figure indicate that a significant part of the material in the chapter at the tail of the arrow is required for understanding the material in the chapter at the head of the arrow. On the other hand, the broken arrows indicate a weaker connection between the corresponding chapters. Indeed, it is possible to follow the material in chapters connected by the solid arrows without prior knowledge of the material in the remaining chapters. Since some readers will already have the requisite knowledge of functional analysis and operator theory, we chose not to precede Chapter 8 with these prerequisites. Instead, we included the latter, presented in a manner best suited to the contents of this book, in the final Chapter 9, which can also serve as a brief, stand-alone introduction to elements of functional analysis.

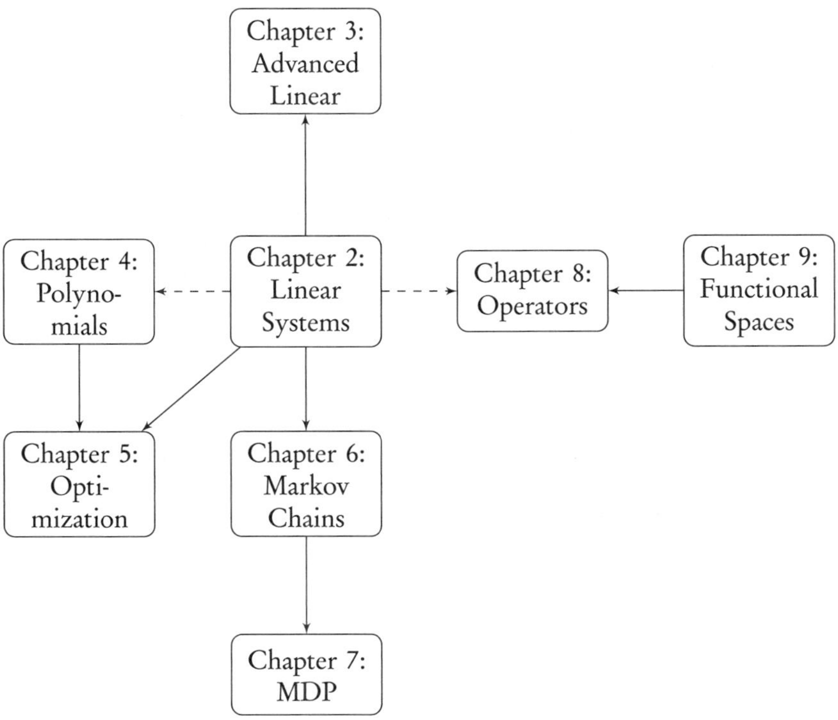

1.4 ▪ Possible Courses with Prerequisites

As mentioned earlier, in addition to the book's research mission, it can also serve as a textbook for at least the following courses.

1. A one-semester introductory course on perturbation theory of finite dimensional linear systems intended for advanced undergraduates or first year graduate students. This course could be based on Sections 2.1–3.2, Section 5.2, and Section 6.2. The only prerequisites for this course are standard undergraduate linear algebra and calculus courses.

2. A one-semester continuation course on perturbation theory intended for graduate students. This course would take the material covered in the preceding introductory course as assumed knowledge and would cover Section 3.3, Chapter 4, Section 5.4, and Sections 8.1–8.6. Prerequisites for this course include complex analysis and very basic functional analysis. In fact, Chapter 9 and Section 8.5 contain accessible review of the necessary material from Fourier and functional analysis.

3. A one-semester course on perturbation theory of Markov chains and Markov decision processes intended for graduate students. This course would cover the material of Chapters 6 and 7 and could be given as a continuation of any of the above listed courses, or it could be made self-contained if it began with Sections 2.1–2.2, possibly at the cost of omitting some of the later sections of Chapters 6 and 7. This course would require some knowledge of basic probability theory and Markov chains.

4. A one-semester course on perturbation theory in infinite dimensional spaces intended for graduate students. This course would cover the material of Chapters 8 and 9.

1.5 ▪ Future Directions

As with most branches of mathematics there is always more to be done. The interested researcher will clearly recognize that there are many opportunities for continuing the various lines of investigation outlined in this book. Below, we mention only a small sample of these.

1. There are many natural extensions to the multiparameter case.

2. Applications of infinite dimensional general results reported in Chapter 8 should be developed in a number of areas, including optimal control, signal processing, and stochastic processes.

3. Efficient numerical implementations for many of the techniques described here are yet to be devised. Much can be done in the way of numerical computation for many of the problems discussed here using standard mathematical packages such as those available within *Mathematica* and MATLAB. Nevertheless, there is much room for development of problem-specific programs that may or may not call on various standard subroutines from existing packages.

Part I

Finite Dimensional Perturbations

Although this may seem a paradox, all exact science is dominated by the idea of approximation.

—Bertrand Russell (1872–1970)

Chapter 2

Inversion of Analytically Perturbed Matrices

2.1 ▪ Introduction and Preliminaries

This chapter and the following one are devoted to a perturbation analysis of the algebraic finite dimensional linear system

$$A(z)x(z) = b(z), \tag{2.1}$$

where the matrix $A(z)$ depends analytically on the parameter z. Namely, $A(z)$ can be expanded as a power series

$$A(z) = A_0 + zA_1 + z^2A_2 + \dots$$

with some nonzero radius of convergence. Mostly in the exposition of the present chapter, z is a complex number and A_i is a matrix with complex elements. If we want to restrict our consideration to the real numbers, we shall use ε instead of z.

In this chapter we study the linear system (2.1) with a square coefficient matrix $A(z)$. (Systems with rectangular matrices $A(z)$ will be studied in Chapter 3.) In particular, we are interested in the case of *singular perturbations* when $A(0)$ is not invertible but $A(z)$ has an inverse for $z \neq 0$, but sufficiently small. We investigate the properties of the matrix-valued function $A^{-1}(z)$. For example, we provide several methods for expanding $A^{-1}(z)$ as a Laurent series:

$$A^{-1}(z) = \frac{1}{z^s}B_{-s} + \dots + \frac{1}{z}B_{-1} + B_0 + zB_1 + \dots. \tag{2.2}$$

The first method is based on the use of augmented block-Toeplitz type matrices and the Moore–Penrose generalized inverse. The second and third methods are based on reduction techniques which allow us to work with spaces of lower dimension. Then, we give specific methods for cases of a linear perturbation $A(z) = A_0 + zA_1$ and a polynomial perturbation $A(z) = A_0 + \dots + z^pA_p$.

It is easier to explain and to understand the techniques of perturbation theory in terms of matrix inversion. However, we are cognizant that numerical analysts would most likely consider algorithms in the context of the solution of a linear system rather than a matrix inversion. The matrix A^{-1} is simply the solution to the linear equation $AX = I$. In that sense calculation of A^{-1} is equivalent to solving the linear system.

Since the methods of this chapter are essentially based on the application of generalized inverse matrices, we briefly review the main definitions and facts from the theory of generalized inverses. The interested reader can find a more detailed discussion in references provided in the bibliographic notes.

There are several types of generalized inverses. The *Moore–Penrose generalized inverse* (or *Moore–Penrose pseudoinverse*) is by far the most commonly used generalized inverse. It can be defined in either geometric or algebraic terms. First we give a "geometric" definition.

Let $A \in \mathbb{C}^{m\times n}$ be the matrix of a linear transformation from $\mathbb{C}^n$ to $\mathbb{C}^m$. And let $N(A) \subseteq \mathbb{C}^n$ and $R(A) \subseteq \mathbb{C}^m$ denote the null space and the range space of this transformation, respectively. The space $\mathbb{C}^n$ can be represented as the direct sum $N(A) \oplus N(A)^\perp$ and the space $\mathbb{C}^m$ can be represented as the direct sum $R(A) \oplus R(A)^\perp$.

Definition 2.1. *The Moore–Penrose generalized inverse of the linear transformation* $A : \mathbb{C}^n \to \mathbb{C}^m$ *is a linear transformation* $A^\dagger : \mathbb{C}^m \to \mathbb{C}^n$ *defined in the following way. Let* $y \in \mathbb{C}^m$*, and write* $y = y_R + y_R^\perp$*, where* $y_R \in R(A)$ *and* $y_R^\perp \in R(A)^\perp$*. Choose* $x \in \mathbb{C}^n$ *such that* $Ax = y_R$*, and write* $x = x_N + x_N^\perp$*, where* $x_N \in N(A)$ *and* $x_N^\perp \in N(A)^\perp$*. Then* $A^\dagger y = x_N^\perp$*.*

Of course, the generalized inverse matrix is just the matrix representation of the corresponding generalized inverse transformation. Next we give an equivalent algebraic definition.

Definition 2.2. *If* $A \in \mathbb{C}^{m\times n}$*, then the Moore–Penrose generalized inverse (or pseudoinverse) is the matrix* $A^\dagger \in \mathbb{C}^{m\times n}$ *uniquely defined by the equations*

$$AA^\dagger A = A, \tag{2.3}$$

$$A^\dagger AA^\dagger = A^\dagger, \tag{2.4}$$

$$(AA^\dagger)^* = AA^\dagger, \tag{2.5}$$

$$(A^\dagger A)^* = A^\dagger A, \tag{2.6}$$

where $(\ )^*$ *denotes a conjugate transpose matrix.*

There are several methods for the computation of Moore–Penrose generalized inverses. The best known and, perhaps, the most computationally stable method is based on the singular value decomposition (SVD). Let $r = r(A)$ be the rank of $A \in \mathbb{C}^{m\times n}$. And let $D = \text{diag}\{\sigma_1, \ldots, \sigma_r\}$ be an invertible diagonal matrix, whose diagonal elements are the positive square roots of the nonzero eigenvalues of A^*A repeated according to multiplicity and arranged in descending order. The numbers $\sigma_1, \ldots, \sigma_r$ are usually referred to as the *singular values* of A. Define also two unitary matrices $U \in \mathbb{C}^{m\times m}$ and $V \in \mathbb{C}^{n\times n}$ as follows: u_k, the kth column of matrix U, is a normalized eigenvector of A^*A corresponding to the eigenvalue σ_k^2 and $v_k = Au_k/\sigma_k$. Then, the SVD is given by

$$A = V\begin{bmatrix} D & 0 \\ 0 & 0 \end{bmatrix} U^*.$$

Now, the generalized inverse $A^\dagger \in \mathbb{C}^{n\times m}$ can be written in the form

$$A^\dagger = U\begin{bmatrix} D^{-1} & 0 \\ 0 & 0 \end{bmatrix} V^*. \tag{2.7}$$

It is easy to check that the above expression for $A^\dagger$ indeed satisfies all four equations (2.3)–(2.6); see Problem 2.1.

The following well-known properties of the Moore–Penrose generalized inverse will be used in what follows:

$$(A^*)^\dagger = (A^\dagger)^*, \tag{2.8}$$
$$A^* = A^*AA^\dagger = A^\dagger AA^*, \tag{2.9}$$
$$(A^*A)^\dagger = A^\dagger A^{*\dagger}, \tag{2.10}$$
$$A^\dagger = (A^*A)^\dagger A^* = A^*(AA^*)^\dagger. \tag{2.11}$$

One can immediately conclude from Definition 2.1 that the generalized inverse is an equation solver. We have the following formal result.

Lemma 2.1. *Consider the assumed feasible system of linear equations*

$$Ax = b, \tag{2.12}$$

where $A \in \mathbb{C}^{m\times n}$ and $b \in \mathbb{C}^m$. Then x is a solution of this system if and only if

$$x = A^\dagger b + v,$$

where $v \in \mathbb{C}^{n\times 1}$ belongs to the null space of A, that is, $Av = 0$.

The next lemma provides a simple condition for the feasibility of linear systems (see Problem 2.2).

Lemma 2.2. *The system of linear equations* (2.12) *is feasible if and only if $w^*b = 0$ for all vectors $w \in \mathbb{C}^{m\times 1}$ that span the null space of the conjugate transpose matrix A^*, that is, $A^*w = 0$.*

An important particular case of the Moore–Penrose generalized inverse is the so-called *group inverse*, defined as follows.

Definition 2.3. *Suppose that A is a square matrix. The group inverse A^g, if it exists, is characterized as the unique matrix satisfying the following three equations:*

$$AA^gA = A, \tag{2.13}$$
$$A^gAA^g = A^g, \tag{2.14}$$
$$AA^g = A^gA. \tag{2.15}$$

Existence of the group inverse of $A \in \mathbb{C}^{n\times n}$ is equivalent to the existence of a decomposition of the space $\mathbb{C}^n$ into a *direct* sum of the null space and the range of A (see Problem 2.3).

We now show that computing the Moore–Penrose generalized inverse reduces to computing the group inverse of a square symmetric matrix. This result seems to be new or, at least, not widely reported.

Lemma 2.3. *The Moore–Penrose generalized inverse of A can be calculated by the formulae*

$$A^\dagger = (A^*A)^gA^* = A^*(AA^*)^g. \tag{2.16}$$

Proof: By (2.11), to prove the above formulae, we need only verify that the Moore–Penrose generalized inverse $(A^*A)^\dagger$ is also the group inverse of A^*A. Thus, we need to verify that $(A^*A)^\dagger$ satisfies (2.13)–(2.15). It is obvious that (2.13) and (2.14) hold, since by definition, the generalized inverse satisfies (2.3) and (2.4). The last identity (2.15) is obtained via

$$\begin{aligned}(A^*A)(A^*A)^\dagger &= A^*AA^\dagger A^{*\dagger} = A^*AA^\dagger A^{\dagger *} = A^*A^{\dagger *}\\ &= (A^\dagger A)^* = A^\dagger A = (A^*A)^\dagger(A^*A),\end{aligned}$$

using (2.10), (2.8), (2.9), (2.6), and (2.11), respectively. Thus, the matrix $(A^*A)^\dagger$ satisfies its analogue of (2.15), and, therefore, $(A^*A)^\dagger = (A^*A)^g$, which immediately yields (2.16). □

Now let us discuss another type of generalized inverse, the so-called *Drazin inverse.* The Drazin inverse can be defined and calculated in the following way: If $A \in \mathbb{C}^{n\times n}$, then it can be represented by the decomposition

$$A = W\begin{bmatrix} S & 0 \\ 0 & N \end{bmatrix} W^{-1}, \tag{2.17}$$

where S is invertible and N is nilpotent. Then, the Drazin inverse is defined by

$$A^\# = W\begin{bmatrix} S^{-1} & 0 \\ 0 & 0 \end{bmatrix} W^{-1}.$$

Note that the Drazin inverse is not an equation solver. However, based on algebraic properties, Drazin inverses have more in common with usual matrix inverses than Moore–Penrose generalized inverses do. In spectral theory of linear operators the Drazin inverse is also known as *reduced resolvent.*

The group inverse is also a particular case of the Drazin inverse. Namely, whenever for a matrix A the group inverse exists, A can be decomposed into (2.17) with $N = 0$. In fact, the group inverse represents the case when the Moore–Penrose generalized inverse and the Drazin inverse coincide.

2.2 ▪ Inversion of Analytically Perturbed Matrices: Algebraic Approach

2.2.1 ▪ Laurent series and fundamental equations

Let $\{A_k\}_{k=0,1,\ldots} \subseteq \mathbb{C}^{n\times n}$ be a sequence of matrices that defines an analytic matrix-valued function

$$A(z) = A_0 + zA_1 + z^2A_2 + \cdots. \tag{2.18}$$

The above series is assumed to converge in some nonempty neighborhood of $z = 0$. In such a case we say that $A(z)$ is an *analytic perturbation* of the matrix $A_0 = A(0)$. Assume the inverse matrix $A^{-1}(z)$ exists in some (possibly punctured) disc centred at $z = 0$. We are primarily interested in the case when A_0 is singular. The next theorem shows that $A^{-1}(z)$ can be expanded as a Laurent series.

Theorem 2.4. *Let $A(z)$ be an analytic matrix-valued function of z in some nonempty neighborhood of $z = 0$ and such that $A^{-1}(z)$ exists in some (possibly punctured) disc centered at*

$z = 0$. *Then, $A^{-1}(z)$ possesses a Laurent series expansion*

$$A^{-1}(z) = \frac{1}{z^s}(X_0 + zX_1 + \cdots), \tag{2.19}$$

where $X_0 \neq 0$ and s is a natural number, known as the order of the pole at $z = 0$.

Proof: Using the Cramer formula, we can write

$$A^{-1}(z) = \frac{\text{adj}A(z)}{\det A(z)}. \tag{2.20}$$

Since the determinant $\det A(z)$ and the elements of the adjugate matrix $\text{adj}A(z)$ are polynomials in $a_{ij}(z)$, $i, j = 1, \ldots, n$, they are analytic functions of z. The division of two analytic functions yields a meromorphic function. Since n is finite, the order of the pole s in the matrix Laurent series (2.19) is finite as well. □

We would like to note that the above proof is essentially based on the finiteness of the dimension of the underlying space. The case of infinite dimensional spaces will be treated in Chapter 8.

Example 2.1. *Let us consider the following example of the analytically perturbed matrix:*

$$A(z) = \begin{bmatrix} 1-z & 1+z \\ 1-2z & 1-z \end{bmatrix}.$$

According to the formula (2.20), the inverse is given by

$$A^{-1}(z) = \frac{1}{-z(1-3z)} \begin{bmatrix} 1-z & -1-z \\ -1+2z & 1-z \end{bmatrix}.$$

Next, to obtain the Laurent series (2.19), we just expand $(\det A(z))^{-1} = 1/(-z(1-3z))$ as a scalar power series, multiply it by $\text{adj}A(z)$, and collect coefficients with the same power of z. In this case, we have

$$\begin{aligned} A^{-1}(z) &= \left(-\frac{1}{z} - 3 - 9z^2 - \ldots\right) \begin{bmatrix} 1-z & -1-z \\ -1+2z & 1-z \end{bmatrix} \\ &= \frac{1}{z}\begin{bmatrix} -1 & 1 \\ 1 & -1 \end{bmatrix} + \begin{bmatrix} -2 & 4 \\ 1 & -2 \end{bmatrix} + z\begin{bmatrix} -6 & 12 \\ 3 & -6 \end{bmatrix} + \ldots. \end{aligned}$$

Of course, the direct application of the Cramer formula (2.20) as in the above example is very inefficient as a method of deriving the Laurent series (2.19). Thus, the main purpose of this section is to provide efficient computational procedures for calculating the Laurent series coefficients X_k, $k \geq 0$.

In fact, we present three methods for computing the coefficients of the Laurent series (2.19) for the inverse of the analytically perturbed matrix (2.18). The first method is based on a direct application of the Moore–Penrose generalized inverse matrix. The other two methods are based on a so-called *reduction technique*. All three methods depend essentially on equating coefficients of powers of z.

By substituting the series (2.18) and (2.19) into the identity $A(z)A^{-1}(z) = I$ and collecting coefficients of the same powers of z, one obtains the following system, which we shall refer to as the *fundamental equations*:

$$\begin{aligned} A_0X_0 &= 0, \\ A_0X_1 + A_1X_0 &= 0 \\ &\vdots \\ A_0X_s + \cdots + A_sX_0 &= I, \\ A_0X_{s+1} + \cdots + A_{s+1}X_0 &= 0 \\ &\vdots \end{aligned}$$

or, in more compact notation,

$$\sum_{i=0}^{k} A_i X_{k-i} = \delta_{ks} I, \quad k = 0, 1, \ldots, \tag{2.21}$$

where δ_{ks} is the Kroneker delta and s is the order of the pole in (2.19). In the next subsection we demonstrate that the infinite system (2.21) of linear equations uniquely determines the coefficients of the Laurent series (2.19). In what follows, if we want to refer to the kth equation (starting from zero) of the above system, we simply write (2.21.k). Note that an analogous system can be derived from the identity $A^{-1}(z)A(z) = I$, that is,

$$\sum_{i=0}^{k} Y_{k-i} A_i = \delta_{ks} I, \quad k = 0, 1, \ldots. \tag{2.22}$$

Since the set of equations (2.21) and the set of equations (2.22) are equivalent (see Problem 2.4), it is sufficient to consider only one of them.

The solution of the fundamental equations in the case of a regular perturbation is straightforward. In that case, A_0^{-1} exists, and hence we solve the fundamental equations (2.21) one by one to obtain

$$X_k = -A_0^{-1} \sum_{i=1}^{k} A_i X_{k-i}, \quad k = 1, 2, \ldots, \tag{2.23}$$

with $X_0 = A_0^{-1}$.

The rest of this section is dedicated to a less obvious analysis of the singular perturbation case. Recall that the latter occurs when A_0 is not invertible but the perturbed matrix $A(z)$ has an inverse for z sufficiently small but different from zero.

2.2.2 ▪ Existence of a solution to the fundamental equations

We begin this subsection with two definitions related to (2.18).

Definition 2.4. *Vectors $\varphi_0, \ldots, \varphi_{r-1}$ are said to form a generalized Jordan chain of the analytic matrix-valued function $A(z)$ at $z = 0$ if $\varphi_0 \neq 0$ and if*

$$\sum_{i=0}^{k} A_i \varphi_{k-i} = 0$$

for each $0 \leq k \leq r-1$. The number r is called the length of the Jordan chain, and φ_0 is called the initial vector.

Let $\{\varphi_0^{(j)}\}_{j=1}^p$ be a system of linearly independent eigenvectors that span the null space of A_0. Then one can construct generalized Jordan chains of $A(z)$ initializing at each of the eigenvectors $\varphi_0^{(j)}$.

Example 2.2. *Consider the linear perturbation $A(z)=A_0+zA_1$ with*

$$A_0=\begin{bmatrix} 1 & 2 \\ 1 & 2 \end{bmatrix}, \quad A_1=\begin{bmatrix} 1 & 3 \\ 0 & 1 \end{bmatrix}.$$

The null space of A_0 is one dimensional and is spanned by

$$\varphi_0=\begin{bmatrix} 2 \\ -1 \end{bmatrix}.$$

We construct the next vector φ_1 by solving

$$A_0\varphi_1=-A_1\varphi_0,$$

$$\begin{bmatrix} 1 & 2 \\ 1 & 2 \end{bmatrix}\begin{bmatrix} \varphi_{11} \\ \varphi_{12} \end{bmatrix}=\begin{bmatrix} 1 \\ 1 \end{bmatrix}.$$

The above system reduces to one equation,

$$\varphi_{11}+2\varphi_{12}=1.$$

Thus, as φ_1 we can take

$$\varphi_0=\begin{bmatrix} 1 \\ 0 \end{bmatrix}.$$

Then, we try to construct the next generalized Jordan vector φ_2 such that $A_0\varphi_2=-A_1\varphi_1$, that is,

$$\begin{bmatrix} 1 & 2 \\ 1 & 2 \end{bmatrix}\begin{bmatrix} \varphi_{21} \\ \varphi_{22} \end{bmatrix}=\begin{bmatrix} -1 \\ 0 \end{bmatrix}.$$

We can see that the above system is infeasible, and consequently, in this example the length of the generalized Jordan chain is equal to two.

Definition 2.5. *Let us define the following augmented matrix $\mathcal{A}^{(t)}\in\mathbb{C}^{(t+1)n\times(t+1)n}$:*

$$\mathcal{A}^{(t)}=\begin{bmatrix} A_0 & 0 & 0 & \cdots & 0 \\ A_1 & A_0 & 0 & \cdots & 0 \\ A_2 & A_1 & A_0 & \cdots & 0 \\ \vdots & \vdots & \vdots & \ddots & \vdots \\ A_t & A_{t-1} & \cdots & A_1 & A_0 \end{bmatrix}.$$

The next lemma relates the order of the pole s to the length of the generalized Jordan chain.

Lemma 2.5. *Let s be the order of the pole at the origin for the inverse matrix function $A^{-1}(z)$. The order of the pole is equal to the maximal length of the generalized Jordan chain of $A(z)$ at $z=0$. Furthermore, any eigenvector $\Phi\in\mathbb{C}^{(s+1)n}$ of $\mathcal{A}^{(s)}$ corresponding to the zero eigenvalue has the property that its first n elements are zero.*

Proof: From the fundamental equations (2.21.0)–(2.21.$s-1$) we can see that any column of $X(z)=z^sA^{-1}(z)=X_0+zX_1+\ldots$ generates a generalized Jordan chain of order s.

Next we show that s is the maximal Jordan chain length. Let us define $\varphi(z)=\varphi_0+z\varphi_1+\cdots+z^{r-1}\varphi_{r-1}$ and multiply it by $A(z)$. We obtain

$$A(z)\varphi(z)=z^r\psi(z),$$

where $\psi(z)$ is an analytic function. Premultiplying the above equation by $X(z)$ and using the identity $X(z)A(z)=z^sI$, we obtain

$$z^s\varphi(z)=z^r\tilde{\psi}(z),$$

where $\tilde{\psi}(z)$ is again an analytic function. As $\varphi_0\neq 0$, we conclude from the above equation that $r\leq s$. Hence, the first statement of the lemma is proved.

Now let us prove the second statement of the lemma. Suppose, on the contrary, that there exists an eigenvector $\Phi\in\mathbb{C}^{(s+1)n}$ such that

$$\mathcal{A}^{(s)}\Phi=0 \tag{2.24}$$

and not all of its first n entries are zero. Then, partition the vector Φ into $s+1$ blocks $\varphi_0,\varphi_1,\ldots,\varphi_s$, and rewrite (2.24) in the form

$$\begin{aligned}A_0\varphi_0&=0,\\ A_0\varphi_1+A_1\varphi_0&=0,\\ &\vdots\\ A_0\varphi_s+\cdots+A_s\varphi_0&=0\end{aligned}$$

with $\varphi_0\neq 0$. This means that we have found a generalized Jordan chain of length $s+1$. Since the maximal length of a generalized Jordan chain of $A(z)$ at $z=0$ is s, we came to a contradiction, and, consequently, $\varphi_0=0$. □

Corollary 2.1. *All vectors $\Phi\in\mathbb{R}^{(s+j+1)n}$ in the null space of the augmented matrix $\mathcal{A}^{(s+j)}$, $j\geq 0$, possess the property that the first $(j+1)n$ elements are zero.*

Example 2.2 (continued from the beginning of Subsection 2.2.2). *Using Cramer's formula, we can calculate*

$$A^{-1}(z)=\frac{1}{z^2}\begin{bmatrix}2+z & -2-3z\\ -1 & 1+z\end{bmatrix}$$

$$=\frac{1}{z^2}\begin{bmatrix}2 & -2\\ -1 & 1\end{bmatrix}+\frac{1}{z}\begin{bmatrix}1 & -3\\ 0 & 1\end{bmatrix}.$$

Indeed, we see that the order of the pole is equal to two, the length of the generalized Jordan chain $\{\varphi_0,\varphi_1\}$.

The following theorem provides a theoretical basis for the recursive solution of the infinite system of fundamental equations (2.21).

Theorem 2.6. *Each coefficient matrix X_k, $k\geq 0$, in the Laurent series expansion (2.19) of $A^{-1}(z)$ is uniquely determined by the previous coefficients $X_0,\ldots,X_{k-1}$ and the set of s fundamental equations (2.21.k)—(2.21.$k+s$).*

Proof: It is obvious that the sequence of Laurent series coefficients $\{X_i\}_{i=0}^{\infty}$ is a solution to the fundamental equations (2.21). Suppose the coefficients X_i, $0 \le i \le k-1$, have been determined. Next, we show that the set of fundamental equations (2.21.k)–(2.21.$k+s$) uniquely determines the next coefficient X_k. Indeed, suppose there exists another solution $\tilde{X}_k$. Since X_k and $\tilde{X}_k$ are both solutions of (2.21.k)–(2.21.$k+s$), we can write

$$\mathscr{A}^{(s)} \begin{bmatrix} \tilde{X}_k \\ \vdots \\ \tilde{X}_{k+s} \end{bmatrix} = \begin{bmatrix} \delta_{s,k} I - \sum_{i=1}^{k} A_i X_{k-i} \\ \vdots \\ \delta_{s,k+s} I - \sum_{i=1}^{k} A_{i+s} X_{k-i} \end{bmatrix} \qquad (2.25)$$

and

$$\mathscr{A}^{(s)} \begin{bmatrix} X_k \\ \vdots \\ X_{k+s} \end{bmatrix} = \begin{bmatrix} \delta_{s,k} I - \sum_{i=1}^{k} A_i X_{k-i} \\ \vdots \\ \delta_{s,k+s} I - \sum_{i=1}^{k} A_{i+s} X_{k-i} \end{bmatrix}, \qquad (2.26)$$

where $\tilde{X}_{k+1}, \dots, \tilde{X}_{k+s}$ are any particular solutions of the nonhomogeneous linear system (2.21.k)–(2.21.$k+s$). Note that (2.25) and (2.26) have identical right-hand sides. Hence, the difference between two solutions, $[\tilde{X}_k - X_k \cdots \tilde{X}_{k+s} - X_{k+s}]^T$, is in the null space of $\mathscr{A}^{(s)}$. Invoking Lemma 2.5, the first n rows of $[\tilde{X}_k - X_k, \dots, \tilde{X}_{k+s} - X_{k+s}]^T$ are zero. In other words, $\tilde{X}_k = X_k$, which proves the theorem. □

Since the first s fundamental equations uniquely determine the leading term of the Laurent series expansion (2.19), we call the first s fundamental equations *determining equations*.

2.2.3 ▪ The determination of the order of the pole

Since some methods proposed in this section depend on the prior knowledge of s, the order of the pole in (2.19), we begin by discussing a procedure for the determination of s. The procedure is based on a rank test of the augmented matrices.

Theorem 2.7. *The order of the pole s is given by the smallest value of t for which* $\operatorname{rank}\mathscr{A}^{(t)} = \operatorname{rank}\mathscr{A}^{(t-1)} + n$, *where $\mathscr{A}^{(t)}$ is as in Definition 2.5 and n is the dimension of $A(z)$.*

Proof: First we note that

$$\dim(N(\mathscr{A}^{t-1})) + \operatorname{rank}(\mathscr{A}^{t-1}) = nt$$

and

$$\dim(N(\mathscr{A}^{t})) + \operatorname{rank}(\mathscr{A}^{t}) = n(t+1).$$

Subtracting the first equation above from the second, we obtain

$$\operatorname{rank}(\mathscr{A}^{t}) = \operatorname{rank}(\mathscr{A}^{t-1}) + n - [\dim(N(\mathscr{A}^{t})) - \dim(N(\mathscr{A}^{t-1}))]. \qquad (2.27)$$

If $\dim(N(\mathscr{A}^{t})) > \dim(N(\mathscr{A}^{t-1}))$, one can construct a generalized Jordan chain $\{\varphi_0, \dots, \varphi_t\}$ of length $t+1$ from a generalized Jordan chain $\{\varphi_0, \dots, \varphi_{t-1}\}$ of length t by solving the equation

$$A_0 \varphi_t = -\sum_{i=1}^{t} A_i \varphi_{t-i}.$$

Since by Lemma 2.5 the maximal length of a generalized Jordan chain is equal to the order s of the pole of $A^{-1}(z)$, $\dim(N(\mathscr{A}^t)) > \dim(N(\mathscr{A}^{t-1}))$ for $t < s$ and $\dim(N(\mathscr{A}^s)) = \dim(N(\mathscr{A}^{s-1}))$. Hence, from (2.27) we conclude that s is the smallest t such that $\operatorname{rank}\mathscr{A}^{(t)} = \operatorname{rank}\mathscr{A}^{(t-1)} + n$. □

The calculation of rank is essentially equivalent to the reduction of $\mathscr{A}^{(t)}$ to a row echelon normal form, and it can be argued that row operations can be used successively in order to calculate the rank of $\mathscr{A}^{(0)}$, $\mathscr{A}^{(1)}$, $\mathscr{A}^{(2)}$, ... and find the minimal value of t for which $\operatorname{rank}\mathscr{A}^{(t)} = \operatorname{rank}\mathscr{A}^{(t-1)} + n$. Note that previous row operations for reducing $\mathscr{A}^{(t-1)}$ to row echelon form are replicated in the reduction of $\mathscr{A}^{(t)}$ and do not need to be repeated. Namely, if a certain combination of row operations reduces $\mathscr{A}^{(t-1)}$ to the row echelon form, the same operations are used again as part of the reduction of

$$\mathscr{A}^{(t)} = \begin{bmatrix} \mathscr{A}^{(t-1)} & 0 \\ * & A_0 \end{bmatrix}$$

to the row echelon form.

Example 2.2 (continued from the beginning of Subsection 2.2.2). *The row echelon form of A_0 is*

$$\begin{bmatrix} 1 & 2 \\ 0 & 0 \end{bmatrix},$$

and hence, $\operatorname{rank}(\mathscr{A}^{(0)}) = 1$. *To determine the rank of $\mathscr{A}^{(1)}$, we augment the block row*

$$\begin{bmatrix} 1 & 2 & 0 & 0 \\ 0 & 0 & 0 & 0 \end{bmatrix}$$

by the block row $[A_1\ A_0]$,

$$\begin{bmatrix} 1 & 2 & 0 & 0 \\ 0 & 0 & 0 & 0 \\ 1 & 3 & 1 & 2 \\ 0 & 1 & 1 & 2 \end{bmatrix}.$$

By subtracting the first row from the third row, and then the third row from the fourth row, we reduce it to the echelon form

$$\begin{bmatrix} 1 & 2 & 0 & 0 \\ 0 & 1 & 1 & 2 \\ 0 & 0 & 0 & 0 \\ 0 & 0 & 0 & 0 \end{bmatrix}.$$

Hence, $\operatorname{rank}(\mathscr{A}^{(1)}) = 2$, *and since* $\operatorname{rank}(\mathscr{A}^{(1)}) - \operatorname{rank}(\mathscr{A}^{(0)}) = 1 < 2$, *we need to continue. Augmenting the above row echelon form by the block row $[0\ A_1\ A_0]$ and interchanging the rows, we obtain*

$$\begin{bmatrix} 1 & 2 & 0 & 0 & 0 & 0 \\ 0 & 1 & 1 & 2 & 0 & 0 \\ 0 & 0 & 1 & 3 & 1 & 2 \\ 0 & 0 & 0 & 1 & 1 & 2 \\ 0 & 0 & 0 & 0 & 0 & 0 \\ 0 & 0 & 0 & 0 & 0 & 0 \end{bmatrix}.$$

Thus, $\operatorname{rank}(\mathscr{A}^{(2)}) - \operatorname{rank}(\mathscr{A}^{(1)}) = 4 - 2 = 2$, *and consequently, the order of the pole is equal to two.*

2.2.4 ▪ Basic generalized inverse method

Here we obtain a recursive formula for the Laurent series coefficients X_k, $k \geq 0$, in the expansion (2.19) of $A^{-1}(z)$ with the help of the Moore–Penrose generalized inverse of the augmented matrix $\mathscr{A}^{(s)}$.

Let $\mathscr{G}^{(s)} \stackrel{def}{=} [\mathscr{A}^{(s)}]^\dagger$ be the Moore–Penrose generalized inverse of $\mathscr{A}^{(s)}$, and define the submatrices $G_{ij}^{(s)} \in \mathbb{C}^{n \times n}$ for $0 \leq i, j \leq t$ by

$$\mathscr{G}^{(s)} = \begin{bmatrix} G_{00}^{(s)} & \cdots & G_{0s}^{(s)} \\ \vdots & \ddots & \vdots \\ G_{s0}^{(s)} & \cdots & G_{ss}^{(s)} \end{bmatrix},$$

where the dimensions and locations of $G_{ij}^{(s)}$ are in correspondence with the block structure of $\mathscr{A}^{(s)}$.

Furthermore, we would like to note that in fact we shall use only the first n rows of the generalized inverse $\mathscr{G}^{(s)}$, namely, $[G_{00}^{(s)} \ \cdots \ G_{0s}^{(s)}]$.

Theorem 2.8. *The coefficients of the Laurent series* (2.19) *can be calculated by the following recursive formula:*

$$X_k = \sum_{j=0}^{s} G_{0j}^{(s)} \left(\delta_{j+k,s} I - \sum_{i=1}^{k} A_{i+j} X_{k-i} \right), \quad k = 1, 2, \ldots, \tag{2.28}$$

initializing with $X_0 = G_{0s}^{(s)}$.

Proof: According to Theorem 2.6, once the coefficients X_i, $0 \leq i \leq k-1$, are determined, the next coefficient X_k can be obtained from the (2.21.k)–(2.21.$k+s$) fundamental equations

$$\mathscr{A}^{(s)} \begin{bmatrix} X_k \\ \vdots \\ X_{k+s} \end{bmatrix} = \begin{bmatrix} \delta_{k,s} I - \sum_{i=1}^{k} A_i X_{k-i} \\ \vdots \\ \delta_{k+s,s} I - \sum_{i=1}^{k} A_{i+s} X_{k-i} \end{bmatrix}.$$

According to Lemma 2.1, the general solution to the above system is given in the form

$$\begin{bmatrix} X_k \\ \tilde{X}_{k+1} \\ \vdots \\ \tilde{X}_{k+s} \end{bmatrix} = \begin{bmatrix} G_{00}^{(s)} & \cdots & G_{0s}^{(s)} \\ G_{10}^{(s)} & \cdots & G_{1s}^{(s)} \\ \vdots & \ddots & \vdots \\ G_{s0}^{(s)} & \cdots & G_{ss}^{(s)} \end{bmatrix} \begin{bmatrix} \delta_{k,s} I - \sum_{i=1}^{k} A_i X_{k-i} \\ \delta_{k+1,s} I - \sum_{i=1}^{k} A_{i+1} X_{k-i} \\ \vdots \\ \delta_{k+s,s} I - \sum_{i=1}^{k} A_{i+s} X_{k-i} \end{bmatrix} + \begin{bmatrix} 0 \\ \Phi_1 \\ \vdots \\ \Phi_s \end{bmatrix},$$

where the first block of the matrix Φ—that is, the last term in the above—is equal to zero according to Lemma 2.5. Thus, we immediately obtain the recursive expression (2.28). Furthermore, applying the same arguments as above to the first $s+1$ fundamental equations, we obtain that $X_0 = G_{0s}^{(s)}$ (see Problem 2.5). □

Note that the terms $\delta_{s,j+k} I$ in the expression (2.28) disappear when the regular coefficients are computed.

Remark 2.1. *The formula* (2.28) *is a generalization of the recursive formula* (2.23) *for the regular case when* A_0 *was invertible.*

Remark 2.2. *From the computational point of view it may be better not to compute the generalized inverse* $\mathcal{G}^{(s)}$ *beforehand, but rather to find the SVD or LU decomposition of* $\mathcal{A}^{(s)}$ *and then use such a decomposition for solving the fundamental equations* (2.21.k)–(2.21.$k+s$). *This is a standard approach for solving linear systems.*

Example 2.3. *Let us consider the perturbed matrix*

$$A(z)=A_0+zA_1=\begin{bmatrix} 1 & 2 & 1 \\ -1 & 1 & 0 \\ 0 & 3 & 1 \end{bmatrix}+z\begin{bmatrix} 1 & -1 & 0 \\ 0 & 1 & -1 \\ -1 & 0 & 1 \end{bmatrix},$$

where $\operatorname{rank}(A_0)=2$. *Construct the augmented matrices*

$$\mathcal{A}^{(0)}=A_0 \quad \text{and} \quad \mathcal{A}^{(1)}=\begin{bmatrix} A_0 & 0 \\ A_1 & A_0 \end{bmatrix},$$

and note that $\operatorname{rank}(\mathcal{A}^{(1)})-\operatorname{rank}(\mathcal{A}^{(0)})=5-2=3$, *which is the dimension of the original coefficients* A_0 *and* A_1. *Therefore, according to the rank test of Theorem 2.7, the Laurent expansion for* $A^{-1}(z)$ *has a pole of order one. Alternatively, we may compute a basis for* $N(\mathcal{A}^{(1)})$, *which in this particular example consists of only one vector,*

$$\Phi=\begin{bmatrix} 0 & 0 & 0 & 1 & 1 & -3 \end{bmatrix}^T.$$

The first three zero elements in Φ *confirm that* X_k *is uniquely determined by the system*

$$\mathcal{A}^{(1)}\begin{bmatrix} X_k \\ X_{k+1} \end{bmatrix}=\begin{bmatrix} \delta_{k,1}I-\sum_{i=1}^{k}A_iX_{k-i} \\ \delta_{k+1,1}I-\sum_{i=1}^{k}A_{i+1}X_{k-i} \end{bmatrix}$$

and hence that the Laurent series (2.19) *has a simple pole. Next, we compute the generalized inverse of* $\mathcal{A}^{(1)}$ *given by*

$$\mathcal{A}^{(1)\dagger}=\mathcal{G}^{(1)}=\begin{bmatrix} G_{00}^{(1)} & G_{01}^{(1)} \\ G_{10}^{(1)} & G_{11}^{(1)} \end{bmatrix}=\begin{bmatrix} 1/3 & -5/12 & -1/12 & 1/8 & 1/8 & -1/8 \\ 0 & 1/4 & 1/4 & 1/8 & 1/8 & -1/8 \\ 1/3 & -5/12 & -1/12 & -3/8 & -3/8 & 3/8 \\ * & * & * & * & * & * \\ * & * & * & * & * & * \\ * & * & * & * & * & * \end{bmatrix}.$$

Consequently,

$$X_0=G_{01}^{(1)}=\frac{1}{8}\begin{bmatrix} 1 & 1 & -1 \\ 1 & 1 & -1 \\ -3 & -3 & 3 \end{bmatrix}, \tag{2.29}$$

and

$$X_1=G_{00}^{(1)}(I-A_1X_0)=\frac{1}{4}\begin{bmatrix} 2 & -1 & -1 \\ 0 & 1 & 1 \\ 2 & -1 & -1 \end{bmatrix}. \tag{2.30}$$

2.2.5 ▪ The case of first order pole

We note that given that the matrix A_0 is singular but the perturbed matrix $A(z)$ is invertible, the case when the Laurent series for $A^{-1}(z)$ has the first order pole ($s = 1$) is *generic*. In other words, if we choose the entries of the perturbation matrices A_k, $k \geq 1$, in a random manner with a sufficiently general distribution, the Laurent series expansion of $A^{-1}(z)$ will, with probability 1, have a pole of order one. At the end of this subsection we shall give a precise mathematical statement of this fact. Of course, there are cases of interest in which the perturbation is not generic and $A^{-1}(z)$ can have higher order poles. We shall thoroughly analyze a general situation in the subsequent sections.

The next theorem provides a computational scheme that is an alternative to the method based on the augmented matrices described in the previous subsection. This scheme allows us to perform computations in lower dimensional spaces. We recommend it when the bases of the null spaces of A_0 and A_0^* are readily available. An example of such a case is given in Chapter 6. This approach will be generalized to the case $s > 1$ in Subsections 2.2.8 and 2.2.9.

Theorem 2.9. *Let the unperturbed matrix A_0 be singular. Let $Q \in \mathbb{C}^{n\times p}$ be a matrix whose columns form a basis for the null space of A_0, and let $M \in \mathbb{C}^{n\times p}$ be a matrix whose columns form a basis for the null space of the conjugate transpose matrix A_0^*. The Laurent series (2.19) has a first order pole if and only if M^*A_1Q is nonsingular. In such a case, the Laurent series coefficients in (2.19) are given by the recursive formula*

$$X_k = (A_0^\dagger - Q[M^*A_1Q]^{-1}M^*A_1A_0^\dagger)\left(\delta_{1,k}I - \sum_{i=1}^{k} A_iX_{k-i}\right)$$
$$+Q[M^*A_1Q]^{-1}M^*\left(\delta_{1,k+1}I - \sum_{i=1}^{k} A_{i+1}X_{k-i}\right), \tag{2.31}$$

where $A_0^\dagger$ is the Moore–Penrose generalized inverse of A_0.

Proof: According to Theorem 2.6, in the case of the first order pole ($s = 1$), the matrix coefficient X_k is uniquely determined by the two equations

$$A_0X_k = R_0, \tag{2.32}$$

$$A_0X_{k+1} + A_1X_k = R_1, \tag{2.33}$$

where $R_0 = \delta_{1,k}I - \sum_{i=1}^{k} A_iX_{k-i}$ and $R_1 = \delta_{1,k+1}I - \sum_{i=1}^{k} A_{i+1}X_{k-i}$. By Lemma 2.1 a general solution to the linear system (2.32) can be written in the form

$$X_k = A_0^\dagger R_0 + QY_k, \tag{2.34}$$

where $Y_k \in \mathbb{C}^{p\times n}$ is some arbitrary matrix. In order for (2.33) to be feasible for X_{k+1}, we require the right-hand side $R_1 - A_1X_k$ to belong to $R(A_0) = N^\perp(A_0^*)$ (see Lemma 2.2), that is,

$$M^*(R_1 - A_1X_k) = 0,$$

where the columns of M form a basis for $N(A_0^*)$. Substituting expression (2.34) for the general solution X_k into the above feasibility condition, one finds that Y_k satisfies the equation

$$M^*(R_1 - A_1(A_0^\dagger R_0 + QY_k)) = 0,$$

which can be rewritten as

$$M^*A_1QY_k = M^*R_1 - M^*A_1A_0^\dagger R_0.$$

Hence, Y_k (and thereby also X_k) is uniquely determined by (2.32) and (2.33) if and only if the matrix M^*A_1Q is nonsingular. Consequently, if M^*A_1Q is invertible, we have

$$Y_k = [M^*A_1Q]^{-1}M^*R_1 - [M^*A_1Q]^{-1}M^*A_1A_0^\dagger R_0.$$

Thus, by substituting the above expression for Y_k into (2.34), we obtain (2.31). □

In particular, for $k=0$ and $k=1$ we have

$$X_0 = Q[M^*A_1Q]^{-1}M^* \tag{2.35}$$

and

$$\begin{aligned} X_1 &= (A_0^\dagger - Q[M^*A_1Q]^{-1}M^*A_1A_0^\dagger)(I - A_1X_0) + Q[M^*A_1Q]^{-1}M^*(-A_2X_0) \\ &= (A_0^\dagger - X_0A_1A_0^\dagger)(I - A_1X_0) - X_0A_2X_0 \\ &= A_0^\dagger - A_0^\dagger A_1X_0 - X_0A_1A_0^\dagger + X_0(A_1A_0^\dagger A_1 - A_2)X_0. \end{aligned} \tag{2.36}$$

Remark 2.3. *A matrix equation $AX = B$ is equivalent to a vector equation $Ax_i = b_i$ for which the solution is formally given by $x_i = A^\dagger b_i + Qy_i$, where $Q \in \mathbb{C}^{n\times p}$. In practice the equations $Ax_i = b_i$ are solved by finding the LU or SVD decomposition of A. Thus we have, for instance, with the SVD, $Ax_i = b_i$ reduced to $\Lambda\xi_i = s_i$, where Λ is a diagonal matrix, $\xi_i = V^*x_i$, and $s_i = U^*b_i$, and the solution is*

$$\xi_i = \Lambda^\dagger s_i + V^*Qy_i,$$

which gives

$$x_i = V\Lambda^\dagger s_i + Qy_i.$$

The next theorem provides a formal justification for the term "generic" in the description of the first order pole.

Theorem 2.10. *Let the unperturbed matrix A_0 be singular. If entries of A_1 are random numbers from $\mathbb{C}$ chosen by a distribution with a continuous density function, the Laurent series* (2.19) *has the first order pole with probability one.*

Proof: From Theorem 2.9, we know that the Laurent series (2.19) has the first order pole if and only if the matrix M^*A_1Q is invertible. In other words, the Laurent series (2.19) has a pole of order larger than one if

$$\det(M^*A_1Q) = 0.$$

The above equation can be regarded as a polynomial whose variables are the n^2 entries of A_1. Thus, it defines a manifold in $\mathbb{C}^{n^2}$ of dimension n^2-1. Since the entries of A_1 have a distribution with a continuous density function, the probability of $\det(M^*A_1Q) \neq 0$ is equal to one. □

Example 2.4. *We consider again*

$$A(z) = A_0 + zA_1 = \begin{bmatrix} 1 & 2 & 1 \\ -1 & 1 & 0 \\ 0 & 3 & 1 \end{bmatrix} + z \begin{bmatrix} 1 & -1 & 0 \\ 0 & 1 & -1 \\ -1 & 0 & 1 \end{bmatrix}.$$

In this example,

$$Q = \begin{bmatrix} 1 \\ 1 \\ -3 \end{bmatrix} \quad \text{and} \quad M = \begin{bmatrix} 1 \\ 1 \\ -1 \end{bmatrix}$$

span the null spaces of A_0 and A_0^*, respectively. Since $s = 1$ in this case (see Example 2.3), this is the generic case and the coefficients X_0 and X_1 can be calculated by the formulae (2.35) *and* (2.36). *Namely,*

$$X_0 = Q(M^* A_1 Q)^{-1} M^* = \begin{bmatrix} 1 \\ 1 \\ -3 \end{bmatrix} \frac{1}{8} \begin{bmatrix} 1 & 1 & -1 \end{bmatrix} = \frac{1}{8} \begin{bmatrix} 1 & 1 & -1 \\ 1 & 1 & -1 \\ -3 & -3 & 3 \end{bmatrix},$$

and

$$X_1 = A_0^\dagger - A_0^\dagger A_1 X_0 - X_0 A_1 A_0^\dagger + X_0 A_1 A_0^\dagger A_1 X_0 = \frac{1}{4} \begin{bmatrix} 2 & -1 & -1 \\ 0 & 1 & 1 \\ 2 & -1 & -1 \end{bmatrix}.$$

We note that the above expressions for X_0 and X_1 are identical to (2.29) *and* (2.30).

2.2.6 ▪ The case of linear perturbation

Let us analyze an important particular case of the linear perturbation

$$A(z) = A + zB. \tag{2.37}$$

First we show that the coefficients of the Laurent series for $A^{-1}(z)$ satisfy an elegant matrix recursion. The reader will observe that coefficients can be readily calculated once Y_{-1} and Y_0 are known. The latter have already been given closed form expressions (2.35) and (2.36) in the generic case of the first order pole. Note that X_0 in (2.35) corresponds to Y_{-1} and X_1 in (2.36) corresponds to Y_0. The general case is covered by formula (2.63), derived later.

Theorem 2.11. *If $A^{-1}(z)$ exists in a punctured neighborhood of $z = 0$, it can be expanded as a Laurent series*

$$A^{-1}(z) = \frac{1}{z^s} Y_{-s} + \cdots + \frac{1}{z} Y_{-1} + Y_0 + zY_1 + \ldots,$$

where the coefficients Y_n, $n = -s, -s+1, \ldots$, satisfy the following recursions:

$$Y_{k+1} = (-Y_0 B) Y_k, \quad k = 0, 1, \ldots, \tag{2.38}$$

$$Y_{-k-1} = (-Y_{-1} A) Y_{-k}, \quad k = 1, \ldots, s-1. \tag{2.39}$$

Moreover, the projections $P := BY_{-1}$ and $\tilde{P} := Y_{-1} B$ separate the regular and singular parts of $A^{-1}(z)$. Namely,

$$A_R^{-1}(z) := \sum_{k=0}^{\infty} z^k Y_k = A^{-1}(z)(I - P) = (I - \tilde{P}) A^{-1}(z), \tag{2.40}$$

$$A_S^{-1}(z) := \sum_{k=-s}^{-1} z^k Y_k = A^{-1}(z)P = \tilde{P}A^{-1}(z). \tag{2.41}$$

Proof: The existence of the Laurent series follows immediately from Theorem 2.4. The regular part of the identity $A(z)A^{-1}(z) = I$ yields

$$A(z)A_R^{-1}(z) + BY_{-1} = I.$$

Premultiplication by $A^{-1}(z)$ and retaining the terms with positive powers of z gives (2.40). It then follows that

$$A_S^{-1}(z) = A^{-1}(z) - A_R^{-1}(z) = A^{-1}(z)P,$$

which yields (2.41). The coefficient of z^{-1} in the above equation is

$$Y_{-1} = Y_{-1}P.$$

Premultiplication of the above equation by B gives $P = P^2$, which shows that P is a projection. The corresponding results for $\tilde{P}$ are obtained in a similar manner using the identity $A^{-1}(z)A(z) = I$. For later reference, we note that the coefficient of z^{-1} in the identity $A(z)A^{-1}(z) = I$ is

$$AY_{-1} + BY_{-2} = 0. \tag{2.42}$$

Next we use the following resolvent identity (see Problem 2.6):

$$A^{-1}(z_2) - A^{-1}(z_1) = (z_1 - z_2)A^{-1}(z_2)BA^{-1}(z_1).$$

By projection of the resolvent identity, with the help of P and $\tilde{P}$, we obtain separate resolvent identities for the regular and singular parts

$$A_R^{-1}(z_2) - A_R^{-1}(z_1) = (z_1 - z_2)A_R^{-1}(z_2)BA_R^{-1}(z_1) \tag{2.43}$$

and

$$A_S^{-1}(z_2) - A_S^{-1}(z_1) = (z_1 - z_2)A_S^{-1}(z_2)BA_S^{-1}(z_1). \tag{2.44}$$

To derive (2.38), we set $z_2 = 0$ in (2.43) to obtain $Y_0 = A_R^{-1}(0) = (I + z_1 Y_0 B)A_R^{-1}(z_1)$. Hence,

$$A_R^{-1}(z_1) = (I + z_1 Y_0 B)^{-1} Y_0 = \sum_{k=0}^{\infty} (-z_1 Y_0 B)^k Y_0,$$

from which (2.38) follows immediately. To derive (2.39), we first note that the coefficient of z_1^{-1} in (2.44) is

$$-Y_{-1} = A_S^{-1}(z_2)BY_{-2} - z_2 A_S^{-1}(z_2)BY_{-1}.$$

Then, we substitute BY_{-2} by the value obtained from (2.42) and replace $A_S^{-1}(z_2)BY_{-1}$ by

$$A_S^{-1}(z_2)BY_{-1} = A_S^{-1}(z_2)P = A_S^{-1}(z_2)$$

to obtain

$$A_S^{-1}(z_2)(z_2 I + AY_{-1}) = Y_{-1}.$$

Thus, for all sufficiently large z we have

$$A_S^{-1}(z) = z^{-1}Y_{-1}(I + z^{-1}AY_{-1})^{-1} = \sum_{k=0}^{\infty} z^{-k-1} Y_{-1}(-AY_{-1})^k.$$

Since we know that $A^{-1}(z)$ has a finite order pole, the above series is finite and hence converges for any nonzero value of z. The recursive formula (2.39) follows immediately from the above expansion. □

It is worth noting that formula (2.38) is in fact a generalization of the formula $Y_{k+1} = (-A^{-1}B)Y_k$, $k = 0, 1, \ldots$, from the regular to the singular case.

Next we show how the order of singularity can be reduced in a successive manner. Let $V = [V_1\ V_2]$ be a unitary matrix such that the columns of V_1 form a basis for the null space of A. In particular, we have

$$\tilde{A} = AV = [0\ \tilde{A}_2], \quad \tilde{B} = BV = [\tilde{B}_1\ \tilde{B}_2].$$

Now let $U = [U_1\ U_2]$ be a unitary matrix such that

$$U^*\tilde{B}_1 = \begin{bmatrix} \bar{B}_{11} \\ 0 \end{bmatrix},$$

where $\bar{B}_{11}$ is square. Then,

$$U^*(A + zB)V = \begin{bmatrix} 0 & \bar{A}_{12} \\ 0 & \bar{A}_{22} \end{bmatrix} + z \begin{bmatrix} \bar{B}_{11} & \bar{B}_{12} \\ 0 & \bar{B}_{22} \end{bmatrix}.$$

We note that if we assume that $A^{-1}(z)$ exists in some punctured neighborhood around $z = 0$, the inverse $\bar{B}_{11}^{-1}$ exists as well. Hence, we can write

$$\begin{aligned} A^{-1}(z) &= V \begin{bmatrix} z\bar{B}_{11} & \bar{A}_{12} + z\bar{B}_{12} \\ 0 & \bar{A}_{22} + z\bar{B}_{22} \end{bmatrix}^{-1} U^* \\ &= V \begin{bmatrix} z^{-1}\bar{B}_{11}^{-1} & -z^{-1}\bar{B}_{11}^{-1}(\bar{A}_{12} + z\bar{B}_{12})(\bar{A}_{22} + z\bar{B}_{22})^{-1} \\ 0 & (\bar{A}_{22} + z\bar{B}_{22})^{-1} \end{bmatrix} U^*. \end{aligned}$$

Thus, the existence of $A^{-1}(z)$ is equivalent to the existence of the inverses $\bar{B}_{11}^{-1}$ and $(\bar{A}_{22} + z\bar{B}_{22})^{-1}$. Of course, now one can again apply the same procedure to the inversion of $\bar{A}(z) = \bar{A}_{22} + z\bar{B}_{22}$. Since the dimension of $\bar{A}_{22}$ is strictly less than the dimension of A, the procedure is terminated with the regular perturbation problem after a finite number of steps. In fact, it is terminated after exactly s steps, where s is the order of the pole of the Laurent series for $A^{-1}(z)$.

In the generic case of the first order pole, we can expand $(\bar{A}_{22} + z\bar{B}_{22})^{-1}$ as follows:

$$(\bar{A}_{22} + z\bar{B}_{22})^{-1} = \bar{A}_{22}^{-1} - z\bar{A}_{22}^{-1}\bar{B}_{22}\bar{A}_{22}^{-1} + \ldots.$$

Consequently, in the generic case the singular part coefficient Y_{-1} and the first coefficient Y_0 of the regular part are given by

$$Y_{-1} = V \begin{bmatrix} \bar{B}_{11}^{-1} & -\bar{B}_{11}^{-1}\bar{A}_{12}\bar{A}_{22}^{-1} \\ 0 & 0 \end{bmatrix} U^* \tag{2.45}$$

and

$$Y_0 = V \begin{bmatrix} 0 & \bar{B}_{11}^{-1}(\bar{A}_{12}\bar{A}_{22}^{-1}\bar{B}_{22} - \bar{B}_{12})\bar{A}_{22}^{-1} \\ 0 & \bar{A}_{22}^{-1} \end{bmatrix} U^*. \tag{2.46}$$

Example 2.4 (continued from Subsection 2.2.5). *Normalizing the vector*

$$\begin{bmatrix} 1 \\ 1 \\ -3 \end{bmatrix}$$

and using the Gram–Schmidt orthogonalization procedure to complete the basis, we obtain

$$V = \begin{bmatrix} 0.3015 & 0.9535 & 0.0 \\ 0.3015 & -0.0953 & 0.9487 \\ -0.9045 & 0.2860 & 0.3162 \end{bmatrix}.$$

Next, to find the matrix U we apply the QR factorization to

$$\tilde{B} = BV = \begin{bmatrix} 0.0 & 1.0488 & -0.9487 \\ 1.2060 & -0.3814 & 0.6325 \\ -1.2060 & -0.6674 & 0.3162 \end{bmatrix}.$$

The factor Q corresponds to U, and the factor R corresponds to $\bar{B}$. Thus, we have

$$\bar{A} = U^* A V = \begin{bmatrix} 0.0 & -0.7416 & -1.5652 \\ 0.0 & -1.2845 & -0.1291 \\ 0.0 & 0.0 & 3.6515 \end{bmatrix},$$

$$\bar{B} = U^* B V = \begin{bmatrix} 1.7056 & 0.2023 & 0.2236 \\ 0.0 & -1.2845 & 1.1619 \\ 0.0 & 0.0 & 0.0 \end{bmatrix}.$$

Consequently, using (2.45) *and* (2.46), *we obtain*

$$Y_{-1} = \begin{bmatrix} 0.125 & 0.125 & -0.125 \\ 0.125 & 0.125 & -0.125 \\ -0.375 & -0.375 & 0.375 \end{bmatrix},$$

$$Y_0 = \begin{bmatrix} 0.5 & -0.25 & -0.25 \\ 0 & 0.25 & 0.25 \\ 0.5 & -0.25 & -0.25 \end{bmatrix}.$$

Then, the subsequent regular coefficients $Y_1, Y_2, \ldots$ can be calculated by the recursion (2.38).

2.2.7 ▪ The case of polynomial perturbation

Here we treat the polynomial perturbation, which is yet another special case of the analytic perturbation, namely,

$$A(z) = A_0 + zA_1 + \cdots + z^p A_p. \tag{2.47}$$

Naturally, $A(z)$ is also referred to as a polynomial matrix. First let us recall the *Smith normal form* for polynomial matrices. There exist unimodular matrices $U(z)$ and $V(z)$ (i.e., the determinants of $U(z)$ and $V(z)$ are nonzero constants) such that

$$U(z)A(z)V(z) = \Lambda(z), \tag{2.48}$$

where $\Lambda(z) = \text{diag}\{0,\ldots,0,\lambda_1(z),\ldots,\lambda_r(z)\}$, r is the generic rank of $A(z)$, and $\lambda_i, i = 1,\ldots,r$, are unique monic polynomials satisfying the divisibility property

$$\lambda_{i+1}(z) \mid \lambda_i(z), \quad i = 1,\ldots,r-1.$$

The matrix $\Lambda(z)$ is called the Smith normal form of $A(z)$.

The Smith normal form is obtained using the elementary row and column operations. By the elementary row and column operations we mean

- interchange of any two columns (or rows);
- addition to any column (row) of a polynomial multiple of any other column (row);
- scaling any column (row) by any nonzero real or complex number.

Example 2.5. *For example, we can obtain the Smith normal form of the matrix*

$$A(z) = \begin{bmatrix} 1+z & 2-z & 1 \\ -1 & 1+z & -z \\ -z & 3 & 1+z \end{bmatrix}$$

by using the Maple command "SmithForm" to obtain

$$\Lambda(z) = \begin{bmatrix} 1 & 0 & 0 \\ 0 & 1 & 0 \\ 0 & 0 & z+z^2 \end{bmatrix},$$

$$U(z) = \begin{bmatrix} 1 & 0 & 0 \\ -\frac{2}{3}-\frac{1}{6}z & \frac{1}{3}-\frac{1}{6}z & \frac{2}{3}-\frac{1}{6}z \\ \frac{1}{8}+\frac{5}{8}z+\frac{1}{8}z^2 & \frac{1}{8}-\frac{1}{8}z+\frac{1}{8}z^2 & -\frac{1}{8}-\frac{3}{8}z+\frac{1}{8}z^2 \end{bmatrix},$$

$$V(z) = \begin{bmatrix} 0 & 0 & 1 \\ 0 & 1 & 1+\frac{4}{3}z \\ 1 & -2+z & -3-\frac{8}{3}z+\frac{4}{3}z^2 \end{bmatrix}.$$

Let us now apply the Smith normal form to the inversion of the polynomial matrices. Suppose, as before, that $A(z)$ has an inverse in some punctured disc around $z = 0$. Then, $r = \dim A(z) = n$, and from (2.48) one can see that

$$A^{-1}(z) = V(z)\Lambda^{-1}(z)U(z). \tag{2.49}$$

From the unimodularity of the matrix polynomials $U(z)$ and $V(z)$, it follows that in the case of singular perturbation, the polynomial $\lambda_r(z)$ has the structure

$$\lambda_r(z) = z^s(z^l + a_{l-1}z^{l-1} + \cdots + a_1 z + a_0),$$

where s is the order of the pole of $A^{-1}(z)$ at $z = 0$. Since $\Lambda(z)$ is diagonal, one easily obtains the Laurent series for its inverse,

$$\Lambda^{-1}(z) = \frac{1}{z^s}[\Lambda_0^{(-1)} + z\Lambda_1^{(-1)} + z^2\Lambda_2^{(-1)} + \ldots]. \tag{2.50}$$

And because all $\lambda_i(z)$ divide $\lambda_r(z)$, the series coefficients $\Lambda_k^{(-1)}$ satisfy the recursion equation

$$\sum_{m=0}^{l} a_m \Lambda_{k-m}^{(-1)} = 0$$

for $k \geq l$.

Next, we show that the same recursion holds for the matrix coefficients X_k of the Laurent series

$$A^{-1}(z) = \frac{1}{z^s}(X_0 + zX_1 + z^2 X_2 + \ldots).$$

Proposition 2.1. *Let p and q be the orders of polynomial matrices $U(z)$ and $V(z)$, respectively. Then, for $k \geq p+q+l$, the Laurent series coefficients X_k satisfy the recursion equation*

$$\sum_{m=0}^{l} a_m X_{k-m} = 0.$$

Proof: Substituting $U(z) = U_0 + zU_1 + \cdots + z^p U_p$, $V(z) = V_0 + zV_1 + \cdots + z^q V_q$, and the Laurent series (2.50) into (2.49), we obtain the formula

$$X_k = \sum_{i=0}^{p+q} \sum_{\mu+\nu=i} V_\mu \Lambda_{k-i}^{(-1)} U_\nu,$$

where the terms with $\nu > p$ and $\mu > q$ are considered to be zero. Using the above expression for X_k, we can write

$$\sum_{m=0}^{l} a_m X_{k-m} = \sum_{m=0}^{l} a_m \sum_{i=0}^{p+q} \sum_{\mu+\nu=i} V_\mu \Lambda_{k-m-i}^{(-1)} U_\nu,$$

$$= \sum_{i=0}^{p+q} \sum_{\mu+\nu=i} V_\mu \sum_{m=0}^{l} a_m \Lambda_{k-i-m}^{(-1)} U_\nu.$$

Since $\sum_{m=0}^{l} a_m \Lambda_{k-i-m}^{(-1)} = 0$ for $k \geq p+q+l$, the above expression is equal to zero as well. □

Example 2.4 (continued from Subsection 2.2.5). *As was noted in the previous section, the regular part coefficients can be calculated by (2.38). Specifically, we have already derived*

$$Y_1 = (-Y_0 B) Y_0 = \frac{1}{8}\begin{bmatrix} -3 & 3 & 3 \\ 1 & -1 & -1 \\ -3 & 3 & 3 \end{bmatrix}$$

and

$$Y_2 = (-Y_0 B) Y_1 = \frac{1}{8}\begin{bmatrix} 3 & -3 & -3 \\ -1 & 1 & 1 \\ 3 & -3 & -3 \end{bmatrix}.$$

It turns out that $Y_2 = -Y_1$, which is not evident given that

$$-Y_0 B = \frac{1}{4}\begin{bmatrix} -3 & 3 & 0 \\ 1 & -1 & 0 \\ -3 & 3 & 0 \end{bmatrix}.$$

However, this fact can be explained with the help of the Smith normal form. First, we note that in this case

$$\lambda_r(z) = z(z+1)$$

(see Example 2.5). The first factor z of $\lambda_r(z)$ implies that $s = 1$, and the second factor $z+1$ implies that the recursion $Y_{k+1} = -Y_k$, for $k \geq 1$, holds.

Next we show that inversion of a polynomial perturbation is equivalent to inversion of a corresponding linear perturbation in augmented space. There might be a number of methods to implement this idea. Here we discuss two schemes.

Theorem 2.12. *Let the polynomial matrix (2.47) have an inverse for $z \neq 0$ and sufficiently small. Consider the linearly perturbed system*

$$[\mathcal{A} + z\mathcal{B}]\mathcal{X}(z) = \mathcal{I}, \tag{2.51}$$

where the matrices $\mathcal{A} \in \mathbb{C}^{np\times np}$, $\mathcal{B} \in \mathbb{C}^{np\times np}$, and $\mathcal{I} \in \mathbb{C}^{np\times n}$ are given by

$$\mathcal{A} := \begin{bmatrix} A_0 & A_1 & \cdots & A_{p-1} \\ 0 & I & \cdots & 0 \\ \vdots & \vdots & \ddots & \vdots \\ 0 & 0 & \cdots & I \end{bmatrix}, \quad \mathcal{B} := \begin{bmatrix} 0 & 0 & \cdots & A_p \\ -I & 0 & \cdots & 0 \\ \vdots & \ddots & \ddots & \vdots \\ 0 & \cdots & -I & 0 \end{bmatrix}, \quad \mathcal{I} := \begin{bmatrix} I \\ 0 \\ \vdots \\ 0 \\ 0 \end{bmatrix}$$

and the matrix $\mathcal{X}(z) = [X_1(z), \ldots, X_p(z)]^T$ has the corresponding block structure. Then, $A^{-1}(z) = X_1(z)$ for $z \neq 0$ and sufficiently small.

Proof: Taking into account the block structure of (2.51), we can write

$$\begin{aligned} A_0X_1(z) + \cdots + A_{p-1}X_p(z) + zA_pX_p(z) &= I, \\ X_2(z) - zX_1(z) &= 0, \\ X_3(z) - zX_2(z) &= 0, \\ &\vdots \\ X_p(z) - zX_{p-1}(z) &= 0, \end{aligned}$$

or, equivalently,

$$\begin{aligned} (A_0 + \cdots + z^pA_p)X_1(z) &= I, \\ X_2(z) &= zX_1(z), \\ X_3(z) &= zX_2(z), \\ &\vdots \\ X_p(z) &= zX_{p-1}(z), \end{aligned}$$

which proves the theorem. □

Theorem 2.13. *Let the polynomial matrix* (2.47) *have an inverse for* $z \neq 0$ *and sufficiently small. Define augmented matrices* $\mathscr{A} \in \mathbb{C}^{np\times np}$, $\mathscr{B} \in \mathbb{C}^{np\times np}$, *and* $\mathscr{I} \in \mathbb{C}^{np\times n}$ *by setting*

$$\mathscr{A} := \begin{bmatrix} A_0 & 0 & \cdots & 0 \\ A_1 & A_0 & \cdots & 0 \\ \vdots & \vdots & \ddots & \vdots \\ A_{p-1} & A_{p-2} & \cdots & A_0 \end{bmatrix}, \quad \mathscr{B} := \begin{bmatrix} A_p & A_{p-1} & \cdots & A_1 \\ 0 & A_p & \cdots & A_2 \\ \vdots & \vdots & \ddots & \vdots \\ 0 & 0 & \cdots & A_p \end{bmatrix}, \quad \mathscr{I} := \begin{bmatrix} I \\ 0 \\ \vdots \\ 0 \end{bmatrix}.$$

Then the solution of the linear perturbation problem

$$(\mathscr{A} + z\mathscr{B})\mathscr{X}(z) = z^{s_*}\mathscr{I}$$

can be expanded as a Laurent series

$$\mathscr{X}(z) = \frac{1}{z^{s_*}} \begin{bmatrix} 0 \\ \vdots \\ 0 \\ X_0 \\ \vdots \\ X_{mod(s,m)} \end{bmatrix} + \frac{1}{z^{s_*-1}} \begin{bmatrix} X_{mod(s,m)+1} \\ \vdots \\ \\ \vdots \\ X_{mod(s,m)+m} \end{bmatrix} + \ldots, \tag{2.52}$$

where $s_* = \lfloor s/m \rfloor + 1$ *and* $\{X_k\}_{k=0}^{\infty}$ *are coefficients of the Laurent series* (2.19), *and where* $\lfloor s/m \rfloor$ *is the integer part and* $mod(s,m)$ *is the remainder of the division of* s *by* m.

Proof: See Problem 2.7. □

Thus, the method of Subsection 2.2.6 for the linear perturbation can be applied to the polynomial perturbation via the transformations described in Theorems 2.12 and 2.13.

Each of the presented augmentation schemes has its own merits. Using the first method to obtain the first k coefficients of the Laurent series for $A^{-1}(z)$, one needs to calculate k augmented coefficients $\mathscr{X}_i, i = 0, \ldots, k-1$, whereas if one utilizes the second augmentation scheme, one needs to compute about m times fewer augmented coefficients. However, in the first method both augmented matrices $\mathscr{A}$ and $\mathscr{B}$ are close to upper triangular form and have a lot of zero elements. Since the procedure of Subsection 2.2.6 is based on simultaneous reduction of matrices $\mathscr{A}$ and $\mathscr{B}$ to upper block triangular form, each iteration of the first method could be more computationally efficient.

Now we show that, in fact, the results on polynomial perturbation can be applied to the general case of analytic perturbation (2.18). Suppose again that the inverse $A^{-1}(z)$ exists in some punctured neighborhood around $z = 0$. Then according to Theorem 2.6 the number of terms in (2.18) that uniquely determine the inversion procedure is finite. Namely, there exists m such that the inverse $(A_0 + \cdots + z^m A_m)^{-1}$ exists for sufficiently small z. Moreover, any $m \geq s$ can be taken. Therefore, we may write

$$\begin{aligned} A^{-1}(z) &= [(A_0 + \cdots + z^s A_s) + z^{s+1}A_{s+1} + \ldots]^{-1} \\ &= [(A_0 + \cdots + z^s A_s)(I + (A_0 + \cdots + z^s A_s)^{-1}(z^{s+1}A_{s+1} + \ldots))]^{-1} \\ &= [I + (A_0 + \cdots + z^s A_s)^{-1}(z^{s+1}A_{s+1} + \ldots)]^{-1}(A_0 + \cdots + z^s A_s)^{-1}. \end{aligned}$$

Thus, we can apply an augmentation approach for polynomial perturbation outlined in either Theorem 2.12 or Theorem 2.13 together with the method for linear perturbation of Subsection 2.2.6.

2.2.8 ▪ One step reduction method

We now present a generalization of the approach given in Subsection 2.2.5 for the case $s=1$. This approach allows us to perform computations in spaces of lower dimensions. In particular, this scheme should be used in the case when it is easy to obtain the bases for the null spaces of A_0 and A_0^*.

The method of this subsection is based on the so-called *reduction technique*. The essence of the reduction technique is captured by the following theorem.

Remark 2.4. *In the next theorem it is important to observe that the reduced system has the same form as the original, but the number of matrix equations is decreased by one and the coefficients are reduced in size to matrices in $\mathbb{C}^{p\times p}$, where p is the dimension of $N(C_0)$ or, equivalently, the number of redundant equations defined by the matrix coefficient C_0. Typically, the dimension of the null space $N(C_0)$ is significantly smaller than m.*

Theorem 2.14. *Let $\{C_k\}_{k=0}^t \subseteq \mathbb{C}^{m\times m}$ and $\{R_k\}_{k=0}^t \subseteq \mathbb{C}^{m\times n}$, with $m \le n$, and suppose that the system of $t+1$ matrix equations*

$$\sum_{i=0}^{k} C_i V_{k-i} = R_k, \quad k=0,\dots,t, \tag{2.53}$$

is feasible. Then the general solution of this system is given by

$$V_k = C_0^{\dagger}\left(R_k - \sum_{i=1}^{k} C_i V_{k-i}\right) + Q W_k, \tag{2.54}$$

where $C_0^{\dagger}$ is the Moore–Penrose generalized inverse of C_0 and $Q \in \mathbb{C}^{m\times p}$ is any matrix whose columns form a basis for the right null space of C_0. Furthermore, the sequence of matrices W_k, $0 \le k \le t-1$, solves a reduced set of t matrix equations

$$\sum_{i=0}^{k} D_i W_{k-i} = S_k, \quad k=0,\dots,t-1, \tag{2.55}$$

where the matrices $D_k \in \mathbb{C}^{p\times p}$ and $S_k \in \mathbb{C}^{p\times n}$, $0 \le k \le t-1$, are computed by the following recursion. Set $U_0 = C_1$ and calculate

$$U_k = C_{k+1} - \sum_{i=1}^{k} C_i C_0^{\dagger} U_{k-i}, \quad k=1,\dots,t-1. \tag{2.56}$$

Then,

$$D_k = M^* U_k Q \quad \text{and} \quad S_k = M^*\left(R_{k+1} - \sum_{i=0}^{k} U_i C_0^{\dagger} R_{k-i}\right), \tag{2.57}$$

where $M^* \in \mathbb{C}^{p\times m}$ is any matrix whose rows form a basis for the left null space of C_0.

Proof: According to Lemma 2.1, the general solution to the matrix equation (2.53) with $k=0$ can be written in the form

$$V_0 = C_0^{\dagger} R_0 + Q W_0, \tag{2.58}$$

where $W_0 \in \mathbb{C}^{p\times n}$ is some arbitrary matrix.

In order for the equation

$$C_0 V_1 = R_1 - C_1 V_0$$

to be feasible, we need that the right-hand side $R_1 - C_1 V_0$ belongs to $R(C_0) = N^{\perp}(C_0^*)$ (see Lemma 2.2), that is,

$$M^*(R_1 - A_1 V_0) = 0,$$

where the rows of M^* form a basis for $N(C_0^*)$. Substituting expression (2.58) for the general solution V_0 into the above feasibility condition, one finds that W_0 satisfies the equation

$$M^*(R_1 - C_1(C_0^{\dagger} R_0 + Q W_0)) = 0,$$

which can be rewritten as

$$M^* C_1 Q W_0 = M^*(R_1 - C_1 C_0^{\dagger} R_0).$$

Thus we have obtained the first reduced fundamental equation (2.55) with $k = 0$ with

$$D_0 := M^* U_0 Q \quad \text{and} \quad S_0 := M^*(R_1 - U_0 C_0^{\dagger} R_0).$$

where $U_0 = C_1$. Next we observed that the general solution of equation (2.53) with $k = 1$ is represented by the formula

$$V_1 = C_0^{\dagger}(R_1 - C_1 V_0) + Q W_1 \tag{2.59}$$

with $W_1 \in \mathbb{C}^{p \times n}$. Moving on and applying the feasibility condition of Lemma 2.2 to (2.53) with $k = 2$, we obtain

$$M^*(R_2 - (C_1 V_1 + C_2 V_0)) = 0,$$

and again the substitution of expressions (2.58) and (2.59) into the above condition yields

$$M^* C_1 (C_0^{\dagger}(R_1 - C_1[C_0^{\dagger} R_0 + Q W_0]) + Q W_1) + M^* C_2 (C_0^{\dagger} R_0 + Q W_0) = M^* R_2,$$

which is rearranged to give

$$M^* C_1 Q W_1 + M^*(C_2 - C_1 C_0^{\dagger} C_1) Q W_0 = M^*(R_2 - C_1 C_0^{\dagger} R_1 - (C_2 - C_1 C_0^{\dagger} C_1) C_0^{\dagger} R_0).$$

The last equation is the reduced equation (2.55) with $k = 1$ with

$$D_1 := M^* U_1 Q \quad \text{and} \quad S_1 := M^*(R_2 - U_0 C_0^{\dagger} R_1 - U_1 C_0^{\dagger} R_0),$$

where $U_1 = C_2 - C_1 C_0^{\dagger} U_0$. Note that this equation imposes restrictions on W_1 as well as on W_0. By proceeding in the analogous way, we eventually obtain the complete system of equations (2.55) with coefficients given by formulae (2.56) and (2.57), each of which can be proved by induction in a straightforward way. □

Now, as in the previous subsection, we suppose that the coefficients X_i, $0 \le i \le k-1$, have already been determined. Then, by Theorem 2.6, the next coefficient X_k is the unique solution to the subsystem of fundamental equations

$$\begin{aligned}
A_0 X_k &= J_k - \textstyle\sum_{i=1}^{k} A_i X_{k-i}, \\
A_0 X_{k+1} + A_1 X_k &= J_{k+1} - \textstyle\sum_{i=1}^{k} A_{i+1} X_{k-i}, \\
&\;\vdots \\
A_0 X_{k+s} + \cdots + A_s X_k &= J_{k+s} - \textstyle\sum_{i=1}^{k} A_{i+s} X_{k-i}.
\end{aligned} \tag{2.60}$$

The above system is like the one given in (2.53) with $C_i = A_i$, $0 \le i \le s$, and with $R_j = J_{k+j} - \sum_{i=1}^{k} A_{i+j} X_{k-i}$, $0 \le j \le s$. Therefore, we can apply the reduction technique described in Theorem 2.14.

Specifically, let $p = \dim(N(A_0))$ be the dimension of the null space of A_0, let $Q \in \mathbb{C}^{n\times p}$ be a matrix whose p columns form a basis for the right null space of A_0, and let $M^* \in \mathbb{C}^{p\times n}$ be a matrix whose p rows form a basis for the left null space of A_0. Of course, although $p=0$ and hence $s=0$ is possible, we are interested in the singular case when $p \ge 1$.

The application of Theorem 2.14 results in the system

$$\begin{aligned} D_0 W_0 &= S_0, \\ D_0 W_1 + D_1 W_0 &= S_1, \\ &\vdots \\ D_0 W_{s-1} + \cdots + D_{s-1} W_0 &= S_{s-1}, \end{aligned} \tag{2.61}$$

where the coefficients D_i and S_i, $i = 0, \ldots, s-1$, are calculated by the recursive formulae (2.56) and (2.57).

It is expected that in many practical applications p is much less than n, and hence the above system (2.61) with $D_i \in \mathbb{C}^{p\times p}$ is much smaller than the original system (2.60).

Now we have two options. We can either apply the reduction technique again (see the next subsection for more details) or solve the reduced system directly by using the generalized inverse approach. In the latter case, we define

$$\mathscr{D}^{(t)} \stackrel{def}{=} \begin{bmatrix} D_0 & 0 & 0 & \cdots & 0 \\ D_1 & D_0 & 0 & \cdots & 0 \\ D_2 & D_1 & D_0 & \cdots & 0 \\ \vdots & \vdots & \vdots & \ddots & \vdots \\ D_t & D_{t-1} & \cdots & D_1 & D_0 \end{bmatrix}$$

and

$$\mathscr{H}^{(t)} \stackrel{def}{=} [\mathscr{D}^{(t)}]^\dagger = \begin{bmatrix} H_{00}^{(t)} & \cdots & H_{0t}^{(t)} \\ \vdots & \ddots & \vdots \\ H_{t0}^{(t)} & \cdots & H_{tt}^{(t)} \end{bmatrix}.$$

Then, by carrying out a computation similar to that presented in the proof of Theorem 2.8, we obtain

$$W_0 = \sum_{i=0}^{s-1} H_{0i}^{(s-1)} S_i.$$

Once W_0 is determined it is possible to obtain X_k from the formula

$$X_k = A_0^\dagger R_0 + Q W_0 = A_0^\dagger R_0 + Q \sum_{i=0}^{s-1} H_{0i}^{(s-1)} S_i.$$

Furthermore, substituting for S_i, $0 \le i \le s-1$, from (2.57) and changing the order of summation gives

$$\begin{aligned} X_k = {} & \left(A_0^\dagger - \sum_{i=0}^{s-1} Q H_{0i}^{(s-1)} M^* U_i A_0^\dagger \right) R_0 \\ & + \sum_{j=1}^{s} \left(Q H_{0j-1}^{(s-1)} M^* - \sum_{i=j}^{s-1} Q H_{0i}^{(s-1)} M^* U_{i-j} A_0^\dagger \right) R_j. \end{aligned} \tag{2.62}$$

Note that, by convention, a sum is set to 0 when the lower limit is greater than the upper limit. Now, substituting $R_j = \delta_{s,k+j} - \sum_{i=1}^{k} A_{i+j} X_{k-i}$, $0 \le j \le s$, into the expression (2.62), we obtain the explicit recursive formula for the Laurent series coefficients

$$X_k = \left(A_0^\dagger - \sum_{i=0}^{s-1} Q H_{0i}^{(s-1)} M^* U_i A_0^\dagger \right) \left(\delta_{s,k} - \sum_{i=1}^{k} A_i X_{k-i} \right) \tag{2.63}$$
$$+ \sum_{j=1}^{s} \left(Q H_{0j-1}^{(s-1)} M^* - \sum_{i=j}^{s-1} Q H_{0i}^{(s-1)} M^* U_{i-j} A_0^\dagger \right) \left(\delta_{s,k+j} - \sum_{i=1}^{k} A_{i+j} X_{k-i} \right)$$

for all $k \ge 1$. In particular, the coefficient of the first singular term in (2.19) is given by the formula

$$X_0 = Q H_{0s-1}^{(s-1)} M^*. \tag{2.64}$$

2.2.9 ▪ Complete reduction method

As was pointed out in the previous subsection, the reduced system has essentially the same structure as the original one, and hence one can again apply the reduction step described in Theorem 2.14. Note that each time the reduction step is carried out, the number of matrix equations is reduced by one. Therefore, one can perform up to s reduction steps. We now outline how these steps can be executed. We start by introducing the sequence of reduced systems. The fundamental matrix equations for the lth reduction step are

$$\begin{aligned} A_0^{(l)} X_0^{(l)} &= R_0^{(l)}, \\ A_0^{(l)} X_1^{(l)} + A_1^{(l)} X_0^{(l)} &= R_1^{(l)}, \\ &\vdots \\ A_0^{(l)} X_{s-l}^{(l)} + \cdots + A_{s-l}^{(l)} X_0^{(l)} &= R_{s-l}^{(l)}. \end{aligned}$$

With $l = 0$, one obtains the original system of fundamental equations, and with $l = 1$ one obtains the reduced system for the first reduction step described in the previous subsection. Initializing with $R_i^{(0)} = 0$, $0 \le i \le s-1$, and $R_s^{(0)} = I$ and with $A_i^{(0)} = A_i$, $0 \le i \le s$, the matrices $A_j^{(l)}$ and $R_j^{(l)}$, $0 \le j \le s-l$, for each reduction step $1 \le l \le s$, can be computed successively by a recursion similar to (2.56) and (2.57). In general we have

$$U_0^{(l)} = A_1^{(l-1)}, \quad U_j^{(l)} = A_{j+1}^{(l-1)} - \sum_{i=1}^{j} A_i^{(l-1)} A_0^{(l-1)\dagger} U_{j-i}^{(l)}, \quad j = 1, \ldots, s-l,$$

$$A_j^{(l)} = M^{(l)*} U_j^{(l)} Q^{(l)}, \quad j = 0, \ldots, s-l,$$

$$R_j^{(l)} = M^{(l)*} \left(-\sum_{i=0}^{j} U_{j-i}^{(l)} A_0^{(l-1)\dagger} R_i^{(l-1)} + R_{j+1}^{(l-1)} \right) \quad j = 0, \ldots, s-l,$$

where $Q^{(l)}$ and $M^{(l)*}$ are the basis matrices for the right and left null spaces, respectively, of the matrix $A_0^{(l-1)}$ and where $A_0^{(l-1)\dagger}$ is the Moore–Penrose generalized inverse of $A_0^{(l-1)}$. After s reduction steps, one obtains the final system of reduced equations

$$A_0^{(s)} X_0^{(s)} = R_0^{(s)}. \tag{2.65}$$

Since X_0 is a unique solution to the subsystem of the first $s+1$ fundamental equations (2.21) and Theorem 2.14 states the equivalence of the lth and $(l+1)$st systems of reduced equations, the system (2.65) possesses a unique solution. Consequently, the matrix $A_0^{(s)}$ is invertible. Thus,

$$X_0^{(s)} = [A_0^{(s)}]^{-1} R_0^{(s)}. \tag{2.66}$$

The original solution $X_0 = X_0^{(0)}$ can be now retrieved by the backward recursive relationship

$$X_0^{(l-1)} = A_0^{(l-1)\dagger} R_0^{(l-1)} + Q^{(l)} X_0^{(l)}, \quad l = s, \dots, 1. \tag{2.67}$$

Now by taking $R_j^{(0)} = J_{k+j} - \sum_{i=1}^{k} A_{i+j} X_{k-i}$, $0 \le j \le s$, we obtain an algorithm for computing the Laurent series coefficients $X_k, k \ge 1$. Of course, recursive formulae similar to (2.28) and (2.63) can be obtained, but they are quite complicated in the general case.

The order s of the pole may also be obtained from the reduction process by continuing the process until $A_0^{(l)}$ becomes nonsingular. The number of reduction steps is equal to the order of the pole. Note also that the sequence of matrices $A_0^{(l)}$, $l \ge 0$, can be computed irrespective of the right-hand sides. Once s is determined, one can compute $R_j^{(l)}$, $1 \le l \le s, 0 \le j \le s - l$.

Of course, an intermediate process with the number of reductions between 1 and s can be used as well. Then, one needs to apply the generalized inverse method to the final step reduced system.

2.3 ▪ Problems

Problem 2.1. Verify that the SVD-based decomposition of the Moore–Penrose generalized inverse

$$A^\dagger = U \begin{bmatrix} D^{-1} & 0 \\ 0 & 0 \end{bmatrix} V^*$$

satisfies equations

$$AA^\dagger A = A,$$
$$A^\dagger A A^\dagger = A^\dagger,$$
$$(AA^\dagger)^* = AA^\dagger,$$
$$(A^\dagger A)^* = A^\dagger A.$$

Problem 2.2. Prove Lemma 2.2. *Hint: The statement of Lemma 2.2 is equivalent to the fact* $N(A^*) = R(A)^\perp$.

Problem 2.3. Prove that the existence of the group inverse of $A \in \mathbb{C}^{n\times n}$ is equivalent to the decomposition of the space $\mathbb{C}^n$ into a *direct* sum of the null space and the range of A.

Problem 2.4. If we substitute the series expansions

$$A(z) = A_0 + zA_1 + z^2 A_2 + \cdots$$

and

$$A^{-1}(z) = \frac{1}{z^s}(X_0 + zX_1 + \cdots)$$

into the equation $A(z)A^{-1}(z)=I$, we obtain the set of equations

$$\sum_{i=0}^{k} A_i X_{k-i} = \delta_{ks} I, \quad k=0,1,\ldots, \tag{2.68}$$

where δ_{ks} is the Kroneker delta, and if we substitute the above series into the equation $A^{-1}(z)A(z)=I$, we obtain the set of equations

$$\sum_{i=0}^{k} X_{k-i} A_i = \delta_{ks} I, \quad k=0,1,\ldots. \tag{2.69}$$

Prove that the sets of equations (2.68) and (2.69) are equivalent.

Problem 2.5. Verify that the initial term X_0 in the recursion (2.28) is indeed equal to $G_{0s}^{(s)}$.

Problem 2.6. Prove that the linear perturbation $A(z) = A + zB$ satisfies the resolvent type identity

$$A^{-1}(z_2) - A^{-1}(z_1) = (z_1 - z_2)A^{-1}(z_2)BA^{-1}(z_1).$$

Problem 2.7. Prove Theorem 2.13. *Hint: The proof is done by collecting and inspecting coefficients in the equation*

$$(\mathscr{A} + z\mathscr{B})\mathscr{X}(z) = z^{s_*}\mathscr{I}.$$

2.4 ▪ Bibliographic Notes

There is a large body of literature on matrix perturbation theory. One can divide the literature into two main groups. In the first group the starting point is the additive componentwise perturbation. Namely, the perturbed matrix is given by $A_0 + \Delta A$, where A_0 is the original matrix and the perturbation matrix ΔA is bounded. In the second group, the starting point is the analytic perturbation. Namely, the perturbed matrix is given by $A(z)$, where $A(0)$ is the original matrix and $A(z)$ is an analytic function of the perturbation parameter z. A good review of the results on componentwise matrix perturbations can be found in [147, 103]. A similarly good review of the results on the analytic matrix perturbations can be found in [99, 22]. Indeed, Kato's seminal treatise [99] inspired many authors, including ourselves. The results on nonlocal properties of the analytic/polynomial matrix perturbations can be found in the books [70, 71]. In the present book we focused on the topics that are not covered in the above-mentioned books. Numerous applications of the matrix perturbation theory to mechanics, physics, dynamic systems, control, and statistics are given in [33, 34, 59, 70, 71, 84, 99, 103, 108, 109, 135, 143, 147].

The inversion of analytically perturbed matrices and operators has been studied in [22, 66, 67, 69, 100, 104, 111, 133, 139, 141, 151, 157]. The inversion of nearly singular operator-valued functions was probably first studied in the paper by Keldysh [100]. An important particular case of the linear perturbation $A_0 + zA_1$ has been treated in [66, 111, 139, 141, 157]. In the case of linear perturbation Vishik and Lyusternik [157] showed that one can express $A^{-1}(z)$ as a Laurent series as long as $A(z)$ is invertible in some punctured neighborhood of the origin and provided an undetermined coefficient method for the computation of the Laurent series coefficients. The linear perturbation is often called a matrix or operator pencil. Gohberg, Goldberg, and Kaashoek [66] gave a detailed account of the spectral theory for linear pencils when the inverse is analytic in an annular region.

Langenhop [111] showed that the coefficients of the regular part of the Laurent series for the inverse of a linear perturbation form a matrix geometric sequence. The proof of this fact was refined later by Schweitzer [139] and by Schweitzer and Stewart [141]. In particular, the authors of [141] proposed an efficient method for computing the Laurent series coefficients. In [86] and [87] the method of [141] has been extended to operator perturbations on Hilbert spaces.

The notion of the generalized Jordan chains has been developed and applied to the inversion of analytically perturbed matrices and operators in [104, 115, 120, 151, 161]. In particular, Gohberg and Sigal [72] used a local Smith form to elaborate on the structure of the principal part of the Laurent series in terms of generalized Jordan chains. Gohberg, Kaashoek, and Van Schagen [69] refined the results of [72]. A comprehensive study of the Smith form and its application to matrix polynomials can be found in [19, 70]. In [67] matrix- and operator-valued functions are considered from the viewpoint of block-Toeplitz operators. Vainberg and Trenogin [151] used the generalized Jordan chains in combination with the Lyapunov–Schmidt operator for the inversion of analytically perturbed operators. Several recent extensions and applications of the Lyapunov–Schmidt operator approach can be found in [143]. Wilkening [161] proposed a fast and numerically stable algorithm for computing generalized Jordan chains with application to inversion of analytic matrix functions.

Sain and Massey [135] have proposed a rank test to determine the order of the pole of the Laurent series. The rank test has been refined by Howlett [84] and extended to the case of meromorphic matrix functions by Zhou [164]. Howlett [84] also proposed a scheme for computing the coefficients of the Laurent series using Gaussian elimination and showed that for polynomial pencils the coefficients satisfied a recursive relationship.

The methods of Sections 2.2.4, 2.2.5, 2.2.8, and 2.2.9 for the inversion of analytically perturbed matrices have been developed in [8, 13]. In particular, the algebraic reduction process of Sections 2.2.5, 2.2.8, and 2.2.9 can be considered as a counterpart of the complex analysis reduction process proposed by Korolyuk and Turbin [104]. We note that Kato [99] developed the reduction process only for the perturbed eigenvalue problem and not for the inversion of the perturbed operators.

A number of linearization methods are available to transform a problem of analytic perturbation or polynomial perturbation to an equivalent problem of linear perturbation. In Section 2.2.7 we have outlined only two schemes. More linearization schemes can be found in [18, 68, 71, 110].

There are a number of excellent books available on the topic of generalized inverses [23, 35, 159]. Details on the SVD and the other computational methods for the generalized inverse can be found in [148, 159]. In particular, a method for the computation of $A^\dagger$ based on elementary row and column operations (LU decomposition) is presented in [148].

Chapter 3

Perturbation of Null Spaces, Eigenvectors, and Generalized Inverses

3.1 ▪ Introduction

In this chapter we continue to investigate the algebraic finite-dimensional linear system

$$A(z)x(z) = b(z), \tag{3.1}$$

where the matrix $A(z)$ depends analytically on the parameter z. Namely, $A(z)$ can be expanded as a power series

$$A(z) = A_0 + zA_1 + z^2A_2 + \dots$$

with some nonzero radius of convergence.

This chapter covers more advanced cases of algebraic linear systems in comparison with the previous chapter. The material is advanced in both problem formulation and employed techniques. In particular, we are interested in the cases when the matrix $A(z)$ is not square or (and) $A(z)$ is not invertible.

As before, we are primarily interested in the case of *singular perturbation*, that is, when $\text{rank}(A(z)) > \text{rank}(A_0)$ for z different from zero and sufficiently small.

In Section 3.2, we analyze the analytic perturbation of null spaces. This problem can be regarded as the linear system (3.1) with $b(z) = 0$. We then apply our results to the perturbation analysis of the eigenvalue problem.

In Section 3.3 we consider the linear system (3.1), where the matrix $A(z)$ is either not square or not invertible or both. This formulation leads to the perturbation analysis of various generalized inverses, such as Drazin generalized inverse or Moore–Penrose generalized inverse. In contrast to the earlier algebraic approach, in Section 3.3 we use a complex analytic approach. In fact, by using the complex analytic approach we derive elegant recursive formulae for the matrix coefficients of the regular part of the Laurent series for matrix inverse (2.2).

Since we extensively use various concepts of generalized inverses, we suggest that the reader review the material about generalized inverses provided in Section 2.1.

3.2 ▪ Perturbation of Null Spaces and the Eigenvalue Problem

3.2.1 ▪ Problem formulation

The primary goal of this section is to analyze the null space of an analytically perturbed matrix

$$A(\varepsilon)=A_0+\varepsilon A_1+\varepsilon^2 A_2+\dots \tag{3.2}$$

with $A_k \in \mathbb{R}^{n\times n}, k=0,1,\dots,$ when the above series converges in a region $0\le|\varepsilon|\le\varepsilon_{\max}$ for some positive $\varepsilon_{\max}$. In this section we restrict ourselves to the real matrices, as we extensively use the orthogonality concept. Of course, analogous results can be obtained for matrices with complex entries. However, to keep the presentation of the material more transparent we have chosen to work with real matrices. We assume that the unperturbed matrix A_0 has eigenvalue zero with geometric multiplicity $m\ge 1$[2] and that the perturbed matrices $A(\varepsilon)$ also have eigenvalue zero with multiplicity $\bar{m}$ for ε sufficiently small but different from zero. In Theorem 3.1 below we show that the dimension of the perturbed null space does not depend on ε in some small punctured neighborhood around $\varepsilon=0$. When the perturbation parameter ε deviates from zero, the zero eigenvalues of the unperturbed matrix may split into zero and nonzero eigenvalues. This fact implies that $\bar{m}\le m$. We assume that $\bar{m}\ge 1$ and (for computational purposes) that the value of $\bar{m}$ should be known in advance. The case when $\bar{m}=0$ and hence $A(\varepsilon)$ is invertible for $\varepsilon\ne 0$ and sufficiently small was dealt with in Section 2.2. A perturbation is said to be *regular* if it is rank-preserving, $\bar{m}=m$; and it is said to be *singular* if it is non–rank-preserving, $\bar{m}<m$. The following examples clarify the distinction between these two types of perturbation.

Example 3.1. Regular perturbation. *Let the perturbed matrix be given by*

$$A(\varepsilon)=A_0+\varepsilon A_1=\begin{bmatrix}0&1\\0&0\end{bmatrix}+\varepsilon\begin{bmatrix}1&0\\0&0\end{bmatrix}.$$

The null spaces of A_0 and $A(\varepsilon)$ are both one dimensional, and they are spanned, respectively, by

$$\tilde{v}=\begin{bmatrix}1\\0\end{bmatrix},\quad v(\varepsilon)=\begin{bmatrix}1\\-\varepsilon\end{bmatrix}=\begin{bmatrix}1\\0\end{bmatrix}+\varepsilon\begin{bmatrix}0\\-1\end{bmatrix}.$$

We can see that $v(\varepsilon)$ is holomorphic and converges to $\tilde{v}$ as ε goes to zero.

Example 3.2. Singular perturbation. *Let*

$$A(\varepsilon)=A_0+\varepsilon A_1=\begin{bmatrix}0&1&0\\0&1&0\\0&0&0\end{bmatrix}+\varepsilon\begin{bmatrix}1&0&0\\0&0&1\\0&0&0\end{bmatrix}.$$

The null space of A_0 is two dimensional and is spanned by

$$\tilde{v}_1=\begin{bmatrix}1/\sqrt{2}\\0\\1/\sqrt{2}\end{bmatrix},\quad \tilde{v}_2=\begin{bmatrix}-1/\sqrt{2}\\0\\1/\sqrt{2}\end{bmatrix}.$$

The null space of $A(\varepsilon)$ is one dimensional and is spanned by the holomorphic vector-valued function

$$v(\varepsilon)=\begin{bmatrix}1\\-\varepsilon\\1\end{bmatrix}=\begin{bmatrix}1\\0\\1\end{bmatrix}+\varepsilon\begin{bmatrix}0\\-1\\0\end{bmatrix}. \tag{3.3}$$

Thus, we can see that as ε goes to zero, $v(\varepsilon)$ converges to a vector which belongs to the unperturbed null space of matrix A_0, but there is a gap between the dimensions of the perturbed and unperturbed null spaces.

[2]Below we will refer only to the geometric multiplicity.

We denote by $\tilde{v}_i, i = 1,\dots,m$, orthonormal eigenvectors of A_0 corresponding to the eigenvalue zero and form the matrix $\tilde{V} := [\tilde{v}_1,\dots,\tilde{v}_m]$. This matrix satisfies the following equations:

$$A_0\tilde{V} = 0, \tag{3.4}$$

$$\tilde{V}^T\tilde{V} = I_m. \tag{3.5}$$

Similarly, let $v_i(\varepsilon), i = 1,\dots,\bar{m}$, be linearly independent eigenvectors of the perturbed matrix $A(\varepsilon)$ corresponding to the eigenvalue zero. Again, one can form the matrix $V(\varepsilon) := [v_1(\varepsilon),\dots,v_{\bar{m}}(\varepsilon)]$, which satisfies the equation

$$A(\varepsilon)V(\varepsilon) = 0. \tag{3.6}$$

Theorem 3.1. *There exists a holomorphic family of vector-valued functions $v_i(\varepsilon)$ which constitute a basis for the null space of $A(\varepsilon)$ for $\varepsilon \neq 0$.*

Proof: We prove the theorem by construction. First, using elementary row and column operations (see Subsection 2.2.7) we transform the perturbed matrix $A(\varepsilon)$ to the form

$$\tilde{A}(\varepsilon) = \begin{bmatrix} A_1(\varepsilon) & A_2(\varepsilon) \\ 0 & 0 \end{bmatrix},$$

where $A_1(\varepsilon) \in \mathbb{R}^{r\times r}$, $r = n - \bar{m}$, and $\det(A_1(\varepsilon))$ is not identically equal to zero. We note that since $A(\varepsilon)$ is transformed into $\tilde{A}(\varepsilon)$ by unimodular transformation, it is enough to prove the theorem statement for the above form. Consider the "candidate" vector-valued functions

$$\tilde{v}_j(\varepsilon) = \begin{bmatrix} \operatorname{adj}(A_1(\varepsilon))A_{2j}(\varepsilon) \\ -\det(A_1(\varepsilon))e_j \end{bmatrix}, \quad j = 1,\dots,\bar{m},$$

where $A_{2j}(\varepsilon)$ is the jth column of $A_2(\varepsilon)$ and $e_j \in R^{\bar{m}\times 1}$ is the jth canonical basis vector of dimension $\bar{m}$. Next, we check that

$$\begin{aligned}\tilde{A}(\varepsilon)\tilde{v}_j(\varepsilon) &= \begin{bmatrix} A_1(\varepsilon)\operatorname{adj}(A_1(\varepsilon))A_{2j}(\varepsilon) - \det(A_1(\varepsilon))A_{2j}(\varepsilon) \\ 0 \end{bmatrix} \\ &= \begin{bmatrix} \det(A_1(\varepsilon))A_{2j}(\varepsilon) - \det(A_1(\varepsilon))A_{2j}(\varepsilon) \\ 0 \end{bmatrix} = \begin{bmatrix} 0 \\ \vdots \\ 0 \end{bmatrix}.\end{aligned}$$

Clearly each vector $\tilde{v}_j(\varepsilon)$ is analytic, and, by their construction, the complete set of $\bar{m}$ of these spans the null space of $\tilde{A}(\varepsilon)$. □

We would like to note that if in the above theorem $\det(A_1(0)) \neq 0$, the perturbation is regular; otherwise it is singular. Furthermore, the fact that $\det(A_1(\varepsilon))$ can have only isolated zeros implies that the dimension of the perturbed null space is constant for all ε sufficiently small but different from zero.

The above theorem also implies that $V(\varepsilon)$ can be expressed as a power series in some neighborhood of zero, namely,

$$V(\varepsilon) = V_0 + \varepsilon V_1 + \varepsilon^2 V_2 + \cdots. \tag{3.7}$$

Of course, one may always obtain an orthonormal basis from an arbitrary basis by applying a Gram–Schmidt-like procedure over the vectors with elements that are power series expansions. This procedure will be discussed in more detail in Section 3.2.5. However, it is more convenient to construct a "quasi-orthonormal" family of eigenvectors described by the condition

$$V_0^T V(\varepsilon) = I_{\bar{m}}, \tag{3.8}$$

where V_0 is the first coefficient of the power series expansion (3.7) (rather than $V^T(\varepsilon)V(\varepsilon) = I_{\bar{m}}$). Note that even though this family of eigenvectors is not orthonormal for $\varepsilon \neq 0$, it is linearly independent when ε is sufficiently small. Also note that (3.8) was introduced in order to make $V(\varepsilon)$ unique once the leading term V_0 is determined. As we show later, there is some freedom in selecting V_0. As mentioned above, we distinguish between two cases: the rank-preserving case when $\bar{m} = m$ and the non-rank-preserving case when $1 \leq \bar{m} < m$. Note that only in the rank-preserving case it is possible to set $V_0 = \tilde{V}$.

Our main goal is to obtain an efficient recursive algorithm for the computation of coefficients $V_k, k = 0, 1, \ldots$. The algorithm for computing $V_k, k = 0, 1, \ldots$, is based on recursively solving a system of *fundamental equations.* Here the fundamental equations are obtained by substituting (3.2) and (3.7) into (3.6) to yield

$$\sum_{i=0}^{k} A_i V_{k-i} = 0, \quad k = 0, 1, \ldots. \tag{3.9}$$

The quasi-normalization condition (3.8) gives rise to another system of equations,

$$V_0^T V_k = \delta_{0k} I_{\bar{m}}, \quad k = 0, 1, \ldots, \tag{3.10}$$

where δ_{0k} is the Kroneker delta. We will refer to the latter system as the system of *normalization equations.*

We treat the cases of regular and singular perturbations separately. In Section 3.2.2 we provide an algorithm for computing the coefficients V_k, $k \geq 0$, in the regular perturbation case. This algorithm is based on a straightforward recursive procedure. The singular perturbation case is treated in Section 3.2.3, where we suggest three algorithms for computing $\{V_i\}_{i=0}^{\infty}$. The first is based on defining an augmented matrix and using its Moore–Penrose generalized inverse. The second algorithm is based on reducing the dimension of the equations to a set of equations whose type coincides with the rank-preserving case. The third algorithm is a combination of the previous two algorithms and is based on an early termination of the reduction process and then solving the resulting system with the help of a generalized inverse. In Section 3.2.5 we show how to transform a "quasi-orthonormal" basis (see (3.8)) into an "orthonormal" one. Finally, in Section 3.2.6 we demonstrate how our results can be applied to a perturbation analysis of the general eigenvalue problem.

3.2.2 ▪ Regular perturbation

The following lemma states a necessary condition for a perturbation to be regular or, in other words, rank-preserving. This condition, of course, can be checked in practice only in the case of a polynomial perturbation.

Lemma 3.2. *If the perturbation is regular, the sequence of matrices $\{A_k\}_{k=0}^{\infty}$ satisfies the following conditions:*

$$\tilde{U}^T \left(\sum_{p=1}^{k+1} (-1)^{p-1} \sum_{\nu_1+\cdots+\nu_p=k+1} A_{\nu_1} A_0^\dagger A_{\nu_2} \cdots A_0^\dagger A_{\nu_p} \right) \tilde{V} = 0, \quad k = 0, 1, \ldots, \tag{3.11}$$

where $\nu_i \geq 1$, and where $\tilde{U}$ and $\tilde{V}$ are bases for the left and right null spaces of the matrix A_0, respectively.

Proof: From equation (3.9) with $k = 0$ we conclude that

$$V_0 = \tilde{V} C_0, \tag{3.12}$$

where C_0 is some coefficient matrix. Since we consider the case of a rank-preserving perturbation, the rank of V_0 is equal to m. This in turn implies that $C_0 \in R^{m\times m}$ and that it is a full rank matrix.

Since $\tilde{U}^T A_0 = 0$, we obtain by Lemma 2.2 the following feasibility condition for equation (3.9.1):

$$\tilde{U}^T A_1 V_0 = 0.$$

Upon substituting (3.12) into the above expression, we obtain

$$\tilde{U}^T A_1 \tilde{V} C_0 = 0.$$

Moreover, since C_0 is a full rank matrix, we conclude that

$$\tilde{U}^T A_1 \tilde{V} = 0, \tag{3.13}$$

which is the first feasibility condition of (3.11).

Since the perturbation is rank-preserving, there exists a holomorphic basis $V(\varepsilon)$ for the perturbed null space such that $V(0) = \tilde{V}$. The coefficients $V_k, k = 0, 1, \ldots,$ of the power series (3.7) satisfy the fundamental equations (3.9). Hence, in particular, the feasibility condition for equation (3.9) with $k = 1$ is satisfied, and using Lemma 2.1 we can write its general solution in the form

$$V_1 = \tilde{V} C_1 - A_0^\dagger A_1 \tilde{V} C_0, \tag{3.14}$$

where $C_1 \in R^{m\times m}$ is some matrix. Next define for each $k = 0, 1, \ldots$

$$D_k = \sum_{p=1}^{k+1} (-1)^{p-1} \sum_{\nu_1+\cdots+\nu_p = k+1} A_{\nu_1} A_0^\dagger A_{\nu_2} \cdots A_0^\dagger A_{\nu_p}.$$

Note that the above formula can be rewritten in the recursive form

$$D_k = A_{k+1} - \sum_{i=1}^{k} A_i A_0^\dagger D_{k-i}, \quad k = 0, 1, \ldots. \tag{3.15}$$

Next we prove by induction that

$$\tilde{U}^T D_k \tilde{V} = 0, \quad k = 0, 1, \ldots, \tag{3.16}$$

and that

$$V_{k+1} = \tilde{V} C_{k+1} - A_0^\dagger \sum_{i=0}^{k} D_i \tilde{V} C_{k-i}, \tag{3.17}$$

where $C_i, i = 0, \ldots, k$, are some coefficient matrices. We recall that (3.16) is condition (3.11). We assume that relations (3.16) and (3.17) hold for $k = 0, \ldots, l$, and then we show that they also hold for $k = l + 1$. Note that we have already proved the induction base.

According to Lemma 2.2, the following feasibility condition for the $(l+2)$nd fundamental equation is satisfied:

$$\tilde{U}^T(A_1V_{l+1}+A_2V_l+\cdots+A_{l+2}V_0)=0.$$

Substituting formula (3.17) for each $V_{k+1}, k=0,\ldots,l$, and rearranging terms, we obtain

$$\tilde{U}^TA_1\tilde{V}C_{l+1}+\tilde{U}^T(A_2-A_1A_0^\dagger D_0)\tilde{V}C_l+\cdots+\tilde{U}^T\left(A_{l+2}-\sum_{i=1}^{l+1}A_iA_0^\dagger D_{l+1-i}\right)\tilde{V}C_0=0.$$

By the inductive hypothesis all terms of the above equation vanish except for the last one. Hence, we have

$$\tilde{U}^T\left(A_{l+2}-\sum_{i=1}^{l+1}A_iA_0^\dagger D_{l+1-i}\right)\tilde{V}C_0=0.$$

Using the recursive formula (3.15) and the fact that C_0 is a full rank matrix, we conclude that $\tilde{U}^TD_{l+1}\tilde{V}=0$.

Next we show that formula (3.17) also holds for $k=l+1$. By Lemma 2.1 the general solution for the $(l+2)$nd fundamental equation is given by

$$V_{l+2}=\tilde{V}C_{l+2}-A_0^\dagger(A_1V_{l+1}+\cdots+A_{l+2}V_0),$$

where C_{l+2} is some coefficient matrix. Substituting (3.17) for $V_{k+1}, k=0,\ldots,l$, into the above equation and rearranging terms yield the formula (3.17) for $k=l+1$. Thus, by induction, relation (3.16) and formula (3.17) hold for any integer k. □

The next theorem provides a recursive formula for the computation of the coefficients $V_k, k=0,1,\ldots$.

Theorem 3.3. *Let the matrix $A(\varepsilon)$ be a regular perturbation of A_0. Then there exists a holomorphic family of eigenvectors $V(\varepsilon)$ corresponding to the zero eigenvalue and satisfying the quasi-normalization condition* (3.8). *Moreover, the coefficients of the power series for $V(\varepsilon)$ can be calculated recursively by the formula*

$$V_k=-A_0^\dagger\sum_{j=1}^{k}A_jV_{k-j},\quad k=1,2,\ldots, \tag{3.18}$$

initializing with $V_0=\tilde{V}$ the right basis of the null space of A_0.

Proof: It follows from the proof of Lemma 3.2 that the general solution of the fundamental equations is

$$V_k=VC_k-A_0^\dagger\sum_{j=1}^{k}A_jV_{k-j},\quad k=1,2,\ldots,$$

with $V_0=\tilde{V}C_0$. By choosing $C_0=I_m$, we obtain $V_0=\tilde{V}$, which satisfies the quasi-normalization condition (3.10) with $k=0$.

Now the coefficients $C_k, k=1,2,\ldots$, are uniquely determined by the quasi-normalization conditions (3.10). Namely, we have

$$V_0^TV_k=V_0^T\left(\tilde{V}C_k-A_0^\dagger\sum_{j=1}^{k}A_jV_{k-j}\right)=0,$$

or, equivalently,

$$\tilde{V}^T \tilde{V} C_k - \tilde{V}^T A_0^\dagger \sum_{j=1}^{k} A_j V_{k-j} = 0,$$

since $V_0 = \tilde{V}$ when $C_0 = I_m$. Recalling that $\tilde{V}^T \tilde{V} = I_m$ and $\tilde{V}^T A_0^\dagger = 0$, since $N(A) = R(A^\dagger)^\perp$ (see Problem 3.1), we obtain that $C_k = 0, k = 1, 2, \ldots$, as required. □

Example 3.1 (continued from Subsection 3.2.1). *First we check that conditions* (3.11) *indeed hold for Example 3.1. For $k = 0$, we have*

$$\tilde{U}^T A_1 \tilde{V} = \begin{bmatrix} 0 & 1 \end{bmatrix} \begin{bmatrix} 1 & 0 \\ 0 & 0 \end{bmatrix} \begin{bmatrix} 1 \\ 0 \end{bmatrix} = 0.$$

Since $A_k = 0, k \geq 2$, the matrices $D_k, k = 1, 2, \ldots$, satisfy the following recursive relationship

$$D_k = -A_1 A_0^\dagger D_{k-1},$$

with $D_0 = A_1$. Next, we calculate

$$A_1 A_0^\dagger = \begin{bmatrix} 1 & 0 \\ 0 & 0 \end{bmatrix} \begin{bmatrix} 0 & 0 \\ 1 & 0 \end{bmatrix} = \begin{bmatrix} 0 & 0 \\ 0 & 0 \end{bmatrix}.$$

Thus, $D_k = 0, k = 1, 2, \ldots$, and hence conditions (3.11) *are indeed satisfied. As the perturbation is rank-preserving, one can take $V_0 = \tilde{V}$. Using the recursive formula* (3.18)*, we compute the terms $V_k, k = 1, 2, \ldots$, by*

$$V_k = -A_0^\dagger A_1 V_{k-1} = -\begin{bmatrix} 0 & 0 \\ 1 & 0 \end{bmatrix} V_{k-1}.$$

This results in

$$V_1 = \begin{bmatrix} 0 \\ -1 \end{bmatrix} \quad \text{and} \quad V_k = \begin{bmatrix} 0 \\ 0 \end{bmatrix}, \quad k = 2, 3, \ldots.$$

Next, we would like to address the issue of the radius of convergence. Above we have implicitly assumed that the series (3.7) has a positive radius of convergence. The next theorem gives a bound on the radius of convergence of the series (3.7) with coefficients as in (3.18).

Theorem 3.4. *Suppose $||A_i|| \leq a r^i$ for some positive constants a and r; then the radius of convergence of the series $V(\varepsilon) = V_0 + \varepsilon V_1 + \cdots$, where V_k is computed by* (3.18)*, is at least $(1 + a||A_0^\dagger||)^{-1} r^{-1}$.*

Proof: First, we prove by induction the inequality

$$||V_k|| \leq ||V_0|| (1 + a||A_0^\dagger||)^k r^k, \tag{3.19}$$

which trivially holds when $k = 0$. Now suppose that inequality (3.19) holds for the coefficients $V_0, \ldots, V_{k-1}$. From (3.18), we obtain

$$||V_k|| \leq ||A_0^\dagger|| \sum_{j=1}^{k} ||A_j|| ||V_{k-j}|| \leq a||A_0^\dagger|| \sum_{j=1}^{k} r^j ||V_{k-j}||.$$

Now using inequality (3.19) for $j = 0, 1, \ldots, k-1$ (which is the inductive hypothesis), we get

$$\begin{aligned} ||V_k|| &\leq a||A_0^\dagger||\sum_{j=1}^{k} r^j ||V_0||(1+a||A_0^\dagger||)^{k-j} r^{k-j} \\ &\leq a||A_0^\dagger||||V_0||r^k \sum_{j=1}^{k}(1+a||A_0^\dagger||)^{k-j}. \end{aligned}$$

Note that

$$\sum_{j=1}^{k}(1+a||A_0^\dagger||)^{k-j} = \frac{(1+a||A_0^\dagger||)^k - 1}{1+a||A_0^\dagger||-1} = \frac{(1+a||A_0^\dagger||)^k - 1}{a||A_0^\dagger||}.$$

Thus,

$$\begin{aligned} ||V_k|| &\leq ||V_0||r^k[(1+a||A_0^\dagger||)^k - 1] \\ &\leq ||V_0||r^k(1+a||A_0^\dagger||)^k, \end{aligned}$$

as required. Consequently, the radius of convergence for the power series $V(\varepsilon) = V_0 + \varepsilon V_1 + \cdots$ is at least $(1+a||A_0^\dagger||)^{-1} r^{-1}$. □

3.2.3 ▪ Singular perturbation: Augmented matrix method

In the next two subsections we deal with the case of singular or non-rank-preserving perturbations, namely, when the dimension of the perturbed null space $N(A(\varepsilon))$ for $0 < \varepsilon \leq \varepsilon_{\max}$ is strictly less than the dimension of $N(A_0)$. We propose two algorithms. The first is based on generalized inverses applied to augmented matrices, whereas the second is based on a reduction technique. Both methods have their own merits. Finally, we also suggest a way to combine these two approaches.

Some definitions are required prior to the introduction of our analysis for the case of singular perturbations, that is, the case when $\bar{m} < m$. First, as in Section 2.2, for any integer t, $t \geq 0$, we define an *augmented matrix* $\mathcal{A}^{(t)} \in \mathbb{R}^{n(t+1)\times n(t+1)}$:

$$\mathcal{A}^{(t)} = \begin{bmatrix} A_0 & 0 & 0 & \cdots & 0 \\ A_1 & A_0 & 0 & \cdots & 0 \\ A_2 & A_1 & A_0 & \cdots & 0 \\ \vdots & \vdots & \vdots & \ddots & \vdots \\ A_t & A_{t-1} & \cdots & A_1 & A_0 \end{bmatrix}.$$

Second, we partition the generalized inverse $\mathcal{G}^{(t)} := [\mathcal{A}^{(t)}]^\dagger$ into a block structure that corresponds to the structure of the augmented matrix $\mathcal{A}^{(t)}$. Namely,

$$\mathcal{G}^{(t)} = \begin{bmatrix} G_{00}^{(t)} & \cdots & G_{0t}^{(t)} \\ \vdots & \ddots & \vdots \\ G_{t0}^{(t)} & \cdots & G_{tt}^{(t)} \end{bmatrix},$$

where $G_{ij}^{(t)} \in \mathbb{R}^{n\times n}$ for $0 \leq i, j \leq t$.

Third, let $M_t \subseteq \mathbb{R}^n$ be the linear subspace of vectors w such that for some vector $v \in N(\mathscr{A}^{(t)}) \subseteq \mathbb{R}^{n(t+1)}$, the first n entries in v coincide with w. Since $\bar{v} \in N(\mathscr{A}^{(t+1)})$ implies that the first $n(t+1)$ entries of $\bar{v}$ form a vector $v \in N(\mathscr{A}^{(t)})$, $M_{t+1} \subseteq M_t$ for any $t \geq 0$, and hence $\dim(M_t)$ is nonincreasing with t. Finally, let $\tau = \arg\min_t\{\dim(M_t)\}$. In other words, τ is the smallest value of t where the minimum of $\dim(M_t)$ is attained. Since $\{\dim(M_t)\}_{t=0}^{\infty}$ is a sequence of nonincreasing integers, the minimum of $\dim(M_t)$ is attained at a finite value of index t.

Theorem 3.5. *For any $V_0 \in M_\tau$, there exists a sequence $\{V_i\}_{i=1}^{\infty}$ which coupled with V_0 solves*

$$\mathscr{A}^{(t)} \begin{bmatrix} V_0 \\ V_1 \\ \vdots \\ V_t \end{bmatrix} = \begin{bmatrix} 0 \\ 0 \\ \vdots \\ 0 \end{bmatrix} \tag{3.20}$$

for any $t \geq 0$. In particular, $\bar{m} = \dim(N(A(z))) = \dim(M_\tau)$.

Proof: A necessary (but not sufficient) condition for V_0 to be a leading term in such a sequence is that $A_0 V_0 = 0$, that is, $V_0 \in M_0$. But what is further required is that for this V_0 there exists a V_1 such that $A_0 V_1 + A_1 V_0 = 0$, that is, $V_0 \in M_1$. Conversely, any $V_0 \in M_1$ (coupled with an appropriate V_1) solves (3.20) for $t = 1$. Similarly, one can see that $V_0 \in M_2$ (coupled with the corresponding V_1 and V_2, which exist by the definition of M_2) if and only if (3.20) holds for $t = 2$. By induction, we conclude that V_0 leads to a solution for (3.20) for any $t \geq 0$ if and only if $V_0 \in M_t$ for any $t \geq 0$, that is, if and only if $V_0 \in M_\tau$. The equality $\bar{m} = \dim(M_\tau)$ follows from the fact that for each $V_0 \in M_\tau$ one can construct an analytically perturbed eigenvector $V(\varepsilon) = V_0 + \varepsilon V_1 + \cdots$. Thus, the dimension of M_τ coincides with the dimension of the perturbed null space. □

The first $\tau + 1$ fundamental equations (3.9) can be written as follows:

$$\mathscr{A}^{(\tau)} \begin{bmatrix} V_0 \\ V_1 \\ \vdots \\ V_\tau \end{bmatrix} = \begin{bmatrix} 0 \\ 0 \\ \vdots \\ 0 \end{bmatrix}. \tag{3.21}$$

Above we argued that any vector in M_τ will lead to a solution of (3.21). Imposing the normalization condition (3.10) with $k = 0$ is now equivalent to requiring that V_0 be an orthonormal basis. Finally, any such orthonormal basis will be appropriate for our purposes.

Once V_0 is determined, the next goal is the determination of the corresponding V_1. Using the augmented matrix notation, we rewrite equations (3.9) with k from 1 to $\tau + 1$ as follows:

$$\mathscr{A}^{(\tau)} \begin{bmatrix} V_1 \\ V_2 \\ \vdots \\ V_{\tau+1} \end{bmatrix} = \begin{bmatrix} -A_1 V_0 \\ -A_2 V_0 \\ \vdots \\ -A_{\tau+1} V_0 \end{bmatrix}, \tag{3.22}$$

which is similar to (3.20) with $t = \tau$ but with a different right-hand side. Note that by definition of τ and by the fact that $V_0 \in M_\tau$, the system (3.22) is solvable. Hence, by Lemma 2.1, we have

$$\begin{bmatrix} V_1 \\ V_2 \\ \vdots \\ V_{\tau+1} \end{bmatrix} = [\mathscr{A}^{(\tau)}]^\dagger \begin{bmatrix} -A_1 V_0 \\ -A_2 V_0 \\ \vdots \\ -A_{\tau+1} V_0 \end{bmatrix} + y$$

for some $y \in N(\mathscr{A}^{(\tau)})$. Note that not any $y \in N(\mathscr{A}^{(\tau)})$ will lead to a solution for the fundamental equations since in (3.22) we have not considered all of them. However, for any $w \in M_\tau$ there exists such a y with w being its first n entries. Moreover, any such w leads to a vector V_1 such that, coupled with V_0, they are the leading two terms in a series expansion for $V(\varepsilon)$. The reason is that whatever was true for V_0 is now true for V_1 since in the latter case one obtains the same set of equations but with a different right-hand side. The normalization condition (3.10) with $k = 1$, coupled with the fact that V_0 is chosen, implies a unique value for the matrix V_1.

Above we have shown how the value of V_0 leads to the value of V_1. Next, we show that this is the case in general. Specifically, once $V_0, \dots, V_k$ are determined, one can compute V_{k+1} by the recursive formula provided in the next theorem.

Theorem 3.6. *The solution of the system of fundamental equations* (3.9) *coupled with the normalization conditions* (3.10) *is given by the recursive formula*

$$V_{k+1} = -(I_n - V_0 V_0^T) \sum_{j=0}^{\tau} G_{0j}^{(\tau)} \sum_{i=1}^{k+1} A_{i+j} V_{k+1-i}, \tag{3.23}$$

where V_0 is any orthogonal basis of the linear subspace M_τ.

Proof: Consider the set of fundamental equations (3.9) from the $(k+1)$st equation to the $(k+1+\tau)$th equation. Since they are feasible, by Lemma 2.1 the general solution is of the form

$$\begin{bmatrix} V_{k+1} \\ \vdots \\ V_{k+1+\tau} \end{bmatrix} = \mathscr{G}^{(\tau)} \begin{bmatrix} -\sum_{i=1}^{k+1} A_i V_{k+1-i} \\ \vdots \\ -\sum_{i=1}^{k+1} A_{i+\tau} V_{k+1-i} \end{bmatrix} + y,$$

where $y \in N(\mathscr{A}^{(\tau)})$. Since the first n entries of y constitute a vector w in M_τ and V_0 is an orthogonal basis of M_τ, the general solution for V_{k+1} can be written as

$$V_{k+1} = -\sum_{j=0}^{\tau} G_{0j}^{(\tau)} \sum_{i=1}^{k+1} A_{i+j} V_{k+1-i} + V_0 C_{k+1}, \tag{3.24}$$

where C_{k+1} is some matrix coefficient that can be determined from the $(k+1)$st normalization condition (3.10). Specifically,

$$-V_0^T \sum_{j=0}^{\tau} G_{0j}^{(\tau)} \sum_{i=1}^{k+1} A_{i+j} V_{k+1-i} + C_{k+1} = 0,$$

and hence

$$C_{k+1} = V_0^T \sum_{j=0}^{\tau} G_{0j}^{(\tau)} \sum_{i=1}^{k+1} A_{i+j} V_{k+1-i}.$$

Substituting the above expression for the coefficient C_{k+1} into the formula (3.24) results in the recursive formula (3.23). This completes the proof. □

Remark 3.1. *We would like to point out that although above we call for* $[A^{(\tau)}]^\dagger$, *only its first* $\bar{m}$ *rows are required in order to carry out the desired computations.*

Example 3.2 (continued from Subsection 3.2.1). *It is easy to check that in this example the subspace* M_1 *is one dimensional and is spanned by the vector* $[c\ 0\ c]^T$, *where* $c \neq 0$ *is an arbitrary constant. Hence,* $\tau = 1$, *and the first term of power series* (3.7) *is given by*

$$V_0 = \frac{1}{\sqrt{2}} \begin{bmatrix} 1 \\ 0 \\ 1 \end{bmatrix}.$$

Then, to compute the terms $V_k, k = 1, 2, \ldots$, *we use the recursive formula* (3.23), *which has the following form for this particular example:*

$$V_{k+1} = -(I - V_0 V_0^T) G_{00} A_1 V_k, \quad k = 0, 1, \ldots.$$

Also,

$$I - V_0 V_0^T = \begin{bmatrix} 0.5 & 0 & -0.5 \\ 0 & 1 & 0 \\ -0.5 & 0 & 0.5 \end{bmatrix}, \quad G_{00} A_1 = \begin{bmatrix} 0 & 0 & 0 \\ 0.5 & 0 & 0.5 \\ 0 & 0 & 0 \end{bmatrix}.$$

Consequently,

$$V_1 = \frac{1}{\sqrt{2}} \begin{bmatrix} 0 \\ -1 \\ 0 \end{bmatrix} \quad \text{and} \quad V_k = \begin{bmatrix} 0 \\ 0 \\ 0 \end{bmatrix}, \ k \geq 2.$$

Note that in both Examples 3.1 and 3.2, we obtained finite expansions for $V(\varepsilon)$ instead of infinite series. Of course, this is due to the simplicity of the examples. However, if one calculates orthonormal bases instead of quasi-orthonormal bases, one will have to deal with *infinite* series even in the case of these simple examples. This fact demonstrates an advantage of using quasi-orthonormal bases instead of orthonormal ones.

3.2.4 ▪ Singular perturbation: Reduction process method

Next we show that by using a reduction process one can transform the system of fundamental equations (3.9) into another system with coefficient matrices of reduced dimensions. Furthermore, the latter system can be solved by the algorithm proposed in Section 3.2.2, for the regular case. Thus, we reduce the singular problem to a regular one. The next theorem is a key to the reduction process.

Theorem 3.7. *A solution of the fundamental equations* (3.9) *together with the normalization conditions* (3.10) *is given by the recursive formula*

$$V_k = \tilde{V} W_k - A_0^\dagger \sum_{j=1}^{k} A_j V_{k-j}, \quad k = 1, 2, \ldots, \tag{3.25}$$

with $V_0 = \tilde{V} W_0$, *and where the sequence of auxiliary matrices* W_k, $k \geq 0$, *is a solution to the next system of reduced fundamental equations,*

$$\sum_{i=0}^{k} B_i W_{k-i} = 0, \quad k = 0, 1, \ldots, \tag{3.26}$$

and reduced normalization conditions

$$W_0^T W_k = \delta_{0k} I_{\bar{m}}, \tag{3.27}$$

where the coefficient matrices B_k, $k \geq 0$, are given by the formula

$$B_k = \tilde{U}^T \left(\sum_{p=1}^{k+1} (-1)^{p-1} \sum_{\nu_1+\cdots+\nu_p=k+1, \nu_i \geq 1} A_{\nu_1} A_0^\dagger A_{\nu_2} A_0^\dagger \cdots A_0^\dagger A_{\nu_p} \right) \tilde{V}, \quad k \geq 0. \tag{3.28}$$

Proof: From the fundamental equation (3.9) with $k = 0$ we conclude that V_0 belongs to the null space of A_0, that is,

$$V_0 = \tilde{V} W_0, \tag{3.29}$$

where $W_0 \in \mathbb{R}^{m \times m_1}$ is some coefficient matrix, and where m_1 is a number to be determined with $\bar{m} \leq m_1 \leq m$. By Lemma 2.2 the equation (3.9.1) is feasible if and only if

$$\tilde{U}^T A_1 V_0 = 0.$$

Substituting the expression given in (3.29) for V_0, we obtain

$$\tilde{U}^T A_1 \tilde{V} W_0 = 0.$$

This is the first equation of the reduced system (3.26) with $B_0 = \tilde{U}^T A_1 \tilde{V}$. Note that m_1 above is the dimension of the null space of B_0. Next we consider the fundamental equation (3.9) with $k = 1$. By Lemma 2.1 its solution has the general form

$$V_1 = \tilde{V} W_1 - A_0^\dagger A_1 V_0, \tag{3.30}$$

where $W_1 \in \mathbb{R}^{m \times m_1}$ is some coefficient matrix, which describes the general solution of the corresponding homogeneous system and where $-A_0^\dagger A_1 V_0$ is a particular solution of (3.9) with $k = 1$. The coefficient matrices W_0 and W_1 have to be chosen so that they satisfy the feasibility condition for the next fundamental equation (3.9) with $k = 2$:

$$\tilde{U}^T (A_1 V_1 + A_2 V_0) = 0.$$

Upon substitution of V_0 (see (3.29)) and V_1 (see (3.30)) into the above condition, one obtains

$$\tilde{U}^T A_1 \tilde{V} W_1 + \tilde{U}^T (A_2 - A_1 A_0^\dagger A_1) \tilde{V} W_0 = 0,$$

which is the reduced fundamental equation (3.26) with $k = 1$, with $B_1 = \tilde{U}^T (A_2 - A_1 A_0^\dagger A_1) \tilde{V}$.

Note that the recursive formula (3.25) is just the general form of the solution of the kth fundamental equation (3.9). The reduced system of equations (3.26) is the set of feasibility conditions for $W_k, k = 0, 1, \ldots$, which are obtained in a way similar to the above considerations. The general formula (3.28) for the coefficients can now be established by an induction argument similar to that given in the proof of Lemma 3.2 (see Problem 3.3).

Next, we show that the new normalization conditions (3.27) also hold. First, consider the normalization condition for W_0. Substituting $V_0 = \tilde{V} W_0$ into (3.10) with $k = 0$, we obtain

$$(\tilde{V} W_0)^T \tilde{V} W_0 = I_{\bar{m}}$$

or

$$W_0^T \tilde{V}^T \tilde{V} W_0 = I_{\bar{m}}.$$

Recall that we have chosen the basis $\tilde{V}$ for the null space of A_0 such that $\tilde{V}^T\tilde{V} = I_m$. The latter implies that

$$W_0^T W_0 = I_{\bar{m}}.$$

Thus, we have obtained the normalization condition (3.27) with $k = 0$. Next we show that the normalization condition (3.27) holds as well for $k = 1,2,\ldots$. Toward this end, substitute the recursive expression (3.25) into (3.10.k) to obtain

$$V_0^T\tilde{V}W_k - V_0^T A_0^\dagger \sum_{j=1}^{k} A_j V_{k-j} = 0.$$

Note that since V_0 belongs to the null space of A_0 and since $N(A) = R(A^\dagger)^\perp$ (see Problem 3.1), $V_0^T A_0^\dagger = 0$. Thus,

$$V_0^T\tilde{V}W_k = 0.$$

By substituting V_0 from (3.29) and taking into account that $\tilde{V}^T\tilde{V} = I_m$, we obtain

$$W_0^T\tilde{V}^T\tilde{V}W_k = W_0^T W_k = 0,$$

which is the normalization condition (3.27). This completes the proof. □

Remark 3.2. *Note that the computation of the coefficient matrices B_k, $k = 0,1,\ldots$, by (3.28) is tedious. Therefore, as in Theorem 2.14, we compute these coefficients in a recursive manner. Specifically, define the sequence of matrices $\{D_k\}_{k=0}^\infty$ as follows:*

$$D_k = \sum_{p=1}^{k+1}(-1)^{p-1} \sum_{\nu_1+\cdots+\nu_p=k+1} A_{\nu_1}A_0^\dagger A_{\nu_2}A_0^\dagger\cdots A_{\nu_p}, \quad k = 0,1,\ldots.$$

These auxiliary matrices can be computed by the recursion

$$D_k = A_{k+1} - \sum_{i=1}^{k} A_i A_0^\dagger D_{k-i}, \quad k = 1,2,\ldots, \tag{3.31}$$

initializing with $D_0 = A_1$ (see Problem 3.4). Then the coefficient matrices B_k, $k = 0,1,\ldots$, are simply given by

$$B_k = U^T D_k V.$$

We would like to point out that the reduced system of equations (3.26) together with the normalization condition (3.27) has exactly the same structure as the initial system of fundamental equations (3.9) with the normalization conditions (3.10). Thus, one has two options as how to proceed from here. The first is to solve it using the augmented matrix method described in the previous subsection. The second is to apply one more reduction step—this time to the system composed of (3.26) and (3.27). If the latter option is pursued, then once again one may face the same alternative, and so on. At first sight, it might seem that one may end up carrying out an infinite number of reduction steps. However, as it turns out, termination is guaranteed after a finite number of steps. The next theorem addresses this issue.

Theorem 3.8. *Suppose that $\{B_k^{(l)}\}_{k=0}^\infty$, $l = 1,2,\ldots$, are the coefficients of the reduced system obtained at the lth reduction step ($B_k^{(1)} = B_k$). Also, let m_l be the dimension of the null space* *of* $B_0^{(l)}$. *Then, the reduction process terminates after a finite number of steps with* $m_l = \bar{m}$, *where* $\bar{m}$ *is the dimension of the null space of the perturbed matrices* $A(\varepsilon)$, $0 < |\varepsilon| < \varepsilon_{\max}$. *Furthermore, the final system of reduced equations (namely, the system of reduced fundamental equations derived at the last reduction step) can be solved by the recursive procedure which was proposed for the case of a regular perturbation described in Subsection 3.2.2 (see formula* (3.18)).

Proof: Note that after each reduction step the dimension of the null space of $B_0^{(l)}$ does not increase. Since we deal with a finite dimensional problem and since the sequence m_l, $l \geq 1$, is of integers, we conclude that the sequence of m_l achieves its limit, say, m_*, in a finite number of steps. Next we argue that this limit m_* equals $\bar{m}$, and once it is reached there is no need to make any further reduction steps. Note also that the solution to the final system of reduced equations (the reduction process terminates when the null space of $B_0^{(l)}$ has dimension m_*) can be obtained by the recursive algorithm proposed in Subsection 3.2.2. The latter means that a basis for the null space of the perturbed matrix $A(\varepsilon)$ is constructed, and this basis is holomorphic with the parameter ε. This basis is formed by m_* linearly independent vectors. However, according to our assumptions the dimension of the null space of $A(\varepsilon)$ is $\bar{m}$. This implies that the limit m_* is equal to $\bar{m}$. □

Finally, we would like to suggest a practical implementation of the above scheme. If one applies the reduction process as described above to calculate V_k, then one needs to compute $B_i, i = 0,\ldots,k+\bar{m}$, $B_i^{(2)}, i = 0,\ldots,k+\bar{m}-1$, and so on. This could result in a large number of calculations, even when the recursive formula given in Remark 3.2 is used. Alternatively, suppose that we have already obtained $V_0,\ldots,V_k, k > r$, where r denotes the number of reduction steps needed to obtain the final system of reduced equations. Then we can rewrite the fundamental equations (3.9) from k to $k+r$ as follows:

$$\begin{aligned}
A_0 V_{k+1} &= -(A_1 V_k + \cdots + A_{k+1} V_0), \\
A_0 V_{k+2} + A_1 V_{k+1} &= -(A_2 V_k + \cdots + A_{k+2} V_0), \\
&\vdots \\
A_0 V_{k+r+1} + \cdots + A_r V_{k+1} &= -(A_{r+1} V_k + \cdots + A_{k+r+1} V_0).
\end{aligned}$$

This system of equations can be effectively solved by the same reduction technique. Moreover, note that the auxiliary matrices such as $B_i^{(l)}$ can be stored and used afterward to compute the next terms $V_{k+2}, V_{k+3}, \ldots$. This suggestion is in line with the approach taken in Section 2.2.

If it is needed, an estimation of the convergence radius can also be obtained for the singular case. This can be done by recursively applying the arguments of Theorem 3.4 (Problem 3.2).

3.2.5 ▪ Orthogonalization of the basis

In the previous subsections we have developed the power series expansion for the eigenvectors corresponding to the zero eigenvalue of $A(\varepsilon)$. The matrix $V(\varepsilon)$ forms a basis for the null space of $A(\varepsilon)$, though it is not necessarily orthogonal. If one needs an orthogonal basis, one can apply the following Gram–Schmidt-like orthogonalization process.

First we perform the Gram–Schmidt procedure (without normalization) over the analytic vector-valued functions $v_i(\varepsilon), i = 1,\ldots,\bar{m}$, which constitute the "quasi-orthogonal" basis $V(\varepsilon)$. Note that summation, multiplication, and division operations used in the orthogonalization procedure need to be carried out on power series (rather than on real numbers). This results in an orthogonal basis for the perturbed null space. Each new basis element is a vector-valued function analytic in the punctured disc: $0 < |\varepsilon| < \varepsilon_{max}$. Next we show that the normalization procedure leads to a basis whose elements are analytic vector-valued functions at $\varepsilon = 0$. Indeed, consider a vector-valued function $a(\varepsilon)$ that is analytic in $0 < |\varepsilon| < \varepsilon_{max}$. It can be expanded as a Laurent series. And let $a_i(\varepsilon) = \varepsilon^m a_{i,m} + \varepsilon^{m+1} a_{i,m+1} + \dots$ with $a_{i,m} \neq 0$ be the largest element (in absolute value and for sufficiently small ε) of the vector $a(\varepsilon)$. Then, clearly

$$||a(\varepsilon)|| = \sqrt{a_1^2(\varepsilon) + \cdots + a_n^2(\varepsilon)} = \varepsilon^m (\nu_0 + \varepsilon \nu_1 + \cdots), \quad \nu_0 > 0.$$

The latter implies that the normalized vector $a(\varepsilon)/||a(\varepsilon)||$ can be expanded as a series with nonnegative powers of ε and with a nonzero leading coefficient. Hence, as a result of the above procedure, we obtain an orthonormal basis.

3.2.6 ▪ Application to the perturbed eigenvalue problem

The results on the perturbation of null spaces can be immediately applied to the general perturbed eigenvalue problem

$$A(\varepsilon)x(\varepsilon) = \lambda(\varepsilon)x(\varepsilon). \tag{3.32}$$

Recall that the perturbed eigenvalue $\lambda(\varepsilon)$ satisfies the characteristic polynomial

$$\det(A(\varepsilon) - \lambda(\varepsilon)I) = 0,$$

that is,

$$(-1)^n \lambda^n + a_{n-1}(\varepsilon)\lambda^{n-1} + \cdots + a_1(\varepsilon)\lambda + a_0(\varepsilon) = 0,$$

where the coefficients $a_i(\varepsilon)$ are analytic functions. Using the method of the Newton polygon (see Section 4.7), it is possible to find a Puiseux expansion for the perturbed eigenvalue:

$$\lambda(\varepsilon) = \lambda_0 + \varepsilon^{1/p}\lambda_1 + \varepsilon^{2/p}\lambda_2 + \dots,$$

where p is some positive integer. Next, introduce an auxiliary variable $\eta := \varepsilon^{1/p}$, and note that the perturbed eigenvalue depends analytically on η. Consequently, the system of equations for the perturbed eigenvectors can be written in the form

$$[A(\eta^p) - \lambda(\eta)I]x(\eta) = 0.$$

Hence, we have reduced the general perturbed eigenvalue problem to the problem of analytic perturbation of the null space, which can be effectively solved by the methods presented in Sections 3.2.2–3.2.4.

3.3 ▪ Perturbation of Generalized Inverses: Complex Analytic Approach

3.3.1 ▪ Problem formulation and some background

In this section we study the generalized inverses of analytically perturbed matrices:

$$A(z) = A_0 + zA_1 + z^2 A_2 + \dots. \tag{3.33}$$

First, we provide the perturbation analysis for the *Drazin generalized inverse*. In such a case we assume that the matrices $A_k, k = 0, 1, \ldots$, are square, of dimension n, with complex entries. Furthermore, since we are interested in the perturbation analysis of the generalized inverse, we assume that the null space of the perturbed matrix $A(z)$ is nontrivial.

Here we also distinguish between regular and singular perturbations. The perturbation is said to be *regular* if it does not change the dimension of null space. Otherwise, the perturbation is said to be *singular*. One of the main advantages of the complex analytic approach is that it allows us to treat both regular and singular perturbations in a unified framework.

If the coefficient matrices $A_k, k = 0, 1, \ldots$, are real and we restrict ourselves to real z, the perturbation analysis of the Drazin generalized inverse can be applied to the perturbation analysis of the *Moore–Penrose generalized inverse*.

The main goals of this section are to prove the existence of the Laurent series expansion for the perturbed Drazin generalized inverse

$$A^{\#}(z) = \sum_{j=-s}^{+\infty} z^j H_j \tag{3.34}$$

and to provide a method for the efficient computation of coefficients $H_j, j = -s, -s+1, \ldots$.

We derive recursive formulae for the (matrix) coefficients of the regular part of the Laurent series expansion (3.34). In addition to their theoretical interest, the recursive formulae are particularly useful when one needs to compute a significant number of terms in the Laurent series expansion. These formulae require knowledge of the singular part coefficients, and the latter are obtained via a reduction technique based on the complex analytic approach. In particular, this reduction technique uses a new notion of *group reduced resolvent*. The order of the pole of (3.34) is obtained as a by-product.

Note that if the matrix $A(z)$ is invertible in some punctured neighborhood around $z = 0$, then its generalized Drazin inverse is just the ordinary inverse. Therefore, the results of the present section are applicable to the problems of Section 2.2. In particular, we obtain the generalization of the recursive formula (2.38) of Subsection 2.2.6.

Last but not least, the limit matrix in the Taylor series expansion of the 0-eigenprojection matrix has a simple expression in terms of (a) the original unperturbed 0-eigenprojection, (b) the perturbation matrices, and (c) the singular part of the Laurent series (3.34). This provides some insight into how the perturbed 0-eigenvectors relate to the original 0-eigenvectors.

In the course of the procedure that we propose, one has to determine the multiplicity of zero eigenvalues and verify whether they are semisimple.[3] Of course, in the general situation neither task is easy in the presence of rounding errors. Nevertheless, we note that in many particular applications this issue can be effectively resolved. For instance, for the perturbation of the Moore–Penrose generalized inverse, the semisimplicity issue is not relevant, as we transform the problem into an equivalent problem with symmetric matrices and the eigenvalues of symmetric matrices are semisimple. Even though these issues should not be underestimated, our primary focus here is on the algebraic structure of the Laurent series (3.34). The details of the practical implementation undoubtedly represent an important direction for future research.

Before proceeding further, let us recall some facts from complex analysis and spectral theory. The interested reader can find more details on spectral theory in the first chapters of Kato's comprehensive book (see also bibliographic notes).

[3] An eigenvalue is said to be *semisimple* if its geometric multiplicity is equal to its algebraic multiplicity.

Any matrix $A \in \mathbb{C}^{n\times n}$ possesses the spectral representation

$$A = \sum_{i=0}^{p} (\lambda_i P_i + D_i), \tag{3.35}$$

where $p+1$ is the number of distinct eigenvalues of A, P_i is the eigenprojection, and D_i is the nilpotent operator corresponding to the eigenvalue λ_i. By convention, λ_0 is the zero eigenvalue of A, that is, $\lambda_0 = 0$. In the case when there is no zero eigenvalue, the eigenvalues are enumerated from $i = 1$. The *resolvent* is another very important object in spectral theory.

Definition 3.1. *The following operator-valued function of the complex parameter ζ is called the resolvent of the operator $A \in \mathbb{C}^{n\times n}$:*

$$R(\zeta) = (A - \zeta I)^{-1}.$$

The resolvent satisfies the *resolvent identity*:

$$R(\zeta_1) - R(\zeta_2) = (\zeta_1 - \zeta_2) R(\zeta_1) R(\zeta_2) \tag{3.36}$$

for all $\zeta_1, \zeta_2 \in \mathbb{C}$ (see Problem 2.6). The resolvent has singularities at the points $\zeta = \lambda_k$, where λ_k are the eigenvalues of A. In a neighborhood of each singular point λ_k the resolvent can be expanded as a Laurent series,

$$R(\zeta) = -\sum_{n=1}^{m_k-1} \frac{1}{(\zeta-\lambda_k)^{n+1}} D_k^n - \frac{1}{\zeta-\lambda_k} P_k + \sum_{n=0}^{\infty} (\zeta-\lambda_k)^n S_k^{n+1}, \tag{3.37}$$

where S_k is the *reduced resolvent* corresponding to the eigenvalue λ_k with geometric multiplicity m_k. In fact, S_k is the Drazin generalized inverse of $(A - \lambda_k I)$. And, in particular, we have $S_0 = A^{\#}$.

The Drazin generalized inverse has the following basic properties:

$$AA^{\#} = I - P_0, \tag{3.38}$$

$$P_0 A^{\#} = 0. \tag{3.39}$$

The above equations show that $A^{\#}$ is the "inverse" of A in the complementary subspace to the generalized null space of A, in the sense that $(AA^{\#})u = u$ for any $u \in R(I - P_0)$. Here by *generalized* null space we mean a subspace which is spanned by all eigenvectors and generalized (Jordan) eigenvectors corresponding to the zero eigenvalue. Note that P_0 is a projection onto this generalized null space.

Moreover, if the underlying space $\mathbb{C}^n$ admits a decomposition into the *direct* sum of the null space and the range of the operator A (recall from Section 2.1 that this is a necessary and sufficient condition for the existence of the group inverse), then the Drazin inverse and the group inverse coincide, and the following Laurent expansion holds:

$$R(\zeta) = -\frac{1}{\zeta} P_0 + \sum_{n=0}^{\infty} \zeta^n (A^{\#})^{n+1}. \tag{3.40}$$

Since the Drazin generalized inverse is the constant term in the Laurent series (3.37) at $\zeta = \lambda_0$, it can be calculated via the Cauchy integral formula

$$A^{\#} = \frac{1}{2\pi i} \int_{\Gamma_0} \frac{1}{\zeta} R(\zeta)\, d\zeta, \tag{3.41}$$

where Γ_0 is a closed positively oriented contour in the complex plane, enclosing 0 but no other eigenvalue of A. The above formula will play a crucial role in what follows.

The Drazin inverse also has a simple expression in terms of eigenprojections, eigenvalues, and nilpotent operators of the original operator A. Namely,

$$A^{\#} = \sum_{i=1}^{p} \left\{ \frac{1}{\lambda_i} P_i + \sum_{j=1}^{m_i-1} (-1)^j \frac{1}{\lambda_i^{j+1}} D_i^j \right\}. \tag{3.42}$$

We emphasize that the above sum is taken over all indices corresponding to nonzero eigenvalues. This expression again demonstrates that the Drazin generalized inverse is the inverse operator in the complementary subspace to the generalized null space. Moreover, this expression exactly represents the inverse operator A^{-1} whenever A has no zero eigenvalue.

3.3.2 ▪ Existence of the Laurent series expansion

In this subsection we prove the existence of a Laurent series expansion (3.34) for the Drazin generalized inverse $A^{\#}(z)$ of the analytically perturbed matrix $A(z)$.

First let us consider the resolvent $R(\zeta,z) := (A(z)-\zeta I)^{-1}$ of the perturbed $A(z)$. One can expand $R(\zeta,z)$ in a power series with respect to the complex variable z near $z=z_0$ as follows:

$$R(\zeta,z) = R(\zeta,z_0) + \sum_{n=1}^{\infty} (z-z_0)^n R^{(n)}(\zeta,z_0), \tag{3.43}$$

where

$$R^{(n)}(\zeta,z_0) := \sum_{\nu_1+\cdots+\nu_p=n} (-1)^p R(\zeta,z_0) A_{\nu_1} R(\zeta,z_0) A_{\nu_2} \cdots R(\zeta,z_0) A_{\nu_p} R(\zeta,z_0),$$

where A_{ν_k} are the coefficients of $A(z)$ and $\nu_k \geq 1$ (see Problem 3.5). The above expansion is called the second Neumann series for the resolvent. It is uniformly convergent for z sufficiently close to z_0 and $\zeta \in \mathcal{D}$, where $\mathcal{D}$ is a compact subset of the complex plane which does not contain the eigenvalues of $A(z_0)$.

Theorem 3.9. *Let $A(z)$ be the analytic perturbation of the matrix A_0 given by* (3.33). *Then, the Drazin generalized inverse $A^{\#}(z)$ of the perturbed operator $A(z)$ can be expanded as a Laurent series* (3.34).

Proof: We first show that there exists a domain $0 < |z| < z_{max}$ such that $A^{\#}(z)$ can be expanded in a Taylor series at any point z_0 in this domain. For a fixed, arbitrary $z > 0$, (3.41) becomes

$$A^{\#}(z) = \frac{1}{2\pi i} \int_{\Gamma_0(z)} \frac{1}{\zeta} R(\zeta,z) d\zeta, \tag{3.44}$$

where $\Gamma_0(z)$ is a closed counterclockwise oriented curve enclosing the origin but no other eigenvalue of $A(z)$.

With z_{max} less than the modulus of any nonzero eigenvalue of A_0, expand the perturbed resolvent in the power series (3.43) around the point z_0 (with $0 < |z_0| < z_{max}$). Then, the substitution of that series in the integral formula (3.44) yields

$$A^{\#}(z) = \frac{1}{2\pi i} \int_{\Gamma_0(z_0)} \frac{1}{\zeta} \left[R(\zeta,z_0) + \sum_{n=1}^{\infty} (z-z_0)^n R^{(n)}(\zeta,z_0) \right] d\zeta.$$

Since the power series for $R(\zeta, z)$ is uniformly convergent for z sufficiently close to z_0, we can integrate the above series term by term,

$$A^{\#}(z) = \frac{1}{2\pi i}\int_{\Gamma_0(z_0)} \frac{1}{\zeta} R(\zeta, z_0)\, d\zeta + \sum_{n=1}^{\infty}(z-z_0)^n \frac{1}{2\pi i}\int_{\Gamma_0(z_0)} \frac{1}{\zeta}[R^{(n)}(\zeta, z_0)]\, d\zeta$$
$$= A^{\#}(z_0) + \sum_{n=1}^{\infty}(z-z_0)^n H_n(z_0), \tag{3.45}$$

where the coefficients are defined by

$$H_n(z_0) := \frac{1}{2\pi i}\int_{\Gamma_0(z_0)} \frac{1}{\zeta}[R^{(n)}(\zeta, z_0)]\, d\zeta.$$

The convergence of power series (3.45) in some nonempty domain $0 < |z| < z_{max}$ can be shown by using the bounds for the contour integrals (see Problem 3.6). From the power series (3.45), we can see that $A^{\#}(z)$ is holomorphic in the domain $0 < |z| < z_{max}$. Consequently, by Laurent's theorem, we conclude that $A^{\#}(z)$ possesses a Laurent series expansion at $z = 0$ (with radius of convergence z_{max}), that is,

$$A^{\#}(z) = \sum_{n=-\infty}^{+\infty} z^n H_n. \tag{3.46}$$

We next show that the pole at $z = 0$ can be at most of finite order. Consider the spectral representation (3.42) for the reduced resolvent of the perturbed operator $A(z)$:

$$A^{\#}(z) = \sum_{i=1}^{p}\left\{\frac{1}{\lambda_i(z)}P_i(z) + \sum_{j=1}^{m_i-1}(-1)^j \frac{1}{\lambda_i^{j+1}(z)} D_i^j(z)\right\}.$$

From the book of Kato, we know that the perturbed eigenvalues $\lambda_i(z)$ are bounded in $|z| \le z_{max}$, and they have at most algebraic singularities. Furthermore, the eigenprojections $P_i(z)$ and nilpotents $D_i(z)$ can also have only algebraic singularities and poles of finite order. Therefore, none of the functions $\lambda_i(z)$, $P_i(z)$, and $D_i(z)$ can have an essential singularity. This latter fact implies that their finite sums, products, or divisions as in $A^{\#}(z)$ do not have an essential singularity as well, and, consequently, the order of pole in (3.46) is finite. This completes the proof. □

3.3.3 ▪ Recursive formula for the regular part coefficients

Here we derive recursive formulae for the coefficients of the regular part of the Laurent series (3.34). We use an analytic technique based on Cauchy contour integrals and resolvent-like identities.

First, observe that the structure of the perturbed Drazin inverse $(A_0 + zA_1 + z^2A_2 + \ldots)^{\#}$ is similar to the structure of the classical resolvent $(A_0 - zI)^{-1}$. Moreover, $A^{\#}(z)$ becomes precisely the resolvent if $A_1 = -I$ and $A_k = 0$ for $k \ge 2$. Therefore, one can expect that these two mathematical objects have some similar features. It turns out that the Drazin inverse of an analytically perturbed matrix $A(z)$ satisfies an identity similar to the resolvent identity (3.36).

Lemma 3.10. *The reduced resolvent $A^{\#}(z)$ of the analytically perturbed operator $A(z) = \sum_{k=0}^{\infty} z^k A_k$ satisfies the resolvent-like identity:*

$$A^{\#}(z_1)-A^{\#}(z_2)=\sum_{k=1}^{\infty}(z_2^k-z_1^k)A^{\#}(z_1)A_kA^{\#}(z_2)+A^{\#}(z_1)P_0(z_2)-P_0(z_1)A^{\#}(z_2), \quad (3.47)$$

where $P_0(z)$ is the eigenprojection matrix corresponding to the zero eigenvalue.

Proof: Consider the following expression:

$$A(z_2)-A(z_1)=\sum_{k=1}^{\infty}(z_2^k-z_1^k)A_k.$$

Premultiplying by $A^{\#}(z_1)$ and postmultiplying by $A^{\#}(z_2)$ yields

$$A^{\#}(z_1)A(z_2)A^{\#}(z_2)-A^{\#}(z_1)A(z_1)A^{\#}(z_2)=\sum_{k=1}^{\infty}(z_2^k-z_1^k)A^{\#}(z_1)A_kA^{\#}(z_2).$$

Then, using (3.38), we obtain

$$A^{\#}(z_1)[I-P_0(z_2)]-[I-P_0(z_1)]A^{\#}(z_2)=\sum_{k=1}^{\infty}(z_2^k-z_1^k)A^{\#}(z_1)A_kA^{\#}(z_2).$$

Equivalently,

$$A^{\#}(z_1)-A^{\#}(z_2)=\sum_{k=1}^{\infty}(z_2^k-z_1^k)A^{\#}(z_1)A_kA^{\#}(z_2)+A^{\#}(z_1)P_0(z_2)-P_0(z_1)A^{\#}(z_2),$$

which is the desired identity (3.47). □

In the next theorem, we obtain a general relation between the coefficients of the Laurent series (3.34).

Theorem 3.11. *Let $H_k, k=-s,-s+1,\ldots,$ be the coefficients of the Laurent series (3.34) and $P_0(z)=\sum_{k=0}^{\infty} z^k P_{0k}$ be a power series for the eigenprojection corresponding to the zero eigenvalue of the perturbed operator. Then the coefficients $H_k, k=-s,-s+1,\ldots,$ satisfy the relation*

$$\sum_{k=1}^{\infty}\sum_{i=0}^{k-1}H_{n-i}A_kH_{m+i-k+1}=-(\eta_n+\eta_m-1)H_{n+m+1}$$

$$-\begin{cases} 0, & m<0, \\ \frac{1}{2\pi i}\int_{\Gamma_1} z_1^{-n-1}A^{\#}(z_1)[P_{0m+1}+z_1P_{0m+2}+\ldots]dz_1, & m\geq 0, \end{cases}$$

$$-\begin{cases} 0, & n<0, \\ \frac{1}{2\pi i}\int_{\Gamma_2} z_2^{-m-1}[P_{0n+1}+z_2P_{0n+2}+\ldots]A^{\#}(z_2)dz_2, & n\geq 0, \end{cases} \quad (3.48)$$

where

$$\eta_m:=\begin{cases} 1, & m\geq 0, \\ 0, & m<0. \end{cases}$$

For the sake of clarity of presentation, the detailed proof is postponed to Subsection 3.3.7. Now the recursive formula for the coefficients of the regular part of the Laurent series (3.34) becomes a corollary of the above general result.

Corollary 3.1. *Suppose that the coefficients $H_k, k=-s,\ldots,-1,0$, and $P_{0k}, k=0,1,\ldots$, are given. Then, the coefficients of the regular part of the Laurent expansion* (3.34) *can be computed by the following recursive formula:*

$$H_{m+1} = -\sum_{i=0}^{m+s}\left(\sum_{j=0}^{s} H_{-j}A_{i+j+1}\right)H_{m-i} - \sum_{i=1}^{m} P_{0m+1-i}H_i$$
$$-(P_{0m+1}H_0+\cdots+P_{0m+1+s}H_{-s})-(H_{-s}P_{0m+1+s}+\cdots+H_0P_{0m+1}) \tag{3.49}$$

for $m=0,1,\ldots$.

Proof: Let us take $n=0$, $m>0$ and then simplify the last two terms in (3.48) by collecting terms in the integrand with z_1^{-1}:

$$\frac{1}{2\pi i}\int_{\Gamma_1} z_1^{-n-1}A^{\#}(z_1)[P_{0m+1}+z_1P_{0m+2}+\ldots]dz_1$$
$$=\frac{1}{2\pi i}\int_{\Gamma_1}\frac{1}{z_1}\left[\frac{1}{z_1^s}H_{-s}+\cdots+\frac{1}{z_1}H_{-1}+H_0+\ldots\right][P_{0m+1}+z_1P_{0m+2}+\ldots]dz_1$$
$$=\frac{1}{2\pi i}\int_{\Gamma_1}\frac{1}{z_1}\frac{1}{z_1^{m+1}}\frac{1}{z_1^s}[H_{-s}P_{0m+1+s}+\cdots+H_0P_{0m+1}]dz_1$$
$$=H_{-s}P_{0m+1+s}+\cdots+H_0P_{0m+1}. \tag{3.50}$$

The last term in (3.48) can be dealt with in a similar fashion:

$$\frac{1}{2\pi i}\int_{\Gamma_2} z_2^{-m-1}[P_{0n+1}+z_2P_{0n+2}+\ldots]A^{\#}(z_2)dz_2$$
$$=\frac{1}{2\pi i}\int_{\Gamma_2} z_2^{-m-1}[P_{01}+z_2P_{02}+\ldots]A^{\#}(z_2)dz_2$$
$$=\frac{1}{2\pi i}\int_{\Gamma_2}\frac{1}{z_2}\frac{1}{Z_2^m}[P_{01}+z_2P_{02}+\ldots]\left[\frac{1}{z_1^s}H_{-s}+\cdots+\frac{1}{z_1}H_{-1}+H_0+\ldots\right]dz_2$$
$$=\frac{1}{2\pi i}\int_{\Gamma_2}\frac{1}{z_2}[P_{01}H_m+P_{02}H_{m-1}+\cdots+P_{0m}H_1+P_{0m+1}H_0+\cdots+P_{0m+1+s}H_{-s}]dz_2$$
$$=\sum_{i=1}^{m}P_{0m+1-i}H_i+(H_{-s}P_{0m+1+s}+\cdots+H_0P_{0m+1}). \tag{3.51}$$

Substituting (3.50) and (3.51) into (3.48) with $n=0$ and $m>0$, we obtain

$$\sum_{k=1}^{\infty}\sum_{i=0}^{k-1}H_{-i}A_kH_{m+i-k+1} = -H_{m+1}-(H_{-s}P_{0m+1+s}+\cdots+H_0P_{0m+1})$$
$$-(P_{0m+1}H_0+\cdots+P_{0m+1+s}H_{-s})-\sum_{i=1}^{m}P_{0m+1-i}H_i.$$

Rearranging terms in the above expression, we obtain

$$H_{m+1} = -\sum_{i=0}^{m+s}\left(\sum_{j=0}^{s} H_{-j}A_{i+j+1}\right)H_{m-i} - \sum_{i=1}^{m} P_{0m+1-i}H_i$$
$$-(P_{0m+1}H_0 + \cdots + P_{0m+1+s}H_{-s}) - (H_{-s}P_{0m+1+s} + \cdots + H_0P_{0m+1}),$$

which is the recursive formula (3.49). □

If the perturbed operator $A(z)$ is invertible for $0 < |z| < z_{max}$, then the inverse $A^{-1}(z)$ can be expanded as a Laurent series,

$$A^{-1}(z) = \frac{1}{z^{-s}}H_{-s} + \cdots + \frac{1}{z}H_{-1} + H_0 + zH_1 + \ldots, \tag{3.52}$$

and the formula (3.49) becomes (Problem 3.7)

$$H_{m+1} = -\sum_{i=0}^{m+s}\left(\sum_{j=0}^{s} H_{-j}A_{i+j+1}\right)H_{m-i}, \quad m = 0, 1, \ldots. \tag{3.53}$$

Furthermore, if the perturbed operator is invertible and the perturbation is linear $A(z) = A_0 + zA_1$, we retrieve the recursive formula (2.38)

$$H_{m+1} = (-H_0A_1)H_m, \quad m = 0, 1, \ldots.$$

3.3.4 ▪ Reduction process

We have seen that the regular terms in the Laurent series expansion (3.34) of $A^{\#}(z)$ can be computed recursively by (3.49). However, to apply (3.49), one first needs to compute $H_{-s}, \ldots, H_{-1}, H_0$, that is, the terms of the singular part.

The complex analytic approach allows us to treat the cases of regular and singular perturbations in a unified framework. In fact, let us first obtain some results on the regular perturbation that will be useful in the reduction process for the singular perturbation.

Regular case. Let us apply analytic function techniques to express the power series for the Drazin inverse of the perturbed operator in the case of *regular* perturbation, that is, when the dimension of the null space of the matrix does not change if the perturbation parameter deviates from zero. In other words, there is no splitting of the zero eigenvalue of the perturbed matrix at $z = 0$. The latter implies that the expansion (3.45) is valid in some neighborhood of $z_0 = 0$ and for any contour $\Gamma_0 := \Gamma_0(0)$ chosen so that it does not enclose eigenvalues other than zero. Namely, the expansion (3.45) takes the form

$$A^{\#}(z) = A_0^{\#} + \sum_{n=1}^{\infty} z^n A_n^{\#}, \tag{3.54}$$

where

$$A_0^{\#} = A^{\#}(0), \quad A_n^{\#} = \frac{1}{2\pi i}\int_{\Gamma_0} \frac{1}{\zeta}[R^{(n)}(\zeta)]\,d\zeta$$

and

$$R^{(n)}(\zeta) = \sum_{\nu_1+\cdots+\nu_p=n} (-1)^p R(\zeta)A_{\nu_1}R(\zeta)A_{\nu_2}\cdots R(\zeta)A_{\nu_p}R(\zeta).$$

It turns out that it is possible to express the coefficients $A_n^{\#}$ in terms of

- the unperturbed Drazin inverse $A^{\#}(0)$,
- the eigenprojection P_0 corresponding to the zero eigenvalue of A_0,
- the perturbation matrices $A_n, n = 1,2,\ldots$.

The next theorem gives the precise statement.

Theorem 3.12. *Suppose that the operator A_0 is perturbed analytically as in (3.33), and assume that the zero eigenvalue of A_0 is semisimple and the perturbation is regular. Then, the matrices $A_n^{\#}, n = 1,2,\ldots$, in the expansion (3.54) are given by the formula*

$$A_n^{\#} = \sum_{p=1}^{n}(-1)^p \sum_{\substack{\nu_1+\cdots+\nu_p=n \\ \mu_1+\cdots+\mu_{p+1}=p+1 \\ \nu_j\geq 1,\mu_j\geq 0}} S_{\mu_1}A_{\nu_1}S_{\mu_2}\ldots A_{\nu_p}S_{\mu_{p+1}}, \tag{3.55}$$

where $S_0 := -P_0$ and $S_k := (A_0^{\#})^k, k = 1,2,\ldots$.

Proof: Since Γ_0 encloses only the zero eigenvalue, we have by (3.54)

$$A_n^{\#} = \frac{1}{2\pi i}\int_{\Gamma_0}\frac{1}{\zeta}R^{(n)}(\zeta)d\zeta = \sum_{\nu_1+\ldots\nu_p=n}(-1)^p\frac{1}{2\pi i}\int_{\Gamma_0}\frac{1}{\zeta}R(\zeta)A_{\nu_1}R(\zeta)\ldots A_{\nu_p}R(\zeta)d\zeta$$

$$= \sum_{\nu_1+\cdots\nu_p=n}(-1)^p\mathrm{Res}_{\zeta=0}\left\{\frac{1}{\zeta}R(\zeta)A_{\nu_1}R(\zeta)\ldots A_{\nu_p}R(\zeta)\right\}.$$

In order to compute the above residue, we replace $R(\zeta)$ by its Laurent series (3.40) in the expression

$$\frac{1}{\zeta}R(\zeta)A_{\nu_1}R(\zeta)\ldots A_{\nu_p}R(\zeta)$$

and collect the terms with $1/\zeta$, that is, the terms

$$\sum_{\sigma_1+\cdots+\sigma_{p+1}=0} S_{\sigma_1+1}A_{\nu_1}S_{\sigma_2+1}\ldots A_{\nu_p}S_{\sigma_p+1}.$$

Next, we change indices $\mu_k := \sigma_k + 1, k = 1,\ldots,p+1$, and rewrite the above sum as

$$\sum_{\mu_1+\cdots+\mu_{p+1}=p+1} S_{\mu_1}A_{\nu_1}S_{\mu_2}\ldots A_{\nu_p}S_{\mu_p},$$

which yields (3.55). □

Remark 3.3. *Of course, formula (3.55) is computationally demanding due to the combinatorial explosion (see Problem 3.12). However, typically only a few terms will be computed by this formula (see the arguments developed below).*

Singular case. We now show that by using a *reduction process*, we can transform the original singular problem into a regular one. We would like to emphasize that the reduction process of this section is different from the algebraic reduction technique proposed in

Sections 2.2 and 3.2. Also this reduction process can be viewed as complimentary to the existing reduction process based on spectral theory (developed in the book of Kato) which is applied to the eigenvalue problem. Moreover, to the best of our knowledge, applying the reduction technique to analytical perturbations of generalized inverses is new.

To develop the reduction technique in the context of the generalized inverses, we need to introduce a new notion of a *group reduced resolvent*. A definition based on spectral representation is as follows.

Definition 3.2. *Let $A:\mathbb{C}^n\to\mathbb{C}^n$ be a linear operator with the spectral representation* (3.35). *Then, the group reduced resolvent $A^{\#\Lambda}$ relative to the group of eigenvalues $\Lambda := \{\lambda_i\}_{i=0}^k$ is defined as follows:*

$$A^{\#\Lambda} \stackrel{def}{=} \sum_{i=k+1}^{p}\left\{\frac{1}{\lambda_i}P_i+\sum_{j=1}^{m_i-1}(-1)^j\frac{1}{\lambda_i^{j+1}}D_i^j\right\},$$

where m_i is the multiplicity of λ_i and D_i is the corresponding nilpotent operator (see (3.35)).

We note that the Drazin generalized inverse (see (3.42)) is a particular case of the group reduced resolvent. In this case, the group of eigenvalues consists only of the zero eigenvalue.

From our definition, the properties of a group reduced resolvent follow easily. In particular, in the next theorem, we will obtain an alternative analytic expression of the group reduced resolvent that will play a crucial role in our perturbation analysis.

Theorem 3.13. *Let A be a linear operator with representation* (3.35). *Then, the group reduced resolvent relative to the eigenvalues $\Lambda=\{\lambda_i\}_{i=0}^k$ is given by*

$$A^{\#\Lambda}=\frac{1}{2\pi i}\int_\Gamma\frac{1}{\zeta}(A-\zeta I)^{-1}d\zeta, \tag{3.56}$$

where Γ is a contour in the complex plane which encloses the set of eigenvalues $\{\lambda_i\}_{i=0}^k$ but none of the other eigenvalues $\{\lambda_i\}_{i=k+1}^p$.

Proof: The resolvent can be represented by (see Problem 3.10)

$$R(\zeta)=-\sum_{i=0}^{p}\left[\frac{1}{\zeta-\lambda_i}P_i+\sum_{j=1}^{m_i}\frac{1}{(\zeta-\lambda_i)^{j+1}}D_i^j\right].$$

Substituting the above expression into the integral of (3.56) yields

$$\frac{1}{2\pi i}\int_\Gamma\frac{1}{\zeta}(A-\zeta I)^{-1}d\zeta=-\frac{1}{2\pi i}\int_\Gamma\sum_{i=0}^{p}\left[\frac{1}{\zeta(\zeta-\lambda_i)}P_i+\sum_{j=1}^{m_i}\frac{1}{\zeta(\zeta-\lambda_i)^{j+1}}D_i^j\right]d\zeta.$$

Using the fact that for every positive integer l

$$\mathrm{Res}_{\zeta=0}\frac{1}{\zeta(\zeta-\lambda)^l}=\frac{1}{(-\lambda)^l}\quad\text{and}\quad\mathrm{Res}_{\zeta=\lambda}\frac{1}{\zeta(\zeta-\lambda)^l}=-\frac{1}{(-\lambda)^l},$$

we obtain

$$\frac{1}{2\pi i}\int_\Gamma \frac{1}{\zeta}(A-\zeta I)^{-1}d\zeta = \sum_{i=0}^{k}\mathrm{Res}_{\zeta=\lambda_i}\left\{-\sum_{i=0}^{p}\left[\frac{1}{\zeta(\zeta-\lambda_i)}P_i+\sum_{j=1}^{m_i}\frac{1}{\zeta(\zeta-\lambda_i)^{j+1}}D_i^j\right]\right\}$$
$$=\sum_{i=k+1}^{p}\left\{\frac{1}{\lambda_i}P_i+\sum_{j=1}^{m_i-1}(-1)^j\frac{1}{\lambda_i^{j+1}}D_i^j\right\}.$$

According to Definition 3.2, the latter expression is equal to the group reduced resolvent, so the proof is complete. □

Lemma 3.14. *Let $P=\sum_{i=0}^{k}P_i$ be the projection corresponding to the group of eigenvalues $\Lambda=\{\lambda_i\}_{i=0}^{k}$; then*

$$A^{\#\Lambda}=(A[I-P])^{\#}.$$

Proof: Since $P_iP_j=\delta_{ij}P_i$ and $D_iP_j=\delta_{ij}D_i$, we have $P_i[I-P]=0$, $D_i[I-P]=0$ for $i=0,\dots,k$ and $P_i[I-P]=P_i$, $D_i[I-P]=D_i$ for $i=k+1,\dots,p$. Then, (3.35) and the above yield

$$(A[I-P])^{\#}=\left(\sum_{i=0}^{p}(\lambda_iP_i+D_i)[I-P]\right)^{\#}=\left(\sum_{i=k+1}^{p}(\lambda_iP_i+D_i)\right)^{\#}.$$

Applying formula (3.42) to the product $A[I-P]$, we obtain

$$(A[I-P])^{\#}=\sum_{i=k+1}^{p}\left\{\frac{1}{\lambda_i}P_i+\sum_{j=1}^{m_i-1}(-1)^j\frac{1}{\lambda_i^{j+1}}D_i^j\right\}.$$

The latter is equal to the group reduced resolvent $A^{\#\Lambda}$ by Definition 3.2. □

Now equipped with this new notion of *group reduced resolvent*, we return to our perturbation analysis. The group of the perturbed eigenvalues $\lambda_i(z)$ such that $\lambda_i(z)\to 0$ as $z\to 0$ is called the 0-group. We denote the 0-group of eigenvalues by Ω. The eigenvalues of the 0-group split from zero when the perturbation parameter differs from zero. Since the eigenvalues of the perturbed operator are algebraic functions of the perturbation parameter, each eigenvalue of the 0-group (other than 0) can be written as

$$\lambda_i(z)=z^{\nu}\lambda_{i\nu}+o(z^{\nu}), \tag{3.57}$$

with $\lambda_{i\nu}\neq 0$, and ν is a positive rational number. The reduction technique is essentially based on the semisimplicity assumption of reduced operators, which will be introduced below. Under that assumption, the power ν in (3.57) must be an integer. The latter implies that we can partition the 0-group into subsets that we call z^l-groups. Namely, we say that the eigenvalue $\lambda_i(z)$ belongs to the z^l-group if $\lambda_i(z)=z^l\lambda_{il}+o(z^l)$, with $\lambda_{il}\neq 0$. We denote the z^l-group by Λ_l.

Consider now the spectral representation of the perturbed reduced resolvent.

$$A^{\#}(z)=\sum_{i=1}^{p}\left\{\frac{1}{\lambda_i(z)}P_i(z)+\sum_{j=1}^{m_i-1}(-1)^j\frac{1}{\lambda_i^{j+1}(z)}D_i^j(z)\right\},$$

where $\{\lambda_i(z)\}_{i=1}^k$ is the 0-group. From the above formula one can see that in this case, the Laurent expansion for the reduced resolvent $A^\#(z)$ will possess terms with negative powers of z. Moreover, it turns out that under our assumptions, the z^k-group eigenvalues contribute to the terms of the Laurent expansion for $A^\#(z)$ with negative powers $-k,-k+1,\dots,-1$ as well as to the regular part of the Laurent expansion.

The basic idea is to first treat the part of the perturbed operator corresponding to the eigenvalues that do not tend to zero as $z \to 0$. Then we subsequently treat the parts of the perturbed operator corresponding to the eigenvalues which belong to the z^1-group, the z^2-group, and so on.

It is helpful to treat the part of $A(z)$ corresponding to the z^{k+1}-group. We have to perform the same algorithm as for the part of the perturbed operator corresponding to the z^k-group. These steps constitute the (finite) reduction process.

Now we implement the above general idea. Consider a fixed contour Γ_0 that encloses only the zero eigenvalue of the unperturbed operator A_0. Note that by continuity of eigenvalues the 0-group of eigenvalues of the perturbed operator $A(z)$ lies inside Γ_0 for z sufficiently small. Therefore, we may define the group reduced resolvent relative to the 0-group of eigenvalues as follows:

$$A^{\#\Omega}(z) = \frac{1}{2\pi i}\int_{\Gamma_0} \frac{1}{\zeta} R(\zeta,z)d\zeta = \frac{1}{2\pi i}\int_{\Gamma_0} \frac{1}{\zeta}(A(z)-\zeta I)^{-1}d\zeta.$$

Since $A^{\#\Omega}(z)$ is an analytic function in some neighborhood of the origin, it can be expanded as a power series

$$A^{\#\Omega}(z) = A_0^{\#\Omega} + \sum_{i=1}^{\infty} z^i A_i^{\#\Omega}. \tag{3.58}$$

Note that $A_0^{\#\Omega} = (A_0)^\#$, and from Theorem 3.12 it follows that the other coefficients $A_i^{\#\Omega}, i = 1,2,\dots$, can be calculated by the formula (3.55). We would like to emphasize that in general the group reduced resolvent $A^{\#\Omega}(z)$ is different from the reduced resolvent $A^\#(z)$. However, we note that $A^{\#\Omega}(z)$ does coincide with $A^\#(z)$ in the case of regular perturbations.

Another operator that is used extensively in the reduction process is the *group projection*,

$$P(z) = \frac{1}{2\pi i}\int_{\Gamma_0} R(\zeta,z)d\zeta,$$

which describes the subspace corresponding to the eigenvalues which split from zero. The group projection is an analytic function in some small neighborhood of the origin (see, e.g., the book of Kato).

Next, as in the classical reduction process, we define the restriction $B(z)$ of the operator $A(z)$ to the subspace determined by the group projection $P(z)$, that is,

$$B(z) := \frac{1}{z}A(z)P(z) = \frac{1}{2\pi i z}\int_{\Gamma_0} \zeta R(\zeta,z)d\zeta,$$

where Γ_0 is some fixed contour enclosing only the zero eigenvalue of the unperturbed operator A_0. For the operator $B(z)$ to be analytic at zero, we need the following assumption.

Assumption S1. *The zero eigenvalue of the operator A_0 is semisimple; that is, the nilpotent operator D_0 corresponding to $\lambda_0 = 0$ is equal to zero.*

Note that this assumption is not too restrictive. For example, in the case of a self-adjoint perturbation operator, the zero eigenvalue of A_0 is semisimple. This is also the case when one studies the Moore–Penrose generalized inverse of an analytically perturbed matrix since it reduces to studying a symmetric perturbation of the Drazin inverse (see Subsection 3.3.5). Whenever Assumption S1 is satisfied, the operator $B(z)$ can be expressed as a power series (Problem 3.11)

$$B(z) = B_0 + \sum_{i=1}^{\infty} z^i B_i,$$

with $B_0 = P(0)A_1P(0)$, and

$$B_n = -\sum_{p=1}^{n+1}(-1)^p \sum_{\substack{\nu_1+\cdots+\nu_p=n+1 \\ \mu_1+\cdots+\mu_{p+1}=p-1 \\ \nu_j\geq 1, \mu_j \geq 0}} S_{\mu_1}A_{\nu_1}S_{\mu_2}\ldots A_{\nu_p}S_{\mu_{p+1}}, \tag{3.59}$$

where $S_0 := -P(0)$ and $S_k := ((A_0)^{\#})^k$.

Since the operator $B(z)$ is analytic in some neighborhood of the origin, we can again construct the expansion for its group reduced resolvent:

$$B^{\#\Omega}(z) = (B_0)^{\#} + \sum_{i=1}^{\infty} z^i B_i^{\#\Omega}. \tag{3.60}$$

The coefficients $B_i^{\#\Omega}, i = 1,2,\ldots$, are calculated by the formula given in Theorem 3.12. This is the first reduction step. To continue, we must distinguish between two cases.

(i) If the splitting of the zero eigenvalue terminates (all branches of the zero eigenvalue have been discovered), and consequently $B(z)$ is a *regular* perturbation of B_0, then $B^{\#\Omega}(z) = B^{\#}(z)$, and the Drazin inverse of the perturbed operator $A(z)$ is given by

$$A^{\#}(z) = A^{\#\Omega}(z) + \frac{1}{z}B^{\#}(z). \tag{3.61}$$

By substituting the series expansions (3.58) and (3.60) for $A^{\#\Omega}(z)$ and $B^{\#}(z)$ into (3.61), we obtain the Laurent series expansion for $A^{\#}(z)$, which has a simple pole at zero.

(ii) If the zero eigenvalue splits further, the expression

$$A^{\#\Omega\backslash\Lambda_1}(z) = A^{\#\Omega}(z) + \frac{1}{z}B^{\#\Omega}(z)$$

represents *only* the *group reduced resolvent* relative to the eigenvalues constituting the 0-group but not the z-group, and we have to continue the reduction process. In fact, we now consider $B(z)$ as a singular perturbation of B_0, and we repeat the procedure with $B(z)$. The 0-group of eigenvalues of B_0 contains all the z^k-groups of $A(0)$ (with $k \geq 2$), but *not* the z-group. Specifically, we construct the next-step reduced operator

$$C(z) = z^{-1}B(z)Q(z),$$

where $Q(z)$ is the eigenprojection corresponding to the 0-group of the eigenvalues of $B(z)$. Again, to ensure that $C(z)$ is an analytic function of z, we require the following assumption.

Assumption S2. *The zero eigenvalue of B_0 is semisimple.*

We would like to emphasize that the subsequent reduction steps are totally analogous to the first one. At each reduction step, we make Assumption Sk that the analogue of B_0 at step k has a semisimple 0-eigenvalue. The final result is stated in the next theorem.

Theorem 3.15. *Let Assumptions Sk hold for $k = 0, 1, \ldots$. Then, the reduction process terminates after a finite number of steps, say, s, and the perturbed Drazin inverse $A^{\#}(z)$ has the following expression:*

$$A^{\#}(z) = A^{\#\Omega}(z) + \frac{1}{z}B^{\#\Omega}(z) + \frac{1}{z^2}C^{\#\Omega}(z) + \cdots + \frac{1}{z^s}Z^{\#}(z). \tag{3.62}$$

Proof: Consider the first reduction step. Since the range spaces $R(P(z))$ and $R(I - P(z))$ represent a direct decomposition of $\mathbb{C}^n$ and the subspace $R(P(z))$ is invariant under the operator $A(z)$, we can write

$$A^{\#}(z) = (A(z)[I - P(z)] + A(z)P(z))^{\#} = (A(z)[I - P(z)])^{\#} + (A(z)P(z))^{\#}$$

$$= A^{\#\Omega}(z) + z^{-1}(z^{-1}A(z)P(z))^{\#},$$

where Lemma 3.14 was used to obtain the first term of the right-hand side. In view of Assumption S1, the operator $B(z) = z^{-1}A(z)P(z)$ is analytic in z, and hence, one can apply the next reduction step. Similarly, Assumptions Sk, $k = 1, 2, \ldots$, guarantee that the reduction process can be carried out. Since the splitting of the zero eigenvalue has to terminate after a finite number of steps, we conclude that the reduction process has to terminate after a finite number of steps as well. Indeed, we successively eliminate the eigenvalues of the z^1-group, the z^2-group, etc. Let $\lambda_i(z) = z^s\lambda_{is} + \ldots$ be the last eigenvalue which splits from zero. Then the corresponding reduced operator $Z(z)$ is regularly perturbed, and the associated reduced resolvent $Z^{\#}(z)$ has the power series defined by Theorem 3.12. This completes the proof. □

Summarizing, to obtain the Laurent series for $A^{\#}(z)$, there are two cases to distinguish. First, if one needs only a few regular terms of $A^{\#}(z)$, then it suffices to replace $A^{\#\Omega}(z), B^{\#\Omega}(z), \ldots$ in (3.62) by their respective power series (3.58) computed during the reduction process. Note that only a few terms of the power series $A^{\#\Omega}(z), B^{\#\Omega}(z), \ldots$ are needed. Otherwise, if one wishes to compute a significant number of regular terms, then compute only $H_{-s}, \ldots, H_{-1}, H_0$ as above (in which case, again, only a few terms of $A^{\#\Omega}(z), B^{\#\Omega}(z), \ldots$ are needed), and then use the recursive formula (3.49). Of course, one needs first to compute the power series expansion of the eigenprojection $P_0(z)$, which can be obtained by several methods, including those described in Section 3.2.

Remark 3.4. *If the operator $A(z)$ has an inverse for $z \neq 0$, then the above algorithm can be used to calculate its Laurent expansion. Hence, the inversion problem $A^{-1}(z)$ is a particular case of the complex analytic approach presented above.*

Example 3.3. *As was mentioned in the introduction, the perturbation analysis of the reduced resolvent can be applied directly to the theory of singularly perturbed Markov chains. More analysis of singularly perturbed Markov chains will follow in Chapter 6. Namely, the reduced resolvent of the generator of a Markov chain is the negative deviation matrix of this chain. The deviation matrix plays a crucial role in the Markov chain theory. For example, it is used to obtain mean first passage times. Taking into account the above remark, we consider an example of a perturbed Markov chain. Let us consider the following perturbed operator:*

$$A(z) = A_0 + zA_1 = \begin{bmatrix} 0 & 0 & 0 \\ 0 & 0.5 & -0.5 \\ 0 & -0.5 & 0.5 \end{bmatrix} + z\begin{bmatrix} 2 & -1 & -1 \\ -3 & 1 & 2 \\ -4 & 3 & 1 \end{bmatrix}.$$

Note that $-A(z)$ is the generator of a Markov chain when z is sufficiently small, real, and positive. The zero eigenprojection and the reduced resolvent of the unperturbed matrix A_0 are given by

$$P(0)=\begin{bmatrix} 1 & 0 & 0 \\ 0 & 0.5 & 0.5 \\ 0 & 0.5 & 0.5 \end{bmatrix}, \quad A_0^{\#}=\begin{bmatrix} 0 & 0 & 0 \\ 0 & 0.5 & -0.5 \\ 0 & -0.5 & 0.5 \end{bmatrix}.$$

In this instance, the Laurent expansion for $A^{\#}(z)$ has a simple pole. Using the method of Hassin and Haviv for the determination of the singularity order of the perturbed Markov chains, one can check that

$$A^{\#}(z)=\frac{1}{z}H_{-1}+H_0+zH_1+\dots.$$

By applying the reduction process, we compute the singular coefficient H_{-1} and the first regular coefficient H_0. Since the zero eigenvalues of the reduced operators are always semisimple in the case of perturbed Markov chains (see the chapter dedicated to Markov chains), we conclude from Theorem 3.15 that

$$H_{-1}=B_0^{\#} \quad \textit{and} \quad H_0=A_0^{\#}+B_1^{\#\Omega}.$$

To compute $B_0^{\#}$ and $B_1^{\#\Omega}$, we need to calculate the first two terms of the expansion for the reduced operator $B(z)$. In particular, from (3.59)

$$B_0=P(0)A_1P(0)=\begin{bmatrix} 2 & -1 & -1 \\ -3.5 & 1.75 & 1.75 \\ -3.5 & 1.75 & 1.75 \end{bmatrix},$$

$$B_1=-(A_0^{\#}A_1P(0)A_1P(0)+P(0)A_1A_0^{\#}A_1P(0)+P(0)A_1P(0)A_1A_0^{\#})$$
$$=\frac{1}{8}\begin{bmatrix} 0 & 4 & -4 \\ -24 & 5 & 19 \\ 20 & -17 & -3 \end{bmatrix},$$

calculated with the help of (3.59)*. Next, we calculate the eigenprojection corresponding to the zero eigenvalue of the operator B_0, that is,*

$$Q(0)=\frac{1}{22}\begin{bmatrix} 14 & 4 & 4 \\ 14 & 15 & -7 \\ 14 & -7 & 15 \end{bmatrix}.$$

Now using formula (3.55) *from Theorem 3.12, we obtain*

$$B_1^{\#\Omega}=Q(0)B_1(B_0^{\#})^2-B_0^{\#}B_1B_0^{\#}+(B_0^{\#})^2B_1Q(0)=\frac{1}{2662}\begin{bmatrix} -16 & 52 & -36 \\ -236 & 41 & 195 \\ 248 & -201 & -47 \end{bmatrix}.$$

Thus, we finally obtain

$$H_{-1}=B_0^{\#}=\frac{1}{5.5^2}\begin{bmatrix} 1 \\ -1.75 \\ -1.75 \end{bmatrix}\begin{bmatrix} 2 & -1 & -1 \end{bmatrix}=\frac{1}{121}\begin{bmatrix} 8 & -4 & -4 \\ -14 & 7 & 7 \\ -14 & 7 & 7 \end{bmatrix}$$

and

$$H_0=B_1^{\#\Omega}+A_0^{\#}=\frac{1}{1331}\begin{bmatrix} -8 & 26 & -18 \\ -118 & 686 & -568 \\ 124 & -766 & 642 \end{bmatrix}.$$

If we have in hand the expansion for the ergodic projection, we can use the recursive formula (3.49) *to compute the regular coefficients. Let us compute by the recursive formula the coefficient H_1 for our example. First, applying the reduction process for the eigenproblem* (*see Chapter* 6), *one can compute the coefficients for the expansion of the ergodic projection associated with z and z^2:*

$$P_{01} = \frac{1}{121}\begin{bmatrix} 2 & -12 & 10 \\ 2 & -12 & 10 \\ 2 & -12 & 10 \end{bmatrix}, \quad P_{02} = \frac{1}{1331}\begin{bmatrix} 32 & -192 & 160 \\ 32 & -192 & 160 \\ 32 & -192 & 160 \end{bmatrix}.$$

Then, according to formula (3.49), *we have*

$$H_1 = -(H_0A_1)H_0 - (P_{01}H_0 + P_{02}H_{-1}) - (H_{-1}P_{02} + H_0P_{01})$$

$$= \frac{1}{14641}\begin{bmatrix} -368 & 1856 & -1488 \\ -2128 & 12416 & -10288 \\ 1744 & -10816 & 9072 \end{bmatrix}.$$

3.3.5 ▪ Perturbation of the Moore–Penrose generalized inverse and the group inverse

Note that the Laurent series for the perturbed group inverse does not always exist. Indeed, the existence of the group inverse of the unperturbed operator does not imply the existence of the group inverse of the perturbed operator.

Example 3.4. *Consider*

$$A(z) = \begin{bmatrix} 0 & z & 0 \\ 0 & 0 & 0 \\ 0 & 0 & 1 \end{bmatrix}.$$

The space $\mathbb{C}^n$ can be decomposed in a direct sum of the null space and range of $A(0)$, but no such decomposition exists if $z \neq 0$. Thus, the unperturbed operator $A(0)$ has a group inverse, and the perturbed operator does not.

The following is a general sufficient condition for the existence of the Laurent series for the perturbed group inverse.

Theorem 3.16. *Let the group inverse $A^g(z)$ of the analytically perturbed matrix $A(z)$ exist in some nonempty (possibly punctured) neighborhood of $z = 0$. Then the group inverse $A^g(z)$ can be expanded as a Laurent series around $z = 0$ with a nonzero radius of convergence.*

In view of previous analyzes, the proof of the theorem is now elementary and is left as an exercise (see Problem 3.9).

As one can see from the following example, even though the Moore–Penrose generalized inverse always exists, it may not be an analytic function of the perturbation parameter.

Example 3.5. *Let*

$$A(z) = \begin{bmatrix} 0 & z \\ 0 & 1 \end{bmatrix}.$$

Its Moore–Penrose generalized inverse is given by

$$A^{\dagger}(z)=\frac{1}{1+z\bar{z}}\begin{bmatrix} 0 & 0 \\ \bar{z} & 1 \end{bmatrix},$$

which is not analytic since it depends on $\bar{z}$.

However, if we restrict $A_k, k = 0,1,\ldots$, to the matrices with real entries and if z belongs to some interval of the real line, we can state the following existence result.

Theorem 3.17. *Let* $A^{\dagger}(\varepsilon)$ *be the Moore–Penrose generalized inverse of the analytically perturbed matrix*

$$A(\varepsilon)=A_0+\varepsilon A_1+\varepsilon^2 A_2+\ldots,$$

where $A_k \in \mathbb{R}^{n\times m}$, $\varepsilon \in \mathbb{R}$, *and the series converges for* $0<|\varepsilon|<\varepsilon_{max}$. *Then,* $A^{\dagger}(\varepsilon)$ *possesses a series expansion*

$$A^{\dagger}(\varepsilon)=\frac{1}{\varepsilon^s}B_{-s}+\cdots+\frac{1}{\varepsilon}B_{-1}+B_0+\varepsilon B_1+\ldots \tag{3.63}$$

in some nonempty punctured vicinity around $\varepsilon = 0$.

Proof: Rewriting (2.16) for the perturbed operator $A(\varepsilon)$ yields

$$A^{\dagger}(\varepsilon)=(A^T(\varepsilon)A(\varepsilon))^g A^T(\varepsilon). \tag{3.64}$$

Note that the group inverse of a symmetric matrix always exists. Hence, by Theorem 3.16, $(A^T(\varepsilon)A(\varepsilon))^g$ has a Laurent series expansion, and so does the Moore–Penrose generalized inverse $A^{\dagger}(\varepsilon)$. □

We would like to emphasize that according to (3.64), computing the series expansion of the perturbed Moore–Penrose generalized inverse $A^{\dagger}(\varepsilon)$ reduces to computing the series expansion of a group inverse. Moreover, $A^T(\varepsilon)A(\varepsilon)$ is a symmetric perturbation; that is, each term of its power series has a symmetric matrix coefficient. This guarantees that the reduction process restricted to the real line is indeed applicable in this case.

3.3.6 ▪ Asymptotics for the perturbed eigenprojection

We know that the eigenprojection $P_0(z)$ of the perturbed matrix corresponding to the identically zero eigenvalue is analytic in some (punctured) neighborhood of $z = 0$ (see Section 3.2), that is,

$$P_0(z) = P_{00}+\sum_{k=1}^{\infty} z^k P_{0k} \tag{3.65}$$

for z sufficiently small but different from zero. For regular perturbations, P_{00} is just $P_0(0)$, and the group projection coincides with the eigenprojection. This is not the case for singular perturbations.

Therefore, an interesting question is how P_{00} in (3.65) relates to the original matrix $P_0(0)$ in the general case and how the power series (3.65) can be computed. The answers to these questions are provided below. This creates an interesting link between this section and Section 3.2.

Proposition 3.1. *The coefficients of the power series* (3.65) *for the perturbed eigenprojection are given by*

$$P_{0k} = -\sum_{i=0}^{s+k} A_i H_{k-i}, \quad k = 1,2,\ldots,$$

where the H_k*'s are as in* (3.34).

Proof: The above formula is obtained by collecting the terms with the same power of z in the identity (3.38) for the perturbed operators. □

Proposition 3.2. *As* $z \to 0$, *the limit eigenprojection matrix* P_{00} *satisfies*

$$P_{00} = P_0(0)\left[I - \sum_{i=1}^{s} A_i H_{-i}\right], \tag{3.66}$$

where $P_0(0)$ *is the* 0*-eigenprojection of the unperturbed matrix* A_0.

Proof: When substituting the Laurent series expansion (3.46) into (3.38)–(3.39) and collecting the terms with the same power, one obtains

$$I - P_{00} = A_0 H_0 + A_1 H_{-1} + \cdots + A_s H_{-s}. \tag{3.67}$$

In addition, from $A(z)P_0(z) = 0$, we immediately obtain

$$A_0 P_{00} = 0 \quad \text{so that } P_{00} = P_0(0)V \tag{3.68}$$

for some matrix V. Moreover, as $P_0(0)^2 = P_0(0)$, we also have $P_0(0)P_{00} = P_0(0)^2 V = P_0(0)V = P_{00}$. Therefore, premultiplying both sides of (3.67) by $P_0(0)$ and using $P_0(0)A_0 = 0$, one obtains (3.66), the desired result. □

Hence, (3.66) relates in a simple manner the limit matrix P_{00} to the original 0-group $P_0(0)$ in terms of the perturbation matrices $A_k, k = 1,\ldots,s$, the original matrix $P_0(0)$, and the coefficients $H_{-k}, k = 1,\ldots,s$, of the singular part of $A(z)^\#$. This shows how the perturbed 0-eigenvectors compare to the unperturbed ones for small z. Observe that in the case of a linear (or first-order) perturbation, only the singular term H_{-1} is involved. Finally, the regular case is obtained as a particular case since then $H_{-k}, k = 1,\ldots,s$, vanish so that $P_{00} = P_0(0)$.

3.3.7 ▪ The proof of the unifying recursive equation

To prove the unifying recursive equation (3.48) contained in Theorem 3.11, we use Cauchy contour integration and the residue technique. First, we present some auxiliary results.

Lemma 3.18. *Let* Γ_1 *and* Γ_2 *be two closed counterclockwise oriented contours in the complex plane around zero, and let* $z_1 \in \Gamma_1, z_2 \in \Gamma_2$. *Furthermore, assume that the contour* Γ_2 *lies inside the contour* Γ_1. *Then the following formulae hold:*

$$\frac{1}{2\pi i}\int_{\Gamma_2} \frac{z_2^{-m-1}}{z_2 - z_1} dz_2 = -\eta_m z_1^{-m-1}, \tag{3.69}$$

$$\frac{1}{2\pi i}\int_{\Gamma_1} \frac{z_1^{-m-1}}{z_2 - z_1} dz_1 = -(1-\eta_n) z_2^{-m-1}, \tag{3.70}$$

with

$$\eta_m := \begin{cases} 0, & m < 0, \\ 1, & m \geq 0, \end{cases}$$

and

$$\frac{1}{2\pi i}\int_{\Gamma_2} \frac{z_2^{-m-1}P_0(z_2)}{z_2 - z_1} dz_2 = \begin{cases} 0, & m < 0, \\ -z_1^{-m-1}[P_{00} + z_1 P_{01} + \cdots + z_1^m P_{0m}], & m \geq 0, \end{cases} \tag{3.71}$$

$$\frac{1}{2\pi i}\int_{\Gamma_1} \frac{z_1^{-n-1}P_0(z_1)}{z_2 - z_1} dz_1 = \begin{cases} -z_2^{-n-1}P_0(z_2), & n < 0, \\ -[P_{0n+1} + z_2 P_{0n+2} + z_2^2 P_{0n+3} + \ldots], & n \geq 0. \end{cases} \tag{3.72}$$

Proof: For the proof of formulae (3.69), (3.70) see Problem 3.13.

Let us establish the auxiliary integral (3.71). If $m < 0$, then the function $\frac{z_2^{-m-1}P_0(z_2)}{z_2 - z_1}$ is analytic inside the area enclosed by the contour Γ_2, and hence the auxiliary integral (3.71) is equal to zero by the Cauchy integral theorem. To deal with the case $m \geq 0$, we first expand the function $\frac{z_2^{-m-1}P_0(z_2)}{z_2 - z_1}$ as a Laurent series:

$$\frac{z_2^{-m-1}P_0(z_2)}{z_2 - z_1} = -\frac{z_2^{-m-1}P_0(z_2)}{z_1(1 - z_2/z_1)}$$

$$= -z_1^{-1} z_2^{-m-1}[P_{00} + z_2 P_{01} + z_2^2 P_{02} + \ldots]\left[1 + \frac{z_2}{z_1} + \frac{z_2^2}{z_1^2} + \ldots\right]$$

$$= z_2^{-m-1}(-z_1^{-1})P_{00} + \cdots + z_2^{-1}(-z_1^{-1})\left[\frac{1}{z_1^m}P_{00} + \frac{1}{z_1^{m-1}}P_{01} + \cdots + P_{0m}\right] + \ldots.$$

Then, according to the residue theorem, we have

$$\frac{1}{2\pi i}\int_{\Gamma_2} \frac{z_2^{-m-1}P_0(z_2)}{z_2 - z_1} dz_2 = (-z_1^{-1})\left[\frac{1}{z_1^m}P_{00} + \frac{1}{z_1^{m-1}}P_{01} + \cdots + P_{0m}\right]$$
$$= -z_1^{-m-1}[P_{00} + z_1 P_{01} + \cdots + z_1^m P_{0m}].$$

Thus, we have calculated the integral (3.71). The same method may be applied to calculate the auxiliary integral (3.72). □

Proof of Theorem 3.11: Each coefficient of the Laurent series (3.34) can be represented by the contour integral formula

$$H_n = \frac{1}{2\pi i}\int_{\Gamma} z^{-n-1} A^{\#}(z) dz, \quad \Gamma \in \mathcal{D}, \tag{3.73}$$

where Γ is a closed positively oriented contour in the complex plane, which encloses zero but no other eigenvalues of A_0. Using (3.73), we can write

$$\sum_{k=1}^{\infty}\sum_{i=0}^{k-1} H_{n-i} A_k H_{m+i-k+1}$$
$$= \sum_{k=1}^{\infty}\sum_{i=0}^{k-1} \frac{1}{2\pi i}\int_{\Gamma_1} z_1^{-n+i-1} A^{\#}(z_1) dz_1 A_k \frac{1}{2\pi i}\int_{\Gamma_2} z_2^{-m-i+k-2} A^{\#}(z_2) dz_2.$$

As in Lemma 3.18, we assume without loss of generality that the contour Γ_2 lies inside the contour Γ_1. Then, we can rewrite the above expressions as double integrals:

$$\sum_{k=1}^{\infty}\sum_{i=0}^{k-1} H_{n-i}A_k H_{m+i-k+1}$$

$$=\sum_{k=1}^{\infty}\sum_{i=0}^{k-1}\left(\frac{1}{2\pi i}\right)^2\int_{\Gamma_1}\int_{\Gamma_2} z_1^{-n+i-1} z_2^{-m-i+k-2} A^{\#}(z_1)A_k A^{\#}(z_2)dz_2dz_1$$

$$=\left(\frac{1}{2\pi i}\right)^2\int_{\Gamma_1}\int_{\Gamma_2} z_1^{-n-1} z_2^{-m-1}\sum_{k=1}^{\infty}\sum_{i=0}^{k-1} z_1^i z_2^{k-i-1} A^{\#}(z_1)A_k A^{\#}(z_2)dz_2dz_1$$

$$=\left(\frac{1}{2\pi i}\right)^2\int_{\Gamma_1}\int_{\Gamma_2} \frac{z_1^{-n-1} z_2^{-m-1}}{z_2-z_1}\sum_{k=1}^{\infty}(z_2^k - z_1^k) A^{\#}(z_1)A_k A^{\#}(z_2)dz_2dz_1.$$

Using the resolvent-like identity (3.47), we obtain

$$\sum_{k=1}^{\infty}\sum_{i=0}^{k-1} H_{n-i}A_k H_{m+i-k+1}$$

$$=\left(\frac{1}{2\pi i}\right)^2\int_{\Gamma_1}\int_{\Gamma_2} z_1^{-n-1} z_2^{-m-1}\left[\frac{A^{\#}(z_1)-A^{\#}(z_2)}{z_2-z_1}-\frac{A^{\#}(z_1)P_0(z_2)}{z_2-z_1}\right.$$
$$\left.+\frac{P_0(z_1)A^{\#}(z_2)}{z_2-z_1}\right]dz_2dz_1.$$

Thus, we obtain

$$\sum_{k=1}^{\infty}\sum_{i=0}^{k-1} H_{n-i}A_k H_{m+i-k+1} = I_1 - I_2 + I_3,$$

where

$$I_1 := \left(\frac{1}{2\pi i}\right)^2\int_{\Gamma_1}\int_{\Gamma_2} z_1^{-n-1} z_2^{-m-1}\frac{A^{\#}(z_1)-A^{\#}(z_2)}{z_2-z_1}dz_2dz_1,$$
$$I_2 := \left(\frac{1}{2\pi i}\right)^2\int_{\Gamma_1}\int_{\Gamma_2} z_1^{-n-1} z_2^{-m-1}\frac{A^{\#}(z_1)P_0(z_2)}{z_2-z_1}dz_2dz_1,$$
$$I_3 := \left(\frac{1}{2\pi i}\right)^2\int_{\Gamma_1}\int_{\Gamma_2} z_1^{-n-1} z_2^{-m-1}\frac{P_0(z_1)A^{\#}(z_2)}{z_2-z_1}dz_2dz_1.$$

Let us separately calculate the integrals I_1, I_2, and I_3. The integral I_1 can be written as

$$I_1 = \left(\frac{1}{2\pi i}\right)^2\int_{\Gamma_1}\int_{\Gamma_2} \frac{z_1^{-n-1} z_2^{-m-1}}{z_2-z_1}A^{\#}(z_1)dz_2dz_1 - \left(\frac{1}{2\pi i}\right)^2\int_{\Gamma_1}\int_{\Gamma_2} \frac{z_1^{-n-1} z_2^{-m-1}}{z_2-z_1}A^{\#}(z_2)dz_2dz_1$$
$$=\left(\frac{1}{2\pi i}\right)^2\int_{\Gamma_1}\left[\int_{\Gamma_2}\frac{z_2^{-m-1}}{z_2-z_1}dz_2\right] z_1^{-n-1}A^{\#}(z_1)dz_1$$
$$-\left(\frac{1}{2\pi i}\right)^2\int_{\Gamma_2}\left[\int_{\Gamma_1}\frac{z_1^{-n-1}}{z_2-z_1}dz_1\right] z_2^{-m-1}A^{\#}(z_2)dz_2.$$

In the last equality we used the Fubini theorem to change the order of integration. Using the auxiliary integrals (3.69) and (3.70), we obtain

$$\begin{aligned} I_1 &= \frac{1}{2\pi i}\int_{\Gamma_1}(-\eta_m z_1^{-m-1})z_1^{-n-1}A^\#(z_1)dz_1 - \frac{1}{2\pi i}\int_{\Gamma_2}(-(1-\eta_n)z_2^{-n-1})z_2^{-m-1}A^\#(z_2)dz_2 \\ &= -\frac{\eta_n+\eta_m-1}{2\pi i}\int_{\Gamma_1} z_1^{-n-m-2}A^\#(z_1)dz_1 = -(\eta_n+\eta_m-1)H_{n+m+1}, \end{aligned}$$

where the second integral can be taken over Γ_1 by the principle of deformation of contours. We calculate the second integral I_2 as follows:

$$\begin{aligned} I_2 &= \frac{1}{2\pi i}\int_{\Gamma_1} z_1^{-n-1}A^\#(z_1)\frac{1}{2\pi i}\int_{\Gamma_2}\frac{z_2^{-m-1}P_0(z_2)}{z_2-z_1}dz_2dz_1 \\ &= \begin{cases} \frac{1}{2\pi i}\int_{\Gamma_1} 0 z_1^{-n-1}dz_1, & m<0, \\ -\frac{1}{2\pi i}\int_{\Gamma_1} z_1^{-n-1}A^\#(z_1)z_1^{-m-1}[P_{00}+z_1P_{01}+\cdots+z_1^m P_{0m}]dz_1, & m\ge 0, \end{cases} \\ &= \begin{cases} 0, & m<0, \\ -\frac{1}{2\pi i}\int_{\Gamma_1} z_1^{-n-m-2}A^\#(z_1)[P_{00}+z_1P_{01}+\cdots+z_1^m P_{0m}]dz_1, & m\ge 0, \end{cases} \\ &= \begin{cases} 0, & m<0, \\ -\frac{1}{2\pi i}\int_{\Gamma_1} z_1^{-n-m-2}A^\#(z_1)[P_0(z_1)-z_1^{m+1}P_{0m+1}-z_1^{m+2}P_{0m+2}-\ldots]dz_1, & m\ge 0, \end{cases} \\ &= \begin{cases} 0, & m<0, \\ \frac{1}{2\pi i}\int_{\Gamma_1} z_1^{-n-m-2}A^\#(z_1)[z_1^{m+1}P_{0m+1}+z_1^{m+2}P_{0m+2}+\ldots]dz_1, & m\ge 0, \end{cases} \\ &= \begin{cases} 0, & m<0, \\ \frac{1}{2\pi i}\int_{\Gamma_1} z_1^{-n-1}A^\#(z_1)[P_{0m+1}+z_1P_{0m+2}+\ldots]dz_1, & m\ge 0, \end{cases} \end{aligned}$$

where, in the above expressions, the auxiliary integral (3.71) and the property $A^\#(z)P_0(z)=0$ have been used. Now, we calculate the last integral I_3 with the help of the auxiliary integral (3.72):

$$\begin{aligned} I_3 &= \frac{1}{2\pi i}\int_{\Gamma_2}\frac{1}{2\pi i}\int_{\Gamma_1}\frac{z_1^{-n-1}P_0(z_1)}{z_2-z_1}dz_1 z_2^{-m-1}A^\#(z_2)dz_2 \\ &= \begin{cases} -\frac{1}{2\pi i}\int_{\Gamma_2} z_2^{-n-m-2}P_0(z_2)A^\#(z_2)dz_2, & n<0, \\ -\frac{1}{2\pi i}\int_{\Gamma_2} z_2^{-m-1}[P_{0n+1}+z_2P_{0n+2}+z_2^2P_{0n+3}+\ldots]A^\#(z_2)dz_2, & n\ge 0, \end{cases} \\ &= \begin{cases} 0, & n<0, \\ -\frac{1}{2\pi i}\int_{\Gamma_2} z_2^{-m-1}[P_{0n+1}+z_2P_{0n+2}+z_2^2P_{0n+3}+\ldots]A^\#(z_2)dz_2, & n\ge 0. \end{cases} \end{aligned}$$

Finally, summing up the three integrals I_1, I_2, and I_3, we obtain the relation (3.48). □

3.4 ▪ Problems

Problem 3.1. Prove that the null space is the orthogonal complement of the range of the Moore–Penrose generalized inverse; namely, prove that $N(A)=R(A^\dagger)^\perp$. *Hint: Use Definition 2.1.*

Problem 3.2. Obtain an estimation of the convergence radius for series (3.7) in the singular perturbation case.

Problem 3.3. Derive formula (3.28) by a formal induction argument.

Problem 3.4. Prove, by induction, the recursion (3.31).

Problem 3.5. Derive expansion (3.43). *Hint: Basically, it is a Taylor power series for the resolvent* $R(\zeta,z)=(A(z)-\zeta I)^{-1}$ *with respect to* z. *See also the book of Kato [99].*

Problem 3.6. Establish convergence of the power series (3.45). *Hint: The convergence is established by making bounds for the contour integrals similarly to [99, Ch.2, Sec.3].*

Problem 3.7. Demonstrate that formula (3.49) becomes (3.53) when $A(z)$ is invertible.

Problem 3.8. For the case of $n=2$ verify (3.55) by tracing the steps of the proof of Theorem 3.12 and calculating the appropriate residues.

Problem 3.9. Prove Theorem 3.16.

Problem 3.10. Prove the following formula for the perturbed resolvent:

$$R(\zeta)=(A-\zeta I)^{-1}=-\sum_{i=0}^{p}\left[\frac{1}{\zeta-\lambda_i}P_i+\sum_{j=1}^{m_i}\frac{1}{(\zeta-\lambda_i)^{j+1}}D_i^j\right].$$

Hint: Use the spectral decomposition. See also the book of Kato [99].

Problem 3.11. Show that, whenever Assumption S1 is satisfied, the operator $B(z)$ can be expressed as a power series,

$$B(z)=B_0+\sum_{i=1}^{\infty}z^iB_i,$$

with $B_0=P(0)A_1P(0)$, and

$$B_n=-\sum_{p=1}^{n+1}(-1)^p\sum_{\substack{\nu_1+\cdots+\nu_p=n+1\\ \mu_1+\cdots+\mu_{p+1}=p-1\\ \nu_j\geq 1,\mu_j\geq 0}}S_{\mu_1}A_{\nu_1}S_{\mu_2}\ldots A_{\nu_p}S_{\mu_{p+1}},$$

where $S_0:=-P(0)$ and $S_k:=((A_0)^{\#})^k$. *Hint: This is similar to the proof of Theorem 3.12. See also the book of Kato [99].*

Problem 3.12. In the regular perturbation case write three first terms for the Taylor series expansion (3.54) of the perturbed Drazin generalized inverse using the general formula for series coefficients (3.55).

Problem 3.13. Establish the expressions

$$\frac{1}{2\pi i}\int_{\Gamma_2}\frac{z_2^{-m-1}}{z_2-z_1}dz_2=-\eta_m z_1^{-m-1},$$

$$\frac{1}{2\pi i}\int_{\Gamma_1}\frac{z_1^{-m-1}}{z_2-z_1}dz_1=-(1-\eta_n)z_2^{-m-1},$$

with

$$\eta_m:=\begin{cases}0, & m<0,\\ 1, & m\geq 0,\end{cases}$$

and where Γ_1 and Γ_2 are two closed counterclockwise oriented contours in the complex plane around zero, and, furthermore, the contour Γ_2 lies inside the contour Γ_1. *Hint: See the book of Kato [99] and the book of Korolyuk and Turbin [104].*

3.5 ▪ Bibliographic Notes

The results of Section 3.2 are based on [12]. We would like to note that even though the problem of the perturbation of null spaces can be regarded as a particular case of the general eigenvalue perturbation problem $A(z)x(z)=\lambda(z)x(z)$, it deserves a special treatment. First, the perturbed null space is always analytic in the perturbation parameter. Second, the results on the perturbation of null spaces can be applied to the perturbation of Markov chains. See Chapter 6 for more on perturbations of Markov chains. And third, the results on the perturbation of null spaces can also be used for the perturbation analysis of the general eigenvalue problem (see Section 3.2.6). The literature on the perturbation of the eigenvalue problem is vast. The interested reader can consult the books [22, 38, 39, 99, 132, 143, 151] and for more recent references the papers [94, 119, 123, 136]. The analysis of the stability properties of the null and range spaces in the context of infinite dimensional operators can be found in [18, 20, 21].

In Section 3.3 we develop a reduction process for generalized inverses. We would like to note that the rank-preserving perturbations of generalized inverses have been well studied (see, e.g., [159] and references therein), whereas the singular perturbations of generalized inverses, to the best of our knowledge, have been studied only in [14, 27] and Section 3.3 of the present book. In particular, Bohl and Lancaster [27] analyzed the perturbation of the group generalized inverse and applied their analysis to chemical networks. One of the key elements of our reduction technique for perturbed generalized inverses is the notion of the group reduced resolvent. A notion similar to the group reduced resolvent was used in [38, 39] to treat clustered eigenvalues. The first analysis of the additive componentwise perturbation of the generalized inverse was provided by Stewart [146]. For the recent development of the additive componentwise perturbation of the generalized inverse the reader is invited to consult [36, 37, 159].

Chapter 4

Polynomial Perturbation of Algebraic Nonlinear Systems

4.1 ▪ Introduction

In the previous chapter we studied the analytic perturbation of linear systems. Even though the class of linear systems is fundamental, many phenomena in nature can only be modeled by a nonlinear system. Typically a model has one or more parameters, and we are interested in how the properties of the system change with changes in a parameter value. The simplest nonlinear model is a polynomial. In fact, polynomials (specifically, characteristic polynomials) are also useful for the analysis of linear systems. Therefore, in the present chapter we study the perturbation of polynomials and polynomial systems. Let us begin with a simple example which demonstrates that the situation in nonlinear algebraic systems is quite different from that in linear algebraic systems.

Example 4.1. *Consider the following polynomial equation:*

$$(1-z)x^2-2x+1=0. \tag{4.1}$$

Since it is a quadratic equation, one can easily find its solutions:

$$x_{1,2}(z)=\frac{1}{1\mp\sqrt{z}}, \quad z\neq 1.$$

These solutions can be expanded as follows:

$$x_1(z)=1+z^{1/2}+z+z^{3/2}+\ldots,$$

$$x_2(z)=1-z^{1/2}+z-z^{3/2}+\ldots.$$

This example shows that the perturbation analysis of nonlinear algebraic systems requires new tools such as *fractional* power series expansions.

Let us consider a general bivariate polynomial equation that can be written in the form

$$Q(x,z)=q_m(z)x^m+q_{m-1}(z)x^{m-1}+\cdots+q_0(z)=0, \tag{4.2}$$

where the coefficients $q_i(z), i=0,\ldots,m$, are also polynomials of the perturbation parameter z. According to the fundamental theorem of algebra, when $z=z_0$, the polynomial equation $Q(x,z_0)=0$ has m roots. We note that some roots can be multiple, and they are counted according to their multiplicity. We are interested in the behavior of roots if the perturbation parameter z deviates slightly from z_0. The following, now classical, result for bivariate polynomials was established by Victor Puiseux in 1850.

Theorem 4.1. *Let $Q(x,z)$ in (4.2) be a bivariate polynomial. Then, a solution of the polynomial equation (4.2) can be expressed in the form of Puiseux fractional power series*

$$x(z)=\sum_{k=k_0}^{\infty} c_k(z-z_0)^{k/d}, \tag{4.3}$$

where k_0 is an integer that can be positive, zero, or negative and d is a natural number greater than or equal to one.

In case k_0 is negative, we refer to such a series as the Laurent–Puiseux series. We also note that the Taylor series is a particular case of the Puiseux series when $d=1$ and $k_0 \geq 0$. The Taylor series case corresponds to the regular perturbation, whereas the Puiseux series case and the Laurent–Puiseux series case correspond to the singular perturbation.

In the present chapter we also consider a system of polynomials which depend on a single perturbation parameter

$$\begin{cases} Q_1(\mathbf{x},z)=0, \\ Q_2(\mathbf{x},z)=0, \\ \quad\vdots \\ Q_n(\mathbf{x},z)=0, \end{cases} \tag{4.4}$$

where $\mathbf{x}=[x_1\ x_2\ \cdots\ x_n]$ and $Q_i(\mathbf{x},z)$ is a multivariate polynomial of $x_1,x_2,\ldots,x_n$ and z. We are interested in studying the behavior of a solution of system (4.4) in a small neighborhood of some point z_0.

In Section 4.3, with the help of a Gröbner bases technique, we substitute the system (4.4) with a decoupled system of bivariate polynomials,

$$\begin{cases} \tilde{Q}_1(x_1,z)=0, \\ \tilde{Q}_2(x_2,z)=0, \\ \quad\vdots \\ \tilde{Q}_n(x_n,z)=0, \end{cases} \tag{4.5}$$

that determines, a priori, a larger variety of solutions than the original system (4.4). But we note that all solutions of (4.4) are included in the solutions of (4.5) and have the same type of analytic expansion as the Puiseux series.

Thus, we reduce the multidimensional problem to the set of one dimensional problems. To make the book self-contained we have added an auxiliary section, Section 4.2, with an introduction to Gröbner bases and Buchberger's algorithm. The reader familiar with these concepts can skip Section 4.2 and continue directly to Section 4.3. For the reader interested in more theoretical and algorithmic details about Gröbner bases, we provide references in the Bibliographic Notes section. For all practical purposes, one can simply use the function `gbasis` from the "Groebner" package of Maple. In Example 4.14 we demonstrate the application of the function `gbasis`. The perturbation of a single polynomial equation is then analyzed in subsequent sections. When the perturbation is regular, we show in Section 4.6 that the coefficients of (4.3) can be computed very efficiently. In Section 4.7 we explain the Newton polygon method for construction of the Puiseux series (4.3) in the singular perturbation case. The Newton diagram method constructs the Puiseux series (4.3) term by term, and in general one cannot determine the integer d. However, if the bivariate polynomial in (4.2) is irreducible, the Newton polygon method leads to the determination of the integer d and hence provides a complete characterization of the behavior of the perturbed solution. Therefore, if one needs to know the value of d, it is recommended to factorize the perturbed polynomial before applying the Newton diagram method. In Section 4.5 we provide a method for the decomposition of a bivariate polynomial into irreducible factors. The method is also based on a Gröbner bases technique. We would like to mention that if d is known, all coefficients c_k of (4.3) can be calculated by simple recursive formulae using the method of undetermined coefficients. In other words, with the help of the Newton polygon method and decomposition into irreducible factors, one determines the integer d and by a change of variables transforms the initial singularly perturbed problem to a regularly perturbed problem.

4.2 ▪ Preliminaries on Gröbner Bases and Buchberger's Algorithm*

To see why Gröbner bases[4] are an essential tool for the analysis of systems of multivariate polynomial equations it is useful to understand some important algebraic concepts. Let us recall some basic definitions. Of course, the reader should feel free to skip familiar definitions.

Briefly, a *ring* is an abelian group with a second binary operation that is associative and is distributive over the abelian group operation. The abelian group operation is called *addition*, and the second binary operation is called *multiplication* in analogy with the integers. One familiar example is the set $\mathbb{Z}$ of integers. The integers are a *commutative* ring, since $ab = ba$. The set of polynomials also forms a commutative ring. The set of 2×2 matrices with integer coefficients is a *noncommutative* ring since, in general, $AB \neq BA$. An *integral domain* is a nontrivial commutative ring with *identity* in which the product of any two nonzero elements is not equal to zero. Examples include the integers and the set of polynomials. Precisely, we have the following definition.

Definition 4.1. *A ring is a set R equipped with two binary operations $+$ and $\cdot$ called addition and multiplication that map every pair of elements of R to a unique element of R. These operations satisfy the following ring axioms (the symbol $\cdot$ is often omitted, and multiplication is denoted simply by juxtaposition), which must be true for all $a, b, c \in R$:*

- *Addition is abelian:*

  1. $(a+b)+c = a+(b+c)$;
  2. *there is an element* $0 \in R$ *such that* $0+a = a$;
  3. $a+b = b+a$; *and*
  4. *for each* $a \in R$ *there exists an element* $-a \in R$ *such that* $a+(-a) = (-a)+a = 0$.

- *Multiplication is associative:*

  5. $(a \cdot b) \cdot c = a \cdot (b \cdot c)$.

[4]Some articles refer to the Groebner bases.

- *Multiplication distributes over addition:*

  6. $a \cdot (b+c) = a \cdot b + a \cdot c$; *and*

  7. $(a+b) \cdot c = a \cdot c + b \cdot c$.

- *Multiplicative identity:*

  8. *there is an element* $1 \in R$ *such that* $a \cdot 1 = 1 \cdot a = a$.

The *ideal* in a ring is an idealized generalization of an element. The study of ideals is central to a structural understanding of a ring.

Definition 4.2. *Let R be a ring. A subset $I \subset R$ is said to be a left ideal in R if $RI \subset I$, where*

$$RI = \left\{ x \mid x = \sum_{i=1}^{n} r_i \cdot x_i \text{ for some } n \in \mathbb{N} \text{ where } r_i \in R \text{ and } x_i \in I \right\}$$

denotes the span of I over R. Similarly a subset $I \subset R$ is said to be right ideal if $IR \subset I$. A subset $I \subset R$ is said to be a two-sided ideal or simply an ideal if it is both a left ideal and a right ideal. A one-sided or two-sided ideal is an additive subgroup of R. If $E \subset R$, then RE is the left ideal generated by E, ER is the right ideal generated by E, and RER is the two-sided ideal generated by E. These are the smallest ideals containing E. If $x \in R$, then Rx and xR are the principal left ideals and right ideals generated by x. The principal ideal RxR is written as $\langle x \rangle$. A ring is said to be simple if it is nonzero and it has no proper nonzero two-sided ideals. A commutative simple ring is precisely a field.

Example 4.2. *The set $M_2(\mathbb{R})$ of 2×2 matrices with real coefficients forms a noncommutative unital ring over the field $\mathbb{R}$ of real numbers. If*

$$E = \begin{bmatrix} 1 & -1 \\ -1 & 1 \end{bmatrix},$$

then the left ideal generated by E is the set of matrices $M_2(\mathbb{R})E = \{F \mid F = AE \text{ for some } A \in M_2(\mathbb{R})\}$. Now since

$$AE = \begin{bmatrix} a & b \\ c & d \end{bmatrix} \begin{bmatrix} 1 & -1 \\ -1 & 1 \end{bmatrix} = \begin{bmatrix} a-b & -a+b \\ c-d & -c+d \end{bmatrix},$$

we can see that $M_2(\mathbb{R})E$ is the set of 2×2 matrices with row sums equal to zero.

A ring in which there is no strictly increasing infinite chain of left ideals is called a left Noetherian ring. A ring in which there is no strictly decreasing infinite chain of left ideals is called a left Artinian ring. The Hopkins–Levitzki theorem states that a left Artinian ring is left Noetherian. The integers form a Noetherian ring which is not Artinian. For commutative rings, the ideals generalize the classical algebraic notion of divisibility and decomposition of an integer into prime numbers. An ideal $P \subset R$ is called a proper ideal if $P \neq R \Leftrightarrow 1 \notin P$. A proper ideal $P \subset R$ is called a prime ideal if, for any elements $x, y \in R$, we have that $xy \in P$ implies either $x \in P$ or $y \in P$. Equivalently, P is prime if for any ideals I, J we have that $IJ \in P$ implies either $I \in P$ or $J \in P$. The latter formulation illustrates the idea of ideals as generalizations of elements.

In general a polynomial ring is a ring formed from the set of polynomials in one or more variables with coefficients in another ring. In order to begin our discussion we will consider a special case in which the set of coefficients is the field of complex numbers. Let $\mathbb{C}$ denote the set of complex numbers, and let $\mathbb{Z}_{\geq 0}$ denote the set of nonnegative integers.

Definition 4.3. *Let $x = (x_1, \ldots, x_n) \in \mathbb{C}^n$ and for each $\alpha \in \mathbb{Z}^n_{\geq 0}$ write $\boldsymbol{x}^\alpha = x_1^{\alpha_1} \cdots x_n^{\alpha_n}$. The function $p : \mathbb{C}^n \to \mathbb{C}$ is called a polynomial in $\boldsymbol{x}$ with complex coefficients if*

$$p(x) = \sum_{\alpha \in A} p_\alpha \boldsymbol{x}^\alpha$$

for some finite subset $A \subset \mathbb{Z}^n_{\geq 0}$ where $p_\alpha \in \mathbb{C}$.

The set of all polynomials in $\boldsymbol{x}$ with complex coefficients forms a commutative *ring* under the operations of addition and multiplication. Addition is defined by

$$\sum_{\alpha \in A} p_\alpha \boldsymbol{x}^\alpha + \sum_{\beta \in B} q_\beta \boldsymbol{x}^\beta = \sum_{\gamma \in C} (p_\gamma + q_\gamma) \boldsymbol{x}^\gamma,$$

where $C = A \cup B$ and where $p_\alpha = q_\beta = 0$ when $\alpha, \beta \notin C$, and multiplication is defined by

$$\sum_{\alpha \in A} p_\alpha \boldsymbol{x}^\alpha \sum_{\beta \in B} q_\beta \boldsymbol{x}^\beta = \sum_{\alpha \in A,\, \beta \in B} p_\alpha q_\beta \boldsymbol{x}^{\alpha+\beta}.$$

The ring is denoted by $\mathbb{C}[\boldsymbol{x}] = \mathbb{C}[x_1, \ldots, x_n]$.

In a univariate polynomial ring, there is a natural ordering on the monomials:

$$\cdots \succ x^{n+1} \succ x^n \succ \cdots \succ x^2 \succ x \succ 1.$$

In a multivariate polynomial ring, there are multiple conventions for ordering monomials, leading to a number of different possible orderings.

Definition 4.4. *A monomial ordering on $\mathbb{C}[\boldsymbol{x}]$ is any relation $\succ$ on $\mathbb{Z}^n_{\geq 0}$ or, equivalently, any relation on the set of monomials $\boldsymbol{x}^\alpha$ for $\alpha \in \mathbb{Z}^n_{\geq 0}$, such that*

1. *$\succ$ is a total ordering on $\mathbb{Z}^n_{\geq 0}$;*
2. *if $\boldsymbol{x}^\alpha \succ \boldsymbol{x}^\beta$ and $\gamma \in \mathbb{Z}^n_{\geq 0}$, then $\boldsymbol{x}^{\alpha+\gamma} \succ \boldsymbol{x}^{\beta+\gamma}$; and*
3. *$\succ$ is a well-ordering on $\mathbb{Z}^n_{\geq 0}$, or, equivalently, every nonempty subset of $\mathbb{Z}^n_{\geq 0}$ has a smallest element.*

It is convenient and sometimes simpler to represent the monomial $\boldsymbol{x}^\alpha = x_1^{\alpha_1} \cdots x_n^{\alpha_n}$ as a *tuple* $\alpha = (\alpha_1, \ldots, \alpha_n) \in \mathbb{Z}^n_{\geq 0}$ where each entry is the degree of the corresponding variable. The purpose of the well-ordering condition is to guarantee that the multivariate polynomial division algorithm will eventually terminate. One example of a commonly used monomial ordering is the lexicographic ordering.

Definition 4.5 (lexicographic ordering). *Let $\alpha, \beta \in \mathbb{Z}^n_{\geq 0}$. We say $\alpha \succ \beta$ if, for some $j \in \{1, \ldots, n\}$, we have $\alpha_i = \beta_i$ for $i < j$ and $\alpha_j > \beta_j$. We say $\boldsymbol{x}^\alpha \succ \boldsymbol{x}^\beta$ if $\alpha \succ \beta$.*

Example 4.3. *The polynomial $p \in \mathbb{C}[x_1, x_2, x_3]$ given by $p = 4x_1^3 - 5x_1^2 x_2^4 x_3 + 3x_1 x_2^6 - 2x_2^6 x_3$ is written using the lexicographic order $x_1 \succ x_2 \succ x_3$ for the terms.*

Definition 4.6. *Let $p = \sum_{\alpha \in A} p_\alpha x^\alpha$ be a polynomial in $\mathbb{C}[x]$, and let $\succ$ be a monomial order on $\mathbb{Z}^n_{\geq 0}$. We define the following:*

1. *the multidegree of p is given by $\overline{\alpha} = \overline{\alpha}(p) = \max\{\alpha \in \mathbb{Z}^n_{\geq 0} \mid p_\alpha \neq 0\} \in \mathbb{Z}^n_{\geq 0}$;*
2. *the leading coefficient of p is given by $\mathrm{LC}(p) = p_{\overline{\alpha}} \in \mathbb{C}$;*
3. *the leading monomial of p is $\mathrm{LM}(p) = x^{\overline{\alpha}}$; and*
4. *the leading term of p is $\mathrm{LT}(p) = \mathrm{LC}(p)\mathrm{LM}(p)$.*

Example 4.4. *For the polynomial $p = 4x_1^3 - 5x_1^2x_2^4x_3 + 3x_1x_2^6 - 2x_2^6x_3$ above, we have, using lexicographic order, $\overline{\alpha} = (3,0,0)$ with $\mathrm{LC}(p) = 4$, $\mathrm{LM}(p) = x_1^3$, and $\mathrm{LT}(p) = 4x_1^3$.*

In the division algorithm for polynomials of one variable, for a given dividend and a given divisor, we are guaranteed a unique quotient and remainder. This is not the case when the polynomials depend on two or more variables. Now the answer depends on both the monomial order and the order of the divisors. We will divide $p \in \mathbb{C}[x]$ by $g_1, \ldots, g_s \in \mathbb{C}[x]$ so that we can write

$$p = q_1 g_1 + \cdots + q_s g_s + r.$$

The strategy is to repeatedly cancel the leading term of p by subtracting an appropriate multiple of the first suitable g_i. The result of the division is not unique because the quotients and remainder depend upon the monomial ordering and the order of the divisors.

Theorem 4.2 (the division algorithm). *Let $\succ$ be a monomial order on $\mathbb{Z}^n_{\geq 0}$, and let $G = (g_1, \ldots, g_s)$ be an ordered s-tuple of polynomials in $\mathbb{C}[x]$. Then every $p \in \mathbb{C}[x]$ can be written as*

$$p = q_1 g_1 + \cdots + q_s g_s + r,$$

where $q_i, r \in \mathbb{C}[x]$ and either $r = 0$ or $r = \sum_{\beta \in B} r_\beta x^\beta$, where $r_\beta \in \mathbb{C}$ and $\beta \in B$ is a finite linear combination of monomials, none of which is divisible by any of $\mathrm{LT}(g_1), \ldots, \mathrm{LT}(g_s)$. We call r the remainder of p on division by G. Furthermore, if $q_i g_i \neq 0$, then we have $\mathbf{deg}(p) \geq \mathbf{deg}(q_i g_i)$.

Example 4.5. *We use lexicographic ordering $x_1 \succ x_2 \succ x_3$. Let $p = x_1^5 x_2^3$, and let $g_1 = x_1^3x_2^2 - x_2^2x_3$ and $g_2 = x_1x_2^2 - x_2x_3$. Since $\mathrm{LT}(g_1) = x_1^3x_2^2$ divides $\mathrm{LT}(p) = x_1^5x_2^3$, we have*

$$p - x_1^2x_2g_1 = x_1^5x_2^3 - x_1^5x_2^3 + x_1^2x_2^3x_3 = x_1^2x_2^3x_3 = r_1.$$

The first term $x_1^2x_2^3x_3$ of r_1 is not divisible by $\mathrm{LT}(g_1) = x_1^3x_2^2$, but it is divisible by $\mathrm{LT}(g_2) = x_1x_2^2$, and so we write

$$r_1 - x_1x_2x_3g_2 = x_1x_2^2x_3^2 = r_2.$$

Again the first term of r_2 is divisible by $\mathrm{LT}(g_2)$. Thus we write

$$r_2 - x_3^2 g_2 = x_2x_3^3 = r.$$

No further divisions are possible, and so we finally obtain

$$p = x_1^2x_2g_1 + (x_1x_2x_3 + x_3^2)g_2 + r.$$

We will soon define the Gröbner basis formally. To motivate the need for such a basis we note that, in general, the division algorithm does not yield a unique remainder. However, if the division of p is a division by the elements G of a Gröbner basis, then we obtain the same remainder r irrespective of the ordering of G. Since it can be shown that every ideal I has a Gröbner basis G, it follows that a polynomial p belongs to an ideal I if and only if division of p by the Gröbner basis G of I returns a remainder of 0.

Definition 4.7. *A subset $I \subset \mathbb{C}[x]$ is a polynomial ideal if it satisfies the following conditions:*

1. $0 \in I$;
2. *if $p, q \in I$, then $p+q \in I$; and*
3. *if $p \in I$ and $q \in \mathbb{C}[x]$, then $pq \in I$.*

There are two commonly used polynomial ideals in $\mathbb{C}[x]I \subset I$. The ideal generated by the finite set of polynomials $\{f_1, \ldots, f_s\} \subset \mathbb{C}[x]$ is defined by

$$\langle f_1, \ldots, f_s \rangle = \left\{ f \mid f = \sum_{i=1}^{s} p_i f_i \text{ where } p_i \in \mathbb{C}[x] \right\}.$$

The ideal consisting of the set of polynomials which vanish everywhere on some given set $S \subset \mathbb{C}^n$ is defined by

$$I(S) = \{f \in \mathbb{C}[x] \mid f(a) = 0 \ \forall a \in S\}.$$

Definition 4.8. *A monomial ideal is an ideal generated by a set of monomials. That is, I is a monomial ideal if there is a subset $A \subset \mathbb{Z}^n_{\geq 0}$ such that I consists of all polynomials of the form $p = \sum_{\alpha \in A} p_\alpha(x) x^\alpha$, where $x \in \mathbb{C}^n$, $p_\alpha(x) \in \mathbb{C}[x]$. We write $I = \langle x^\alpha \mid \alpha \in A \rangle$.*

If I is a monomial ideal generated by the set $A \subset \mathbb{Z}^n_{\geq 0}$ and if $x^\beta \in I$, then x^β is divisible by x^α for some $\alpha \in A$. Furthermore, for every polynomial $p \in I$ we can say that every term of p lies in I and that p is a $\mathbb{C}$-linear combination of the monomials in I.

Example 4.6. *The set $I = \langle x_1^3, x_1^2 x_2 x_3, x_1 x_2^2 x_3^5 \rangle$ is a monomial ideal.*

Definition 4.9. *Let $x \in \mathbb{C}^n$, and let $I \subset \mathbb{C}[x]$ be a nonzero ideal.*

1. *Let $\mathrm{LT}(I)$ be the set of leading terms of elements of I;*

$$\mathrm{LT}(I) = \{c x^\alpha \mid \exists\ p \in I \text{ with } \mathrm{LT}(p) = c x^\alpha\}.$$

2. *We denote by $\langle \mathrm{LT}(I) \rangle$ the ideal generated by the elements of $\mathrm{LT}(I)$. Note that*

$$\langle \mathrm{LM}(g_1), \ldots, \mathrm{LM}(g_s) \rangle = \langle \mathrm{LT}(g_1), \ldots, \mathrm{LT}(g_s) \rangle \text{ and } \langle \mathrm{LM}(I) \rangle = \langle \mathrm{LT}(I) \rangle.$$

The characteristic property of the ideal $\langle \mathrm{LT}(\langle g_1, \ldots, g_s \rangle) \rangle$ is that every element is divisible by $\mathrm{LT}(g_i)$ for some $i \in \{1, \ldots, s\}$, and so

$$\langle \mathrm{LT}(g_1), \ldots, \mathrm{LT}(g_s) \rangle \subset \langle \mathrm{LT}(\langle g_1, \ldots, g_s \rangle) \rangle.$$

However, the opposite inclusion may not be true, and so the monomial ideals $\langle \mathrm{LT}(\langle g_1, \ldots, g_s \rangle) \rangle$ and $\langle \mathrm{LT}(g_1), \ldots, \mathrm{LT}(g_s) \rangle$ are not always the same. We make the following definition.

Definition 4.10 (Gröbner basis). *Let $\succ$ be a fixed monomial ordering on $\mathbb{C}[x]$, where $x \in \mathbb{C}^n$. A finite subset $G = g_1, \ldots, g_t$ is a Gröbner basis if*

$$\langle LT(g_1), \ldots, LT(g_t) \rangle = \langle LT(\langle g_1, \ldots, g_t \rangle) \rangle$$

or, equivalently, if

$$\langle LM(g_1), \ldots, LM(g_t) \rangle = \langle LM(\langle g_1, \ldots, g_t \rangle) \rangle.$$

In the theory of commutative algebra the Hilbert basis theorem states that every ideal in the ring of multivariate polynomials over a Noetherian ring is finitely generated. Equivalently we may say that every algebraic set over a field can be described as the set of common roots of finitely many polynomial equations. Hilbert proved the theorem for the special case of polynomial rings over a field in the course of his proof of finite generation of rings of invariants. As a corollary to the Hilbert basis theorem applied to $\langle \mathrm{LT}(\langle g_1, \ldots, g_t \rangle) \rangle$ we have the following result.

Corollary 4.1. *Let $I = \langle g_1, \ldots, g_t \rangle$ be a nonzero polynomial ideal in $\mathbb{C}[x]$ with an ordering $\succ$, where $x \in \mathbb{C}^n$. Then I has a Gröbner basis.*

We will not discuss the inductive proof proposed by Hilbert but rather will focus on the generation of a finite Gröbner basis for I using the Buchberger algorithm. We wish to obtain a generating set such that all the leading terms of the polynomials in the set generate the leading terms of the ideal I. This fails when there is a cancellation of leading terms. To avoid unwanted subsequent cancellations we construct new polynomials by applying a simple cancellation procedure to each pair of existing polynomials.

Definition 4.11 (S-polynomial). *Let $x \in \mathbb{C}^n$, and suppose $g, h \in \mathbb{C}[x]$ with an ordering $\succ$ are nonzero polynomials.*

1. *If* $\mathbf{deg}(g) = \overline{\alpha}$ *and* $\mathbf{deg}(h) = \overline{\beta}$, *then let* $\gamma = (\gamma_1, \ldots, \gamma_n)$ *where* $\gamma_i = \max[\overline{\alpha}_i, \overline{\beta}_i]$ *for each* $i = 1, \ldots, n$. *We call* x^γ *the least common multiple of* $\mathrm{LT}(g)$ *and* $\mathrm{LT}(h)$, *written as* $x^\gamma = \mathrm{LCM}(\mathrm{LT}(g), \mathrm{LT}(h))$.
2. *The S-polynomial of g and h is defined by the formula*

$$S(g,h) = \frac{x^\gamma}{\mathrm{LT}(g)} \cdot g - \frac{x^\gamma}{\mathrm{LT}(h)} \cdot h.$$

Example 4.7. *Let $g = 3x^2yz^3 - x^2z^3 + 2y$, and let $h = xy^2z + xy^2 - 2x$, where we use the lexicographic order $x > y > z$. Now $\overline{\alpha} = (2,1,3)$ and $\overline{\beta} = (1,2,1)$, and so $\gamma = (2,2,3)$ and we have*

$$S(g,h) = \frac{x^2y^2z^3}{3x^2yz^3} \cdot g - \frac{x^2y^2z^3}{xy^2z} \cdot h = \frac{y}{3} \cdot g - xz^2 \cdot h = -x^2y^2z^2 - \frac{x^2yz^3}{3} + 2x^2z^2 + \frac{2xy}{3}.$$

Note the cancellation of the leading terms in the construction of the S-polynomial. Once a basis contains all necessary S-polynomials defined from the polynomial pairs in the generating set, then it follows that

$$\langle \mathrm{LT}(\langle g_1, \ldots, g_s \rangle) \rangle \subset \langle \mathrm{LT}(g_1), \ldots, \mathrm{LT}(g_t) \rangle,$$

and hence the ideals are equal.

Theorem 4.3 (Buchberger's criterion). *Let I be a polynomial ideal. Then a basis $G = \{g_1,\ldots,g_s\}$ for I is a Gröbner basis for I if and only if, for all pairs $i \neq j$, the remainder on division of $S(g_i,g_j)$ by G is zero.*

We make the following definition.

Definition 4.12. *We write $\overline{p}^G$ for the remainder on division of p by the list of polynomials $G = \{g_1,\ldots,g_s\}$. That is, we write*

$$p = q_1 g_1 + \cdots + q_s g_s + \overline{p}^G.$$

Example 4.8. *Reinterpreting Example 4.5, using the ordering $x_1 \succ x_2 \succ x_3$ with $p = x_1^5 x_2^3$ and $G = \{x_1^3x_2^2 - x_2^2x_3, x_1x_2^2 - x_2x_3\}$, we can write*

$$p = x_1^2x_2 \cdot (x_1^3x_2^2 - x_2^2x_3) + (x_1x_2x_3 + x_3^2)\cdot(x_1x_2^2 - x_2x_3) + x_2x_3^3,$$

and hence $\overline{p}^G = x_2x_3^3$.

Theorem 4.4 (Buchberger's algorithm). *Let $I = \langle g_1,\ldots,g_s\rangle \neq \{0\}$ be a polynomial ideal. Then a Gröbner basis for I can be constructed in a finite number of steps.*

Let $G = \{g_1,g_2\}$, and let $S_{i,j} = S(g_i,g_j)$ be the S-polynomial for the pair $\{g_i,g_j\}$. Note that $S_{i,j} = q_i \cdot g_i + q_j \cdot g_j + \overline{S_{i,j}}^G$, and if we replace G by $G \cup \{\overline{S_{i,j}}^G\}$, then we have $S_{i,j} = q_i \cdot g_i + q_j \cdot g_j + 1 \cdot \overline{S_{i,j}}^G$, and hence the new remainder is zero. This observation is the basis for Buchberger's algorithm, which proceeds as follows. Let $G = \{g_1,\ldots,g_s\}$ be a list of the polynomials defining I. For each pair of polynomials (g_i,g_j) in G calculate their S-polynomial $S_{i,j}$, and divide it by G, obtaining the remainder $\overline{S_{i,j}}^G$. If $\overline{S_{i,j}}^G \neq 0$, add $\overline{S_{i,j}}^G$ to G and start again with $G = G \cup \{\overline{S_{i,j}}^G\}$. Repeat the process until all S-polynomials defined by polynomial pairs in G have remainder 0 after division by G.

Example 4.9. *Consider the ring $\mathbb{C}[x]$, where $x \in \mathbb{Z}^2_{\geq 0}$ with lexicographic order. Let $I = \langle -2x_1x_2 + x_1, x_1^3x_2 - 2x_1^2 + x_2\rangle$. Let $G = \{-2x_1x_2 + x_1, x_1^3x_2 - 2x_1^2 + x_2\}$. Now*

$$S_{1,2} = \frac{x_1^3x_2}{-2x_1x_2}\cdot g_1 - \frac{x_1^3x_2}{x_1^3x_2}\cdot g_2 = -\frac{x_1^2}{2}\cdot g_1 - g_2 = -\frac{x_1^3}{2} + 2x_1^2 - x_2,$$

and since $S_{1,2}$ is not divisible by $G = (g_1,g_2)$, it follows that $\overline{S_{1,2}}^G = -x_1^3/2 + 2x_1^2 - x_2$. Thus we redefine $G = \{-2x_1x_2 + x_1, x_1^3x_2 - 2x_1^2 + x_2, -x_1^3/2 + 2x_1^2 - x_2\}$. Now we repeat the construction process. We know that $S_{1,2} = g_3$, and so division by $G = \{g_1,g_2,g_3\}$ gives $\overline{S_{1,2}}^G = 0$. Now we calculate

$$S_{1,3} = \frac{x_1^3x_2}{-2x_1x_2}\cdot g_1 - \frac{x_1^3x_2}{-\frac{x_1^3}{2}}\cdot g_3 = -\frac{x_1^2}{2}\cdot g_1 + 2x_2\cdot g_3 = -\frac{x_1^3}{2} + 4x_1^2x_2 - 2x_2^2,$$

from which it follows on division by $G = \{g_1,g_2,g_3\}$ that

$$S_{1,3} = -2x_1 \cdot g_1 + 1\cdot g_3 - 2x_1^2 + x_2.$$

Hence $\overline{S_{1,3}}^G = -2x_1^2 + x_2$. *Thus we redefine* $G = \{-2x_1x_2 + x_1, x_1^3x_2 - 2x_1^2 + x_2, -x_1^3/2 + 2x_1^2 - x_2, -2x_2^2 + x_2\}$. *Now we know* $\overline{S_{1,2}}^G = 0$ *and* $\overline{S_{1,3}}^G = 0$. *We compute*

$$S_{2,3} = \frac{x_1^3x_2}{x_1^3x_2} \cdot g_2 - \frac{x_1^3x_2}{-\frac{x_1^3}{2}} \cdot g_3 = g_2 + 2x_2 \cdot g_3 = 4x_1^2x_2 - 2x_1^2 - 2x_2^2 + x_2,$$

from which it follows on division by $G = \{g_1, g_2, g_3, g_4\}$ *that*

$$S_{2,3} = -2x_1 \cdot g_1 + 1 \cdot g_4,$$

and so $\overline{S_{2,3}}^G = 0$. *We have*

$$S_{2,4} = \frac{x_1^3x_2^2}{x_1^3x_2} \cdot g_2 - \frac{x_1^3x_2^2}{-2x_2^2} \cdot g_4 = x_2 \cdot g_2 + \frac{x_1^3}{2} \cdot g_4 = \frac{x_1^3x_2}{2} - 2x_1^2x_2 + x_2^2,$$

from which it follows on division by $G = \{g_1, g_2, g_3, g_4\}$ *that*

$$S_{2,4} = \left(-\frac{x_1^2}{4} + x_1\right) \cdot g_1 - \frac{1}{2} \cdot g_2 - \frac{1}{4} \cdot g_4$$

and hence $\overline{S_{2,4}}^G = 0$. *Finally, we have*

$$S_{3,4} = \frac{x_1^3x_2^2}{-\frac{x_1^3}{2}} \cdot g_3 - \frac{x_1^3x_2^2}{-2x_2^2} \cdot g_4 = -2x_2^2 \cdot g_3 + \frac{x_1^3}{2} \cdot g_4 = \frac{x_1^3x_2}{2} - 4x_1^2x_2^2 + 2x_2^3,$$

and on division by $G = \{g_1, g_2, g_3, g_4\}$ *we obtain*

$$S_{3,4} = \left(-\frac{x_1^2}{4} + 2x_1x_2 + x_1\right) \cdot g_1 - \frac{1}{2} \cdot g_3 - \left(x_2 + \frac{1}{2}\right) \cdot g_4,$$

from which it follows that $\overline{S_{3,4}}^G = 0$. *Thus* $G = \{-2x_1x_2 + x_1, x_1^3x_2 - 2x_1^2 + x_2, -x_1^3/2 + 2x_1^2 - x_2, -2x_2^2 + x_2\}$ *is a Gröbner basis for* I.

The observant reader may have noticed in Example 4.9 that the final polynomial in the Gröbner basis was a univariate polynomial. This is no accident. We have the following important result.

Theorem 4.5 (the elimination property). *Let* I *be a polynomial ideal in* $\mathbb{C}[\mathbf{x}]$, *where* $\mathbf{x} = (x_1, \ldots, x_n) \in \mathbb{C}^n$. *We call* $I \cap \mathbb{C}[x_{\ell+1}, \ldots, x_n]$ *the* ℓ*th elimination ideal in* $\mathbb{C}[x_{\ell+1}, \ldots, x_n]$. *Note that if* $\ell = 0$, *we just get* I. *If* G *is a Gröbner basis for* I *with respect to lexicographic order with* $x_1 \succ \cdots \succ x_n$, *then for all* $0 \le \ell \le n$ *we have that* $G_\ell = G \cap \mathbb{C}[x_{\ell+1}, \ldots, x_n]$ *is a Gröbner basis for the* ℓ*th elimination ideal. Note that a polynomial* $g \in \mathbb{C}[x_{\ell+1}, \ldots, x_n]$ *if and only if the leading term* $\mathrm{LT}(g) \in \mathbb{C}[x_{\ell+1}, \ldots, x_n]$.

Example 4.10. *Let* $I = \langle x^2 + y + z - 1, x + y^2 + z - 1, x + y + z^2 - 1\rangle$ *be a polynomial ideal in* $\mathbb{C}[x, y, z]$ *where we use the lexicographis order* $x > y > z$. *We can use the Buchberger* *algorithm to generate the Gröbner basis*

$$G=\{x+y+z^2-1, y^2-y-z^2+z, 2yz^2+z^4-z^2, z^6-4z^4+4z^3-z^2\}.$$

From Theorem 4.5 it follows that the Gröbner bases for $I\cap\mathbb{C}[y,z]$ *and* $I\cap\mathbb{C}[z]$ *are given by*

$$G_1=\{y^2-y-z^2+z, 2yz^2+z^4-z^2, z^6-4z^4+4z^3-z^2\}$$

and

$$G_2=\{z^6-4z^4+4z^3-z^2\},$$

respectively.

One must be careful to interpret the elimination property correctly. The spaces in Theorem 4.5 may sometimes be trivial, as the following example shows.

Example 4.11. *Let* $I=\langle x^2-y, y^2-z\rangle\subset\mathbb{C}[x,y,z]$ *where we use the lexicographic order* $x>y>z$*. It is easy to see that the S-polynomial for* $g_1=x^2-y$ *and* $g_2=y^2-z$ *is given by*

$$S_{1,2}=\frac{x^2y^2}{x^2}g_1-\frac{x^2y^2}{y^2}g_2=y^2(x^2-y)-x^2(y^2-z)=x^2z-y^3=zg_1-yg_2,$$

and hence $\overline{S_{1,2}}^G=0$*. Thus* $G=\{g_1,g_2\}$ *is a Gröbner basis for* I*. This means that* $G_1=G\cap\mathbb{C}[y,z]=\{y^2-z\}$ *is a basis for* $I_1=I\cap\mathbb{C}[y,z]$*, as expected. However,* $G_2=G\cap\mathbb{C}[z]=\{0\}$ *is trivial. The apparent dilemma is resolved by noting that* $I_2=I\cap\mathbb{C}[z]=\{0\}$ *is also trivial. Thus the statement of the theorem remains true.*

The apparent dilemma in Example 4.11 can be resolved in a more practical way. Imagine we wish to solve the equations $x^2-y=0$ and $y^2-z=0$. Since we have only two equations in three unknowns, the system is underdetermined, and we might reasonably expect an infinite number of solutions. Since the second equation gives $y=\pm\sqrt{z}$ and the first equation gives $x=\pm\sqrt{y}=\pm\sqrt{\pm\sqrt{z}}$, it is clear that $z\in\mathbb{C}$ plays the role of a parameter rather than a variable. The previous example would be less confusing if we let $I=\langle x^2-y, y^2-c\rangle\subset\mathbb{C}[x,y]$, where $c\in\mathbb{C}$ is an arbitrary parameter. Now it is clear that $G=\{g_1,g_2\}$ with $g_1=g_1(x,y)$ and $g_2=g_2(y)$. Theorem 4.5 now tells us that $G_1=G\cap\mathbb{C}[y]=\{g_2\}$ is a Gröbner basis for $I_1=I\cap\mathbb{C}[y]$.

The elimination property is very useful for solving systems of multivariate polynomial equations. Consider a system of polynomial equations $f_1(\boldsymbol{x})=0,\ldots,f_s(\boldsymbol{x})=0$, where $\boldsymbol{x}\in\mathbb{C}^n$. Our initial aim will be to use a Gröbner basis $G=\{g_1,\ldots,g_t\}$ for the ideal $I=\langle f_1,\ldots,f_s\rangle=\langle g_1,\ldots,g_t\rangle\subset\mathbb{C}[\boldsymbol{x}]$ to replace this system by a reduced system of polynomial equations in the form $g_1(\boldsymbol{x})=0,\ldots,g_t(\boldsymbol{x})=0$ where the elimination property shows that the set G contains a subset $H=\{h_1,\ldots,h_n\}$ such that $h_\ell(\boldsymbol{x})=h_\ell(x_\ell,\ldots,x_n)$ for each $\ell=1,\ldots,n$. Thus we can replace the original system of polynomial equations by a truncated triangular system of polynomial equations $h_1(x_1,\ldots,x_n)=0,\ldots,h_n(x_n)=0$, the solutions of which contain all the solutions of the original system. The zero set of H can now be determined by back substitution. We add two important provisos. First, as we saw in the previous example, we must have a sufficient number of equations to ensure the system has only a finite number of solutions. Second, we must understand that by using the truncated system and possibly omitting some equations from the reduced system we may introduce additional solutions that do not satisfy the original system. These solutions will be termed *ghost* solutions.

Example 4.12. *Solve the system of equations*

$$x^2+y+z-1=0, \quad x+y^2+z-1=0, \quad x+y+z^2-1=0.$$

We define an ideal $I=\langle f_1,f_2,f_3\rangle \subset \mathbb{C}[x,y,z]$, *where* $f_1=x^2+y+z-1$, $f_2=x+y^2+z-1$, *and* $f_3=x+y+z^2-1$, *and we use the lexicographic ordering* $x\succ y\succ z$. *We have already seen in Example 4.10 that* $I=\langle g_1,g_2,g_3,g_4\rangle$, *where*

$$g_1=x+y+z^2-1,\ g_2=y^2-y-z^2+z,\ g_3=2yz^2+z^4-z^2,\ g_4=z^6-4z^4+4z^3-z^2,$$

and so we may replace the original system by a reduced system

$$x+y+z^2-1=0,\ y^2-y-z^2+z=0,\ 2yz^2+z^4-z^2=0,\ z^6-4z^4+4z^3-z^2=0.$$

Now we choose a triangular subset

$$h_1=g_1(x,y,z),\ h_2=g_2(y,z),\ h_3=g_4(z)$$

and solve instead the truncated system

$$x+y+z^2-1=0,\ y^2-y-z^2+z=0,\ z^6-4z^4+4z^3-z^2=0.$$

The last equation can be rewritten as

$$z^2(z-1)^2(z^2+2z-1)=0,$$

from which we deduce $z=0,1,-1\pm\sqrt{2}$. *When* $z=0$ *the second equation gives* $y(y-1)=0$, *from which it follows that* $y=0,1$. *When* $(y,z)=(0,0)$ *the first equation gives* $x=1$, *and when* $(y,z)=(1,0)$ *the first equation gives* $x=0$. *Thus we have two solutions* $(1,0,0)$ *and* $(0,1,0)$. *When* $z=1$ *the second equation once again gives* $y(y-1)=0$, *and so* $y=0,1$. *When* $(y,z)=(0,1)$ *the first equation gives* $x=0$, *and when* $(y,z)=(1,1)$ *the first equation gives* $x=-1$. *Thus we have two more solutions* $(0,0,1)$ *and* $(-1,1,1)$ *to the truncated system. Now it turns out that* $(-1,1,1)$ *does not satisfy the original system, and hence it is a so-called ghost solution. When* $z=-1+\sqrt{2}$ *the second equation gives*

$$y^2-y-4+3\sqrt{2}=0,$$

from which it follows that

$$y=\frac{1}{2}\pm\frac{3-2\sqrt{2}}{2}=-1+\sqrt{2} \text{ or } 2-\sqrt{2}.$$

If $(y,z)=(-1+\sqrt{2},-1+\sqrt{2})$, *then the first equation gives* $x=-1+\sqrt{2}$. *Thus* $(-1+\sqrt{2},-1+\sqrt{2},-1+\sqrt{2})$ *is also a solution to the truncated system. If* $(y,z)=(2-\sqrt{2},-1+\sqrt{2})$, *then the first equation gives* $x=-4+3\sqrt{2}$. *The solution* $(-4+3\sqrt{2},2-\sqrt{2},-1+\sqrt{2})$ *to the truncated system turns out to be a ghost solution. When* $z=-1-\sqrt{2}$, *similar arguments show that* $(-1-\sqrt{2},-1-\sqrt{2},-1-\sqrt{2})$ *is also a valid solution to the original system.*

To study perturbations in the solutions to multivariate polynomial equations when the coefficients change, it is convenient to consider systems of equations in the form

$f_1(x,z) = 0,\dots,f_s(x,z) = 0$, where $x \in \mathbb{C}^n$ is the variable and where $z \in \mathbb{C}$ is a complex parameter. To this end we consider the ring of multivariate polynomials over the field $F_z^\#$ of rational functions in the parameter z. Thus we consider elements of the form

$$p(x,z) = \sum_{\alpha \in A} p_\alpha(z) x^\alpha,$$

where $A \subset \mathbb{Z}^n_{\geq 0}$ is a finite set and where the coefficient $p_\alpha(z)$ is a quotient of polynomials in $\mathbb{C}[z]$. The properties of this more general polynomial ring $F_z^\#[x]$ where $x \in \mathbb{C}^n$ are obvious extensions of the properties for the polynomial ring $\mathbb{C}[x]$ studied in the previous sections.

Example 4.13. *Solve the perturbed polynomial equations* $z^2x_1^2x_2 + 2(z^2+1)x_1 - x_2 = 0$ *and* $(z^2+1)x_1x_2 - (z+2) = 0$ *near* $z = 0$. *Note that when* $z = 0$, *the system reduces to* $2x_1 - x_2 = 0$ *and* $x_1x_2 - 2 = 0$, *which has two solutions* $(x_1,x_2) = \pm(1,2)$. *Let* $G = \{g_1, g_2\}$ *where* $g_1 = z^2x_1^2x_2 + 2(z^2+1)x_1 - x_2$ *and* $g_2 = (z^2+1)x_1x_2 - (z+2)$. *We have*

$$S_{1,2} = \frac{x_1^2x_2}{z^2x_1^2}g_1 - \frac{x_1^2x_2}{(z^2+1)x_1x_2}g_2 = \frac{1}{z^2}g_1 - \frac{x_1}{z^2+1}g_2 = \frac{2z^4+z^3+6z^2+2}{z^2(z^2+1)}x_1 - \frac{1}{z^2}x_2.$$

Thus $\overline{S_{1,2}}^G = S_{1,2}$. *It is convenient to add a multiple of* $\overline{S_{1,2}}^G$ *to the basis. Thus we define*

$$g_3 = (2z^4+z^3+6z^2+2)x_1 - (z^2+1)x_2.$$

Now

$$\begin{aligned}
S_{1,3} &= \frac{x_1^2x_2}{z^2x_1^2x_2}g_1 - \frac{x_1^2x_2}{(2z^4+z^3+6z^2+2)x_1}g_3 \\
&= \frac{1}{z^2}g_1 - \frac{x_1x_2}{(2z^4+z^3+6z^2+2)}g_3 \\
&= \frac{(z^2+1)}{2z^4+z^3+6z^2+2}x_1x_2^2 + \frac{2(z^2+1)}{z^2}x_1 - \frac{1}{z^2}x_2 \\
&= \frac{x_2}{2z^4+z^3+6z^2+2}g_2 + \frac{2(z^2+1)}{z^2}x_1 - \frac{2(z^2+1)^2}{z^2(2z^4+z^3+6z^2+2)}x_2 \\
&= \frac{x_2}{2z^4+z^3+6z^2+2}g_2 + \frac{2(z^2+1)}{z^2(2z^4+z^3+6z^2+2)}g_3,
\end{aligned}$$

and hence $\overline{S_{1,3}}^G = 0$. *We also have*

$$\begin{aligned}
S_{2,3} &= \frac{x_1x_2}{(z^2+1)x_1x_2}g_2 - \frac{x_1x_2}{(2z^4+z^3+6z^2+2)x_1}g_3 \\
&= \frac{1}{z^2+1}g_2 - \frac{x_2}{2z^4+z^3+6z^2+2}g_3 \\
&= \frac{z^2+1}{2z^4+z^3+6z^2+2}x_2^2 - \frac{z+2}{z^2+1}.
\end{aligned}$$

Thus $\overline{S_{2,3}}^G = S_{2,3}$. *It is convenient to add a multiple of* $\overline{S_{2,3}}^G$ *to the basis. Thus we define*

$$g_4 = (z^2+1)^2x_2^2 - (z+2)(2z^4+z^3+6z^2+2).$$

In a similar fashion we can show that

$$S_{1,4} = \left[\frac{(z+2)x_1}{(z^2+1)^2} - \frac{x_2}{z^2(z^2+1)} \right] g_3, \quad S_{2,4} = \frac{z+2}{(z^2+1)^2} g_3,$$

and

$$S_{3,4} = \frac{z+2}{(z^2+1)^2} g_3 - \frac{x_2}{(z^2+1)(2z^4+z^3+6z^2+2)} g_4.$$

Thus $\overline{S_{1,4}}^G = \overline{S_{2,4}}^G = \overline{S_{3,4}}^G = 0$, *and hence* $G = \{g_1, g_2, g_3, g_4\}$ *is a Gröbner basis for* I. *Therefore, the reduced set of equations becomes* $g_1 = 0, g_2 = 0, g_3 = 0, g_4 = 0$. *From this set we can select a truncated system,*

$$(2z^4+z^3+6z^2+2)x_1 - (z^2+1)x_2 = 0 \text{ and } (z^2+1)^2 x_2^2 - (z+2)(2z^4+z^3+6z^2+2) = 0,$$

to solve. Thus we eventually obtain

$$(x_1, x_2) = \pm \left(\sqrt{\frac{z+2}{2z^4+z^3+6z^2+2}}, \frac{\sqrt{(z+2)(2z^4+z^3+6z^2+2)}}{z^2+1} \right).$$

One can now check that both solutions verify the original system of equations $g_1 = 0$ *and* $g_2 = 0$.

The approach presented in the above example will be generalized in the next section.

4.3 ▪ Reduction of the System of Perturbed Polynomials

Here we show that the system of multivariate polynomials (4.4) which depend on a single perturbation parameter can be transformed into the system of bivariate polynomials (4.5). This can be done with the help of a Gröbner bases technique. Thus, we reduce the multidimensional problem to the set of one dimensional problems. To each one dimensional problem in (4.5) we can then apply the Newton diagram method described in Section 4.7.

To perform such a transformation, we note that the set of solutions of (4.4) can be considered as an algebraic variety W_1 given as a null set of the ideal I_{W_1} generated by polynomials $Q_i(\mathbf{x}, z)$, $i = 1, 2, \ldots, n$. Throughout this section we assume that W_1 is zero dimensional for any fixed z in some small neighborhood around z_0. We adopt the term order $T_1 := x_1 \prec x_2 \prec \cdots \prec x_n$, keeping z as a parameter, and then we find the *reduced Gröbner basis* of I_{W_1}. To keep the notation simple we denote this basis by $GB^{(1)}(W_1) = \{g_1(\mathbf{x}, z), \ldots, g_t(\mathbf{x}, z)\}$.

Lemma 4.6.

(i) *One can order* $g_1, \ldots, g_t$ *so that* g_1 *is a* univariate *polynomial in the variable* x_1, *polynomial* g_2 *contains only the variables* x_1, x_2, *polynomial* g_3 *contains only* x_1, x_2, x_3, *and so forth until the polynomial* g_n, *containing* $x_1, \ldots, x_n$. *In particular,* $t = n$.

(ii) *The coefficients of* g_i, $i = 1, \ldots, t$, *are rational functions in the variable* z.

Proof: The proof of the first part we leave as an exercise (see Problem 4.1). The Buchberger algorithm for Gröbner bases involves a construction of *S-polynomials* from pairs (g_i, g_j) and their further *reduction* with respect to current generators. All such steps involve a division by a leading coefficient (this may produce rational functions in z), multiplication by a monomial in $\mathbf{x}$, and taking linear combinations of such objects that clearly produce only polynomials in $\mathbf{x}$ with rational coefficients in z. □

Building upon the results of Lemma 4.6, we have the following theorem.

Theorem 4.7. *In a neighborhood of* $(\mathbf{x}_0, z_0)$ *the variety* W_1 *belongs to, a priori, a larger variety* $\tilde{W}_1$ *defined as a union of zero-sets of* τ *systems of* n *irreducible bivariate polynomials* $p_1^i(x_1, z), p_2^i(x_2, z), \ldots, p_n^i(x_n, z)$, $i = 1, \ldots, \tau$.

Proof: Consider polynomial g_1 in the reduced Gröbner basis described in Lemma 4.6. Having multiplied by the least common multiple of the denominators of its coefficients, we obtain a bivariate polynomial in x_1 and z that we denote by $\tilde{g}_1(x_1, z)$. This polynomial can be factorized into prime (*irreducible*) factors (see Section 4.5):

$$\tilde{g}_1(x_1, z) = \prod_1^{\ell_1} p_j(x_1, z). \tag{4.6}$$

Without loss of generality we assume that the initial point $(\mathbf{x}_0, z_0)$ belongs to the zero-set of $p_1(x_1, z)$, the first factor in (4.6). We note that $(\mathbf{x}_0, z_0)$ might also belong to the zero-set of some other $p_j(x_1, z)$, and a branch of $\{p_j(x_1, z) = 0\}$ variety could provide an actual solution for x_1 related to the original system. We now add $p_1(x_1, z)$ to the $GB^{(1)}(W_1)$, change the term order to $T_2 := x_2 \prec x_1 \prec \cdots \prec x_n$, and construct the reduced Gröbner basis $GB^{(2)}(W_1)$ initiated by the set of generators $GB^{(1)}(W_1)$ and the term order T_2. By Lemma 4.6 the first element of $GB^{(2)}(W_1)$ will be a univariate polynomial $g_2(x_2)$ with rational coefficients in z. Again, multiplying by the least common multiple of the coefficients' denominators and taking the irreducible factor $p_2(x_2, z)$ such that $(\mathbf{x}_0, z_0)$ belongs to its zero-set, we obtain the second irreducible bivariate (in x_2 and z) polynomial that we add to $GB^{(2)}$ to continue with the process. □

Remark 4.1. *Variety* $\tilde{W}_1$ *might contain some ghost solutions* $x_j = x_j(z)$ *that are solutions of the system of chosen bivariate polynomials but are not the solutions of the original system* (4.4). *The "ghost" solutions arise as a result of solving irreducible bivariate polynomials* $\{p_j^{(i)}\}$ *without consideration of the remaining polynomials in the Gröbner bases (see Problem 4.2).*

Note that the superscript i in the above refers to a selection of one irreducible component from each of the product expressions of the type (4.6) for $\tilde{g}_k(x_k, z)$, $k = 1, 2, \ldots, n$. Thus, τ could be a large number.

The benefit of the preceding theorem is that the zero-sets of irreducible bivariate polynomials can be represented as Taylor or Puiseux or Laurent–Puiseux series in z, regardless of whether $x_j = x_j(z)$ is a "ghost" or an actual solution of the original system. This allows us to describe solution-set W_1 separately for each variable x_j as solutions of bivariate polynomial equations. In the next section we discuss how we can carry out the classification of the expansion types.

To construct the reduced Gröbner bases $GB^{(1)}(W_1), \ldots, GB^{(n)}(W_1)$, one can use, for instance, the function `gbasis` from the "Groebner" package of Maple.

Example 4.14. *Consider the following system of perturbed polynomials:*

$$\begin{cases} x_1^2+x_2^2+zx_1x_2+z^2x_1+z=0, \\ z^2x_1^2+zx_2^2+x_1x_2+zx_2+z^2=0. \end{cases} \tag{4.7}$$

Using the Maple commands

```
with(Groebner) : gbasis(S,plex(x1,x2));gbasis(S,plex(x2,x1));
```

where S denotes the system of perturbed equations (4.7)*, we transform* (4.7) *into the following set of bivariate polynomial equations:*

$$\begin{cases} (z^5+z^4-3z^3+1)x_1^4+(-2z^5-z^3+z^2+2z)x_1^3 \\ +(z^6+z^2+z)x_1^2+(-z^4+2z^2)x_1+z^3=0, \\ \\ (z^5+z^4-3z^3+1)x_2^4+(2z^6+z^5-z^4-3z^3+z^2)x_2^3 \\ +(z^7+2z^5+3z^6-5z^4+z^2+z)x_2^2 \\ +(3z^7-z^5-3z^4+2z^3)x_2+(z^8+z^6-2z^5+z^4)=0. \end{cases}$$

4.4 ▪ Classification of Expansion Types

In this section we address the fundamental question of classification of the series expansions of irreducible bivariate polynomials. In the preceding section we have seen that the variety determining the solution set of a system of n perturbed polynomials in $n+1$ variables (including the perturbation parameter) belongs to the variety determining the solution set of a system of n bivariate polynomials constructed by repeated application of Buchberger's algorithm. The latter are products of irreducible bivariate polynomials which, in turn, determine whether solutions, as functions of the perturbation, take the form of Taylor, Laurent, or Puiseux series in the neighborhood of a point of interest. Indeed, it should be said that while most of the perturbation analysis is carried out in a neighborhood of $z_0=0$, in general, one can consider a perturbation of the parameter from a nonzero value, that is, $z_0 \neq 0$. And hence, perhaps one of the most interesting questions in this area is the following: *Is there anything special about the value of z_0 that determines the form of the series expansion?* We provide a complete answer to this question.

Consequently, we return to the irreducible bivariate polynomial of the same form as in (4.2). Namely, let

$$Q(x,z)=q_m(z)x^m+q_{m-1}(z)x^{m-1}+\cdots+q_0(z)=\sum_{k=0}^{m} q_k(z)x^k \tag{4.8}$$

be an irreducible polynomial with complex coefficients in two complex variables (x,z) that defines the algebraic variety $W_Q:=\{(x,z):Q(x,z)=0\}$.

By Puiseux's theorem the solution $x=x(z)$ of $Q(x,z)=0$ may be an m'-valued algebraic function with branching points of order $m'\le m$. The zeros of the discriminant $\mathrm{Dis}(Q,z)$ of $Q(x,z)$ with respect to the variable z, that is, the values of z where $Q(x,z)$ has multiple roots, play a critical role in the determination of branching points. In Theorem 4.8 below, we provide a complete classification of different types of series expansions for the perturbed solution of (4.8). Before proceeding to the statement of the theorem, we recall that polynomial $Q(x,z_0)$ has multiple roots if and only if its discriminant is equal to zero. We recall that the discriminant is given by the formula

$$\mathrm{Dis}(Q,z_0) = q_m(z_0)\prod_{i<j}(r_i - r_j),$$

where $r_i, i = 1,\dots,n$, are (possibly multiple) roots of the polynomial $Q(x,z_0)$ with respect to x. The discriminant of a polynomial can also be expressed in terms of the polynomial's coefficients (see Problem 4.3).

Due to the irreducibility of Q the set of zeros $\mathscr{Z}(Q)$ of $\mathrm{Dis}(Q,z)$ is finite. Decompose $\mathscr{Z}(Q) = \mathscr{Z}_m(Q)\bigcup\mathscr{Z}'(Q)$, where the set $\mathscr{Z}_m(Q)$ stands for the zeros of $q_m(z)$, and, respectively, $\mathscr{Z}'(Q)$ stands for the zeros of $\mathrm{Dis}(Q,z)$, that are not zeros of $q_m(z)$. The following theorem provides the algebraic analytic form of the function $x(z)$ in various situations with respect to the nature of the point z_0.

Theorem 4.8. *The following classification takes place:*

(i) *If $z_0 \notin \mathscr{Z}(Q)$ and is not a zero of $q_m(z)$, then in a neighborhood of z_0 every one of the m branches of the solution $x(z)$ is holomorphic, and so it has the vector analytic representation*

$$x(z) = \sum_{k=0}^{\infty} c_k (z - z_0)^k.$$

(ii) *If $z_0 \in \mathscr{Z}'(Q)$, then z_0 is a branching point of some order $m' \leq m$ for every branch $f(z)$ of the solution $x(z)$ and also $\lim_{z\to z_0}(z - z_0)f(z) = 0$. In this case the solution $x(z)$ has a Puiseux series representation*

$$x(z) = \sum_{k=0}^{\infty} c_k (z - z_0)^{k/m'}.$$

(iii) *If $z_0 \in \mathscr{Z}_m(Q)$ and is a zero of multiplicity $m_0 > 0$ of $q_m(z)$, then for any branch $f(z)$ of $x(z)$ the point z_0 is a branching point of some order $m' \leq m$ and $\lim_{z\to z_0}(z - z_0)^{m_0+\delta} f(z) = 0$ for all $\delta > 0$. In this situation the solution $x(z)$ has a Laurent–Puiseux series representation*

$$x(z) = \sum_{k=-k_0}^{\infty} c_k (z - z_0)^{k/m'}.$$

(iv) *If $z_0 \notin \mathscr{Z}(Q)$ and is the zero of multiplicity $m_0 > 0$ of $q_m(z)$, then z_0 is a pole of order m_0 for every branch $f(z)$ of the solution $x(z)$, and in this situation the solution $x(z)$ has a Laurent series representation*

$$x(z) = \sum_{k=-m_0}^{\infty} c_k (z - z_0)^k.$$

Proof: (i) Denote by x_0 one of the m roots of $Q(x,z_0)$. Choose a closed neighborhood $U_{x_0} := \{x : |x - x_0| \leq \rho\}$ that does not contain any other roots of $Q(x,z_0)$, and set $\mu := \min_{\{|x-x_0|=\rho\}} |Q(x,z_0)| > 0$. By the uniform continuity of Q on compact sets, there exists a closed neighborhood $U_{z_0} := \{z : |z - z_0| \leq \delta\}$ such that $|Q(x,z) - Q(x,z_0)| < \mu$

for all $(x,z) \in U_{x_0} \times U_{z_0}$. Representing $Q(x,z)$ as the sum $Q(x,z) = Q(x,z_0) + (Q(x,z) - Q(x,z_0))$, we apply the Rouché theorem to $Q(x,z_0)$ and to $Q(x,z) - Q(x,z_0)$ on U_{x_0} to obtain that for every $z \in U_{z_0}$ there is only one zero of $Q(x,z)$ in U_{x_0} that equals

$$f(z) = \frac{1}{2\pi i} \oint_{\{|\zeta - x_0| = \rho\}} \frac{\zeta \frac{\partial Q}{\partial \zeta}(\zeta, z)}{Q(\zeta, z)} d\zeta. \tag{4.9}$$

This function $f(z)$ extends analytically from U_{z_0} to $\mathbb{C}_z \setminus \mathscr{Z}(Q)$, and by the uniqueness theorem for holomorphic functions, $Q(f(z), z) = 0$ on $\mathbb{C}_z \setminus \mathscr{Z}(Q)$. All points in $\mathscr{Z}(Q)$ are branching points of the same order for the extended $f(z)$. This order cannot exceed m, as there are at most m roots of $Q(x,z)$ for any fixed z. The extended $f(z)$ is some algebraic function that, therefore, satisfies some polynomial equation $\tilde{Q}(f(z), z) = 0$ for some irreducible polynomial $\tilde{Q}(x,z) = 0$. Since every value $x = f(z)$ is also a root of $Q(x,z)$, polynomial Q must divide $\tilde{Q}$; this is only possible in case $Q = c\tilde{Q}$ for some constant c due to the irreducibility of Q. Hence, the branching order for $f(z)$ at every point in $\mathscr{Z}(Q)$ equals m.

(ii)–(iii) Let $z_0 \in \mathscr{Z}(Q)$ be a zero of $q_m(z)$ of multiplicity $m_0 \geq 0$. Assume $m_0 = 0$ in case $z_0 \in \mathscr{Z}'(Q)$. Fix δ such that $0 < \delta < \frac{1}{m}$. Write $q_m = q_m(z) = (z - z_0)^{m_0} \tilde{q}_m$, and substitute $f(z) = (z - z_0)^{-(m_0 + \delta)} g(z)$ in the identity $Q(f(z), z) \equiv 0$. Multiplying by $(z - z_0)^{(m-1)m_0 + m\delta}$, we obtain that $x = g(z)$ satisfies the polynomial equation in x:

$$x^m + (z - z_0)^{\delta} \frac{q_{m-1}}{\tilde{q}_m} x^{m-1} + \cdots + (z - z_0)^{(m-1)m_0 + m\,\delta} \frac{q_0}{\tilde{q}_m} = z^m$$

$$+ \sum_{k=0}^{m-1} (z - z_0)^{(m-k-1)m_0 + m\,\delta} \frac{q_k}{\tilde{q}_m} x^k = 0. \tag{4.10}$$

As the leading coefficient (at x^m) is 1 and all other coefficients approach 0 as $z \to z_0$, for sufficiently small $|z - z_0|$ all zeros of (4.10) are in absolute value less than any given small number. As $g(z)$ is a zero of (4.10), for every z, it follows that $\lim_{z \to z_0} g(z) = 0$. Now, the proofs of (ii) for $m_0 = 0$ and (iii) for $m_0 > 0$ follow from Lemmas 4.9–4.10 established below. The proof of part (iv) is left as an exercise (see Problem 4.4). □

By analogy with Chapters 2 and 3, we can call the case (i) in Theorem 4.8 *regular perturbation* and the other cases *singular perturbations.*

Note that, for a regular point $(x_0, z_0) \in W_Q$, the coefficients of the Taylor series of $x(z)$ can be effectively computed by a contour integral applied to the formula (4.9), as stated in the following lemma.

Lemma 4.9. *If* $z_0 \notin \mathscr{Z}(Q)$, *then in a neighborhood* $\{|z - z_0| \leq \mu\}$ *of* z_0 *each of the* m *branches of* $x(z)$ *is a Taylor series*

$$x(z) = \sum_{k=0}^{\infty} c_k (z - z_0)^k,$$

where

$$c_k = -\frac{1}{4\pi^2} \oint_{\{|\eta - z_0| = \mu\}} \oint_{\{|\zeta - x_0| = \rho\}} \frac{\zeta \frac{\partial Q}{\partial \zeta}(\zeta, \eta)}{(\eta - z_0)^{k+1} Q(\zeta, \eta)} d\zeta\, d\eta \tag{4.11}$$

for some positive ρ *and* μ.

Proof: This follows immediately from the standard Cauchy formula for the coefficients of the Taylor series coefficients for the function $f(z)$ holomorphic in $\{|z-z_0| \leq \mu\}$. □

For $z_0 \in \mathscr{Z}(Q)$ it is convenient to introduce a variable ω as $z = z_0 + \omega^{m'}$ and then to represent the Laurent–Puiseux series in $z - z_0$ as a Laurent series in ω.

Lemma 4.10. *Let $z_0 \in \mathscr{Z}(Q)$ be a zero of $q_m(z)$ of multiplicity $m_0 \geq 0$. Then $x(z)$ admits a Puiseux series representation*

$$x(z) = \sum_{k=-m_0 m'}^{\infty} c_k (z - z_0)^{\frac{k}{m'}}$$

for

$$c_k = \frac{1}{2\pi i} \oint_{\{|\omega|=\delta^{\frac{1}{m'}}\}} \frac{\phi(\omega)}{\omega^{k+1}} d\omega, \; k = -m_0 m', \ldots, \infty, \; m' \leq m, \tag{4.12}$$

where the function $\phi(\omega)$ locally admits an integral representation

$$\phi(\omega) = \frac{1}{2\pi i} \oint_{\{|\zeta - x_k| = \rho\}} \frac{\zeta \frac{\partial Q}{\partial \zeta}(\zeta, z_0 + \omega^{m'})}{Q(\zeta, z_0 + \omega^{m'})} d\zeta \tag{4.13}$$

for some x_k such that $Q(x_k, z_0 + \omega_k^{m'}) = 0$, $\frac{\partial Q}{\partial x}(x_k, z_0 + \omega_k^{m'}) \neq 0$, and $|\omega - \omega_k| < \delta'$ for some $\delta' > 0$.

Proof: Since z_0 is a branching point for $x(z)$ of order $m' \leq m$, the function $\phi(\omega) := x(z_0 + \omega^{m'})$ is a holomorphic function in ω in a punctured neighborhood of 0 and therefore admits a Laurent series representation

$$\phi(\omega) = \sum_{k=-\infty}^{\infty} c_k \omega^k,$$

and the coefficients c_k can be evaluated as stated in Lemma 4.9 for all k. In particular, we obtain $c_k = 0$ for $k < -m_0 m'$. The contour $\gamma_\delta := \{|\omega| = \delta^{\frac{1}{m'}}\}$ can be chosen so that in some δ'-strip neighborhood of γ_δ there are no points ω such that $z_0 + \omega^{m'} \in \mathscr{Z}(Q)$, and so part (i) of Theorem 4.8 is applicable. □

We would like to note that once the classification of the type of series is carried out with the help of Theorem 4.8, the series coefficients can be obtained by formulae (4.11) and (4.12). However, in Sections 4.6 and 4.7 we discuss more efficient methods for computing the series coefficients.

4.5 ▪ Irreducible Factorization of Bivariate Polynomials

Theorem 4.8 applies only to irreducible polynomials. In general, it is not easy to check if a polynomial is irreducible. Below we provide a procedure based on Gröbner bases to check whether a polynomial is irreducible. In the case of reducible polynomials, the procedure eventually produces a factorization into irreducible factors.

Let

$$Q(z,w)=\sum_{\alpha+\beta\le m} c_{\alpha\beta}z^\alpha w^\beta$$

be a polynomial in (z,w) of degree $m>1$ with complex coefficients $c_{\alpha\beta}$. Without loss of generality we assume that $Q(0,0)\neq 0$, that is, $c_{00}\neq 0$; this can be achieved by moving the origin away from the zero set of Q. Having fixed two positive integers m_1 and $m_2=m-m_1$, we would like to find out if it is possible to represent $Q(z,w)$ as a product

$$Q(z,w)=Q_1(z,w)Q_2(z,w) \tag{4.14}$$

for some polynomials $Q_1=\sum_{\alpha+\beta\le m_1} a_{\alpha\beta}z^\alpha w^\beta$ and $Q_2=\sum_{\alpha+\beta\le m_2} b_{\alpha\beta}z^\alpha w^\beta$ of degrees m_1 and m_2, respectively. Without loss of generality we assume that $a_{00}=1$. Equating coefficients in (4.14) at each power product $z^\alpha w^\beta$, we obtain the following system of, at most, quadratic equations in the coefficients $(a_{\alpha\beta},b_{\alpha\beta})$ that determines all possible factorizations of Q into two factors of prescribed degrees m_1 and m_2,

$$\sum_{\substack{\gamma+\lambda=\alpha\\ \delta+\mu=\beta}} a_{\gamma\delta}b_{\lambda\mu}=\begin{cases} c_{\alpha\beta} & : \ c_{\alpha\beta}\neq 0,\\ 0 & : \ \text{otherwise},\end{cases} \tag{4.15}$$

that we denote by $\mathcal{F}(Q,m_1,m_2)$. Any solution $\{a_{\alpha\beta}\}$ and $\{b_{\alpha\beta}\}$ of $\mathcal{F}(Q,m_1,v_2)$ provides a factorization of Q into factors of degrees m_1 and m_2. Under the assumption $a_{00}=1$, system (4.15) has, at most, finitely many solutions. If the solution set of (4.15) in $\mathcal{F}(Q,m_1,m_2)$ is empty, then Q cannot be factorized into polynomials of degrees m_1 and m_2. Consider the ideal I_{m_1,m_2} of the polynomials in the variables $\{a_{\alpha\beta}\}$ and $\{b_{\alpha\beta}\}$ generated by $\mathcal{F}(Q,m_1,m_2)$. The system $\mathcal{F}(Q,m_1,m_2)$ has no solutions if and only if any Gröbner basis of I_{m_1,m_2} consists of just a unit. Because Q has at most finitely many factors of the prescribed degrees, the only alternative case is when the solution set of $\mathcal{F}(Q,m_1,m_2)$ is finite. Then it follows from the Buchberger algorithm that if we adopt a pure lexicographic term order, then the first element in the corresponding Gröbner basis will be univariate, the second will be bivariate, and so forth, which enables us to find the solutions $a_{\alpha\beta},b_{\alpha\beta}$ precisely. Running this algorithm for $m_1=1,\dots,[\frac{m}{2}]$, we either verify that Q is irreducible or come across the smallest value m_1 that provides a factorization. Polynomial Q_1 of the degree m_1 then has to be irreducible. Applying the same algorithm to Q_2 and so on, we eventually obtain all other irreducible factors of Q.

4.6 ▪ Computing Series Coefficients for Regularly Perturbed Polynomials

Let us consider the regular case when a solution of the perturbed polynomial equation (4.2) can be expressed as a Taylor series

$$x(z)=\sum_{k=0}^{\infty} c_k(z-z_0)^k. \tag{4.16}$$

Invoking the implicit function theorem, we can identify the regular case by a simple condition

$$\frac{\partial Q}{\partial x}(c_0,z_0)\neq 0.$$

The zero order term c_0 of the Taylor expansion is a solution of the unperturbed equation $Q(x,z_0)=0$. To calculate the higher order terms, one can use the formula (4.11) (see Problem 4.5 on application of this formula). However, a simpler way is to differentiate the perturbed equation several times. Namely, to obtain the first order term, one needs to differentiate the left-hand side of (4.2) once with respect to z. That is,

$$\frac{\partial Q}{\partial x}(x(z),z)x'(z)+\frac{\partial Q}{\partial z}(x(z),z)=0.$$

Then, we set $z=z_0$, and noting that $x'(z_0)=c_1$, we obtain

$$\frac{\partial Q}{\partial x}(c_0,z_0)c_1+\frac{\partial Q}{\partial z}(c_0,z_0)=0,$$

and, consequently,

$$c_1=-\frac{\frac{\partial Q}{\partial z}(c_0,z_0)}{\frac{\partial Q}{\partial x}(c_0,z_0)}.$$

To obtain the coefficient c_2, we need to differentiate $Q(x,z)$ twice with respect to z. That is,

$$\left[\frac{\partial^2 Q}{\partial x^2}(x(z),z)x'(z)+\frac{\partial^2 Q}{\partial x\partial z}(x(z),z)\right]x'(z)+\frac{\partial Q}{\partial x}(x(z),z)x''(z)$$

$$+\frac{\partial^2 Q}{\partial x\partial z}(x(z),z)x'(z)+\frac{\partial^2 Q}{\partial z^2}(x(z),z)=0,$$

which results in a formula for $c_2=x''(z_0)/2$:

$$c_2=-\frac{\frac{\partial^2 Q}{\partial x^2}(c_0,z_0)c_1^2+2\frac{\partial^2 Q}{\partial x\partial z}(c_0,z_0)c_1+\frac{\partial^2 Q}{\partial z^2}(c_0,z_0)}{2\frac{\partial Q}{\partial x}(c_0,z_0)}.$$

Example 4.15. *Consider a polynomial equation*

$$Q(x,z)=x^2+(z+2)x+z=0.$$

When $z_0=0$, the polynomial equation reduces to $x^2+2x=0$, which has two solutions, $x_0=0$ and $x_0=-2$. Let us consider the point $(x_0,z_0)=(0,0)$. Since

$$\frac{\partial Q}{\partial x}(0,0)=2\neq 0,$$

the perturbation is regular, and the perturbed solution can be expanded as a Taylor series $x(z)=zc_1+z^2c_2+\dots$ $(c_0=x_0=0)$, where the coefficients c_1 and c_2 are given by

$$c_1=-\frac{\frac{\partial Q}{\partial z}(c_0,z_0)}{\frac{\partial Q}{\partial x}(c_0,z_0)}=-\frac{1}{2}$$

and

$$c_2 = -\frac{\frac{\partial^2 Q}{\partial x^2}(c_0,z_0)c_1^2 + 2\frac{\partial^2 Q}{\partial x \partial z}(c_0,z_0)c_1 + \frac{\partial^2 Q}{\partial z^2}(c_0,z_0)}{2\frac{\partial Q}{\partial x}(c_0,z_0)}$$

$$= -\frac{2\cdot(-1/2)^2 + 2\cdot(-1/2)}{2\cdot 2} = \frac{1}{8}.$$

Continuing to take derivatives, one can obtain any number of coefficients $c_k, k = 1,2,\ldots$. However, as is now apparent, this approach is still quite cumbersome. Next, we describe a very efficient approach for the coefficient computation based on the application of the Newton method directly to the power series. First, we recall that in order to numerically find a solution of the equation $q(x) = 0$ one may apply the Newton method as follows:

$$x^{(i+1)} = x^{(i)} - q(x^{(i)})/q'(x^{(i)}) \tag{4.17}$$

from some initial point $x^{(0)}$ which should not be far from the solution. Denote the solution by x^* and the error of the ith iteration by $e^{(i)} = x^{(i)} - x^*$. Then, from the Taylor series expansions

$$q(x^{(i)}) = q'(x^*)e^{(i)} + \frac{q''(x^*)}{2}(e^{(i)})^2 + \ldots,$$

$$q'(x^{(i)}) = q'(x^*) + q''(x^*)e^{(i)} + \ldots,$$

we have

$$e^{(i+1)} = \frac{q''(x^*)}{2q'(x^*)}(e^{(i)})^2 + \sum_{j=3}^{\infty} d_j (e^{(i)})^2, \tag{4.18}$$

where the coefficients $d_j, j = 3,4,\ldots$, are rational expressions of the derivatives of $q(x)$ evaluated at x^* with powers of $q'(x^*)$ as the denominators.

Now let us formally apply the Newton method to the perturbed equation $Q(x,z) = 0$ over the field of rational functions starting with $X^{(0)} = c_0$. Namely, we perform the following iterations:

$$X^{(i+1)}(z) = X^{(i)}(z) - Q(X^{(i)}(z),z)/Q'_x(X^{(i)}(z),z). \tag{4.19}$$

Note that $X^{(i)}(z)$ admits a Taylor series expansion. Then, from (4.18) we conclude that if we start with $X^{(0)} = c_0$, as a result of the ith iteration we generate correctly the first 2^i terms of the Taylor expansion (4.16).

We would also like to mention that the above method can easily be generalized for the solution of a regularly perturbed *polynomial system* (see Problem 4.6).

4.7 ▪ Newton Polygon Method for Singularly Perturbed Polynomials

In the previous section an efficient method for computing series coefficients of a solution of a regular perturbed polynomial equation was suggested. Now let us describe how a singularly perturbed polynomial equation can be transformed into a regular perturbed polynomial equation. The transformation is based on the Newton polygon technique.

We say that a polynomial equation is *singularly perturbed* if $Q'_x(x_0,z_0)=0$. Without loss of generality we assume that $z_0=0$ and $q_m(0)\neq 0$. In particular, if $q_m(0)=0$, one can always choose nonnegative integers λ and μ which satisfy the following conditions:

$$\mu+\text{ord}(q_m(z))=m\lambda \tag{4.20}$$

and

$$\mu+\text{ord}(q_j(z))\geq j\lambda, \quad j=1,\dots,m-1, \tag{4.21}$$

where $\text{ord}(f(z))$ denotes the degree of the lowest degree term of the (possibly fractional) power series expansion of $f(z)$. Then, the leading coefficient of the polynomial $z^\mu Q(x/z^\lambda,z)$ does not vanish at zero, its solutions can be expanded in series with nonnegative powers, and they correspond to the solutions of the original polynomial multiplied by z^λ. Let us illustrate the above change of variable with an example.

Example 4.16. *Consider the perturbed polynomial equation*

$$zx^2-(1+z)x+1=0.$$

One can check that 1 *and* $1/z$ *are solutions of the above equation. According to part* (iv) *of Theorem 4.8, there should be a solution with a pole. To remove the singularity, we make the transformation*

$$z^\mu Q(x/z^\lambda,z)=0,$$

with $\lambda=1$ *and* $\mu=1$. *A reader can check that these* λ *and* μ *satisfy conditions* (4.20) *and* (4.21). *The transformed equation takes the form*

$$x^2-(1+z)x+z=0.$$

Its solutions are z *and* 1, *corresponding to the solutions of the original equation multiplied by* z.

The Newton polygon process makes a series of transformations that lead to a regular perturbation problem. Let us formally describe it.

Newton Polygon Process:

1. Set k to 1 and $Q_1(x,z)=Q(x,z)$.

2. Let $q_{i,k}(z)$ be the coefficient of x^i in the polynomial $Q_k(x,z)$, and let $r_{i,k}z^{\rho_{i,k}}$ be the lowest degree term of $q_{i,k}(z)$ if $q_{i,k}(z)\not\equiv 0$. Construct a Newton polygon associated with the polynomial $Q_k(x,z)$. That is, plot the points $(0,\rho_0),\dots(i,\rho_i),\dots(m,\rho_m)$ on a plane for $i\in\{0,1,\dots,m\}$ and $q_{i,k}(z)\not\equiv 0$. Next, let us draw a line through the point $(0,\rho_0)$ which coincides with the ordinate axis, and rotate this line counterclockwise around $(0,\rho_0)$ until it touches one of the other points, say, (l,ρ_l). In fact, several points may fall on the line. Then, we choose the point on the line with the largest abscissa, draw a line through it, parallel to the ordinate axis, and again rotate it counterclockwise until it touches another point. Continuing in the same fashion, we obtain the lower envelope of the convex hull for the points $(0,\rho_0),\dots(i,\rho_i),\dots(m,\rho_m)$. This lower envelope is called the Newton polygon.

3. If $k=1$, choose any segment $y+\gamma_i x=\beta_i$ of the Newton polygon. If $k>1$, choose a segment with $\gamma_k>0$ (such a segment always exists). Denote by S_k a set of indices

for which point (i,ρ_i) lies on the chosen segment. Solve the following polynomial equation:

$$\sum_{i\in S_k} r_{i,k}x^i = 0.$$

Let c_k be any of the nonzero roots (such a nonzero solution always exists).

4. If c_k is a simple root, go to Step 6; else go to Step 5.

5. Increment k, set $z^{-\beta_k}Q_k(z^{\gamma_k}(x+c_k),z)$ as $Q_{k+1}(x,z)$, and go to Step 2.

6. Stop with $t=k$, the number of stages taken by the Newton polygon process, and assign

$$\hat{Q}(x,z)=z^{-\beta_t}Q_t(z^{\gamma_t}x,z), \quad \bar{Q}(x,z)=\hat{Q}(x,z^d),$$

where d is the smallest common denominator of $\gamma_1,\dots,\gamma_t$ (if $\gamma_1=0$, take 1 as the denominator of γ_1).

We note that by construction the coefficients of $\bar{Q}(x,z)$ are polynomials in z. Next, let us show that the Newton polygon process indeed terminates with a regularly perturbed polynomial.

Theorem 4.11. *Upon the termination of the Newton polygon process, c_t is a simple root of the polynomial $\bar{Q}(x,0)$.*

Proof: It follows from the last step of the Newton polygon process that c_t is a simple root of the equation $\sum_{i\in S_t} r_{i,t}x^i = 0$. Let us show that in fact $\bar{Q}(x,0)=\sum_{i\in S_t} r_{i,t}x^i$. To simplify the notation, let $\rho_{i,t}=\rho_i$, $r_{i,t}=r_i$, $\beta_t=\beta$, $\gamma_t=\gamma$, $S_t=S$, and $S^c=\{1,\dots,m\}\backslash S_t$. We have

$$Q_t(x,z)=(r_m z^{\rho_m}+r_m(z))x^m+\cdots+(r_0 z^{\rho_0}+r_0(z)),$$

where $\text{ord}(r_i)>\rho_i$. Then, the polynomial $\hat{Q}(x,z)$ takes the form

$$\hat{Q}(x,z)=\sum_{i\in S} r_i x^i+\sum_{j\in S^c} r_j z^{\alpha_j+j\gamma-\beta}x^j+\sum_{i=0}^{m} z^{i\gamma-\beta}r_i(z)x^i.$$

Since $\beta=\rho_i+i\gamma<\rho_j+j\gamma$ for all $i\in S$ and for all $j\in S^c$, we have

$$\bar{Q}(x,0)=\hat{Q}(x,0)=\sum_{i\in S} r_i x^i. \qquad \square$$

The above theorem implies that $\bar{Q}(x,z)=0$ is a regularly perturbed polynomial equation. Now we can formally state a connection between the regularly perturbed polynomial equation $\bar{Q}(x,z)=0$ and the original singularly perturbed polynomial equation $Q(x,z)=0$.

Theorem 4.12. *Computing the Puiseux series expansion for $x(z)$, a root of singularly perturbed polynomial equation $Q(x,z)=0$ has been transformed into the following regular* *perturbation problem: Compute the Taylor series expansion for $\bar{x}(z)$ starting from c_t corresponding to a perturbed solution of $\bar{Q}(x,z)=0$. The Puiseux series expansion for the original singular perturbation problem can be retrieved by*

$$x(z)=\sum_{i=1}^{t-1} c_i z^{\gamma_1+\cdots+\gamma_i}+z^{\gamma_1+\cdots+\gamma_t}\bar{x}(z^{1/d}), \tag{4.22}$$

where d is the smallest common denominator of $\gamma_1,\ldots,\gamma_t$ (if $\gamma_1=0$, take 1 as the denominator of γ_1). Moreover, the initial segment of the Puiseux expansion $\sum_{i=1}^{t} c_i z^{\gamma_1+\cdots+\gamma_i}$ defines a unique root of the polynomial equation $Q(x,z)=0$.

Proof: Theorem 4.11 states that the problem of finding a power expansion for the perturbed solution of the equation $\bar{Q}(x,z)=0$ starting with c_t is regular. Formula (4.22) follows from the following transformation which summarizes the Newton polygon process:

$$\hat{Q}(x,z)=z^{-(\beta_1+\cdots+\beta_t)}Q\left(\sum_{i=1}^{t-1} c_i z^{\gamma_1+\cdots+\gamma_i}+z^{\gamma_1+\cdots+\gamma_t}x,z\right).$$

The uniqueness of the power expansion follows from the fact that the Newton polygon process does not terminate until c_k is a simple root. □

The next theorem provides a condition for the finite number of stages of the Newton polygon process.

Theorem 4.13. *If the discriminant of the perturbed polynomial (4.2) is not identically equal to zero, the Newton polygon process has a finite number of stages. Furthermore, the number of stages is bounded above, as follows:*

$$t\le \operatorname{ord}(\operatorname{Dis}(Q))+1. \tag{4.23}$$

Proof: We are interested in the case $t\ge 2$. There are at least two cycles of solutions $x_{1,j}(z)$ and $x_{2,j}(z)$ whose series expansions have the same first $t-1$ nonzero terms. We can write them in the form

$$x_{1,j}(z)=\sum_{i=1}^{\infty} c_{1,a_i}\xi_1^{ja_i}z^{a_i/d_1}, \quad j=0,\ldots,d_1-1,$$

and

$$x_{2,j}(z)=\sum_{i=1}^{\infty} c_{2,b_i}\xi_2^{jb_i}z^{b_i/d_2}, \quad j=0,\ldots,d_2-1,$$

where $\{a_i\}$, $\{b_i\}$ are strictly increasing nonnegative integer sequences such that none of c_{1,a_i}, c_{2,b_i} vanish and $c_{1,a_i}=c_{2,b_i}$, $a_i/d_1=b_i/d_2$ for $i=1,\ldots,t-1$, and

$$\xi_1=e^{2\pi\sqrt{-1}/d_1}, \quad \xi_2=e^{2\pi\sqrt{-1}/d_2}.$$

Without loss of generality, we assume that $d_1\le d_2$. Since the series expansions for $x_1(z)$ and $x_2(z)$ agree in the first $t-1$ terms for $j=0,\ldots,d_1-1$, we have

$$\operatorname{ord}(x_1(z)-x_2(z))\ge (b_{t-1}+1)/d_2=a_{t-1}/d_1+1/d_2.$$

Consequently, we obtain

$$\text{ord}(\text{Dis}(Q)) \geq \text{ord}\left(\prod_{j=0}^{d_1-1}(x_1(z)-x_2(z))\right) \geq d_1(a_{t-1}/d_1 + 1/d_2) \geq a_{t-1}+1.$$

Since $\{a_i\}$ is a strictly increasing nonnegative integer sequence, $a_{t-1} \geq t-2$. Thus, we establish the bound (4.23), from which it follows that if the discriminant does not vanish identically, the number of stages of the Newton polygon process is finite. □

The next corollary provides a simpler, rough, bound on the number of stages of the Newton polygon process.

Corollary 4.2. *The number of stages of the Newton polygon process satisfies the following bound:*

$$t \leq p(2m-1)+1,$$

where $p = \max_{0\leq i\leq m}(deg(q_i(z)))$.

Proof: The discriminant is a determinant of order $2m-1$ (see Problem 4.3) whose elements are polynomials of degree at most p. Since by assumption the discriminant cannot vanish identically, $\text{ord}(\text{Dis}(Q)) \leq p(2m-1)$. □

Let us demonstrate the application of the Newton polygon method continuing Example 4.1.

Example 4.1 (continued from Section 4.1). *Let us apply the Newton polygon method to construct Puiseux series expansions for the perturbed polynomial equation*

$$Q(x,z) = (1-z)x^2 - 2x + 1 = 0.$$

For this equation, we have $q_2(z) = 1-z$, $q_1(z) = -2$, *and* $q_0(z) = 1$. *Since* $q_2(0) = 1 \neq 0$, *we set* $Q_1(x,z) = Q(x,z)$. *The Newton polygon corresponding to the first iteration is shown in Figure 4.1. There is only one horizontal segment which corresponds to the equation*

$$x^2 - 2x + 1 = 0$$

or

$$(x-1)^2 = 0.$$

We can see that 1 *is a multiple root of the above equation, and we have to continue the process. The horizontal segment lies on the line* $y + 0x = 0$. *Hence,* $\gamma_1 = 0$ *and* $\beta_1 = 0$.

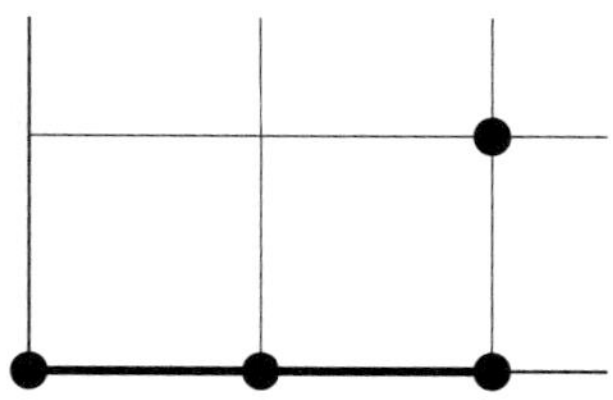

Figure 4.1. *The first Newton polygon for Example 4.1*

To continue the process, we make a change of variable $x \to x+1$. *Thus, we have*

$$Q_2(x,z) = Q_1(x+1,z) = (1-z)x^2 - 2zx - z.$$

The Newton polygon corresponding to the second iteration is shown in Figure 4.2. The endpoints of the segment determine the equation

$$x^2 - 1 = 0,$$

which has two simple roots $+1$ *and* -1. *Thus, we stop the process* ($t = 2$). *Since the segment lies on the line* $y - 1/2x = 1$, *we have* $\gamma_2 = 1/2$ *and* $\beta_2 = 1$.

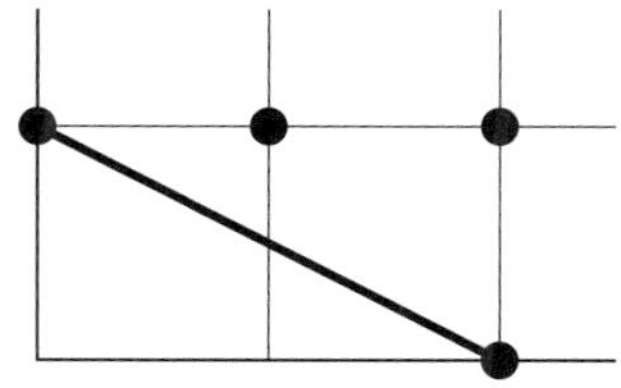

Figure 4.2. *The second Newton polygon for Example 4.1*

We make the following two transformations:

$$\hat{Q}(x,z) = z^{-\beta_2} Q_2(z^{\gamma_2} x, z) = z^{-1} Q_2(z^{1/2} x, z) = (1-z)x^2 - 2z^{1/2}x - 1,$$

$$\bar{Q}(x,z) = \hat{Q}(x,z^d) = \hat{Q}(x,z^2) = (1-z^2)x^2 - 2zx - 1.$$

We note that the above equation is a regularly perturbed polynomial equation which admits the following two solutions:

$$\bar{x}_1(z) = \frac{1}{1-z} = 1 + z + z^2 + \dots$$

and

$$\bar{x}_2(z) = -\frac{1}{1+z} = -1 + z - z^2 + \dots.$$

Then, the Puiseux series for the original perturbed polynomial equation can be retrieved by the formula (4.22). *Namely, we obtain*

$$\begin{aligned} x_1(z) &= 1 + z^{1/2}\bar{x}_1(z^{1/2}) \\ &= 1 + z^{1/2}(1 + z + z^2 + \dots) \\ &= 1 + z^{1/2} + z + z^{3/2} + \dots, \end{aligned}$$

and

$$\begin{aligned} x_2(z) &= 1 + z^{1/2}\bar{x}_2(z^{1/2}) \\ &= 1 + z^{1/2}(-1 + z - z^2 + \dots) \\ &= 1 - z^{1/2} + z - z^{3/2} + \dots. \end{aligned}$$

4.8 ▪ An Example of Application to Optimization

Let us demonstrate that the theoretical results from the preceding subsections can be applied to polynomial equations arising in constrained nonlinear optimization. This example creates a connection with the next chapter.

Consider an instance of a constrained minimization problem

$$\min_{x,y} \quad f(x,y,\varepsilon)$$

$$\text{subject to} \quad h(x,y,\varepsilon)=0,$$

where

$$\begin{aligned} f(x,y,\varepsilon) &= \frac{x^4}{4}+\frac{y^4}{4}+\frac{\varepsilon}{3}x^3y+\varepsilon x, \\ h(x,y,\varepsilon) &= 2x^2+xy-y^2+2\varepsilon x^2+\varepsilon xy-\varepsilon y^2, \end{aligned}$$

and ε is a parameter. To distinguish the parameter from the main variables we use the Greek letter ε and emphasize that we are interested only in a valued real parameter.

We observe that the vanishing gradient in variables (x,y,λ) of the Lagrangian $f+\lambda h$, that is, $\frac{\partial(f+\lambda h)}{\partial x}=\frac{\partial(f+\lambda h)}{\partial y}=\frac{\partial(f+\lambda h)}{\partial \lambda}=0$, requires the solution of simultaneous polynomial equations

$$\begin{aligned} f_1 &= x^3+\varepsilon x^2y+\varepsilon+4\lambda x+\lambda y+4\lambda\varepsilon x+\lambda\varepsilon y=0, \\ f_2 &= y^3+1/3\,\varepsilon x^3+\lambda x-2\lambda y+\lambda\varepsilon x-2\lambda\varepsilon y=0, \\ h &= 2x^2+xy-y^2+2\varepsilon x^2+\varepsilon xy-\varepsilon y^2=0. \end{aligned}$$

The generator polynomials $\{f_1,f_2,h\}$ do not comprise a *Gröbner basis.* To obtain a Gröbner basis for $F=\{f_1,f_2,h\}$ we first choose the *pure lexicographic term order* $y \prec x \prec \lambda$. Using Maple command `with(Groebner):gbasis` we find a Gröbner basis in this case, that is,

$$\begin{aligned} &[72\varepsilon^2+120y^3\varepsilon^2+32\varepsilon^2y^6+156y^6\varepsilon+9\varepsilon y^3-306y^6, \\ &-306y^4+156\varepsilon y^4-45\varepsilon y+32\varepsilon^2y^4+72\varepsilon^2y-198\varepsilon x+48\varepsilon^2x, \\ &-8100\lambda\varepsilon^2-10692\lambda\varepsilon+2592\lambda\varepsilon^3+160\varepsilon^3y^5-3672y^5 \\ &+1164y^5\varepsilon^2+342y^5\varepsilon-999\varepsilon^2y^2+5400\varepsilon y^2+408\varepsilon^3y^2]. \end{aligned} \tag{4.24}$$

The first element of this basis is a bivariate polynomial

$$S(y,\varepsilon)=72\varepsilon^2+120y^3\varepsilon^2+32\varepsilon^2y^6+156y^6\varepsilon+9\varepsilon y^3-306y^6,$$

which admits an irreducible factorization $S(y,\varepsilon)=S_1(y,\varepsilon)S_2(y,\varepsilon)$ with

$$S_1(y,\varepsilon)=51y^3+8\varepsilon y^3+24\varepsilon, \quad S_2(y,\varepsilon)=-6y^3+4\varepsilon y^3+3\varepsilon.$$

Solutions of $S_1(y,\varepsilon)=0$ and $S_2(y,\varepsilon)=0$ are Puiseux series that, in the closed form, can be written as

$$\begin{aligned} y_1 &= 2\frac{\sqrt[3]{-3\varepsilon\,(51+8\varepsilon)^2}}{51+8\varepsilon}, \\ y_2 &= 1/2\frac{\sqrt[3]{-12\varepsilon\,(-3+2\varepsilon)^2}}{-3+2\varepsilon}. \end{aligned} \tag{4.25}$$

Analogously, adopting lexicographic order $x \prec y \prec \lambda$, we obtain the Gröbner basis

$$[-9\,\varepsilon^2 - 12\,\varepsilon^2 x^3 + 32\,\varepsilon^2 x^6 - 171\,\varepsilon\,x^3 + 156\,\varepsilon\,x^6 - 306\,x^6, \tag{4.26}$$
$$156\,\varepsilon\,x^4 - 12\,\varepsilon^2 x - 306\,x^4 - 108\,\varepsilon\,x + 45\,\varepsilon\,y + 32\,\varepsilon^2 x^4 + 12\,\varepsilon^2 y,$$
$$1539\,\lambda\,\varepsilon^2 + 1215\,\lambda\,\varepsilon + 324\,\lambda\,\varepsilon^3 + 64\,\varepsilon^3 x^5$$
$$+984\,\varepsilon^2 x^5 - 6426\,x^5 + 2664\,x^5 \varepsilon - 12\,\varepsilon^3 x^2 - 783\,\varepsilon^2 x^2 - 3618\,\varepsilon\,x^2].$$

The first element of this basis is a bivariate polynomial

$$R(x,\varepsilon) = -9\,\varepsilon^2 - 12\,\varepsilon^2 x^3 + 32\,\varepsilon^2 x^6 - 171\,\varepsilon\,x^3 + 156\,\varepsilon\,x^6 - 306\,x^6,$$

which admits an irreducible factorization $R(x,\varepsilon) = R_1(x,\varepsilon)R_2(x,\varepsilon)$ with

$$R_1(x,\varepsilon) = 51\,x^3 + 8\,\varepsilon\,x^3 + 3\,\varepsilon, \quad R_2(x,\varepsilon) = -6\,x^3 + 4\,\varepsilon\,x^3 - 3\,\varepsilon.$$

Solutions of $R_1(x,\varepsilon) = 0$ and $R_2(x,\varepsilon) = 0$ are Puiseux series that in the closed form can be written as

$$x_1 = \frac{\sqrt[3]{-3\,\varepsilon\,(51+8\,\varepsilon)^2}}{51+8\,\varepsilon}, \tag{4.27}$$
$$x_2 = 1/2\,\frac{\sqrt[3]{12}\sqrt[3]{\varepsilon\,(-3+2\,\varepsilon)^2}}{-3+2\,\varepsilon}.$$

Hence, a solution of our optimization program must be one of the pairs of $(x_i(\varepsilon), y_j(\varepsilon),$ $i,j = 1,2)$ from (4.27), (4.25).

Direct substitution into h shows that only two pairs, namely, $(x = \frac{\sqrt[3]{-3\,\varepsilon\,(51+8\,\varepsilon)^2}}{51+8\,\varepsilon}, y = 2\,\frac{\sqrt[3]{-3\,\varepsilon\,(51+8\,\varepsilon)^2}}{51+8\,\varepsilon})$ and $(x = 1/2\,\frac{\sqrt[3]{12\varepsilon\,(-3+2\,\varepsilon)^2}}{-3+2\,\varepsilon}, y = 1/2\,\frac{\sqrt[3]{-12\,\varepsilon\,(-3+2\,\varepsilon)^2}}{-3+2\,\varepsilon})$, satisfy the constraint $h(x,y,\varepsilon) = 0$ since $h(x,y) = (1+\varepsilon)(y+x)(2x-y)$, and for the above expressions $y = 2x$ and $y = -x$, respectively.

Therefore, solutions for the Karush–Kuhn–Tucker conditions (disregarding λ) could only be

$$x = \frac{\sqrt[3]{-3\,\varepsilon\,(51+8\,\varepsilon)^2}}{51+8\,\varepsilon}, \quad y = 2\,\frac{\sqrt[3]{-3\,\varepsilon\,(51+8\,\varepsilon)^2}}{51+8\,\varepsilon}; \tag{4.28}$$
$$x = 1/2\,\frac{\sqrt[3]{12\varepsilon\,(-3+2\,\varepsilon)^2}}{-3+2\,\varepsilon}, \quad y = 1/2\,\frac{\sqrt[3]{-12\,\varepsilon\,(-3+2\,\varepsilon)^2}}{-3+2\,\varepsilon}.$$

In this simple example the above solutions could also have been derived by eliminating the constraint and applying elementary calculus. For instance, substituting $y = 2x$ back into f, we obtain that

$$f(x, y = 2x, \varepsilon) = \frac{17}{4}\,x^4 + 2/3\,\varepsilon\,x^4 + \varepsilon\,x,$$

and the zero of the derivative of this function in x occurs at exactly $x = \frac{\sqrt[3]{-3\,\varepsilon\,(51+8\,\varepsilon)^2}}{51+8\,\varepsilon}$, while the second derivative at this point is $\frac{3^{2/3}\left(\varepsilon\,(8\,\varepsilon+51)^2\right)^{2/3}}{8\,\varepsilon+51} > 0$.

Similarly, substituting $y = -x$ into f, we obtain that

$$f(x, y = -x, \varepsilon) = \frac{1}{2}x^4 - \frac{1}{3}\varepsilon x^4 + \varepsilon x$$

with the zero of its derivative occurring at $x = 1/2\,\frac{\sqrt[3]{12\varepsilon(-3+2\varepsilon)^2}}{-3+2\varepsilon}$. The second derivative of f at this point is $\frac{12^{2/3}(\varepsilon(2\varepsilon-3)^2)^{2/3}(8\varepsilon+51)}{4(2\varepsilon-3)^2} > 0$.

Hence, (4.28) indeed provides the two solutions of the original optimization problem.

Finally, we note that the closed form expressions in (4.28) are indeed Puiseux series. For instance, one can readily verify that the first few terms of an expansion of x, y from the first pair (4.28) are

$$x(\varepsilon) = \frac{1}{51}\sqrt[3]{-7803}\sqrt[3]{\varepsilon} - \frac{8}{7803}\sqrt[3]{-7803}\varepsilon^{4/3} + O(\varepsilon^{7/3}),$$
$$y(\varepsilon) = \frac{2}{51}\sqrt[3]{-7803}\sqrt[3]{\varepsilon} - \frac{16}{7803}\sqrt[3]{-7803}\varepsilon^{4/3} + O(\varepsilon^{7/3}).$$

4.9 ▪ Problems

Problem 4.1. Prove the first part of Lemma 4.6. Specifically, prove that one can order the elements of the reduced Gröbner basis $g_1, \dots, g_t$ so that g_1 is a univariate polynomial in the variable x_1, polynomial g_2 contains only the variables x_1, x_2, polynomial g_3 contains only x_1, x_2, x_3, and so forth until the polynomial g_n containing $x_1, \dots, x_n$. *Hint: See Theorem 4.5 in Section 4.2 and reference [4], which provides an excellent introduction to Gröbner bases techniques.*

Problem 4.2. Find "ghost" solutions in the following system of polynomials:

$$\begin{cases} x_2(x_1^2 - z) = 0, \\ x_1 - x_2^2 = 0. \end{cases}$$

Problem 4.3. Show that the discriminant of the polynomial

$$Q(x) = q_m x^m + q_{m-1}x^{m-1} + \cdots + q_1 x + q_0$$

can be expressed in terms of its coefficients as follows:

$$\mathrm{Dis}(Q)$$
$$= \det \begin{bmatrix} 1 & 0 & \dots & 0 & m & 0 & \dots & 0 & 0 \\ q_{m-1} & q_m & \dots & 0 & (m-1)q_{m-1} & mq_m & \dots & 0 & 0 \\ \vdots & \vdots & \ddots & \vdots & \vdots & \vdots & \ddots & \vdots & \vdots \\ q_2 & q_3 & \dots & q_m & 2q_2 & 3q_3 & \dots & mq_m & 0 \\ q_1 & q_2 & \dots & q_{m-1} & q_1 & 2q_2 & \dots & (m-1)q_{m-1} & mq_m \\ q_0 & q_1 & \dots & q_{m-2} & 0 & q_1 & \dots & (m-2)q_{m-2} & (m-1)q_{m-1} \\ \vdots & \vdots & \ddots & \vdots & \vdots & \vdots & \ddots & \vdots & \vdots \\ 0 & 0 & \dots & q_1 & 0 & 0 & \dots & q_1 & 2q_2 \\ 0 & 0 & \dots & q_0 & 0 & 0 & \dots & 0 & q_1 \end{bmatrix}.$$

Problem 4.4. Prove part (iv) of Theorem 4.8. Namely, show that if $z_0 \notin \mathscr{Z}(Q)$ and is the zero of multiplicity $m_0 > 0$ of $q_m(z)$, then z_0 is a pole of order m_0 for every branch $f(z)$ of the solution $x(z)$ to the polynomial equation $Q(x, z_0) = 0$ and that, in this situation, the solution $x(z)$ has a Laurent series representation

$$x(z) = \sum_{k=-m_0}^{\infty} c_k (z - z_0)^k.$$

Hint: Follow arguments similar to those in the proof of part (iii) *of Theorem 4.8.* An alternative approach is described in the beginning of Section 4.7.

Problem 4.5. For the polynomial equation

$$Q(x,z)=x^2+(z+2)x+z=0$$

show that the perturbation is regular around the point $(x_0,z_0)=(-1,0)$. Then, calculate the first four terms for the series of the solution $x(z)=c_0+zc_1+z^2c_2+z^3c_3+\dots$ by the following methods:

(a) using formula (4.11);

(b) by differentiation $Q(x,z)$ with respect to z;

(c) using Newton-like method (4.19).

Verify that the three methods give the same answer.

Problem 4.6. Generalize the Newton-like method from the case of a single regularly perturbed polynomial to a regularly perturbed system of n polynomials. *Hint: Try it for $n=2$ in the first instance.*

Problem 4.7. Consider the perturbed polynomial equation

$$(1-z)x^2-2zx+z^2=0.$$

(a) Use the Newton polygon method to transform the problem into a regular perturbation problem.

(b) To the regular perturbation problem from (a) apply the Newton-like method (4.19) to calculate the first four terms of the solution series.

(c) Use formula (4.22) to obtain the first four coefficients of the Puiseux series expansion for the original singularly perturbed problem.

Problem 4.8. Find the first three terms of the series of solutions around the point $(0,0,0)$ of the polynomial equations in Example 4.14.

4.10 ▪ Bibliographic Notes

Puiseux published a key theorem about fractional power series expansion for a solution of a bivariate polynomial equation in 1850 in [129]. That result, in some form, was probably known even before 1850. Newton (as reported in [124]) proposed using a geometric method based on convex polygons to determine the powers of the fractional power series expansions of the polynomial equation solutions. Later this method was named in Newton's honor. The interested reader can find a comprehensive treatment of bivariate polynomials in [158]. Surprisingly, it is not easy to find a thorough exposition of the Newton method in the literature. We can point to [151, 158], which describe the Newton diagram method to some extent. In [151] one can find a number of applications of perturbed nonlinear systems. In this book we have adopted the presentation of the Newton polygon method from [105].

To the best of our knowledge the application of Gröbner bases to the perturbation analysis of polynomial systems was first proposed in [53]. The results of [53] were refined in [10]. In [53] and [10] the interested reader can find more theoretical details. Theorem 4.7 is analogous to the Remmert–Stein lemma [160] for complex analytic varieties. The material of Sections 4.3 and 4.5 is heavily based on the Gröbner bases technique. The book of Adams and Lostaunau [4] provides a comprehensive and accessible introduction to the Gröbner bases. Another short and accessible introduction to Gröbner bases and their applications is given by their discoverer Bruno Buchberger [32]. The application of the results of Sections 4.3 and 4.5 does not require a deep knowledge of the Gröbner bases theory. For all practical purposes, the reader can simply use the function `gbasis` from the "Groebner" package of Maple. The Newton-like method for the computation of the series of a solution of a perturbed polynomial equation was proposed by Kung and Traub in [105].

Part II

Applications to Optimization and Markov Processes

Mathematics is the language with which God has written the universe.

—Galileo Galilei (1564–1642)

Chapter 5

Applications to Optimization

5.1 ▪ Introduction and Motivation

In this chapter we focus on a range of optimization problems where some (or all) of the data/parameters of the problem are perturbed. We will be interested in the behavior of "solutions" as the perturbation parameter tends to zero. This class of problems is closely related to the well-established topics of sensitivity or postoptimality and parametric analyses of mathematical programs. However, the approach we are proposing (based on series expansions) is perhaps more generic than what has been done hitherto.

In particular, we propose describing the asymptotic behavior of solutions[5] to a generic, perturbed, mathematical program[6]

$$\begin{aligned} &\max f(x,\varepsilon) \\ \text{s.t.}\quad &\text{(i)}\quad g_i(x,\varepsilon)=0,\ i=1,\dots,m, \\ &\text{(ii)}\quad h_j(x,\varepsilon)\le 0,\ j=1,\dots,p, \end{aligned} \tag{MP(ε)}$$

where $x \in \mathbb{R}^n, \varepsilon \in [0,\infty)$, and f, g_i's, h_j's are functions on $\mathbb{R}^n \times [0,\infty)$. In particular, they can be analytic functions or polynomials in ε. The case $\varepsilon = 0$ corresponds to the underlying *unperturbed program* that will be denoted by (MP(0)). The parameter, ε, will be called the *perturbation*. We will be especially concerned with characterizing solutions, $x^{op}(\varepsilon)$, of (MP(ε)) as functions of the perturbation parameter, ε, and in their limiting behavior as $\varepsilon \downarrow 0$.

Before proceeding further, we would like to motivate the context in which problems such as MP(ε) arise naturally in practice. Let us suppose that we have a given engineering maximization problem similar in structure to MP(ε) except that it has no perturbation parameter ε, but, instead, its equality constraints $\tilde{g}_i(x,p)=0$, $i=1,\dots,m$, depend in a known way on some physical parameter p. It is natural to assume that a "default" value p^* of that parameter is given. If, as functions of p, the constraints are twice differentiable, they can be replaced by their Taylor series approximations which, to the second order, have the form

$$\tilde{g}_i(x,p^*)+\tilde{g}_i'(x,p^*)(p-p^*)+\frac{1}{2}\tilde{g}_i''(x,p^*)(p-p^*)^2=0,\ i=1,\dots,m.$$

[5]The word "solution" is used in a broad sense at this stage. In some cases the solution will indeed be a global optimum, while in other cases it will be only a local optimum or a stationary point.

[6]Clearly, the theory for minimization parallels that for maximization.

Now, by setting $\varepsilon := (p - p^*)$ and defining

$$g_i(x,\varepsilon) := \tilde{g}_i(x,p^*) + \tilde{g}_i'(x,p^*)\varepsilon + \frac{1}{2}\tilde{g}_i''(x,p^*)\varepsilon^2 = 0, \; i = 1,\dots,m,$$

we obtain equality constraints of the form given in MP(ε) with $g_i(x,\varepsilon)$ as a polynomial in ε.

A closely related and also natural situation is where the "default" value p^* is actually the average of, say, N observations of a random variable P that has an unknown mean μ and a known variance σ^2. If, for instance, P were normally distributed $N(\mu,\sigma^2)$, then the interval $[p^* - \frac{2}{\sqrt{N}}\sigma, p^* + \frac{2}{\sqrt{N}}\sigma]$ is approximately the 95% confidence interval for μ. In performing a sensitivity analysis on the value of the physical parameter p, it is thus natural to also consider the constraints

$$g_i(x,\varepsilon) := \tilde{g}_i(x, p^* \pm \varepsilon 2\sigma) = 0, \; i = 1,\dots,m,$$

where the perturbation parameter is now directly related to the number of observations taken to estimate μ, namely,

$$\varepsilon = \frac{1}{\sqrt{N}}.$$

In this case, the behavior of a solution $x^{op}(\varepsilon)$ of (MP(ε)) as $\varepsilon \downarrow 0$ is directly related to the value (if any) of additional observations.

Of course, this reasoning extends to the case where two or more parameters are being estimated with the help of a statistical procedure involving the same number, N, of observations. The perturbation, $\varepsilon = \frac{1}{\sqrt{N}}$, will still be the same but the known standard deviations σ_s of the parameters p_s will enter the constraint functions without changing the possible analyses in any essential way.

The obvious limitation in the formulation of (MP(ε)) is that we are considering small perturbations with only a *single* perturbation parameter at a time. However, by the end of this chapter it will be clear that even this case can yield interesting and even counterintuitive results. For instance, with the interpretation presented just above it is possible to construct examples where gaining extra information by taking more observations yields no improvements to the quality of the solution $x^{op}(\varepsilon)$.

We shall consider the (MP(ε)) problem at three levels of generality.

A. **Asymptotic linear programming.** Here all functions are linear in x, and the problem (MP(ε)) can be converted to an essentially equivalent perturbed linear program:

$$\begin{aligned} &\max[c(\varepsilon)x] \\ \text{s.t.}\quad &A(\varepsilon)x = b(\varepsilon), \\ &x \geq 0. \end{aligned} \qquad (\text{LP}(\varepsilon))$$

B. **Asymptotic polynomial programming.** Here all functions $f(x,\varepsilon)$, $g_i(x,\varepsilon)$, and $h_j(x,\varepsilon)$ are polynomials in x and ε.

C. **Asymptotic analytic programming.** Here all functions $f(x,\varepsilon)$, $g_i(x,\varepsilon)$, and $h_j(x,\varepsilon)$ are analytic functions in x and ε.

Our aims, in order of increasing complexity (and generality), are as follows:

1. To demonstrate that in the case (A) the asymptotic simplex method working in the space of Laurent series (described in detail in Section 5.2) can be effectively implemented to find an asymptotically optimal solution of (LP(ε)).

2. To demonstrate that under very mild conditions there exist a neighborhood $(0,\varepsilon^*)$, an integer $M>0$, and a solution $x^{op}(\varepsilon)$ of (MP(ε)) in that neighborhood that is expressible as a Puiseux series of the form

$$x^{op}(\varepsilon)=\sum_{\nu=K}^{\infty}\varepsilon^{\frac{\nu}{M}}\mathbf{c}_\nu. \tag{PS}$$

   See Section 5.4 for more details.

There is an enormous number of problems that are formulated as either linear or nonlinear programs. In a vast majority of cases it is assumed that the objective function and the constraints are fully and precisely known. However, that is rarely the case in applications. Hence a fundamental question that arises concerns the stability (or instability) of a solution when the problem is slightly perturbed. It will be seen below that this can be a very difficult question.

Even in the simplest case of linear programming, standard operations research textbooks discuss only the most straightforward cases and scrupulously avoid the general issue of how to analyze the effect of a perturbation when the whole coefficient matrix is affected by it. The next, well-known, example illustrates that even in the "trivial" case of linear programming the effect of a small perturbation can be "nontrivial."

Example 5.1. *Consider the linearly perturbed linear program*

$$\begin{aligned} &\max\{(1+\varepsilon)x_1+2x_2\}=F(\varepsilon)\\ s.t.\quad &x_1(1-\varepsilon)+x_2(1+\varepsilon)\le 1+\tfrac{1}{2}\varepsilon,\\ &-x_1-x_2\le -1,\\ &x_1\ge 0, x_2\ge 0. \end{aligned} \tag{MP$^1(\varepsilon)$}$$

The unperturbed problem (MP^1(0)) *is*

$$\max\{x_1+2x_2 \mid x_1+x_2=1,\ x_i\ge 0, i=1,2\}=F(0),$$

which has a solution: $x_1=0, x_2=1, F(0)=2$. *As* (MP^1(ε)) *is a linear program, the solution can be easily checked to be*

$$x_1(\varepsilon)=\frac{1}{4},\quad x_2(\varepsilon)=\frac{3}{4},\quad F(\varepsilon)=(1+\varepsilon)\frac{1}{4}+\frac{6}{4}=\frac{7}{4}+\varepsilon\frac{1}{4}.$$

Hence,

$$\lim_{\varepsilon\to 0}F(\varepsilon)=\frac{7}{4}\neq 2=F(0). \tag{5.1}$$

Thus the optimal objective function value has a discontinuity at $\varepsilon=0$ *even though* $x_1(\varepsilon)$ *and* $x_2(\varepsilon)$ *are continuous (actually constant) for* $\varepsilon>0$.

Example 5.1 does not demonstrate how fractional powers present in (PS) can naturally arise in mathematical programming. This is illustrated in the next simple example.

Example 5.2. *Let*

$$f(x_1,x_2,\varepsilon)=\frac{x_1^4}{4}+\frac{x_2^4}{4}+\frac{\varepsilon}{3}x_1^3x_2+\varepsilon x_1,$$

and consider its stationary points satisfying

$$\frac{\partial f}{\partial x_1}=x_1^3+\varepsilon x_1^2x_2+\varepsilon=0;\quad \frac{\partial f}{\partial x_2}=x_2^3+\frac{\varepsilon}{3}x_1^3=0.$$

It is easy to check that the solutions $(x_1(\varepsilon),x_2(\varepsilon))$ *satisfy* $x_2(\varepsilon)=-[x_1(\varepsilon)\varepsilon^{\frac{1}{3}}]/\sqrt[3]{3}$ *and* $x_1^3(\varepsilon)$ $[1-\varepsilon^{\frac{4}{3}}/\sqrt[3]{3}]=-\varepsilon$ *and hence that*

$$x_1(\varepsilon)=-\varepsilon^{\frac{1}{3}}-\varepsilon^{\frac{5}{3}}/3\sqrt[3]{3}\cdots;\quad x_2(\varepsilon)=-\varepsilon^{\frac{2}{3}}/\sqrt[3]{3}+\varepsilon^2/3\sqrt[3]{9}\cdots.$$

Despite the fractional powers, the above solution is better behaved than the solution of Example 5.1, because here $(x_1(\varepsilon),x_2(\varepsilon))\longrightarrow(x_1(0),x_2(0))$ *as* $\varepsilon\downarrow 0$.

Examples 5.1–5.2 suggest that the understanding of the expansion (PS) is, in many cases, the key to understanding the asymptotic behavior of solutions to the mathematical program MP(ε). Indeed, this approach promises to offer a unified analytic perspective of quite a diverse range of asymptotic behaviors.

Of course, there is more than one kind of desirable asymptotic behavior that the solutions $\mathbf{x}^{op}(\varepsilon)$ of MP(ε) may exhibit. To illustrate this, we informally define an *asymptotically optimal* (a-optimal) solution as one that is "uniformly" optimal for all $\varepsilon\in(0,\bar{\varepsilon}]$; let us denote such a solution by $x^{ap}(\varepsilon)$. This is stronger than the notion of a *limiting optimal* solution that can be thought of as "δ-optimal" (for $\delta>0$) in MP(ε_k) for any sequence $\varepsilon_k\to 0$ that we shall denote by x^{lim}. Alternatively, one could have defined x^{op} as being *sequentially* a-optimal if there exists a sequence $\varepsilon_k\to 0$ such that $x^{op}=\lim_{k\to\infty}x^{op}(\varepsilon_k)$, where $x^*(\varepsilon_k)$ is optimal in MP(ε_k) for each k. This last definition is restricted by the requirement that the sequence of optimal solutions needs to be selected in such a way as to be convergent.

The examples below demonstrate some of the differences between these different notions of asymptotic optimality for the simplest case of a perturbed linear program LP(ε).

Example 5.3. *This example shows that a-optimality not only is different from limiting optimality but also gives the user a solution that is, in a natural sense, more robust. Consider the perturbed linear program LP(ε):*

$$\min\{10x_1+10x_2\}$$

subject to

$$\begin{aligned}\varepsilon x_2-x_3&=0,\\ x_1+x_2+x_3&=1,\\ x_1,x_2,x_3&\geq 0.\end{aligned}$$

For each $\varepsilon>0$ *this linear program possesses two basic feasible solutions* $(1,0,0)$ *and* $(0,1/(1+\varepsilon),\varepsilon/(1+\varepsilon))$ *(see Figure 5.1). Clearly,* $x^{ap}(\varepsilon):=(0,1/(1+\varepsilon),\varepsilon/(1+\varepsilon))$ *is an optimal solution for any positive value of* ε*; that is,* $x^{ap}(\varepsilon)$ *is an a-optimal solution.*

Note that the point $(0,1,0)$ *is a* limiting *optimal; that is, the optimal value* $10/(1+\varepsilon)$ *of the perturbed linear programming program converges to* 10 *as* ε *goes to zero. Thus, we can see*

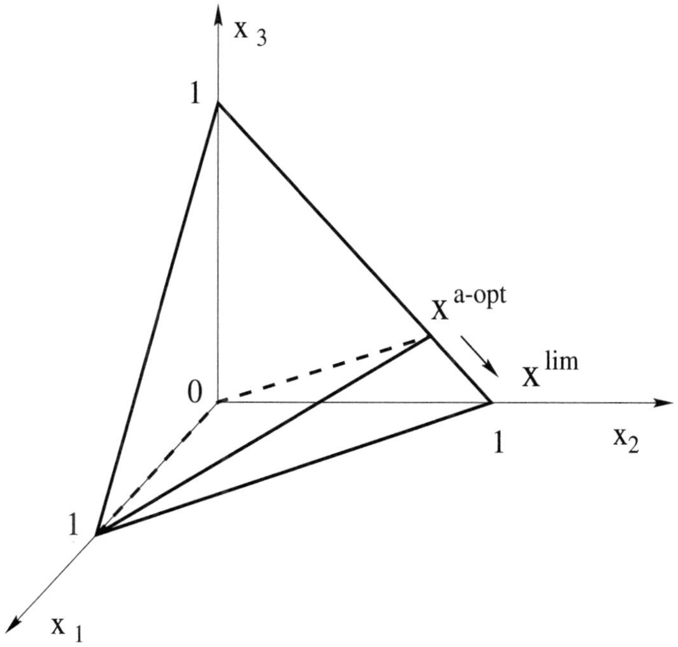

Figure 5.1. *Comparison between a-optimal and limiting optimal solutions [58]*

that the notion of an a-optimal solution is more "robust" than the notion of a limiting optimal solution in the sense that it is optimal (not just approximately optimal) for some interval of values of ε.

Example 5.4. *Again, consider the perturbed linear program $LP(\varepsilon)$:*

$$\max_{x_1,x_2}\{x_2\}$$

subject to

$$\begin{cases} x_1+x_2=1, \\ (1+\varepsilon)x_1+(1+2\varepsilon)x_2=1+\varepsilon, \end{cases} \tag{5.2}$$

$$x_1 \geq 0, \quad x_2 \geq 0.$$

It is obvious that the system of constraints (5.2) *has the unique feasible solution $x^{ap}=(1,0)$ when $\varepsilon>0$, which is also an a-optimal solution (see Figure 5.2). However, the optimal solution of the original unperturbed ($\varepsilon=0$) problem is $(0,1)$, which is not anywhere near the previous solution.*

Example 5.5. *Now, consider just a slightly modified perturbed linear program $LP(\varepsilon)$:*

$$\max_{x_1,x_2}\{x_1\}$$

subject to

$$\begin{cases} x_2=\frac{1}{2}, \\ \varepsilon x_1+x_2=1, \end{cases} \tag{5.3}$$

$$x_1 \geq 0, \quad x_2 \geq 0.$$

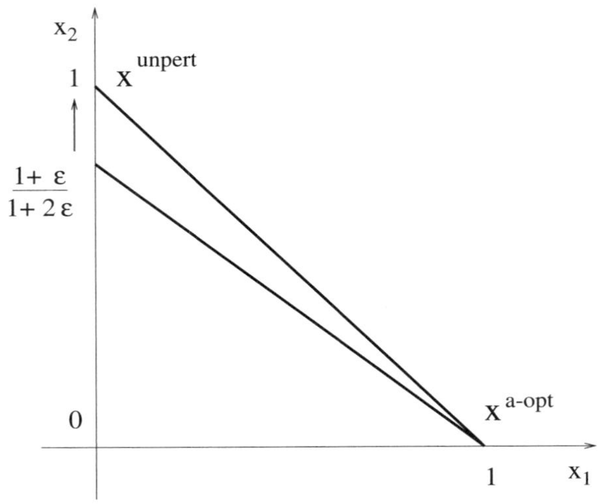

Figure 5.2. *An example of a singularly perturbed linear program [58]*

It can now be easily checked that, when $\varepsilon > 0$, (5.3) *has the unique feasible solution*

$$x^{ap}(\varepsilon) = \frac{1}{\varepsilon}\left(\frac{1}{2}, 0\right) + \left(0, \frac{1}{2}\right),$$

which is also an a-optimal solution and is of the form of a Laurent series with a pole of order one. Thus $x^{ap}(\varepsilon) \uparrow (\infty, \frac{1}{2})$ *as* $\varepsilon \downarrow 0$, *and yet the feasible region is empty at* $\varepsilon = 0$.

5.2 ▪ Asymptotic Simplex Method

5.2.1 ▪ Preliminaries

Let us consider the perturbed linear program

$$\min_x \{(c^{(0)} + \varepsilon c^{(1)})x\} \tag{5.4}$$

subject to

$$(A^{(0)} + \varepsilon A^{(1)})x = b^{(0)} + \varepsilon b^{(1)}, \quad x \geq 0. \tag{5.5}$$

This is the case of a linear perturbation. Later we show how our method can be generalized to the case of a polynomial perturbation, where the coefficient matrix is of the form

$$A(\varepsilon) = A^{(0)} + \varepsilon A^{(1)} + \cdots + \varepsilon^p A^{(p)},$$

and similarly for $b(\varepsilon)$ and $c(\varepsilon)$. As was mentioned in the introductory Section 4.1, we are interested in the determination of an *asymptotically* optimal solution. For linear programming we can define it as follows.

Definition 5.1. *The set of basic indices B is said to be asymptotically optimal (or a-optimal) for the perturbed linear program* (5.4),(5.5) *if it is optimal for the linear program* (5.4),(5.5) *with any given* $\varepsilon \in (0, \bar{\varepsilon}]$, *where* $\bar{\varepsilon} > 0$.

The effect of perturbations (for small values of ε) can be either small or large. Typically the effect of a perturbation is large when the dimension of the perturbed feasible set is different from the dimension of the original feasible set. This underlies the classification of problems into either *regular* or *singular* perturbation problems. More precisely, we have the following definition.

Definition 5.2. *Let $B = \{j_1, j_2, \dots, j_m\}$ be a subset of m indices selected from $\{1,2,\dots,n\}$, and $A_B(\varepsilon)$ be the $m \times m$ submatrix of $A(\varepsilon)$ whose columns correspond to the index set B for $\varepsilon \geq 0$. In all cases it will be assumed that* $\text{rank}[A(\varepsilon)] = m$ *for $\varepsilon > 0$ and sufficiently small. There are now three cases:*

(i) Regular perturbation: $A_B^{-1}(0)$ *exists whenever $A_B^{-1}(\varepsilon)$ exists for $\varepsilon > 0$ and sufficiently small.*

(ii) Weakly singular (or pseudo-singular) perturbation: $\text{rank}[A(0)] = m$, *but there exits at least one B such that* $\text{rank}[A_B(0)] < m$ *and* $\text{rank}[A_B(\varepsilon)] = m$ *for $\varepsilon > 0$ and sufficiently small.*

(iii) Strongly singular perturbation: $\text{rank}[A(0)] < m$.

It can be shown that an a-optimal solution of the regularly perturbed linear program is always the optimal solution of the original unperturbed linear program (see Problem 5.1). However, in the case of singular perturbations the latter is often not true. Let us demonstrate this phenomenon with the help of the following elegant example:

$$\max_{x_1, x_2} \{x_2\}$$

subject to

$$\begin{cases} x_1 + x_2 = 1, \\ (1+\varepsilon)x_1 + (1+2\varepsilon)x_2 = 1+\varepsilon, \end{cases} \tag{5.6}$$

$$x_1 \geq 0, \quad x_2 \geq 0.$$

It is obvious that the system of constraints (5.6) has the unique feasible solution $x_1^{op}(\varepsilon) = 1, x_2^{op}(\varepsilon) = 0$ when $\varepsilon > 0$. Of course, this is also an optimal solution if ε is not equal to zero. However, the optimal solution of the original ($\varepsilon = 0$) problem is $x_1^{op} = 0, x_2^{op} = 1$, which is not anywhere near the previous solution. Thus we can see that in the singularly perturbed linear programs the gap between the solution of the original problem and $\lim_{\varepsilon \to 0} x^{op}(\varepsilon)$ may arise.

5.2.2 ▪ Operations with Laurent series and lexicographic ordering

All algebraic operations are naturally defined for the Laurent series expansions. Let $g(\varepsilon)$ and $h(\varepsilon)$ be analytic functions in some nonempty punctured neighborhood around $\varepsilon = 0$ with the corresponding Laurent series expansions

$$g(\varepsilon) = \sum_{k=-s_1}^{\infty} \varepsilon^k g^{(k)},$$

$$h(\varepsilon) = \sum_{k=-s_2}^{\infty} \varepsilon^k h^{(k)}.$$

Without loss of generality, let us assume that $s_1 \le s_2$. Then, the sum of $g(\varepsilon)$ and $h(\varepsilon)$ is given by

$$g(\varepsilon)+h(\varepsilon)=\sum_{k=-s_2}^{\infty}\varepsilon^k(g^{(k)}+h^{(k)}),$$

the multiplication $g(\varepsilon)h(\varepsilon)$ is given by

$$g(\varepsilon)h(\varepsilon)=\sum_{k=-(s_1+s_2)}^{\infty}\varepsilon^k\sum_{l_1+l_2=k}g^{(l_1)}h^{(l_2)},$$

and the division $g(\varepsilon)/h(\varepsilon)$ is given by

$$f(\varepsilon)=\frac{g(\varepsilon)}{h(\varepsilon)}=\sum_{k=s_2-s_1}^{\infty}\varepsilon^k f^{(k)},$$

where the coefficients $\{f^{(k)}\}_{k=s_2-s_1}^{\infty}$ are calculated by the following recursive relation (see Problem 5.2):

$$f^{(k)}=\frac{1}{h^{(-s_2)}}\left[g^{(-s_1)}-\sum_{l=0}^{k-1}h^{(-s_2+k-l)}x^{(s_2-s_1+l)}\right]. \tag{5.7}$$

Next we define the *lexicographic ordering* that allows us to compare two functions in some small neighborhood of zero.

Definition 5.3. *A vector a (possibly infinite) is called* lexicographically nonnegative, *written* $a \succeq 0$, *if the first nonzero element (if any) in a is positive, and a is called* lexicographically positive, *written* $a \succ 0$, *if* $a \succeq 0$ *and* $a \neq 0$. *For two vectors a and b we say that a is* lexicographically greater (strictly greater) *than b, if* $a-b\succeq 0$ $(a-b\succ 0)$.

Suppose that we have an analytic function $g(\varepsilon)$ that is expanded as a Laurent series at $\varepsilon=0$ with a finite singular part

$$g(\varepsilon)=\sum_{k=-s}^{\infty}\varepsilon^k g^{(k)}.$$

We construct from the coefficients of the above series the infinite vector

$$\gamma=[g^{(-s)},\dots,g^{(0)},g^{(1)},\dots].$$

It is easy to see that $g(\varepsilon)>0$ for ε sufficiently small and positive if and only if $\gamma\succ 0$. Moreover, if $g(\varepsilon)$ is a rational function, then only a finite number of elements in γ needs to be checked (see Lemma 5.2). The comparison (in a neighborhood of 0) between two functions $g(\varepsilon)$ and $h(\varepsilon)$ possessing Laurent series expansions with finite order poles can be carried out similarly by considering the infinite vector of coefficients associated with $g(\varepsilon)-h(\varepsilon)$.

5.2.3 ▪ Revised simplex method

To ensure that this chapter is self-contained and has consistent notation, we briefly outline the now classical revised simplex method. More detailed explanations and proofs can be found in any standard book on linear programming. Consider the linear program

$$\min_x\{cx\} \tag{5.8}$$

subject to

$$Ax = b, \quad x \geq 0, \tag{5.9}$$

where A is an $m \times n$ matrix ($m \leq n$), b is an $m \times 1$ vector, and c is a $1 \times n$ vector of objective function coefficients.

Denote by B the index set of the current basic variables, and denote by N the index set of all other variables. Let A_B be the corresponding basis and $x_B = A_B^{-1} b$ be the corresponding vector of basic variables. Suppose that all elements of this vector are positive (nondegeneracy assumption). We can now partition c into c_B and c_N.

Recall that when applying the revised simplex method the following quantities need to be computed or updated:

(i) Simplex multipliers:

$$\lambda = c_B A_B^{-1}.$$

(ii) Reduced cost coefficients:

$$r_N = \lambda A_N - c_N,$$

where N corresponds to the set of nonbasic column indices, namely, $N := \{1, 2, \ldots, n\} \backslash B$.

(iii) Entering column:

$$y_k = A_B^{-1} a_k,$$

where

$$k = \arg\max_{j \in N} \{r_j | r_j > 0\}.$$

(iv) Exiting column index:

$$p = \arg\min_{l} \left\{ \frac{[x_B]_l}{[y_k]_l} | [y_k]_l > 0 \right\}.$$

Of course, most of the above depends on the knowledge of A_B^{-1}, and it is a well-known fact that this inverse need not be computed from scratch but can be updated at each iteration.

In particular, if B' is the next set of basic variable indices and we construct the vector and a matrix

$$\xi = \left[-\frac{y_1}{y_p}, \ldots, -\frac{y_{p-1}}{y_p}, \frac{1}{y_p}, -\frac{y_{p+1}}{y_p}, \ldots, -\frac{y_m}{y_p} \right]^T,$$

$$E = [e_1, \ldots, e_{p-1}, \xi, e_{p+1}, \ldots, e_m],$$

then the next inverse basis is given by

$$A_{B'}^{-1} = E A_B^{-1}.$$

In the above, e_i denotes the ith element of the standard unit basis.

5.2.4 ▪ Laurent series for the basis matrix

From the preceding description of the classical revised simplex method we see that everything depends on the inverse basis matrix A_B^{-1}, even if the latter is updated rather than explicitly computed. In the perturbed case it is necessary to realize that the analogous matrix $A_B^{-1}(\varepsilon)$ is a Laurent series whose structure and coefficients determine the asymptotic behavior, as $\varepsilon \to 0$. In particular,

$$A_B^{-1}(\varepsilon) = \frac{1}{\varepsilon^s}U^{(-s)} + \cdots + \frac{1}{\varepsilon}U^{(-1)} + U^{(0)} + \cdots. \tag{5.10}$$

If $A_B(\varepsilon)$ becomes singular at $\varepsilon = 0$, then the above series will have a pole of order s at $\varepsilon = 0$ and will contain a nontrivial *singular part* defined by

$$A_B^S(\varepsilon) = \frac{1}{\varepsilon^s}U^{(-s)} + \cdots + \frac{1}{\varepsilon}U^{(-1)}. \tag{5.11}$$

Similarly, a *regular part* of (5.10) is defined by

$$A_B^R(\varepsilon) = U^{(0)} + \varepsilon U^{(1)} + \cdots. \tag{5.12}$$

Clearly, if $A_B^S(\varepsilon) \neq 0$ and ε is small, standard simplex operations could result in unstable numerical behavior. The methods developed in this section overcome this difficulty by working with the coefficients of (5.10).

At first sight it might appear that computations involving the series expansion (5.10) would be too difficult. Fortunately, recursive formulae developed in Section 2.2 (see (2.38)–(2.39)) provide tools that can be adapted to the revised simplex method. A key observation here is that if $U^{(0)}$ and $U^{(-1)}$ are known, then the other coefficients of (5.10) can be obtained according to

$$U^{(k)} = (-U^{(0)}A_B^{(1)})^k U^{(0)} = U^{(0)}(-A_B^{(1)}U^{(0)})^k, \quad k = 1, 2, \ldots, \tag{5.13}$$

$$U^{(-k)} = (-U^{(-1)}A_B^{(0)})^{k-1} U^{(-1)} = U^{(-1)}(-A_B^{(0)}U^{(-1)})^{k-1}, \quad k = 2, \ldots, s, \tag{5.14}$$

where $A_B^{(i)}$ are the coefficients of ε^i in the basis matrix $A_B(\varepsilon)$. Consequently, if $U^{(0)}$ and $U^{(-1)}$ can be efficiently updated, when moving from one basis to another, then any coefficient of (5.10) can also be calculated for the next basis via (5.13), (5.14) if needed.

In general, we need to compute $U^{(0)}$ and $U^{(-1)}$ for the initial step of the asymptotic simplex method. There are two natural approaches to this problem. The first approach is to compute the singular part and the first regular coefficient of asymptotic expansion (5.10) by using methods presented in Section 2.2. The other approach is to start the asymptotic simplex method with an analogue of the phase 1 method for the linear programming. Note that when we introduce artificial variables in phase 1, the linear program becomes weakly singular, even if it was strongly singular before modification. The latter enables us to start from a basis matrix that possesses a Taylor series expansion instead of a Laurent series. This significantly facilitates computations. In addition, if we use the phase 1 method, we need not be concerned about the choice of an initial basic feasible solution.

Example 5.6. *We illustrate the formulae* (5.13) *and* (5.14) *with the help of the following perturbed matrix:*

$$A(\varepsilon) = A^{(0)} + \varepsilon A^{(1)} = \begin{bmatrix} 1 & 1 \\ 1 & 1 \end{bmatrix} + \varepsilon \begin{bmatrix} 2 & -1 \\ -1 & 2 \end{bmatrix}.$$

The inverse of $A(\varepsilon)$ can be easily calculated:

$$A^{(-1)}(\varepsilon) = \frac{1}{3\varepsilon(2+\varepsilon)} \begin{bmatrix} 1+2\varepsilon & -1+\varepsilon \\ -1+\varepsilon & 1+2\varepsilon \end{bmatrix}.$$

Then we expand each element of $A^{(-1)}(\varepsilon)$ as a Laurent series:

$$A^{(-1)}(\varepsilon) = \begin{bmatrix} \frac{1}{6\varepsilon} + \sum_{k=0}^{\infty}(-1)^k \frac{\varepsilon^k}{2^{k+2}} & -\frac{1}{6\varepsilon} + \sum_{k=0}^{\infty}(-1)^k \frac{\varepsilon^k}{2^{k+2}} \\ -\frac{1}{6\varepsilon} + \sum_{k=0}^{\infty}(-1)^k \frac{\varepsilon^k}{2^{k+2}} & \frac{1}{6\varepsilon} + \sum_{k=0}^{\infty}(-1)^k \frac{\varepsilon^k}{2^{k+2}} \end{bmatrix}. \tag{5.15}$$

In particular, we have

$$U^{(-1)} = \frac{1}{6}\begin{bmatrix} 1 & -1 \\ -1 & 1 \end{bmatrix}, \quad U^{(0)} = \frac{1}{4}\begin{bmatrix} 1 & 1 \\ 1 & 1 \end{bmatrix},$$

and for $k \geq 1$

$$U^{(k)} = (-1)^k \frac{1}{2^{k+2}} \begin{bmatrix} 1 & 1 \\ 1 & 1 \end{bmatrix}.$$

Next we check the formulae (5.13) *and* (5.14). *Since*

$$-U^{(-1)}A^{(0)} = -\frac{1}{6}\begin{bmatrix} 1 & -1 \\ -1 & 1 \end{bmatrix}\begin{bmatrix} 1 & 1 \\ 1 & 1 \end{bmatrix} = \begin{bmatrix} 0 & 0 \\ 0 & 0 \end{bmatrix},$$

all matrices $U^{(k)}$ for $k < -1$ are equal to zero matrices, which is consistent with (5.15). *Now we calculate $U^{(k)}$, $k \geq 1$, by the formula* (5.13)*:*

$$U^{(k)} = \left(-\begin{bmatrix} 1/4 & 1/4 \\ 1/4 & 1/4 \end{bmatrix}\begin{bmatrix} 2 & -1 \\ -1 & 2 \end{bmatrix}\right)^k \begin{bmatrix} 1/4 & 1/4 \\ 1/4 & 1/4 \end{bmatrix}$$

$$= (-1)^k \begin{bmatrix} 1/4 & 1/4 \\ 1/4 & 1/4 \end{bmatrix}^{k+1} = (-1)^k \frac{1}{2^{k+1}} \begin{bmatrix} 1/2 & 1/2 \\ 1/2 & 1/2 \end{bmatrix}^{k+1}$$

$$= (-1)^k \frac{1}{2^{k+1}} \begin{bmatrix} 1/2 & 1/2 \\ 1/2 & 1/2 \end{bmatrix} = \begin{bmatrix} (-1)^k \frac{1}{2^{k+2}} & (-1)^k \frac{1}{2^{k+2}} \\ (-1)^k \frac{1}{2^{k+2}} & (-1)^k \frac{1}{2^{k+2}} \end{bmatrix}.$$

As one can see, the last expression coincides with the coefficients of regular terms of the power series expansions in (5.15).

5.2.5 ▪ Asymptotic simplex method—the algorithm

The notation of the algorithm presented below is unavoidably complex. However, the basic ideas are a natural generalization of the revised simplex method to the field of Laurent series. These may be easier to understand, on first reading, by following Subsection 5.2.3 prior to reading the detailed statement of the algorithm.

As before, let the initial set of basic indices B be given, and let $N := \{1,\ldots,n\}\backslash B$ be the set of nonbasic indices.

Step 1: Obtain or update only the singular and the first regular coefficients of the Laurent series expansion (5.10) for the inverse basis matrix $A_B^{-1}(\varepsilon)$ with the order of a pole at $\varepsilon = 0$ denoted by s. The implementation of this step is discussed at the end of this section.

Step 2: Set $i := -s$ and $\lambda^{(-s-1)} := 0$.
Step (2a) Calculate the ith term of the Laurent expansion for the vector of simplex multipliers

$$\lambda^{(i)} := c_B^{(0)} U^{(i)} + c_B^{(1)} U^{(i-1)}.$$

Step (2b) Calculate the ith term of the Laurent expansion for the vector of nonbasic reduced cost coefficients,

$$r_N^{(i)} := \lambda^{(i)} A_N^{(0)} + \lambda^{(i-1)} A_N^{(1)} - \delta_{0i} c_N^{(0)} - \delta_{1i} c_N^{(1)},$$

where δ_{0i} and δ_{1i} are the Kronecker deltas. Let $N^{(-s-1)} := N$ and $N^{(i)} = \{j \in N^{(i-1)} | r_j^{(i)} = 0\}$. If $r_j^{(i)} < 0$ for all $j \in N^{(i-1)}$, STOP; the current solution is a-optimal. If there is an index k such that

$$k := \arg \max_{j \in N^{(i-1)}} \{r_j^{(i)} | r_j^{(i)} > 0\},$$

then k identifies the entering nonbasic variable; go to Step 3.
Step (2c) If the algorithm has not stopped and k has not been identified in Step (2b), and $i < m+1$, then $N^{(i)}$ is not empty. Increment the index i, and return to Step (2a) to consider the higher order approximation of the reduced cost coefficients $r_N(\varepsilon)$.

Step 3: Set $i := -s$, $U^{(-s-1)} := 0$, and $P = \emptyset$.
Step (3a) Calculate the ith term of the Laurent expansion for the entering column $y_k(\varepsilon)$:

$$y_k^{(i)} = U^{(i)} a_k^{(0)} + U^{(i-1)} a_k^{(1)}.$$

Step (3b) Let $Q^{(-s-1)} := \{1, \dots, m\}$ and $Q^{(i)} := \{j \in Q^{(i-1)} | [y_k^{(i)}]_j = 0\}$. Add the index $j \in Q^{(i-1)}$ to the set P if $[y_k^{(i)}]_j > 0$. If $Q^{(i)} = \emptyset$, then go to Step (3d).
Step (3c) If $Q^{(i)} \neq \emptyset$ and $i < m$, then increment i by one, and return to Step (3a). If $i = m$, go to Step (3d). Lemma 5.2 guarantees that $[y_k(\varepsilon)]_j \equiv 0, j \in Q^{(m)}$.
Step (3d) Stop. At this point the set P of candidate row indices is determined.

Step 4: Set $i := 0$.
Step (4a) Form the set of indices corresponding to the maximal powers of the leading coefficients in (5.21):

$$S^{(-1)} := \left\{ j | j = \arg\max_l \{t_l - q_l | l \in P\} \right\}.$$

Step (4b) Calculate the $(q_l + i)$th and $(t_l + i)$th terms of expansions (5.19), (5.20), respectively,

$$[y_k^{(q_l+i)}]_l = [U^{(q_l+i)}]_l a_k^{(0)} + [U^{(q_l+i+1)}]_l a_k^{(1)}, \quad l \in S^{(i-1)},$$

and

$$x_{Bl}^{(t_l+i)} = [U^{(t_l+i)}]_l b^{(0)} + [U^{(t_l+i+1)}]_l b^{(1)}, \quad l \in S^{(i-1)}.$$

Step (4c) Calculate the ith coefficient of expansion (5.21):

$$\Delta_l^{(i)} = \left(x_{Bl}^{(t_l+i)} - \sum_{j=0}^{i-1} [y_k^{(q_l+i-j)}]_l \Delta_l^{(j)} \right) \Big/ [y_k^{(q_l)}]_l, \quad l \in S^{(i-1)}.$$

Step (4d) Form the following set of indices:

$$S^{(i)} := \left\{ j | j = \arg\min_{l} \{\Delta_l^{(i)} | l \in S^{(i-1)}\} \right\}.$$

If $S^{(i)}$ consists of a unique index p, go to Step 5. If $S^{(i)}$ is not a singleton and $i < 2m+1$, we should take into account a higher order approximation of $\Delta_l(\varepsilon)$. Namely, increment i by one, and return to Step (4b). However, if $i = 2m+1$, choose any $p \in S^{(i)}$, and go to Step 5.

Step 5: Construct a new basis $A_{B'}(\varepsilon)$ obtained from $A_B(\varepsilon)$ by replacing $a_p(\varepsilon)$ with $a_k(\varepsilon)$. Go to Step 1.

This completes the algorithm.

Remark 5.1. *Note that if we know the first regular and the first singular terms of the Laurent expansion* (5.10)*, then the computation of Laurent series coefficients for simplex quantities* $\lambda(\varepsilon), x_B(\varepsilon)$*, and* $y(\varepsilon)$ *is easily performed by the recursive formulae*

$$\lambda^{(t)} = \lambda^{(t-1)} D_1, \quad y_k^{(t)} = D_2 y_k^{(t-1)},$$

$$x_B^{(t)} = D_2 x_B^{(t-1)}, \quad t \geq 2,$$

where $D_1 := -A_B^{(1)} U^{(0)}$, $D_2 := -U^{(0)} A_B^{(1)}$, *and*

$$\lambda^{(-t)} = \lambda^{(-t+1)} F_1, \quad y_k^{(-t)} = F_2 y_k^{(-t+1)},$$

$$x_B^{(-t)} = F_2 x_B^{(-t+1)}, \quad t \geq 3,$$

where $F_1 := -A_B^{(0)} U^{(-1)}$, $F_2 := -U^{(-1)} A_B^{(0)}$.

Remark 5.2. *As in the revised simplex method, it is possible to update the expansion* (5.10) *for the inverse of the new basis matrix* $A_{B'}(\varepsilon)$ *via the multiplication of the series by*

$$E(\varepsilon) = [e_1, \ldots, e_{p-1}, \xi(\varepsilon), e_{p+1}, \ldots, e_m],$$

where $\xi(\varepsilon) = [-\frac{y_1(\varepsilon)}{y_p(\varepsilon)}, \ldots, -\frac{y_{p-1}(\varepsilon)}{y_p(\varepsilon)}, \frac{1}{y_p(\varepsilon)}, -\frac{y_{p+1}(\varepsilon)}{y_p(\varepsilon)}, \ldots, -\frac{y_m(\varepsilon)}{y_p(\varepsilon)}]^T$. *Since the division of two Laurent series in the scalar case is not a problem (see the recursive formula* (5.7)*), one can easily obtain the Laurent series for* $E(\varepsilon)$*:*

$$E(\varepsilon) = \frac{1}{\varepsilon^t} E^{(-t)} + \cdots + \frac{1}{\varepsilon} E^{(-1)} + E^{(0)} + \ldots. \tag{5.16}$$

Let s' *be the order of the pole of the updated basis* B'*; then the coefficients* $U'^{(k)}, k = -s', -s'+1, \ldots,$ *of the Laurent series for* $A_{B'}^{(-1)}(\varepsilon)$ *are calculated by the following formula:*

$$U'^{(k)} = \sum_{i+j=k} E^{(i)} U^{(j)}, \quad k = -s', -s'+1, \ldots. \tag{5.17}$$

However, we would like to emphasize that we need to update only the coefficients $U'^{(-1)}, U'^{(0)}$ *by the above formula. The other coefficients, if needed, can be restored by iterative formulae* (5.13), (5.14) *in a more efficient way. The computational complexity for this updating procedure is analyzed in the next subsection.*

5.2.6 ▪ Basic ideas illustrated by an example

Let us give a basic idea of each step of the asymptotic simplex method and illustrate it with the following example of the singularly perturbed linear program.

Example 5.7.

$$\min\{-10x_1 - 10x_2 - 10x_3\}$$

subject to

$$\begin{array}{rrrrcl} \varepsilon x_1 & +\varepsilon x_2 & & -0.5x_4 & = & 0, \\ & -\varepsilon x_2 & +\varepsilon x_3 & -0.5x_4 & = & 0, \\ x_1 & +x_2 & +x_3 & +x_4 & = & 1, \end{array}$$

$$x_1, x_2, x_3, x_4 \geq 0.$$

In this example the perturbed coefficient matrix is $A(\varepsilon) = A^{(0)} + \varepsilon A^{(1)}$ *with*

$$A^{(0)} = \begin{bmatrix} 0 & 0 & 0 & -0.5 \\ 0 & 0 & 0 & -0.5 \\ 1 & 1 & 1 & 1 \end{bmatrix}, \quad A^{(1)} = \begin{bmatrix} 1 & 1 & 0 & 0 \\ 0 & -1 & 1 & 0 \\ 0 & 0 & 0 & 0 \end{bmatrix}.$$

Basic idea of Step 2: We have to decide which column enters the basis. Namely, among the nonbasic elements of the reduced cost vector

$$r_N(\varepsilon) := \lambda(\varepsilon)A_N(\varepsilon) - c_N(\varepsilon), \quad \text{where} \quad \lambda(\varepsilon) := c_B(\varepsilon)A_B^{-1}(\varepsilon), \tag{5.18}$$

we need to find k such that

$$k \in \arg\max_{j \in N}\{r_j(\varepsilon) | r_j(\varepsilon) > 0, \varepsilon \in (0, \bar{\varepsilon}]\}.$$

Substituting (5.10) into (5.18), we obtain the next asymptotic expansion,

$$r_N(\varepsilon) = \frac{1}{\varepsilon^s} r_N^{(-s)} + \frac{1}{\varepsilon^{s-1}} r_N^{(-s+1)} + \ldots,$$

where

$$r_N^{(i)} := \lambda^{(i)} A_N^{(0)} + \lambda^{(i-1)} A_N^{(1)} - \delta_{0i} c_N^{(0)} - \delta_{1i} c_N^{(1)} \quad \text{and} \quad \lambda^{(i)} := c_B^{(0)} U^{(i)} + c_B^{(1)} U^{(i-1)}.$$

Let us consider the (possibly semi-infinite) matrix

$$R = \begin{bmatrix} r_N^{(-s)} \\ r_N^{(-s+1)} \\ \ldots \end{bmatrix}$$

and denote its ith column by R_i.

As mentioned above, the lexicographic ordering can be used to compare functions in the "small" neighborhood of zero. In particular, it is easy to see that

$$\arg\max_{j \in N}\{r_j(\varepsilon) | r_j(\varepsilon) > 0, \varepsilon \in (0, \bar{\varepsilon}]\} = \arg\text{lex-}\max_{j \in N}\{R_j | R_j \succ 0\},$$

where "lex-max" is a maximum with respect to the lexicographical ordering of the columns of R and "arg lex-max" is an index at which "lex-max" is attained. Note that to compare two reduced cost coefficients $r_i(\varepsilon)$ and $r_j(\varepsilon)$ for sufficiently small ε we need only check a *finite* number of elements of the vectors R_i and R_j. This follows from the fact that $r_i(\varepsilon)$ and $r_j(\varepsilon)$ are rational functions (see Lemma 5.2 and also Problem 5.8). In practical implementation of the lexicographical entering rule we calculate the rows of matrix R one by one.

Example 5.7 (continued from the beginning of Subsection 5.2.6). *We start with the set of basic indices $B=\{1,3,4\}$. Since 2 is the only nonbasic index, we just have to check the sign of $r_2(\varepsilon)$. We find that $r_2^{(-1)}=0$ and $r_2^{(0)}=9$ (see Problem 5.4). Hence, $r_2(\varepsilon)>0$, and column 2 enters a new basis.*

Basic idea of Step 3: Now, as in the revised simplex method, we have to find out which elements of the vector $y_k(\varepsilon)=A_B^{-1}(\varepsilon)a_k(\varepsilon)$ are positive for $\varepsilon>0$ and sufficiently small. Namely, we have to identify the set of indices $P:=\{l|[y_k(\varepsilon)]_l>0,\varepsilon\in(0,\bar{\varepsilon}]\}$. Toward this, as in Step 2, we first expand $y_k(\varepsilon)$ as a Laurent series,

$$y_k(\varepsilon)=\frac{1}{\varepsilon^s}y_k^{(-s)}+\frac{1}{\varepsilon^{s-1}}y_k^{(-s+1)}+\dots,$$

and then define an auxiliary semi-infinite matrix

$$Y=[y_k^{(-s)},y_k^{(-s+1)},\dots].$$

Let Y_l denote the lth row of matrix Y. The set P is given by $P=\{l|Y_l\succ 0\}$. For a practical implementation of Step 3 we introduced the set $Q^{(i)}$ of indices corresponding to components of vector function $y_k(\varepsilon)$ with the first i coefficients of the Laurent series equal to zero.

Example 5.7 (continued from the beginning of Subsection 5.2.6). *We start Step 3 with $P=\{\emptyset\}$ and $Q^{(-2)}=\{1,3,4\}$. Then we calculate $y_2^{(-1)}$, which is $[0\ 0\ 0]^T$. Since all elements of $y_2^{(-1)}$ are zeros, $Q^{(-1)}=Q^{(-2)}=\{1,3,4\}$. Next, we calculate $y_2^{(0)}$, which is $[1.5\ -0.5\ 0]^T$ (see Problem 5.4). Thus, $Q^{(0)}=\{4\}$, and we add index 1 to set P. Since $[y_2^{(1)}]_4=1$, we add index 4 to set P. We finish Step 3 with $P=\{1,4\}$ and $Q=\{\emptyset\}$.*

Basic idea of Step 4: Now we have to choose a basic variable which exits the basis; namely, we have to find

$$p\in\arg\min_l\left\{\frac{[x_B(\varepsilon)]_l}{[y_k(\varepsilon)]_l}\mid l\in P,\varepsilon\in(0,\bar{\varepsilon}]\right\}.$$

To find such a p we again use the lexicographical ordering.

According to the previous step the functions $[y_k(\varepsilon)]_l, l\in P$, are expressed as a Laurent series

$$[y_k(\varepsilon)]_l=\varepsilon^{q_l}[y_k^{(q_l)}]_l+\varepsilon^{q_l+1}[y_k^{(q_l+1)}]_l+\dots,\tag{5.19}$$

with $y_l^{(q_l)}>0$. Under the nondegeneracy assumption, Assumption 5.2, and Lemma 5.3, $[x_B(\varepsilon)]_l$ can be expressed as a power series with a positive leading coefficient

$$[x_B(\varepsilon)]_l=\varepsilon^{t_l}x_{Bl}^{(t_l)}+\varepsilon^{t_l+1}x_{Bl}^{(t_l+1)}+\dots,\qquad x_{Bl}^{(t_l)}>0.\tag{5.20}$$

Then, the quotient $\Delta_l(\varepsilon):=[x_B(\varepsilon)]_l/[y_k(\varepsilon)]_l$ is written in terms of the Laurent series

$$\Delta_l(\varepsilon)=\varepsilon^{t_l-q_l}(\Delta_l^{(0)}+\varepsilon\Delta_l^{(1)}+\varepsilon^2\Delta_l^{(2)}+\dots),\tag{5.21}$$

where the coefficients $\Delta_l^{(i)}$ are calculated by simple recursive formulae (5.7). As in the previous steps, we introduce an auxiliary index set $S^{(i)}$ to perform the comparison according to the lexicographical ordering in an efficient recursive manner.

Example 5.7 (continued from the beginning of Subsection 5.2.6). *Since* $\Delta_1(\varepsilon)=1/3+o(\varepsilon)$ *and* $\Delta_4(\varepsilon)=1+o(\varepsilon)$ *(see Problem 5.4), the maximal power of* ε *of leading terms in the series for* $\Delta_1(\varepsilon)$ *and* $\Delta_4(\varepsilon)$ *is zero, and therefore* $S^{(-1)}=\{1,4\}$. *As the leading coefficient of* $\Delta_1(\varepsilon)$ *is smaller than the leading coefficient of* $\Delta_4(\varepsilon)$, $S^{(0)}=\{1\}$. *Since* $S^{(0)}$ *is a singleton, we terminate Step 4 with column 1 exiting the basis.*

The set of new basic indices is $B'=\{2,3,4\}$. *Since* $r_1^{(-1)}=r_1^{(0)}=0$ *and* $r_1^{(1)}=-20/3$, $r_2(\varepsilon)<0$ *for all sufficiently small* $\varepsilon>0$. *Thus, the new basis is a-optimal.*

5.2.7 ▪ Asymptotic simplex method—convergence and computational complexity

First let us introduce some rather mild assumptions that will be relaxed in Subsection 5.2.8. Let M_ε denote the feasible region of (5.4), (5.5) and M_0 be the feasible region of the unperturbed problem.

Assumption 5.1. *The region* M_0 *is bounded.*

The above assumption ensures that basic feasible solutions of the perturbed program (5.4), (5.5) can be expanded as Taylor series (see Lemma 5.3 for details).

Assumption 5.2. *The perturbed problem is nondegenerate; namely, every element of the basic feasible vector* $x_B(\varepsilon)=A_B^{-1}(\varepsilon)b(\varepsilon), \varepsilon\in(0,\bar{\varepsilon}]$ *is positive.*

We now prove the finite convergence of the asymptotic simplex method. Note that this theorem states that the asymptotic simplex method finds an a-optimal basic feasible solution that is stable in the sense of having a power series expansion in ε.

Theorem 5.1. *Let Assumptions 5.1 and 5.2 hold. Then the asymptotic simplex method finds an a-optimal basic index set for perturbed linear program* (5.4), (5.5), $0<\varepsilon<\bar{\varepsilon}$, *in a finite number of steps. Furthermore, if we let* B^* *denote this a-optimal basic index set, then the basic variables of the a-optimal solution are expressed by the power series*

$$x_{B^*}(\varepsilon)=x_{B^*}^{(0)}+\varepsilon x_{B^*}^{(1)}+\ldots, \qquad \varepsilon<\min\{\bar{\varepsilon},1/||D_2||\}, \tag{5.22}$$

where

$$x_{B^*}^{(0)}=U^{(0)}b^{(0)}+U^{(-1)}b^{(1)}, \quad x_{B^*}^{(1)}=U^{(1)}b^{(0)}+U^{(0)}b^{(1)},$$

and the subsequent coefficients are calculated by the recurrent formula

$$x_{B^*}^{(k)}=D_2x_{B^*}^{(k-1)}, \quad k\geq 2,$$

with $D_2=-U^{(0)}A_B^{(1)}$. *The matrices* $U^{(0)}$ *and* $U^{(-1)}$ *are the coefficients of* ε^0 *and* ε^{-1} *in Laurent series* (5.10) *corresponding to the basic set* B^*, *that is,* $B=B^*$.

Proof: At each iteration of the asymptotic simplex method it is necessary to determine the column that enters the basis and the column that exits it. Namely, it is necessary to determine the following numbers:

$$k \in \arg\max_{j\in N}\{r_j(\varepsilon)|r_j(\varepsilon)>0\}, \quad p \in \arg\min_{l}\left\{\frac{[x_B(\varepsilon)]_l}{[y_k(\varepsilon)]_l}|[y_k(\varepsilon)]_l>0\right\}.$$

According to Lemma 5.2, in the asymptotic simplex method k and p are determined in a finite number of steps by a recursive procedure analogous to the lexicographic ordering of the coefficients of Laurent/power series expansions.

Next let us show that the asymptotic simplex method has a finite number of iterations. Note that after each iteration the objective function $c(\varepsilon)x$ is decreased in the lexicographic sense by the subtraction of the function

$$r_k(\varepsilon)\frac{[x_B(\varepsilon)]_p}{[y_k(\varepsilon)]_p}$$

for all $\varepsilon \in (0,\bar{\varepsilon}]$. Since all quantities in the above expression are positive for small $\varepsilon > 0$ (in particular, $[x_B(\varepsilon)]_p > 0$ for $\varepsilon \in (0,\bar{\varepsilon}]$ due to the nondegeneracy assumption), after each iteration the objective function is strictly decreased for all $\varepsilon \in (0,\bar{\varepsilon}]$. Hence, cycling is impossible, and the asymptotic simplex method converges in a finite number of iterations.

The series expansion (5.22) is obtained by substituting the Laurent expansion (5.10) into $x_{B^*}(\varepsilon) = A_{B^*}^{-1}(\varepsilon)(b^{(0)} + \varepsilon b^{(1)})$ and observing that $x_{B^*}(\varepsilon)$ cannot have a singular part because of Lemma 5.3 (proved in Subsection 5.2.9). The inequality $\varepsilon < 1/||D_2||$ is the standard convergence condition for a Neumann series. □

Once an a-optimal basis is found by the asymptotic simplex method, one may exactly calculate the optimal solution for any sufficiently small value of the perturbation parameter.

Corollary 5.1. *The following is an exact formula for the optimal solution of the perturbed linear program:*

$$x_{B^*}(\varepsilon) = x_{B^*}^{(0)} + \varepsilon[I-\varepsilon D_2]^{-1}x_{B^*}^{(1)}, \quad \varepsilon < \min\left\{\bar{\varepsilon}, \frac{1}{||D_2||}\right\}, \tag{5.23}$$

where $x_{B^*}^{(0)}$, $x_{B^*}^{(1)}$, *and* D_2 *are as in Theorem 5.1.*

Note that the above updating formula is computationally stable even in the case of singular perturbations, since one needs only to invert the matrix that is close to the identity.

Proposition 5.1. *The updating procedure for terms* $U^{(-1)}$ *and* $U^{(0)}$ *of the Laurent series expansion* (5.10) *requires* $O(\bar{s}m^2)$ *operations, where* $\bar{s}$ *is the maximal order of poles of the Laurent expansions for basis matrices.*

Proof: Note that for our updating procedure we need to compute $\bar{s}$ terms of the Laurent series (5.16). To calculate the Laurent series for $E(\varepsilon)$, we need to calculate m scalar Laurent expansions for elements of $\xi(\varepsilon)$. This can be done by applying the recursive formula (5.7). Since the computation of each scalar expansion requires $O(\bar{s}^2)$ flops, the computation of first $\bar{s}$ terms of Laurent series (5.16) requires $O(\bar{s}^2 m)$ operations. Note that since matrix $E^{(i)}$ has a special structure, the matrix multiplication $E^{(i)}U^{(j)}$ requires only $O(m^2)$ operations. Then the calculation of $U^{(-1)}$ and $U^{(0)}$ by formula (5.17) demands $O(\bar{s}m^2)$

flops. Consequently, the whole updating procedure requires $O(\bar{s}^2 m + \bar{s} m^2)$ operations, which is equivalent to $O(\bar{s} m^2)$ as $\bar{s} \le m$. □

In practice, since the case of $\bar{s} \ge 2$ is nongeneric, we expect that $\bar{s} \ll m$ in many applications. If $\bar{s} = 1$, which we expect will be a typical case, the complexities of basis updating in the asymptotic simplex method and in the standard revised simplex methods are comparable.

5.2.8 ▪ Computational aid and generalizations

Note that the main computational difficulties of the asymptotic simplex method arise in Steps 2 and 3 if the functions $r_j(\varepsilon)$ and $y_l(\varepsilon)$ are identically zero, namely, if $r_j(\varepsilon) = 0$, $y_l(\varepsilon) = 0$ for any ε. In this case we are forced to calculate all terms in corresponding expansions up to the $(m+1)$st term. Of course, we are interested in identifying such identically zero elements by an efficient method.

One simple heuristic solution for the above problem is proposed here. Note that if $r_j(\varepsilon) = 0$ or $y_l(\varepsilon) = 0$ for $\varepsilon \in (0, \bar{\varepsilon}]$, then these equalities hold for any $\varepsilon \in \mathbb{R}$. This fact follows from the observation that $r_j(\varepsilon)$ and $y_l(\varepsilon)$ are rational functions of ε and every rational function has either no zero or isolated zeros, or it is identically zero (see Problem 5.3). Therefore, we can detect elements that are identically zero not only in the neighborhood of $\varepsilon = 0$ but also at any point $\varepsilon \in \mathbb{R}$.

For instance, choose an ε^* such that the basis matrix $A_B(\varepsilon^*)$ is well conditioned. Then calculate $r_j(\varepsilon^*)$ and $y_l(\varepsilon^*)$ (now we can directly use formulae $r_N(\varepsilon^*) = \lambda(\varepsilon^*) A_N(\varepsilon^*) - c_N(\varepsilon^*)$ and $y(\varepsilon^*) = A_B^{-1}(\varepsilon^*) a_k(\varepsilon^*)$ instead of their expansions). If $r_j(\varepsilon^*) \ne 0$ and $y_l(\varepsilon^*) \ne 0$, then the functions are certainly not identically zero. If we obtain some zero elements, then we should add to ε^* arbitrary small variation and check whether this is an isolated zero or an identical zero. Of course, ε^* and its small variation (if necessary) can be chosen according to the features of the specific problem.

Next, we show how the asymptotic simplex method can be modified to solve problems more general than linear program (5.4), (5.5) under Assumptions 5.1 and 5.2.

We first point out that our method can be readily generalized to a linear program with polynomial perturbations. Namely, suppose that the coefficients $A(\varepsilon)$, $b(\varepsilon)$, and $c(\varepsilon)$ in the perturbed linear program are polynomials of ε. In particular, a basis matrix has the form

$$A_B(\varepsilon) = A_B^{(0)} + \varepsilon A_B^{(1)} + \cdots + \varepsilon^p A_B^{(p)}. \tag{5.24}$$

Clearly, the lexicographic entry and exit rules will need only some "technical" changes. Note that in the case of polynomial perturbations one needs at worst to check $(m+1)p$ terms of Laurent expansions for the entry rule and $2mp$ terms for the exit rule. The main difficulties that we face now are in the inversion of the polynomial matrix (5.24). It turns out that our methods for calculating the Laurent series

$$A_B^{-1}(\varepsilon) = \frac{1}{\varepsilon^s} U^{(-s)} + \frac{1}{\varepsilon^{s-1}} U^{(-s+1)} + \ldots \tag{5.25}$$

can be generalized for the case of polynomial perturbations. In particular, in Chapter 3 (see (3.53)) we showed that

$$U^{(k+1)} = -\sum_{i=0}^{p-1} \left(\sum_{j=0}^{p-1-i} U^{(-j)} A^{(j+i+1)} \right) U^{(k-i)}, \quad k = 0, 1, \ldots. \tag{5.26}$$

Thus, only some singular and the first regular terms of Laurent expansion (5.25) have to be obtained or updated. The other terms, if needed, can be computed in an efficient way by the recursive formula (5.26). Again, on the first iteration of the asymptotic simplex method one may use an analogue of the phase 1 method to obtain the initial Laurent expansion. For the following iterations one may use the generalized version of the updating algorithm that we introduced in Remark 5.2.

Note that Assumption 5.1 guarantees that an a-optimal solution can be expanded in Taylor series. We have introduced this assumption in order to restrict ourselves to the most common and interesting case where the a-optimal solution differs from the optimal solution of the unperturbed problem but both solutions are finite. In this case there exists a computationally stable updating formula (5.23). Of course, one can consider a perturbed linear program without this restriction. Then one will need to deal with the basic solutions in the form of general Laurent series with singular terms. Again, the asymptotic algorithm for that case would not be much different from that presented in Section 5.2.5.

5.2.9 ▪ Auxiliary results

Here we present several auxiliary theoretical results that help us develop the asymptotic simplex method.

Lemma 5.2. *Suppose $c(\varepsilon)=a(\varepsilon)/b(\varepsilon)$ is a rational function with the degrees of the polynomials $a(\varepsilon)$ and $b(\varepsilon)$ being m and n, respectively. Then the function $c(\varepsilon)$ can be expanded as a Laurent series,*

$$c(\varepsilon)=\frac{1}{\varepsilon^s}c^{(-s)}+\frac{1}{\varepsilon^{s-1}}c^{(-s+1)}+\dots,$$

in some punctured neighborhood of zero with the order of pole s that is at most n. Moreover, if $c^{(-s)}=c^{(-s+1)}=\cdots=c^{(m)}=0$, then $c(\varepsilon)\equiv 0$.

Proof: Since polynomials are analytic functions, the division of two polynomials is a meromorphic function. Next we show that the pole order of $c(\varepsilon)$ cannot be larger than n. Let us consider the equation

$$b(\varepsilon)c(\varepsilon)=a(\varepsilon),$$

$$(b^{(n)}\varepsilon^n+\cdots+b^{(0)})(c^{(-s)}\varepsilon^{-s}+c^{(-s+1)}\varepsilon^{-s+1}+\dots)=a^{(m)}\varepsilon^m+\cdots+a^{(0)}. \tag{5.27}$$

If we suppose that $s>n$, there are terms with negative powers of ε on the left-hand side of the above equation and no terms with negative powers of ε on the right-hand side. This leads to a contradiction, and hence the order of the pole s cannot exceed n. Finally, if $c^{(-s)}=\cdots=c^{(m)}=0$, equation (5.27) takes the form

$$\varepsilon^{m+1}(b^{(n)}\varepsilon^n+\cdots+b^{(0)})(c^{(m+1)}+c^{(m+2)}\varepsilon+\dots)=a^{(m)}\varepsilon^m+\cdots+a^{(0)}.$$

Collecting terms with the same powers of ε, we obtain $a^{(0)}=\cdots=a^{(m)}=0$, that is, $a(\varepsilon)\equiv 0$, and hence $c(\varepsilon)=a(\varepsilon)/b(\varepsilon)\equiv 0$. □

Lemma 5.3. *Let Assumption 5.1 hold. Then, any basic feasible solution of the perturbed program* (5.4), (5.5) *can be expanded as the Taylor series*

$$x_B(\varepsilon)=x_B^{(0)}+\varepsilon x_B^{(1)}+\dots$$

for $\varepsilon\in(0,\bar{\varepsilon}]$, where $\bar{\varepsilon}>0$ is sufficiently small.

Proof: Recall that any basic feasible solution can be given by the formula

$$x_B(\varepsilon) = A_B^{-1}(\varepsilon) b(\varepsilon). \tag{5.28}$$

According to Theorem 2.4 from Chapter 2, the inverse basis matrix $A_B^{-1}(\varepsilon)$ possesses, in general, a Laurent series expansion. Thus one can see from the formula (5.28) that $x_B(\varepsilon)$ possesses a Laurent series expansion in some punctured neighborhood of $\varepsilon = 0$ as well.

Now we shall show that the Laurent series for $x_B(\varepsilon)$ does not have a singular part. Suppose this is not the case, and some basic feasible solution has a Laurent series with a nontrivial singular part. The latter implies that there exists a sequence $\{\varepsilon_k\}_{k=0}^{\infty}$ such that $\varepsilon_k \to 0$ and $||x_B(\varepsilon_k)|| \to \infty$ as $k \to \infty$.

Next we define the following auxiliary sequence:

$$y_k := \frac{x_B(\varepsilon_k)}{||x_B(\varepsilon_k)||}.$$

Note that $||y_k|| = 1,\ k = 1, 2, \ldots$. According to the Bolzano–Weierstrass principle on bounded sequences in finite dimensional spaces, there exists a convergent subsequence $\{y_{k_l}\}$ with the limit $\bar{y}$ as $k_l \to \infty$, and $||\bar{y}|| = 1$.

From the definition of the basic feasible solution one obtains

$$A(\varepsilon_{k_l}) y_{k_l} = \frac{b(\varepsilon_{k_l})}{||x(\varepsilon_{k_l})||}$$

for each k_l. Now let k_l tend to the infinity to obtain

$$A^{(0)} \bar{y} = 0 \quad \text{and} \quad \bar{y} \geq 0.$$

Take any feasible solution of the unperturbed problem x_f such that

$$A^{(0)} x_f = b^{(0)}, \quad x_f \geq 0.$$

It is easy to see that $x_f + \lambda \bar{y}$ is also a feasible solution for any $\lambda \geq 0$. Since $||\bar{y}|| = 1$, the latter means that the original feasible region M_0 is unbounded, which is a contradiction to Assumption 5.1.

Thus, every basic feasible solution of the perturbed program can be expanded as a Taylor series. □

Remark 5.3. *Note that the first term $x_B^{(0)}$ of the Taylor expansion for $x_B(\varepsilon)$ might not be a* basic *feasible solution for the original program. This may occur in the case of singular perturbations.*

5.3 ▪ Asymptotic Gradient Projection Methods

5.3.1 ▪ Preliminaries

In this section we analyze a mathematical program with nonlinear strictly convex objective function and linear constraints that depend on a small perturbation parameter ε. Namely, consider the program (PR_ε),

$$\min_x f(x)$$

subject to

$$A(\varepsilon)x = b(\varepsilon), \tag{5.29}$$

where $x \in \mathbb{R}^n$, $f \in C^1$, $A(\varepsilon) \in \mathbb{R}^{m\times n}$ is an analytic matrix-valued function, $b(\varepsilon) \in \mathbb{R}^m$ is an analytic vector-valued function, and the level sets $L_c = \{x | f(x) \leq c\}$ are assumed to be compact (or empty) for every $c \in \mathbb{R}$. The corresponding *unperturbed* program (PR_0) is given by

$$\min_x f(x)$$

subject to

$$A(0)x = b(0). \tag{5.30}$$

Suppose that $x^{op}(\varepsilon)$ and $x^{op}(0)$ are optimal solutions of the perturbed and unperturbed problems, respectively. In particular, we are interested in the *singular perturbation* case, when $x^{op}(\varepsilon) \not\to x^{op}(0)$ as $\varepsilon \to 0$. This situation takes place when some of the perturbed constraints (5.29) become linearly dependent as the perturbation parameter goes to zero. Here we provide a modification of the gradient projection method to solve the auxiliary well-defined mathematical program. The key element of our method is the use of an asymptotic expansion of the projection operator onto the perturbed feasible set. As will be shown in the next section, this expansion can be readily calculated with the help of results from Chapter 2.

5.3.2 ▪ Perturbations of linear manifolds and projections

Singular perturbations occur when the rank of the perturbed constraint matrix $A(\varepsilon)$ is strictly greater than the rank of the unperturbed matrix $A(0)$. To investigate the behavior of the linear manifold described by (5.29) with respect to the perturbation parameter, let us consider an auxiliary quadratic program (AP)

$$\min (x - x_0)^T (x - x_0)$$

subject to

$$A(\varepsilon)x = b(\varepsilon),$$

where x_0 is some arbitrary point in $\mathbb{R}^n$. The above (convex) quadratic program has the Lagrangian function of the form

$$L(\lambda, x) = x^T x - 2x_0^T x + x_0^T x_0 + \lambda(A(\varepsilon)x - b(\varepsilon)), \tag{5.31}$$

where λ is an m-dimensional vector of Lagrange multipliers. Now the first order optimality condition $\nabla L = 0$ reduces to

$$2x - 2x_0 + A^T(\varepsilon)\lambda = 0. \tag{5.32}$$

Under the standard assumption that, for $\varepsilon > 0$ and sufficiently small, $A(\varepsilon)$ has full row rank, the inverse $(A(\varepsilon)A^T(\varepsilon))^{-1}$ exists, and premultiplication of (5.32) by $A(\varepsilon)$ leads to

$$\lambda = 2(A(\varepsilon)A^T(\varepsilon))^{-1}[-b(\varepsilon) + A(\varepsilon)x_0]. \tag{5.33}$$

Now substitution of (5.33) into (5.32) yields the following simple solution for (AP):

$$x(\varepsilon) = d(\varepsilon) + P(\varepsilon)x_0, \tag{5.34}$$

where

$$d(\varepsilon) = A^T(\varepsilon)(A(\varepsilon)A^T(\varepsilon))^{-1}b(\varepsilon) \tag{5.35}$$

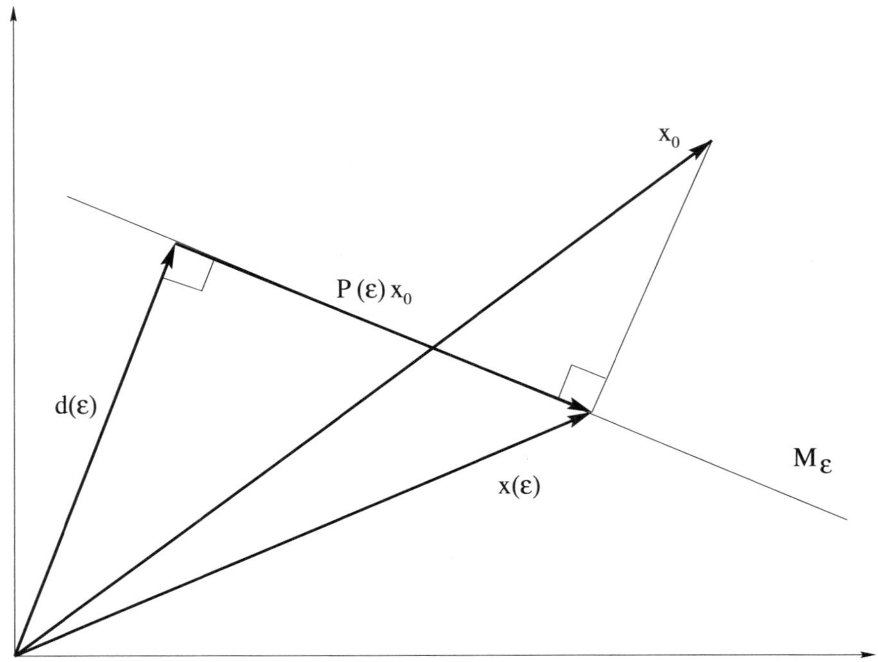

Figure 5.3. *Projection onto the perturbed manifold [8]*

is orthogonal to the perturbed linear manifold described by (5.29) and

$$P(\varepsilon) = I - A^T(\varepsilon)(A(\varepsilon)A^T(\varepsilon))^{-1}A(\varepsilon) \tag{5.36}$$

is a projection operator onto this manifold (see Figure 5.3). In what follows we will show that the perturbed projection $P(\varepsilon)$ always possesses a Taylor series expansion around $\varepsilon = 0$, even though the matrix $(A(\varepsilon)A^T(\varepsilon))^{-1}$ may have a singularity at this point. The perpendicular $d(\varepsilon)$ can be either finite or infinite when the perturbation parameter tends to zero. This motivates a further refinement.

Note that the unperturbed problem may be infeasible, even if the perturbed problem is feasible. In this case, the perturbed linear manifold defined by (5.29) moves to infinity (and $||d(\varepsilon)|| \to \infty$) when the perturbation parameter tends to zero. However, in the case of a feasible unperturbed problem it is possible that the perturbed manifold does not move away when $\varepsilon \to 0$; that is, $||d(\varepsilon)||$ is bounded when $\varepsilon \to 0$ (these two cases are demonstrated in Example 5.9).

Therefore, it is sufficient for our purposes to demand the boundedness of this quantity in some small neighborhood of $\varepsilon = 0$.

Now let us analyze the dependence of the projection matrix $P(\varepsilon)$ on the small parameter ε. First we need the following auxiliary lemma.

Lemma 5.4. *Let an orthogonal projection $Q(\varepsilon)$ depend on a parameter ε; then its Euclidean norm and its elements are uniformly bounded with respect to ε.*

Proof: The proof follows immediately from the fact that $||Q(\varepsilon)x||_2 \le ||x||_2$, because $Q(\varepsilon)$ is an orthogonal projection matrix. □

As we demonstrate in the next example, the above statement need not be true in general for a nonorthogonal projection.

Example 5.8. *Let us consider*

$$Q(\varepsilon)=\begin{bmatrix} 1/2 & 1/(2\varepsilon) \\ \varepsilon/2 & 1/2 \end{bmatrix}.$$

Since $Q^2=Q$, it is a projection. However, it is not an orthogonal projection $Q^T \neq Q$ and $q_{12}(\varepsilon)=1/(2\varepsilon)\to\infty$ as $\varepsilon\to 0$.

The matrix $P(\varepsilon)$ defined in (5.36) is an orthogonal projection (see Problem 5.14). Now we are able to formulate and prove the main result of this subsection.

Theorem 5.5. *The projection matrix $P(\varepsilon)$ defined in* (5.36) *possesses a Maclaurin series expansion at $\varepsilon=0$. Namely,*

$$P(\varepsilon)=P_0+\varepsilon P_1+\varepsilon^2 P_2+\dots$$

for $\varepsilon>0$ and sufficiently small.

Proof: The proof for the regular case follows immediately from the Neumann expansion of $[A(\varepsilon)A^T(\varepsilon)]^{-1}$ and is left to the reader to verify in Problem 5.15. Consequently we consider the more complicated singular case.

Since in the singular case the rows of the matrix $A(\varepsilon)$ become linearly dependent when the perturbation parameter tends to zero, the matrix $A(\varepsilon)A^T(\varepsilon)$ does not have a full rank when $\varepsilon=0$ (see Problem 5.13). However, for $\varepsilon>0$ and sufficiently small, $[A(\varepsilon)A^T(\varepsilon)]^{-1}$ exists and hence, by Theorem 2.4 of Chapter 2, possesses a Laurent series expansion in some neighborhood of $\varepsilon=0$. Namely,

$$[A(\varepsilon)A^T(\varepsilon)]^{-1}=\frac{1}{\varepsilon^s}C_{-s}+\frac{1}{\varepsilon^{s-1}}C_{-s+1}+\dots+\frac{1}{\varepsilon}C_{-1}+C_0+\varepsilon C_1+\dots \tag{5.37}$$

for $0<|\varepsilon|<\varepsilon_*$. This implies that the projection $P(\varepsilon)=I-A^T(\varepsilon)(A(\varepsilon)A^T(\varepsilon))^{-1}A(\varepsilon)$ can also be expanded as a Laurent series. However, $P(\varepsilon)$ is an orthogonal projection, and hence it is uniformly bounded for $0<|\varepsilon|<\varepsilon_*$. Consequently, the Laurent expansion for $P(\varepsilon)$ cannot have any terms with negative powers of ε; that is, $P(\varepsilon)$ possesses a Maclaurin series at $\varepsilon=0$. □

Example 5.9. *Let us consider the following linear constraints:*

$$(1-\varepsilon)x_1+x_2+x_3=1,$$

$$x_1+(1-\varepsilon^2)x_2+x_3=1.$$

Thus, we have

$$A(\varepsilon)=A_0+\varepsilon A_1+\varepsilon^2 A_2,\quad b=b_0=\begin{bmatrix}1\\1\end{bmatrix},$$

with

$$A_0=\begin{bmatrix}1&1&1\\1&1&1\end{bmatrix},\quad A_1=\begin{bmatrix}-1&0&0\\0&0&0\end{bmatrix},\quad A_2=\begin{bmatrix}0&0&0\\0&-1&0\end{bmatrix}.$$

First, we calculate $A(\varepsilon)A^T(\varepsilon)$ and its inverse

$$A(\varepsilon)A^T(\varepsilon)=\begin{bmatrix} 3-2\varepsilon+\varepsilon^2 & 3-\varepsilon-\varepsilon^2 \\ 3-\varepsilon-\varepsilon^2 & 3-2\varepsilon^2+\varepsilon^4 \end{bmatrix},$$

$$(A(\varepsilon)A^T(\varepsilon))^{-1}=\frac{1}{\varepsilon^2(2+2\varepsilon-2\varepsilon^3+\varepsilon^4)}\begin{bmatrix} 3-2\varepsilon^2+\varepsilon^4 & -3+\varepsilon+\varepsilon^2 \\ -3+\varepsilon+\varepsilon^2 & 3-2\varepsilon+\varepsilon^2 \end{bmatrix}$$

$$=\frac{1}{\varepsilon^2}\begin{bmatrix} 1.5 & -1.5 \\ -1.5 & 1.53 \end{bmatrix}+\frac{1}{\varepsilon}\begin{bmatrix} -1.5 & 2 \\ 2 & -2.5 \end{bmatrix}+\begin{bmatrix} 0.5 & -1.5 \\ -1.5 & 3 \end{bmatrix}+\varepsilon\begin{bmatrix} 1 & 0 \\ 0 & -1.5 \end{bmatrix}.$$

Using the above expansion, we may also expand $A^T(\varepsilon)(A(\varepsilon)A^T(\varepsilon))^{-1}$ and $A^T(\varepsilon)(A(\varepsilon)\cdot A^T(\varepsilon))^{-1}A(\varepsilon)$:

$$A^T(\varepsilon)(A(\varepsilon)A^T(\varepsilon))^{-1}=\frac{1}{\varepsilon}\begin{bmatrix} -1 & 1 \\ 0.5 & -0.5 \\ 0.5 & -0.5 \end{bmatrix}+\begin{bmatrix} 0.5 & -0.5 \\ 0.5 & 0 \\ -1 & 1.5 \end{bmatrix}+\varepsilon\begin{bmatrix} 0.5 & 0 \\ -1 & 1 \\ 1 & -1.5 \end{bmatrix}+\ldots,$$

$$A^T(\varepsilon)(A(\varepsilon)A^T(\varepsilon))^{-1}A(\varepsilon)=\begin{bmatrix} 1 & 0 & 0 \\ 0 & 0.5 & 0.5 \\ 0 & 0.5 & 0.5 \end{bmatrix}+\varepsilon\begin{bmatrix} 0 & -0.5 & 0.5 \\ -0.5 & 0.5 & 0 \\ 0.5 & 0 & -0.5 \end{bmatrix}+\ldots.$$

Note how the singularity is subsequently reduced. Thus, the perturbed projection is given by

$$\begin{aligned} P(\varepsilon)&=I-A^T(\varepsilon)(A(\varepsilon)A^T(\varepsilon))^{-1}A(\varepsilon) \\ &=P_0+\varepsilon P_1+\ldots \\ &=\begin{bmatrix} 0 & 0 & 0 \\ 0 & 0.5 & -0.5 \\ 0 & -0.5 & 0.5 \end{bmatrix}+\varepsilon\begin{bmatrix} 0 & 0.5 & -0.5 \\ 0.5 & -0.5 & 0 \\ -0.5 & 0 & 0.5 \end{bmatrix}+\ldots. \end{aligned}$$

In particular, we note that

$$P(0)=I-\begin{bmatrix} 1 \\ 1 \\ 1 \end{bmatrix}\frac{1}{3}\begin{bmatrix} 1 & 1 & 1 \end{bmatrix}=\begin{bmatrix} 2/3 & -1/3 & -1/3 \\ -1/3 & 2/3 & -1/3 \\ -1/3 & -1/3 & 2/3 \end{bmatrix}\neq P_0.$$

Let us next use (5.35) *to calculate the orthogonal vector to the perturbed manifold*

$$d(\varepsilon)=\begin{bmatrix} 0 \\ 0.5 \\ 0.5 \end{bmatrix}+\varepsilon\begin{bmatrix} 0.5 \\ 0 \\ -0.5 \end{bmatrix}+\ldots.$$

Now, if we take

$$b(\varepsilon)=b_0=\begin{bmatrix} 2 \\ 1 \end{bmatrix},$$

the unperturbed constraints will be infeasible, and the norm of

$$d(\varepsilon)=\frac{1}{\varepsilon}\begin{bmatrix} -1 \\ 0.5 \\ 0.5 \end{bmatrix}+\begin{bmatrix} 0.5 \\ 1 \\ -0.5 \end{bmatrix}+\varepsilon\begin{bmatrix} 1 \\ -1 \\ -0.5 \end{bmatrix}+\ldots$$

goes to infinity as ε goes to zero.

Even though by using results from Chapter 2 one can obtain Taylor series for $P(\varepsilon)$ at $\varepsilon = 0$, in the general case of analytic perturbations $A(\varepsilon) = A_0 + \varepsilon A_1 + \varepsilon^2 A_2 + \dots$, $b(\varepsilon) = b_0 + \varepsilon b_1 + \varepsilon^2 b_2 + \dots$ the calculations become much easier and transparent in the case of linear perturbations.

Linear perturbation: Here we assume that $A(\varepsilon) = A_0 + \varepsilon A_1$ and $b(\varepsilon) = b_0 + \varepsilon b_1$. This implies that we need to obtain the Laurent series (5.37) for the inverse of quadratic perturbation $A_0 A_0^T + \varepsilon(A_0 A_1^T + A_1 A_0^T) + \varepsilon^2 A_1 A_1^T$. Recall that if we have in hand the coefficients C_{-1}, C_0, and C_1, then the other coefficients of the regular part of (5.37) can be efficiently computed by the recursive formula (3.52) from Section 3.3. In this particular setting, the formula (3.52) has the form

$$C_{k+1} = -[C_0(A_0 A_1^T + A_1 A_0^T) + C_{-1} A_1 A_1^T] C_k - [C_0 A_1 A_1^T] C_{k-1} \tag{5.38}$$

for $k = 1, 2, \dots$. The coefficients C_{-1}, C_0, and C_1 can be computed by any method of Section 2.2. For instance, in the generic case of the pole of order one, the basic generalized inverse method gives

$$C_{-1} = G_{01},$$

$$C_0 = G_{00}[I - (A_0^T A_1^T + A_1 A_0^T) C_{-1}] + G_{01}[-A_1 A_1^T C_{-1}],$$

$$C_1 = G_{00}[-(A_0^T A_1^T + A_1 A_0^T) C_0 - A_1 A_1^T C_{-1}] + G_{01}[-A_1 A_1^T C_0],$$

where $G_{ij} \in \mathbb{R}^{m \times m}$, $i, j = 0, 1$, are the blocks of the generalized inverse

$$\mathcal{G} = \begin{bmatrix} G_{00} & G_{01} \\ G_{10} & G_{11} \end{bmatrix} = \begin{bmatrix} A_0 A_0^T & 0 \\ A_0^T A_1^T + A_1 A_0^T & A_0 A_0^T \end{bmatrix}^{\dagger}.$$

Next, upon substituting the Laurent series (5.37) into (5.36) and equating coefficients of like powers of ε, we obtain the power series for the projection matrix

$$P(\varepsilon) = \sum_{k=0}^{\infty} \varepsilon^k P_k, \tag{5.39}$$

with

$$P_k = \delta_{0k} I - A_0^T C_k A_0 - A_1^T C_{k-1} A_0 - A_0^T C_{k-1} A_1,$$

where $k = 0, 1, \dots$ and δ_{0k} is the Kroneker delta. In what follows we will also need a Laurent series expansion for $d(\varepsilon)$. Again, upon the substitution of (5.37) into (5.35) and equating coefficients with the same powers of ε, we obtain the Laurent series

$$d(\varepsilon) = \sum_{k=-s}^{\infty} \varepsilon^k d_k, \tag{5.40}$$

where

$$d_{-s} = A_0^T C_{-s} b_0$$

and

$$d_k = A_0^T C_k b_0 + A_0^T C_{k-1} b_1 + A_1^T C_{k-1} b_0$$

for $k = -s + 1, -s + 2, \dots$.

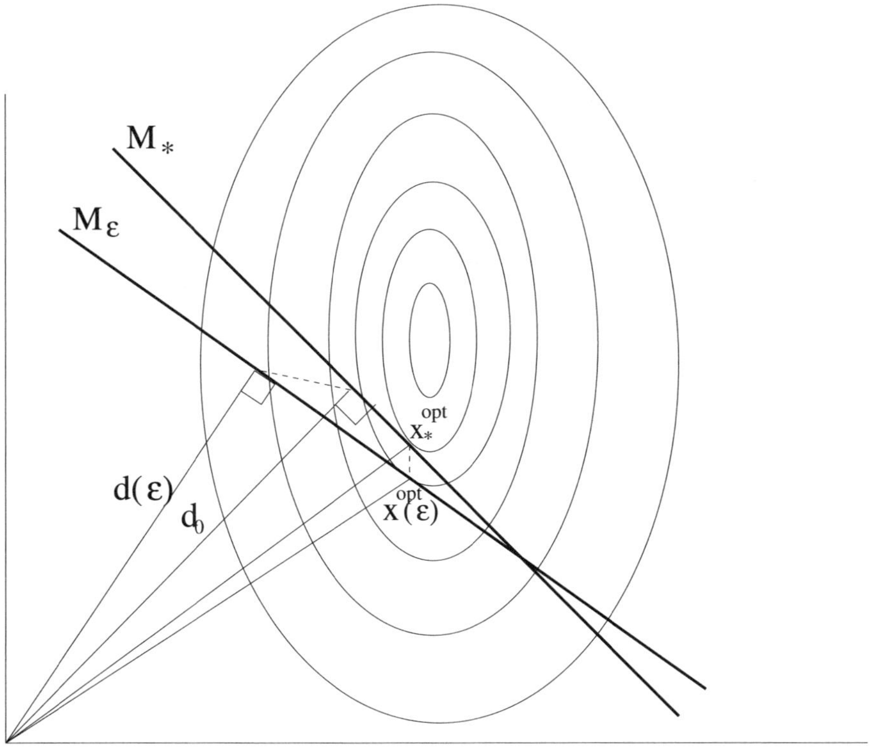

Figure 5.4. *The limiting manifold and the auxiliary mathematical program [8]*

5.3.3 ▪ Application of the gradient projection method to the limiting mathematical program

Let M_ε denote the perturbed linear manifold described by the linear system of constraints (5.29), let M_0 denote the linear manifold corresponding to unperturbed constraints (5.30), and finally let M_* denote (when it exists) a *limiting linear manifold*, that is, the limit of set M_ε as ε goes to zero (see Figure 5.4). More precisely, this means that $d(\varepsilon) \to d_0$ and $P(\varepsilon) \to P_0$ as $\varepsilon \to 0$. We are interested in the case where the distance from the origin to the limiting manifold M_* is finite. Equivalently, this is the case when the power series expansion (5.40) for $d(\varepsilon)$ does not have any terms with negative powers of ε, that is,

$$d(\varepsilon) = d_0 + \varepsilon d_1 + \varepsilon^2 d_2 + \dots.$$

Note that in this case $||d_0|| = \text{dist}\{0, M_*\}$ and P_0 (the first term of (5.39)) is an orthogonal projection operator onto M_* (see Problem 5.16). In fact, M_* is uniquely characterized by d_0 and P_0; that is, any vector y from M_* can be written in the form

$$y = d_0 + P_0 x$$

for some $x \in \mathbb{R}^n$.

Let us now briefly review the well-known gradient projection method with linear equality constraints. Suppose we want to find an optimal solution to the following mathematical program:

$$\min_x f(x)$$

subject to

$$Ax = b.$$

Here we assume that $f(x)$ is strictly convex with compact level sets. Then, it is known that a unique optimal solution exists, and it can be found by the iterative *gradient projection method*. First we construct the projection matrix $P = I - A^T(AA^T)^{-1}A$ onto the feasible region and find any feasible solution x_0. Then, the gradient projection method is performed according to the iteration

$$x_{k+1} = x_k - \alpha_k P g_k, \tag{5.41}$$

where $g_k := \nabla f(x_k)$ and $\alpha_k := \arg\min_\alpha \{f(x_k - \alpha P g_k)\}$. The Lagrangian function corresponding to the above convex program is

$$L(\lambda, x) = f(x) - \lambda^T(Ax - b),$$

and the necessary and sufficient condition, $\nabla L = 0$, for optimality takes the form

$$\nabla f(x) - \lambda^T A = 0.$$

By an argument analogous to that used to derive (5.33) we can check that $\lambda = (AA^T)^{-1}A\nabla f(x)$ and hence that the necessary and sufficient optimality condition takes the form

$$[I - A^T(AA^T)^{-1}A]\nabla f(x) = P\nabla f(x) = 0. \tag{5.42}$$

Now suppose that we need to solve the perturbed mathematical program (PR_ε) when the perturbation is singular. If one tries to apply the above gradient projection method directly to (PR_ε), one will face at least two problems. The first problem is that the matrix $A(\varepsilon)A^T(\varepsilon)$ becomes ill-conditioned when ε is close to zero. This will lead to an incorrect computation of the projection matrix $P(\varepsilon)$. The latter in turn implies that the sequence $\{x_k\}$ generated by the gradient projection method may leave the feasible set. The second problem is that, since ε may not be known exactly or may even be unknown, it is very difficult to find the first feasible solution $x_0 \in M_\varepsilon$. In other words, the perturbed feasible set described by (5.29) could be very sensitive to changes in ε in the case of singular perturbations.

Example 5.10. *For instance, let the precision of our calculations be* 10^{-3}*. Suppose we want to find a feasible vector with minimal length for the constraints of Example 5.9 for* $\varepsilon = 0.01$*. If a numerical error occurs in the first element of* $b(\varepsilon)$*, that is, instead of the vector* $b(\varepsilon) = [1\ 1]^T$ *we consider the vector* $b(\varepsilon) = [1.001\ 1]^T$*, and we use directly the formula* (5.35) *for the calculation of* $d(\varepsilon)$*, we obtain*

$$d(0.01) = \begin{bmatrix} -0.0944 \\ 0.5504 \\ 0.5441 \end{bmatrix}.$$

The above vector has about 10% *error in the Euclidean norm with respect to the reference vector* $d_0 = [0\ 0.5\ 0.5]^T$*. However, from the original optimization problem formulation, we might know that the solution should be finite. Hence, we use only the regular part of the Laurent series expansion*

$$d(\varepsilon) = \frac{1}{\varepsilon}\begin{bmatrix} -0.001 \\ 0.0005 \\ 0.0005 \end{bmatrix} + \begin{bmatrix} 0.0005 \\ 0.5005 \\ 0.499 \end{bmatrix} + \dots.$$

Despite the fact that the terms of the above series have also been calculated with the error, the first regular term

$$\tilde{d}_0 = \begin{bmatrix} 0.0005 \\ 0.5005 \\ 0.499 \end{bmatrix}$$

produces an answer with only 0.05% *error in norm.*

As a solution to the outlined problems, we propose to substitute the ill-posed program (PR_ε) with a specially constructed auxiliary *limiting mathematical program* (PR_*):

$$\min_x \{f(x) | x \in M_*\},$$

where M_* is a limiting feasible set of (5.29). Equivalently, (PR_*) can be thought of as the problem

$$\min_x f(x)$$

subject to

$$x = d_0 + P_0 z,$$
$$z \in \mathbb{R}^n.$$

Of course, the above is equivalent to the unconstrained problem

$$\min_z f(d_0 + P_0 z),$$

and hence the necessary and sufficient optimality condition for (PR_*) is

$$P_0 \nabla f(x) = 0.$$

Now we can solve the above limiting program by the gradient projection method. As an initial feasible solution one may take $x_0 = d_0$. Then, the iterative procedure takes the form

$$x_{k+1} = x_k - \alpha_k P_0 g_k, \tag{5.43}$$

where P_0 is the first term in (5.39). As a result, we obtain an approximation to the optimal solution x_*^{op} of the auxiliary limiting program (PR_*), which we shall show is close to the optimal solution of the perturbed problem for small values of the perturbation parameter. Namely, we can state the following result that has a geometric interpretation illustrated in Figure 5.4.

Theorem 5.6. *Suppose the distance from the origin to the limiting manifold M_* is finite and f is strictly convex with compact level sets. Then, the optimal solution $x^{op}(\varepsilon)$ of the perturbed mathematical program (PR_ε) converges to the optimal solution x_*^{op} of the limiting program (PR_*) as ε tends to zero.*

Proof: Let us consider the optimality equations for the perturbed program:

$$\nabla f(x^{op}) + \lambda^T A(\varepsilon) = 0, \tag{5.44}$$

$$A(\varepsilon) x^{op} = b(\varepsilon). \tag{5.45}$$

In the case of regular perturbations, that is, when $A(\varepsilon)$ does not change the rank at $\varepsilon = 0$, we can apply the implicit function theorem to show that $x^{op}(\varepsilon)$ is continuous at $\varepsilon = 0$ and converges to the optimal solution of (PR_0) (and hence to the optimal solution of (PR_*), which in this case coincides with (PR_0)) as $\varepsilon \to 0$.

The case of singular perturbations requires a more detailed argument. Let us choose $\bar{\varepsilon} > 0$ such that $A(\varepsilon)$ has a constant rank on the interval $(0, \bar{\varepsilon}]$. The latter is always possible in the finite dimensional case (see Theorem 3.1 for a similar statement). For any $0 < \underline{\varepsilon} < \bar{\varepsilon}$ the optimal solution of the perturbed problem $x^{op}(\varepsilon)$ is continuous on the closed interval $[\underline{\varepsilon}, \bar{\varepsilon}]$. The justification of this continuity is the same as in the preceding case of regular perturbations.

Now let us prove, by contradiction, that $x^{op}(\varepsilon)$ is bounded at $\varepsilon = 0$. Note that premultiplication (5.34) by $A(\varepsilon)$ yields $A(\varepsilon)d(\varepsilon) = b(\varepsilon)$; thus $d(\varepsilon)$ is feasible for (PR_ε). Since $f(x)$ is strictly convex with compact level sets and $d(\varepsilon) \to d_0$ as $\varepsilon \to 0$, there exists a constant c such that $d(\varepsilon)$ belongs to the set $L_c = \{x \in \mathbb{R}^n | f(x) \leq c\}$ for $\varepsilon \in [0, \bar{\varepsilon}]$. Suppose, on the contrary, that $||x^{op}(\varepsilon)|| \to \infty$ as $\varepsilon \to 0$. Then, there exists some ε' for which $f(x^{op}(\varepsilon')) > c$. On the other hand, $f(d(\varepsilon')) \leq c$, since $d(\varepsilon') \in L_c$. Consequently, $f(x^{op}(\varepsilon')) > f(d(\varepsilon'))$, contradicting the optimality of $x^{op}(\varepsilon')$. Hence $x^{op}(\varepsilon)$ is bounded on $[0, \bar{\varepsilon}]$.

Next we show that in fact $x^{op}(\varepsilon)$ has a finite limit, say, x_*, as ε tends to zero. Suppose, on the contrary, that there is no limit; then, since $x^{op}(\varepsilon)$ is bounded on $[0, \bar{\varepsilon}]$, there exist at least two sequences $\{\varepsilon'_k, k = 0, 1, \ldots | \varepsilon'_k \to 0\}$ and $\{\varepsilon''_k, k = 0, 1, \ldots | \varepsilon''_k \to 0\}$ such that $x^{op}(\varepsilon'_k) \to x'_* \in M_*$, $x^{op}(\varepsilon''_k) \to x''_* \in M_*$ as $k \to \infty$, and $x'_* \neq x''_*$. Since $x^{op}(\varepsilon'_k)$ and $x^{op}(\varepsilon''_k)$ are optimal solutions of the perturbed problems $(PR_{\varepsilon'_k})$ and $(PR_{\varepsilon''_k})$, respectively, by (5.42) we write the following optimality conditions for all k:

$$P(\varepsilon'_k)\nabla f(x^{op}(\varepsilon'_k)) = 0, \quad P(\varepsilon''_k)\nabla f(x^{op}(\varepsilon''_k)) = 0.$$

Recall that $P(\varepsilon) \to P_0$ as $\varepsilon \to 0$ and $f \in C^1$. Next we let k approach infinity. By continuity of the gradient, we obtain

$$P_0 \nabla f(x'_*) = 0, \quad P_0 \nabla f(x''_*) = 0,$$

with $x'_*, x''_* \in M_*$ and $x'_* \neq x''_*$. This contradicts the fact that the limiting mathematical program has a unique solution. The latter must hold as the objective function is strictly convex with compact level sets and the limiting constraints form a linear manifold. Consequently, the limit $x_* = \lim_{\varepsilon \to 0} x^{op}(\varepsilon)$ exists. Moreover, from the optimality condition $P_0 \nabla f(x_*) = 0$ we conclude that it is an optimal solution to (PR_*). This completes the proof. □

5.4 ▪ Asymptotic Analysis for General Nonlinear Programming: Complex Analytic Perspective

In this section we return to a more general perturbed mathematical program

$$\min f(\varepsilon, x) \tag{5.46}$$

subject to

$$\begin{aligned} h_i(\varepsilon, x) &= 0; & i &= 1, 2, \ldots, p \\ g_j(\varepsilon, x) &\leq 0; & j &= 1, 2, \ldots, m, \end{aligned}$$

where all functions may now depend on the perturbation parameter ε in addition to the original decision variables $x_1, x_2, \ldots, x_n$. Our analysis will be based on a belief that an essential understanding of the asymptotic behavior of the solutions as $\varepsilon \downarrow 0$ can be gained from determining what type of functions $x_k(\varepsilon)$'s are for each $k = 1,2,\dots,n$ and that this applies to both regular and singular perturbations.

Of course, an explicit functional form cannot be hoped for at the level of generality considered below. Consequently, if it were possible to characterize $x_k(\varepsilon)$'s in terms of series expansion in appropriate powers of ε, that would already provide a lot of insight into the asymptotic behavior of solutions as $\varepsilon \downarrow 0$.

We shall claim that Puiseux series are the natural mathematical objects to use in this context, and we first observe that the class of Puiseux series

$$G(\varepsilon) = \sum_{\nu=K}^{\infty} c_\nu \varepsilon^{\frac{\nu}{M}},$$

where M is a positive integer and K is an arbitrary (fixed) integer, includes both Laurent and power series.

In fact, the perturbed mathematical program introduced in the previous section can be viewed as a special case of a slightly more general problem:

$$\min_{x} f(\varepsilon, x) \tag{5.47}$$

subject to

$$(\varepsilon, x) \in \Omega \subset \mathbb{R}^{n+1},$$

where the feasible region Ω is viewed as a subset of $\mathbb{R}^{n+1}$ rather than $\mathbb{R}^n$ because of the inclusion of the perturbation parameter ε, even though the minimization is with respect to x only. Since the objective is to characterize solutions x of (5.47) as functions of ε and since this may involve solving simultaneous equations of a finite number of nonlinear functions, it is reasonable to expect that the complex space $\mathbb{C}^{n+1}$ may be the natural space to work in. Of course, at the end of the analysis, we shall consider the intersection of the solution sets with $\mathbb{R}^{n+1}$.

Toward this end we assume that, in $\mathbb{C}^{n+1}$, the most general "feasible region" that we shall consider will be a *complex analytic variety* $W \subset \mathscr{W}$, where $\mathscr{W}$ is some open set in $\mathbb{C}^{n+1}$. Recall (see also Bibliographic Notes) that W is an analytic variety in $\mathscr{W}$ if for each $p \in W$ there exists a neighborhood U of p and holomorphic functions $\theta_1, \theta_2, \dots, \theta_s$ such that $\theta_i(z) = 0$ for all $z \in W \cap U$ and $i = 1,2,\dots,s$, and W is closed in $\mathscr{W}$.

We begin by fixing some analytic variety W that we shall view as the extension of the feasible region Ω into $\mathbb{C}^{n+1}$. That is, W contains all the points (η, z) of interest and defines $\Omega = W \cap \mathbb{R}^{n+1}$. We adopt the convention that points in Ω will be denoted by (ε, x) rather than (η, z) whenever it is necessary to emphasize that they are real-valued. Similarly, we define $W_\eta = \{z \in \mathbb{C}^n \mid (\eta, z) \in W\}$ when $\eta \in \mathbb{C}$, $W_\varepsilon = \{z \in \mathbb{C}^n \mid (\varepsilon, z) \in W\}$ when $\varepsilon \in \mathbb{R}$, and $W_\varepsilon \cap \mathbb{R}^n = \{x \in \mathbb{R}^n \mid (\varepsilon + 0i, x_1 + 0i, \dots, x_n + 0i) \in W\}$. Finally, we postulate that our objective function in (5.47) derives from a holomorphic function $f : \mathscr{W} \to \mathbb{C}$ such that

$$f(\Omega) \subset \mathbb{R}.$$

We may now define the minimization problem (5.47) as a minimization problem with respect to the analytic variety W. That is,

$$\min_{x} f(\varepsilon, x)$$

subject to

$$x \in W_\varepsilon \cap \mathbb{R}^n \tag{5.48}$$

for any $\varepsilon \in \mathbb{R}$ such that $W_\varepsilon \cap \mathbb{R}^n \neq \phi$.

It is now possible to define the *solution set* of (5.48) for any $\varepsilon > 0$ as $S_\varepsilon = \{x \in W_\epsilon \cap \mathbb{R}^n \mid x \text{ attains the minimum in (5.48)}\}$ and the corresponding set in $\mathbb{R}^{n+1}$, namely, $\mathsf{S}_\varepsilon = \{(\varepsilon, x) \in \Omega \mid x \in S_\varepsilon\}$.

Next we introduce the field of Puiseux series with real coefficients. The elements of this field are functions $G(\varepsilon)$ of the form

$$G(\varepsilon) = \sum_{k=K}^{\infty} c_k \varepsilon^{\frac{k}{M}}, \tag{5.49}$$

where K is some integer, M is a positive integer, and the real coefficients $\{c_k\}_{k=K}^{\infty}$ are such that the above series converges for all ε sufficiently small. Of course, c_k's and hence $G(\varepsilon)$ can be vector-valued.

Our goal is to establish that, under weak conditions, there exists a Puiseux series $G(\varepsilon)$ such that

$$x(\varepsilon) = G(\varepsilon) \in S_\varepsilon \tag{5.50}$$

for all $\varepsilon > 0$ and sufficiently small. In the remainder of this section we introduce some of the notation that will be used later on.

For any holomorphic function $g : \mathcal{W} \to \mathbb{C}$ we define the *gradient* of $g(\eta, z)$ at $z = (z_1, z_2, \dots, z_n)$ such that $(\eta, z) \in \mathcal{W}$ by

$$\nabla g(\eta, z) = \left(\frac{\partial g}{\partial z_1}, \frac{\partial g}{\partial z_2}, \dots, \frac{\partial g}{\partial z_n} \right),$$

where $\frac{\partial g}{\partial z_i}$ is evaluated at (η, z). Similarly, the *Hessian matrix* of $g(\eta, z)$ at z is defined by

$$\nabla^2 g(\eta, z) = \left(\frac{\partial^2 g(\eta, z)}{\partial z_i \partial z_j} \right)_{i,j=1}^{n,n}.$$

If $v, v' \in \mathbb{C}^m$, then $v.v'$ is the *holomorphic inner product* of v and v', that is, the plain inner product which does not involve conjugation. Finally, if $E \subset \mathbb{C}^m$, the *orthogonal complement* of E is given by

$$E^\perp = \{v \in \mathbb{C}^m \mid e.v = 0 \ \forall e \in E\}.$$

5.4.1 ▪ Minimization under constraints

We now return to the original mathematical programming problem (5.46), but with the simplification that there are only p equality constraints: $h_i(\varepsilon, x) = 0;\ i = 1, 2, \dots, p$. We shall return to the case of both equality and inequality constraints at the end of this section. To cast the problem in our setting we assume that $\mathcal{W}$ is an open set in $\mathbb{C}^{n+1}$ and $h_1, h_2, \dots, h_p, f$ are all holomorphic functions mapping $\mathcal{W} \to \mathbb{C}$ such that $\mathcal{W} \cap \mathbb{R}^{n+1}$ is mapped by these functions into $\mathbb{R}$. We consider the perturbed minimization problem:

$$\min f(\varepsilon, x)$$

subject to

$$h_i(\varepsilon, x) = 0;\ i = 1, 2, \dots, p. \tag{5.51}$$

Let $h = (h_1, \dots, h_p)$; $\mathcal{W} \to \mathbb{C}^p$, and define the set

$$W = h^{-1}(0, \dots, 0) = \{(\eta, z) \mid h_i(\eta, z) = 0;\ i = 1, 2, \dots, p\}.$$

Clearly, as the zero set of p holomorphic functions, W is a complex analytic variety. For a fixed η, let

$$\nabla h_i(\eta,z)=\left(\frac{\partial h_i}{\partial z_1}(\eta,z),\ldots,\frac{\partial h_i}{\partial z_n}(\eta,z)\right)$$

for all z such that $(\eta,z)\in W$ and $i=1,2,\ldots,p$. Let $\Gamma(\eta,z)$ be the subspace of $\mathbb{C}^n$ spanned by $\nabla h_i(\eta,z)$ for $i=1,2,\ldots,p$. We are now ready to generalize a standard "second order optimality condition" to this new setting.

Definition 5.4. *We shall say that a point $(\varepsilon,x)\in \mathcal{W}\cap\mathbb{R}^{n+1}$ satisfies optimality conditions of the second order (or is a strict* stationary point*) if*

(i) *the gradients of the constraints are independent, that is,* $\dim\Gamma(\varepsilon,x)=p$,

(ii) $\nabla f(\varepsilon,x)\in\Gamma(\varepsilon,x)$, *that is, there exist Lagrange multipliers (dependent on ε) $\lambda_1,\lambda_2,\ldots,\lambda_p\in\mathbb{R}$, not all zero, such that*

$$\sum_{i=1}^{p}\lambda_i\nabla h_i(\varepsilon,x)+\nabla f(\varepsilon,x)=0,$$

(iii) *the Hessian $L(\varepsilon,x,\lambda)$ of the Lagrangian of* (5.51) *is positive definite on $\Gamma^{\perp}(\varepsilon,x)$, that is,*

$$L(\varepsilon,x,\lambda)=\sum_{i=1}^{p}\lambda_i\nabla^2 h_i(\varepsilon,x)+\nabla^2 f(\varepsilon,x)$$

is a positive definite matrix.

Note that conditions (i)–(iii) are analogous to the standard 2nd order necessary conditions for a strict local minimum. Let $\mathcal{S}$ denote the set strict stationary points in $\mathcal{W}\cap\mathbb{R}^{n+1}$, and let $\bar{\mathcal{S}}$ be the closure of $\mathcal{S}$.

Motivated by the Karush–Kuhn–Tucker-type condition (ii), we shall now consider the subset of the feasible region W defined by

$$W_1=\left\{(\varepsilon,x)\in W\mid \operatorname{rank}\left[\nabla h_1(\varepsilon,x),\ldots,\nabla h_p(\varepsilon,x),\nabla f(\varepsilon,x)\right]\le p\right\},$$

where $[\nabla h_1(\cdot),\ldots,\nabla h_p(\cdot),\nabla f(\cdot)]$ is an $n\times(p+1)$ matrix whose columns are the above gradient vectors. Since the rank condition defining W_1 consists of certain determinants being equal to zero, W_1 is clearly a complex analytic variety. Furthermore, since (ii) holds at any $(\varepsilon,x)\in\mathcal{S}$, we have that $\mathcal{S}\subset W_1$.

Lemma 5.7. *Let $\mathcal{V}\subset\mathcal{W}$ be the open set of points (η,z) satisfying the independent gradient condition* (i)*. Suppose, in addition, that $(\eta,z)\in\mathcal{V}\cap W_1$. There exists a unique set of holomorphic functions: $\mathcal{V}\to\mathbb{C}$ such that $\lambda_i=\lambda_i(\eta,z)$, $i=1,\ldots,p$, are the unique Lagrange multipliers satisfying*

$$\sum_{i=1}^{p}\lambda_i\nabla h_i(\eta,z)+\nabla f(\eta,z)=0 \tag{5.52}$$

for $(\eta,z)\in\mathcal{V}\cap W_1$.

Proof: Multiplying (5.52) by the transpose of $\nabla h_j(\eta, z)$ for $j = 1,2,\ldots,p$ yields a set of p equations

$$\sum_{i=1}^{p}\left[\nabla h_j(\eta,z)\cdot\nabla h_i(\eta,z)\right]\lambda_i = -\nabla h_j(\eta,z)\cdot\nabla f(\eta,z)$$

for $j = 1,2,\ldots,p$. We can think of the above system of equations, with the argument (η, z) suppressed, as simply the linear system

$$A\lambda = b,$$

where the (i,j)th element of A is $a_{ij} = \nabla h_j(\eta,z).\nabla h_i(\eta,z)$ for $i,j = 1,2,\ldots,p$, and $b_i = -\nabla h_i(\eta,z)\nabla f(\eta,z)$ for $i = 1,2,\ldots,p$.

It is now easy to check that the independent gradient condition (i) implies that A is nonsingular. Hence $\lambda = A^{-1}b$ defines the unique set Lagrange multiplier solutions $\lambda_i(\eta,z)$, $i = 1.2,\ldots,p$, satisfying (ii). Clearly, these functions are holomorphic. □

Theorem 5.8. *The complex analytic variety W_1 is one dimensional near any $(\varepsilon,x)\in\mathcal{S}$.*

Proof: Consider a holomorphic function $F : \mathcal{W}\times\mathbb{C}^p \to \mathbb{C}^{p+n}$ defined by

$$F(\eta,z,\lambda) = \left(h_1(\eta,z),\ldots,h_p(\eta,z),\sum_{i=1}^{p}\lambda_i\nabla h_i(\eta,z)+\nabla f(\eta,z)\right),$$

where $z = (z_1,\ldots,z_n)$ and $\lambda = (\lambda_1,\ldots,\lambda_p)$.

Note that the zero set of F, namely,

$$W_2 := F^{-1}(0),$$

is a complex analytic variety in $\mathcal{W}\times\mathbb{C}^p$. Let $A(\eta,z) = (\nabla h_1(\eta,z),\ldots,\nabla h_p(\eta,z))$ be the $n\times p$ matrix of gradients of h_i's, and let $L(\eta,z,\lambda) = \sum_{i=1}^{p}\lambda_i\nabla^2 h_i(\eta,z)+\nabla^2 f(\eta,z)$ be the Hessian of the Lagrangian as in (iii). Hence for $(\eta,z,\lambda)\in W_2$ the Jacobian of F with respect to (z,λ) is given by the $(p+n)\times(p+n)$ matrix

$$\frac{\partial F}{\partial(z,\lambda)} = \begin{pmatrix} A^T(\eta,z) & 0 \\ L(\eta,z,\lambda) & A(\eta,z)\end{pmatrix}.$$

We claim that $\frac{\partial F}{\partial(z,\lambda)}$ is nonsingular at $(\eta,z,\lambda) = (\varepsilon,x,\lambda)$, satisfying (i), (ii), and (iii). To verify this suppose that there exists (u,v) (not equal to 0) such that $\left[\frac{\partial F}{\partial(z,\lambda)}\right](u,v)^T = 0$. That is,

$$A^T(\varepsilon,x)u^T = 0,$$
$$L(\varepsilon,x,\lambda)u^T + A(\varepsilon,x)v^T = 0.$$

However, the first of the above equations implies that $uA(\varepsilon,x) = 0$, so multiplying the second equation by u, on the left, yields

$$uL(\varepsilon,x,\lambda)u^T = 0.$$

However, the positive definiteness of $L(\varepsilon,x,\lambda)$ implies that $u = 0$, which in turn leads to $A(\varepsilon,x)v^T = 0$, which contradicts (i). We can now apply the implicit function theorem to show that in a neighborhood $U_2(\subset \mathbb{C}^{n+1} \times \mathbb{C}^p)$ of $(\varepsilon, x, \lambda)$, $W_2 \cap U_2$ is a one dimensional manifold. Define a map $\pi : \mathcal{V} \cap W_1 \to W_2$ by

$$\pi(\eta, z) = (\eta, z, \lambda(\eta, z)),$$

where $\mathcal{V}$ is as in Lemma 5.7. For some sufficiently small neighborhood $U_1(\subset \mathscr{C}^{n+1})$ of (ε, x)

$$\pi(W_1 \cap U_1) \subset W_2 \cap U_2.$$

However, since $W_2 \cap U_2$ is a one dimensional manifold, the z and λ coordinates of $\pi(\eta, z)$ can be parameterized by η via holomorphic functions. That is, $\pi(\eta, z) = (\eta, z(\eta), \lambda(\eta))$, where $\lambda(\eta) = \lambda(\eta, z(\eta))$, $z = z(\eta)$ for $(\eta, z) \in W_1 \cap U_1$. Hence $(\eta, z) = (\eta, z(\eta))$ on $W_1 \cap U_1$, and therefore $W_1 \cap U_1$ is also a one dimensional manifold. □

We are now in a position to state and prove the main theorem of this section.

Theorem 5.9. *Given any $(0, x) \in \bar{\mathscr{S}}$, there exist an n-vector of Puiseux series in ε (with real coefficients), $G(\varepsilon) = (G_1(\varepsilon), G_2(\varepsilon), \ldots, G_n(\varepsilon))$ such that for $\varepsilon > 0$ and sufficiently small*

$$(\varepsilon, G(\varepsilon)) \in \mathscr{S},$$

and

$$G(0) = \lim_{\varepsilon \downarrow 0} G(\varepsilon) = x.$$

Proof: Let $\bar{Q}$ be a compact neighborhood of $(0, x)$. Take a sequence $\{(\varepsilon_q, x_q)\}_{q=1}^{\infty}$ in $(W_1 \cap \bar{Q}) \cap \mathscr{S}$ such that $\varepsilon_q \downarrow 0$ and $x_q \to x$ as $q \to \infty$. Since $\bar{Q}$ is compact, only finitely many of the one dimensional components of W_1 intersect $\bar{Q}$. By Theorem 5.8, infinitely many of the points (ε_q, x_q) must lie in at least one such component. Let $\bar{W}_1$ be such an irreducible, one dimensional component, and assume, without loss of generality, that $\{(\varepsilon_q, x_q))\}_{q=1}^{\infty} \subset \bar{W}_1$.

Because $\bar{W}_1$ is one dimensional the Remmert–Stein representation theorem ensures that there exists an n-vector of Puiseux series $G(\varepsilon) = (G_1(\varepsilon), \ldots, G_n(\varepsilon))$ with real coefficients such that for $\varepsilon > 0$ and sufficiently small

$$(\varepsilon, x') = (\varepsilon, G(\varepsilon)) \in \bar{W}_1. \tag{5.53}$$

In particular, for members of the sequence in $\bar{W}_1$

$$x_q = G(\varepsilon_q) \to G(0) = x. \tag{5.54}$$

Note also that while we know that $(\varepsilon_q, x_q) = (\varepsilon_q, G(\varepsilon_q)) \in \mathscr{S}$ for all $q = 1, 2, \ldots$, we need to prove that this is also the case for all $\varepsilon > 0$ and sufficiently small. That is, we need to verify that (i)–(iii) are satisfied at $(\varepsilon, G(\varepsilon))$ for all $\varepsilon > 0$ and sufficiently small. These can be verified by recalling that for any Puiseux series $H(\varepsilon)$, with real coefficients, if a statement $H(\varepsilon) = (or \ \geq \ or \ \leq)$ *constant* is valid for all $\varepsilon_q \downarrow 0$, then it is valid for all $\varepsilon > 0$ and sufficiently small. This is a consequence of the fact that $H(\varepsilon_q) = 0$ for all $\epsilon_q \downarrow 0$ implies $H(\varepsilon) = 0$ for all $\varepsilon > 0$ and sufficiently small.

Further, since x_q is real for every $q = 1, 2, \ldots$, we have from (5.54) that

$$\mathbb{I}_m(G(\varepsilon_q)) = 0$$

infinitely often in the neighborhood of $\varepsilon = 0$. Hence $\mathbb{I}_m(G(\varepsilon)) \equiv 0$, and $G(\varepsilon) \in \mathbb{R}^n$ in that neighborhood. Now, verification of (i)–(iii) at $(\varepsilon, G(\varepsilon))$ for $\varepsilon > 0$ and sufficiently small becomes a simple matter. For instance, if (i) were not satisfied for such ε, then the matrix

$$A(\varepsilon) = \left(a_{ij}(\varepsilon)\right)_{i,j=1}^{p},$$

where $a_{ij}(\varepsilon) = \nabla h_j(\varepsilon, G(\varepsilon)).\nabla h_i(\varepsilon, G(\varepsilon))$ for all $i,j = 1,2,\ldots,p$, is singular at $\varepsilon = \varepsilon_q$ for $q = 1,2,\ldots,\infty$. Thus the Puiseux series $H(\varepsilon) := \det[A(\varepsilon)] = 0$ for all $\varepsilon = \varepsilon_q$. Hence $H(\varepsilon) \equiv 0$ for all $\varepsilon > 0$ and sufficiently small, yielding the desired contradiction. (ii) and (iii) can be verified similarly. This completes the proof. □

Example 5.11. *Consider the perturbed nonlinear program*

$$\min\{-x_1^2\}$$

subject to

$$x_1^2 - x_2^2 + \varepsilon x_2^4 \quad = \quad 1.$$

It is easy to check that the first order optimality conditions for the problem are

$$\begin{aligned} -2x_1 \quad +2x_1\lambda \quad\quad\quad\quad\quad &= 0, \\ -2\lambda x_2 \quad +4\varepsilon\lambda x_2^3 &= 0, \\ x_1^2 \quad -x_2^2 \quad\quad +\varepsilon x_2^4 &= 1. \end{aligned}$$

In Problem 5.17 the reader is invited to verify that there are three parameterized families of solutions of these optimality conditions, namely,

$$\left(4\varepsilon x^2 - 4\varepsilon - 1 = 0, y = \frac{\sqrt{2}}{2\sqrt{\varepsilon}}, \lambda = 1\right), \left(4\varepsilon x^2 - 4\varepsilon - 1 = 0, y = -\frac{\sqrt{2}}{2\sqrt{\varepsilon}}, \lambda = 1\right),$$

and

$$(y = 0, x^2 - 1 = 0, \lambda = 1).$$

The quadratic equation for x and ε leads to a solution (for $\varepsilon > 0$ and sufficiently small)

$$x(\varepsilon) = \frac{1}{2}\frac{\sqrt{\varepsilon(4\varepsilon+1)}}{\varepsilon} = \frac{1}{2}\frac{1}{\varepsilon} + \sqrt{\varepsilon} - \varepsilon^{\frac{3}{2}} + 2\varepsilon^{\frac{5}{2}} - 5\varepsilon^{\frac{7}{2}} + O(\varepsilon^{\frac{9}{2}}).$$

Remark 5.4. *It is easy to check that the results of this section frequently can be extended to the case where* (5.51) *is replaced by*

$$\min f(\varepsilon, x)$$

subject to

$$\begin{aligned} h_i(\varepsilon, x) &= 0, & i &= 1,2,\ldots,p, \\ g_j(\varepsilon, x) &\leq 0, & j &= 1,2,\ldots,m. \end{aligned}$$

In this case, by considering at each feasible point (ε, x) the combined set of equality and "active" inequality constraints, the problem is effectively reduced to (5.51). *Of course, active inequalities are those that are equal to 0 at the point (ε, x) in question.*

5.5 ▪ Problems

Problem 5.1. Prove that an a-optimal solution of the regularly perturbed linear program is always the optimal solution of the original unperturbed linear program (e.g., see [126]).

Problem 5.2. Verify the validity of recursion (5.7) for the coefficients of the expansion of the ratio of two Laurent series.

Problem 5.3. Let $r_j(\varepsilon)$ and $y_l(\varepsilon)$ be rational functions of ε. Prove that these functions either have no zero or isolated zeros or that they are identically zero. In particular, prove that if $r_j(\varepsilon)=0$ or $y_l(\varepsilon)=0$ for $\varepsilon \in (0,\bar{\varepsilon}]$, then these equalities hold for any $\varepsilon \in \mathbb{R}$.

Problem 5.4. Consider the example discussed in Section 5.2.6, namely,

$$\min\{-10x_1 - 10x_2 - 10x_3\}$$

subject to

$$\begin{array}{cccccc} \varepsilon x_1 & +\varepsilon x_2 & & -0.5x_4 & = & 0, \\ & -\varepsilon x_2 & +\varepsilon x_3 & -0.5x_4 & = & 0, \\ x_1 & +x_2 & +x_3 & +x_4 & = & 1, \end{array}$$

$$x_1, x_2, x_3, x_4 \geq 0.$$

Verify the results of the calculations reported in Section 5.2.6—particularly the following:

1. Show that if we begin with a basis corresponding to $B = \{1,3,4\}$, then in the expansion of $r_2(\varepsilon)$: $r_2^{(-1)} = 0$ and $r_2^{(0)} = 9$ so that $r_2(\varepsilon) > 0$ for $\varepsilon > 0$ and sufficiently small and hence that column 2 enters the new basis.

2. Show that if we start Step 3 with $P = \{\emptyset\}$ and $Q^{(-2)} = \{1,3,4\}$, then $y_2^{(-1)} = [0\,0\,0]^T$, the elements of $y_2^{(-1)}$ are all zeros, and that $Q^{(-1)} = Q^{(-2)} = \{1,3,4\}$. Next, verify that $y_2^{(0)} = [1.5\ \ -0.5\ 0]^T$ and hence that $Q^{(0)} = \{4\}$, resulting in the index 1 being added to the set P. Finally, verify that $[y_2^{(1)}]_4 = 1$, resulting in the index 4 being added to the set P. Show that Step 3 finishes with $P = \{1,4\}$ and $Q = \{\emptyset\}$.

3. Show that Step 4 terminates with column 1 exiting the basis.

4. For the new basis $B = \{2,3,4\}$ show that $r_1^{(-1)} = r_1^{(0)} = 0$ and $r_1^{(1)} = -20/3$ and hence that $r_1(\varepsilon) < 0$ for all sufficiently small $\varepsilon > 0$. Conclude that the new basis is a-optimal.

Problem 5.5. Prove that if index $i = m+1$ is reached in Step (2c) of Section 5.2.5 and $N^{(m+1)}$ is still nonempty, then $r_j(\varepsilon) \equiv 0$ for $j \in N^{(m+1)}$ and ε sufficiently small. Hence prove that $r_N(\varepsilon) \leq 0$ for all ε sufficiently small; that is, the current solution is a-optimal.

Problem 5.6. Use Lemma 5.3 to prove that the feasible region for the perturbed problem is bounded. Hence, or otherwise, prove that in our setting the set P introduced in Step 3 of Section 5.2.5 when Step (3d) is reached.

Problem 5.7.

1. Find an a-optimal solution in the regularly perturbed problem

$$\min\{10x_1 + 8x_2 - 4x_3\}$$

subject to

$$\begin{array}{rcrcrcl} (5+2\varepsilon)x_1 & + & \varepsilon x_2 & + & (1-\varepsilon)x_3 & = & 0, \\ x_1 & + & x_2 & & +x_3 & = & 1, \end{array}$$

$$x_1, x_2, x_3, x_4 \geq 0.$$

2. Now, change the above problem to a singularly perturbed one by replacing the coefficients $(5+\varepsilon)$, ε, $(1-\varepsilon)$ in the first equation by $(5+\varepsilon)$, $(5+\varepsilon)$, $(5-\varepsilon)$. Apply the a-simplex method again, and comment on the computational difficulty encountered.

Problem 5.8. Use Lemma 5.2 to prove that $\Delta_p(\varepsilon) \equiv \Delta_q(\varepsilon)$ if $p, q \in R^{(2m+1)}$.

Problem 5.9. Let analytic functions $a(\varepsilon)$ and $b(\varepsilon)$ be represented by the power series $a(\varepsilon) = \varepsilon^t a^{(t)} + \varepsilon^{t+1} a^{(t+1)} + \ldots, a^{(t)} \neq 0$, and $b(\varepsilon) = \varepsilon^q b^{(q)} + \varepsilon^{q+1} b^{(q+1)} + \ldots, b^{(q)} \neq 0$, respectively. Prove that the quotient of these analytic functions may be expressed as a power series (for sufficiently small ε)

$$c(\varepsilon) = \frac{a(\varepsilon)}{b(\varepsilon)} = \varepsilon^{t-q} c^{(0)} + \varepsilon^{t-q+1} c^{(1)} + \ldots, \quad c^{(0)} \neq 0,$$

whose coefficients are calculated by the recurrent formula

$$c^{(k)} = \left[a^{(t+k)} - \sum_{i=0}^{k-1} b^{(q+k-i)} c^{(i)} \right] \Big/ b^{(q)}, \quad k = 0, 1, 2, \ldots.$$

Hint: See the book of Markushevich [121] on complex analysis.

Problem 5.10. Prove the validity of the updating formula for the optimal solution of the perturbed linear program

$$x_{B^*}(\varepsilon) = x_{B^*}^{(0)} + \varepsilon[I - \varepsilon D_2]^{-1} x_{B^*}^{(1)}, \quad \varepsilon < \min\left\{\bar{\varepsilon}, \frac{1}{||D_2||}\right\},$$

where $x_{B^*}^{(0)}$ and $x_{B^*}^{(1)}$ are as in Theorem 5.1.

Problem 5.11. Assume a basis matrix has the form

$$A_B(\varepsilon) = A_B^{(0)} + \varepsilon A_B^{(1)} + \cdots + \varepsilon^p A_B^{(p)}.$$

Prove the generalized recursive formula (due to Korolyuk and Turbin [104])

$$B^{(k+1)} = -\sum_{i=0}^{p-1} \left(\sum_{j=0}^{p-1-i} B^{(-j)} A^{(j+i+1)} \right) B^{(k-i)}, \quad k = 0, 1, \ldots.$$

Show that the above formula is a particular case of the more general formula (3.49) from Section 3.3.

Problem 5.12. Follow the discussion in Subsection 5.2.8 to prove that the asymptotic simplex method can be readily generalized to a linear program with polynomial perturbations. Namely, suppose that the coefficients $A(\varepsilon)$, $b(\varepsilon)$, and $c(\varepsilon)$ in the perturbed linear program are polynomials of ε. In particular, a basis matrix has the form

$$A_B(\varepsilon)=A_B^{(0)}+\varepsilon A_B^{(1)}+\cdots+\varepsilon^p A_B^{(p)}.$$

Note that in the case of polynomial perturbations one needs at worst to check $(m+1)p$ terms of Laurent expansions for the entry rule and $2mp$ terms for the exit rule.

Problem 5.13. Prove that the matrix AA^T has full rank if and only if the matrix A has full row rank.

Problem 5.14. Prove that $P(\varepsilon)$ defined in (5.36) for $\varepsilon > 0$ and sufficiently small is an orthogonal projection matrix.

Problem 5.15. Suppose that $A(\varepsilon)=\sum_{k=0}^{\infty}\varepsilon^k A_k$ and that A_0 has full row rank. Prove that $P(\varepsilon)$ defined in (5.36) possesses a Maclaurin series expansion at $\varepsilon=0$.

Problem 5.16. Let $P(\varepsilon)$ be as in (5.36), and consider its power series expansion

$$P(\varepsilon)=P_0+\varepsilon P_1+\varepsilon^2 P_2+\ldots.$$

Also let $d(\varepsilon)$ be as in (5.35), and assume that it also has the power series expansion only with nonnegative powers of ε:

$$d(\varepsilon)=d_0+\varepsilon d_1+\varepsilon^2 d_2+\ldots.$$

Define $M_*=\{x|x=d_0+P_0 z, z\in\mathbb{R}^n\}$.

1. Prove that $||d_0||_2=\text{dist}\{0,M_*\}=\min\{||x||_2|x\in M_*\}$.

2. Prove that P_0 is an orthogonal projection operator onto M_*.

Problem 5.17. Consider the example mentioned in Section 5.4:

$$\min\{-x_1^2\}$$

subject to

$$x_1^2 \quad -x_2^2 \quad +\varepsilon x_2^4 \quad = \quad 1.$$

1. Verify that the first order optimality conditions for this problem are

$$\begin{array}{rrrcl} -2x_1 & +2x_1\lambda & & = & 0, \\ & -2\lambda x_2 & +4\varepsilon\lambda x_2^3 & = & 0, \\ x_1^2 & -x_2^2 & +\varepsilon x_2^4 & = & 1, \end{array}$$

where λ is the Lagrange multiplier corresponding to the single constraint.

2. Verify that there are three parameterized families of solutions of these optimality conditions, namely,

$$\left(4\varepsilon x^2 - 4\varepsilon - 1 = 0, y = \frac{\sqrt{2}}{2\sqrt{\varepsilon}}, \lambda = 1\right), \left(4\varepsilon x^2 - 4\varepsilon - 1 = 0, y = -\frac{\sqrt{2}}{2\sqrt{\varepsilon}}, \lambda = 1\right),$$

and

$$(y = 0, x^2 - 1 = 0, \lambda = 1).$$

3. Hence show that the quadratic equation for x and ε leads to a Puiseux series solution (for $\varepsilon > 0$ and sufficiently small) of the form

$$x(\varepsilon) = \frac{1}{2}\frac{\sqrt{\varepsilon(4\varepsilon+1)}}{\varepsilon} = \frac{1}{2}\frac{1}{\varepsilon} + \sqrt{\varepsilon} - \varepsilon^{\frac{3}{2}} + 2\varepsilon^{\frac{5}{2}} - 5\varepsilon^{\frac{7}{2}} + O(\varepsilon^{\frac{9}{2}}).$$

5.6 ▪ Bibliographic Notes

Note that most literature on sensitivity analysis and parametric programming (see, e.g., [55, 56, 64, 65]) concerns the perturbation of the objective function and the right-hand side. Past investigations of the perturbation of the entire coefficient matrix are quite limited. Moreover, the majority of authors restrict themselves to the case of regular perturbations. Jeroslow [95, 96] was, perhaps, the first who studied the general case. He considered the elements of matrices $A(\varepsilon), b(\varepsilon)$, and $c(\varepsilon)$ as arbitrary rational functions. To deal with such perturbed LPs, Jeroslow [95, 96] proposed a simplex-like method which works directly over the field of rational functions. The main disadvantage of that method is that the polynomials involved in calculation can have a high degree. For instance, the inversion of a basis matrix takes $O(m^4 \log(m))$ flops. Jeroslow's method can be viewed as an instance of a more general theory of extending algorithms for parametric problems (see Eaves and Rothblum [52]).

The existence of an a-optimal set of basic indices in the case when $A(\varepsilon), b(\varepsilon)$, and $c(\varepsilon)$ are rational functions was shown in [95, 96, 106]. Note that in [95, 96, 106] a semi-infinite interval $[\bar{t}, +\infty)$ is considered instead of the interval $(0, \bar{\varepsilon}]$. It is clear that the formulation with coefficients that are rational functions on $[\bar{t}, +\infty)$ is equivalent to a formulation with coefficients that are rational functions of ε on $(0, \bar{\varepsilon}]$. Furthermore, by multiplying by a common denominator the coefficients that are rational functions, the problem formulation can be converted to the polynomial setting.

The notion of the asymptotic optimal solution is stronger than the notion of a *limiting* optimal solution [6, 126]. Our definitions of singular perturbations are more general than those proposed in [63, 126]. In particular, the notions of weakly singular and strongly singular perturbations appear to have been introduced in [58]. Interestingly, perhaps, the asymptotic behavior of solutions to mathematical programming problems can be quite subtle, even in the case of linear programs. Recently, Avrachenkov et al. [9] showed that a discontinuity in the limit is possible even when the rank of the coefficient matrix does not change. The latter phenomenon, described as a "pseudosingular" or "weakly singular" perturbation, is discussed in detail in [9].

The asymptotic simplex method of Section 5.2 is related to the techniques introduced by Lamond [106, 107] and Huang [89]; however, it exploits the structure of Laurent series to a greater degree. In particular, Lamond [106] proposed a method for the expansion of the inverse of the basis matrix which demands $O(\bar{s}m^3)$ flops, where $\bar{s}$ is the maximal order of poles of the basis matrices. Huang [89] further improved the expansion of the perturbed basis matrix by proposing an algorithm which demands only $O(m^3)$ flops. In another paper [107] Lamond proposed updating the asymptotic expansion for the inverse of the perturbed basis matrix rather than computing it anew. However, his approach applies only to some particular cases, that is, when the inverse of the perturbed basis matrix has the pole of order one. This updating procedure demands $O(m^2)$ operations, which is comparable with the standard simplex method. In this chapter we proposed an updating procedure which deals with the general case and demands only $O(\bar{s}m^2)$. Moreover, our procedure is simpler than the inversion technique of Huang [89] and the updating algorithm of Lamond [107]. It is based on the elegant recursive formulae of Langenhop [111] and Schweitzer and Stewart [141]. In Section 5.2, if $\bar{s} \ll m$ (as can be expected in practice), then the estimated number of operations $O(\bar{s}m^2)$ needed in our updating procedure could be significantly less than $O(m^3)$, which is required in Huang's method [89].

The main difficulties faced when considering the inversion of the polynomially perturbed matrix (5.24) are that we cannot directly apply the methods of Lamond [106, 107] and Huang [89] for calculating the Laurent series. This is because these methods are heavily dependent on the linearity of the perturbation. Note that the iterative formulae (5.13), (5.14) that we use in our analysis were also derived for the case of linear perturbations. However, they can be generalized for the case of polynomial perturbations.

Note that prior proofs of Lemma 5.2 can be found in the papers of Lamond [106, 107] and Huang [89].

The material of Section 5.3, especially the result about the Maclaurin series expansion of the perturbed projection matrix, clearly has applications in many practical problems where the projection matrix plays a key role. One such application in the context of a statistical problem involving "systematic bias" was developed in Filar et al. [59].

Section 5.4 is based on the articles [10], [53], and [43]. The latter work is a generalization of the complex analytic approach in stochastic games [150].

It is important to note the comprehensive treatment of perturbed optimization presented in Bonnans and Shapiro [29] and in their preceding survey paper [28]. Indeed, these authors formulate their perturbed optimization problems in more general Banach spaces but also discuss perturbed mathematical programs in the finite dimensional case. They mostly concentrate on the case of regular perturbations. Our discussion in this chapter, mostly concentrated on the case of singular perturbations, can be seen as complementing parts of the comprehensive development presented in [28, 29].

For background on linear and nonlinear programming, we recommend the excellent books by Boyd and Vandenberghe [31], Cottle and Lemke [42], and Luenberger [118].

Chapter 6
Applications to Markov Chains

6.1 ▪ Introduction, Motivation, and Preliminaries

Finite state Markov chains (MCs) are among the most widely used probabilistic models of discrete event stochastic phenomena. Named after A.A. Markov, a famous Russian mathematician, they capture the essence of the existentialist "here and now" philosophy in the so-called Markov property, which, roughly speaking, states that probability transitions to a subsequent state depend only on the current state and time. This property is less restrictive than it might first appear because there is a great deal of flexibility in the choice of what constitutes the "current state."

The theory of MCs today constitutes a classical topic in the wider subjects of probability and stochastic processes. This theory has been applied in a wide spectrum of contexts ranging from weather prediction, to signal processing, to telecommunications. It is the ubiquity of MCs as a preferred modeling paradigm that lends importance to the analysis of their asymptotic behavior under perturbations. This chapter is devoted to the latter subject.

In this introductory section we briefly review some known facts from MC theory. Since MC theory is covered in most undergraduate curricula, we state some known results without proofs. The reader interested in more details is referred to some of the references mentioned in the bibliographical notes at the end of the chapter. In the present study we are concerned only with the finite state space MCs.

Definition 6.1. *A sequence of random variables* $\{X_t\}_{t\geq 0}$, *whose values belong to a finite set* $\mathcal{S}=\{1,\ldots,N\}$, *is said to be a (homogeneous finite)* Markov chain *(MC) with state space* $\mathcal{S}$, *initial distribution* $\alpha=\{\alpha_i\}_{i=1}^{n}$, *and transition matrix* $P=[p_{ij}]_{i,j=1}^{N}$ *if and only if* $P\{X_0=i\}=\alpha_i, i\in\mathcal{S}$, *and*

$$P\{X_{t+1}=i_{t+1}|X_t=i_t,\ldots,X_0=i_0\}=P\{X_{t+1}=i_{t+1}|X_t=i_t\}=p_{i_t,i_{t+1}} \tag{6.1}$$

for all $t\geq 0$ *and* $i_0,\ldots,i_{t+1}\in\mathcal{S}$. *The above equation is called the* Markov property.

Homogeneity is introduced by the second equality in (6.1), which shows that the conditional probability of state i_{t+1} at time $t+1$ given state i_t at time t has a prescribed value independent of t. If we denote the distribution of a discrete random variable X_t by

$x_t \in R^{1\times N}$, then the evolution of the process is given by the matrix equation

$$x_{t+1} = x_t P = \alpha P^{t+1}, \tag{6.2}$$

where the elements of matrix P are given by (6.1). If the MC is aperiodic, then the powers of the transition matrix P converge to a limit. In general, however, one has to consider the *Cesaro limit*, or the *stationary distribution matrix*, or the *ergodic projection*

$$\Pi = \lim_{T\to\infty} \frac{1}{T+1} \sum_{t=0}^{T} P^t, \tag{6.3}$$

which is known to be well defined.

A well-known example illustrates a phenomenon that can arise naturally when perturbations are present. Indeed, this example shows the essence of the so-called *singular perturbation*. Let us consider the perturbed MC with the following transition matrix:

$$P(\varepsilon) = \begin{bmatrix} 1-\varepsilon & \varepsilon \\ \varepsilon & 1-\varepsilon \end{bmatrix}.$$

Then, it is easy to see that the ergodic projection is given by

$$\Pi(\varepsilon) = \begin{cases} \begin{bmatrix} 1 & 0 \\ 0 & 1 \end{bmatrix}, & \varepsilon = 0, \\ \begin{bmatrix} 0.5 & 0.5 \\ 0.5 & 0.5 \end{bmatrix}, & 0 < \varepsilon \le 1. \end{cases}$$

From the above example, we can see that the ergodic projection has a discontinuity at $\varepsilon = 0$. The explanation for this fact is that the perturbed chain has fewer ergodic classes than the original chain. Hence, the stationary distribution matrix corresponding to the unperturbed chain has a larger rank than the one corresponding to the perturbed MC.

More generally, we shall consider situations where the probability transition matrix $P(\varepsilon)$ of an MC depends on ε in a prescribed way (e.g., linearly, polynomially, or analytically) and study the asymptotic behavior of important characteristics of the MC as $\varepsilon \to 0$. Of course, the case $\varepsilon = 0$ corresponds to the unperturbed chain.

Next, we review in a little more detail some structural properties of an arbitrary, finite state MC with a probability transition matrix P and its associated ergodic projection Π, as introduced in (6.3).

The name ergodic projection for Π stems from the fact that Π is the eigenprojection of the transition matrix P corresponding to its maximal eigenvalue 1. We call the MC *irreducible* if for any two states there is a positive probability of moving from one state to another in a finite number of transitions. In the case of an irreducible MC, the Cesaro limit can be easily constructed. Namely, we first determine the *stationary distribution* or the *invariant measure* as a solution of the linear system

$$\begin{cases} \mu P = \mu, \\ \mu \mathbf{1} = 1, \end{cases}$$

where, in this instance, $\mathbf{1} = [1 \cdots 1]^T \in R^{n\times 1}$. Elsewhere, the vector $\mathbf{1}$ will denote the vector of all ones of whatever dimension is needed to make a given equation consistent.

Now, for such an irreducible MC, the ergodic projection is given by

$$\Pi = \mathbf{1}\mu. \tag{6.4}$$

Note that Π has identical rows. This demonstrates that in the irreducible case the starting state has no influence on the long-run behavior of the chain.

However, the above is not the case in general. In the general *multichain* case one can always relabel the states in such an order that the transition matrix will take the following *canonical* form:

$$P = \begin{bmatrix} P_1 & \cdots & 0 & 0 \\ \vdots & \ddots & \vdots & \vdots \\ 0 & \cdots & P_n & 0 \\ R_1 & \cdots & R_n & S \end{bmatrix} \begin{matrix} \}\,\Omega_1 \\ \vdots \\ \}\,\Omega_n \\ \}\,\Omega_T, \end{matrix}$$

where the set of states Ω_i represents the ith ergodic class with transition matrix P_i and Ω_T represents the set of *transient* states. Let $N = |\Omega_1| + \cdots + |\Omega_n| + |\Omega_T|$ denote the total number of states in the MC. Note that the elements of submatrix S are transition probabilities inside the transient set, and the elements of R_i represent the one step probabilities of transition from the transient states to the ergodic states of class Ω_i.

It can be easily checked that the ergodic projection matrix Π inherits most of its structure from the above. Namely,

$$\Pi = \begin{bmatrix} \Pi_1 & \cdots & 0 & 0 \\ \vdots & \ddots & \vdots & \vdots \\ 0 & \cdots & \Pi_n & 0 \\ R_1^* & \cdots & R_n^* & 0 \end{bmatrix} \begin{matrix} \}\,\Omega_1 \\ \vdots \\ \}\,\Omega_n \\ \}\,\Omega_T, \end{matrix}$$

where the zero matrix in the bottom right corner replaces S because, in the long run, the transient will not be observed any more.

Often it is more convenient to use the MC generator $G := P - I$ rather than the transition matrix itself. We will use the following notation for the generator in the canonical form:

$$G = \begin{bmatrix} A_1 & \cdots & 0 & 0 \\ \vdots & \ddots & \vdots & \vdots \\ 0 & \cdots & A_n & 0 \\ R_1 & \cdots & R_n & T \end{bmatrix} \begin{matrix} \}\,\Omega_1 \\ \vdots \\ \}\,\Omega_n \\ \}\,\Omega_T, \end{matrix}$$

where $A_i = P_i - I$ and $T = S - I$.

In the multichain case, the ergodic projection Π can still be given mainly in terms of invariant measures of the ergodic classes Ω_i. However, the expression is more involved than the formula (6.4). First, we form the matrix of invariant measures

$$M = \begin{bmatrix} m_1 \\ \cdots \\ m_n \end{bmatrix} \in R^{n \times N}, \tag{6.5}$$

where $m_i = [0 \cdots 0 \ \mu_i \ 0 \cdots 0]$ with invariant measure μ_i placed in correspondence with the ergodic class Ω_i. Of course, $\mu_i \in R^{1 \times |\Omega_i|}$ can be found from the solution of the system

$$\begin{cases} \mu_i A_i = 0, \\ \mu_i \mathbf{1} = 1, \end{cases}$$

where $\mathbf{1}$ is a vector of ones with length $|\Omega_i|$. Next, we form the matrix of probabilities of absorption in one of the ergodic classes,

$$Q = [q_1 \cdots q_n] \in R^{N \times n}, \tag{6.6}$$

where

$$q_i = \begin{bmatrix} 0 \\ 1 \\ 0 \\ \varphi_i \end{bmatrix} \begin{matrix} \\ \}\,\Omega_i \\ \\ \}\,\Omega_T, \end{matrix} \tag{6.7}$$

where the jth element of the vector q_i represents the probability that the process initiated in state j will be absorbed in the ith ergodic class. The subvector φ_i can be calculated by

$$\varphi_i = (I - S)^{-1} R_i \mathbf{1} = -T^{-1} R_i \mathbf{1}. \tag{6.8}$$

Then, the ergodic projection is given by

$$\Pi = QM. \tag{6.9}$$

Note that $MQ = I$. Clearly, (6.9) can be seen as a generalization of (6.4).

Now probabilities of absorption in Ω_i are contained in the $|\Omega_T| \times |\Omega_i|$ matrix

$$R_i^* = \varphi_i \mu_i. \tag{6.10}$$

That is, R_i^* has the same dimension as R_i, and every element (k, j) constitutes the probability of absorption in Ω_i through state j, starting in state k. Now, let us illustrate the above theoretical development with the help of an example.

Example 6.1. *Consider an MC with the following transition matrix:*

$$P = \begin{bmatrix} 0 & 1 & 0 & 0 & 0 \\ 1/2 & 1/2 & 0 & 0 & 0 \\ 0 & 0 & 1/2 & 1/2 & 0 \\ 0 & 0 & 1/3 & 2/3 & 0 \\ 1/10 & 2/10 & 2/10 & 2/10 & 3/10 \end{bmatrix} \begin{matrix} \}\,\Omega_1 \\ \}\,\Omega_1 \\ \}\,\Omega_2 \\ \}\,\Omega_2 \\ \}\,\Omega_T. \end{matrix}$$

There are two ergodic classes, Ω_1 and Ω_2, and there is one transient state in Ω_T. In this case $R_1 = [\frac{1}{10}\ \frac{2}{10}]$, $R_2 = [\frac{2}{10}\ \frac{2}{10}]$, and $S = [\frac{3}{10}]$. First we need to use (6.8) *to calculate φ_1 and φ_2, which happen to be scalars because there is only one transient state:*

$$\varphi_1 = \left(1 - \frac{3}{10}\right)^{-1} \begin{bmatrix} \frac{1}{10} & \frac{2}{10} \end{bmatrix} \begin{bmatrix} 1 \\ 1 \end{bmatrix} = \frac{3}{7},$$

$$\varphi_2 = \left(1 - \frac{3}{10}\right)^{-1} \begin{bmatrix} \frac{2}{10} & \frac{2}{10} \end{bmatrix} \begin{bmatrix} 1 \\ 1 \end{bmatrix} = \frac{4}{7}.$$

Now we can construct the matrix Q. Note that if the process starts in Ω_i, then, naturally, the probability of absorption in Ω_i is 1. There are two ergodic classes with two states each, so we have

$$Q = \begin{bmatrix} 1 & 0 \\ 1 & 0 \\ 0 & 1 \\ 0 & 1 \\ 3/7 & 4/7 \end{bmatrix}.$$

After we calculate the stationary distributions $\mu_1 = [\frac{1}{3}\ \frac{2}{3}]$ and $\mu_2 = [\frac{2}{5}\ \frac{3}{5}]$ of the irreducible subchains corresponding to Ω_1 and Ω_2, we can construct the matrix of invariant measures.

Note that the "stationary distribution" of the transient state is zero.

$$M = \begin{bmatrix} 1/3 & 2/3 & 0 & 0 & 0 \\ 0 & 0 & 2/5 & 3/5 & 0 \end{bmatrix}.$$

Also, using (6.10) *we find that* $R_1^* = \frac{3}{7}[\frac{1}{3}\ \frac{2}{3}] = [\frac{1}{7}\ \frac{2}{7}]$ *and, similarly,* $R_2^* = [\frac{8}{35}\ \frac{12}{35}]$*. If the process starts in the transient state, the probability of absorption in the first ergodic class is the sum of the probabilities of going into state* 1 *and state* 2*. So the probability of absorbing in* Ω_1 *is* $\frac{1}{7} + \frac{2}{7} = \frac{3}{7} = \varphi_1$*. And the probability of absorbing in* Ω_2 *is* $\frac{8}{35} + \frac{12}{35} = \frac{4}{7} = \varphi_2$*. One can now check that, indeed,* $MQ = \left[\begin{smallmatrix} 1 & 0 \\ 0 & 1 \end{smallmatrix}\right]$*, and after constructing the stationary distribution matrix*

$$\Pi = QM = \begin{bmatrix} 1/3 & 2/3 & 0 & 0 & 0 \\ 1/3 & 2/3 & 0 & 0 & 0 \\ 0 & 0 & 2/5 & 3/5 & 0 \\ 0 & 0 & 2/5 & 3/5 & 0 \\ 1/7 & 2/7 & 8/35 & 12/35 & 0 \end{bmatrix}.$$

Note that the columns q_i *of* Q *are also right eigenvectors of* Π *corresponding to the eigenvalue* 1*. This follows immediately from the identity* $\Pi Q = QMQ = Q$*. We can also easily check that* $M\Pi = M$ *and* $\Pi P = P\Pi = \Pi$ *hold.*

Before proceeding further, let us briefly review a few of the known facts about the fundamental matrix and the mean first passage times. Let P be a transition matrix of an MC, and let Π be the associated ergodic projection; then the *fundamental matrix* is defined as follows:

$$Z := [I - P + \Pi]^{-1}.$$

Another equivalent definition of the fundamental matrix can be given in the form of matrix series

$$Z := \lim_{T\to\infty}(c)\left[\sum_{t=0}^{T}(P-\Pi)^t\right] = \lim_{T\to\infty}\frac{1}{T+1}\sum_{t=0}^{T}\left[\sum_{n=0}^{t}(P-\Pi)^n\right],$$

where lim(c) denotes a Cesaro limit. Of course, if the chain is aperiodic, we have the convergence in the usual sense. If Π expresses the ergodic (long-run) behavior of the chain, then, according to the second definition, matrix Z represents the transient (short-run) behavior of the MC. The fundamental matrix is very useful in the perturbation analysis of MCs. Another important application of the fundamental matrix is to the mean first passage times.

Definition 6.2. *If an ergodic MC is initiated in state* i*, the expected number of transitions to reach state* j *for the first time is called the* mean first passage time *from* i *to* j*. It is denoted by* $\mathcal{M}_{ij}$*.*

Obviously, the mean first passage time has a sensible definition only for the *ergodic* chains. Once we have in hand the fundamental matrix Z, the mean first passage time can be immediately computed by the simple formula

$$\mathcal{M}_{ij} = \frac{z_{jj} - z_{ij}}{\mu_j}, \tag{6.11}$$

where $\mu = [\mu_1 \cdots \mu_N]$ is an invariant measure of the MC. By convention, $\mathcal{M}_{ii} = 1/\mu_i$, which is the expected return time.

It has been known since at least the 1980s that the *fundamental matrix* of a singularly perturbed MC can be expanded as a Laurent series:

$$Z(\varepsilon) = [I - P(\varepsilon) + \Pi(\varepsilon)]^{-1} = \frac{1}{\varepsilon^s} Z_{-s} + \cdots + \frac{1}{\varepsilon} Z_{-1} + Z_0 + \varepsilon Z_1 + \ldots. \tag{6.12}$$

In our development, we prefer to first obtain the Laurent series for the *deviation matrix*

$$H(\varepsilon) := Z(\varepsilon) - \Pi(\varepsilon)$$

rather than straight away for the fundamental matrix $Z(\varepsilon)$. There are several reasons for this. In particular, it is easier to implement the reduction process for the deviation matrix. Of course, once the Laurent series for the deviation matrix is obtained, we can immediately calculate the Laurent series for the fundamental matrix.

We conclude this introduction by stating the following well-known formulae (see Problem 6.1) for fundamental matrix Z and the deviation matrix H, of an MC:

$$Z = [\Pi + I - P]^{-1} = [\Pi - G]^{-1}, \tag{6.13}$$

$$H = Z - \Pi = [\Pi + I - P]^{-1} - \Pi = [\Pi - G]^{-1} - \Pi. \tag{6.14}$$

In Section 6.3 we obtain several results on the asymptotic behavior of the fundamental matrix of the singularly perturbed MC. Then we apply these results to the perturbation analysis of mean first passage times.

6.2 ▪ Asymptotic Analysis of the Stationary Distribution Matrix

In this section we consider the general case of an analytic perturbation of a finite MC. That is, the probability transition matrix of the perturbed MC is $P(\varepsilon)$, an analytic function of a perturbation parameter ε. Hence, it can be represented by power series

$$P(\varepsilon) = P_0 + \varepsilon P_1 + \varepsilon^2 P_2 + \ldots, \tag{6.15}$$

where it is assumed that the coefficient matrices P_k are known. Even though the above power series may converge in some complex neighborhood around $\varepsilon = 0$, we will consider only some real interval $[0, \varepsilon_{max}]$, where the elements of the matrix $P(\varepsilon)$ are nonnegative reals whose values are less than or equal to one. We make no assumption at all about the structure of the unperturbed and perturbed MCs.

It will be shown that the stationary distribution matrix $\Pi(\varepsilon)$ of this perturbed MC has an analogous power series expansion

$$\Pi(\varepsilon) = \Pi_0 + \varepsilon \Pi_1 + \varepsilon^2 \Pi_2 + \ldots. \tag{6.16}$$

Of course, the asymptotic behavior of $\Pi(\varepsilon)$ as $\varepsilon \to 0$ is determined by the coefficient matrices Π_k, for $k = 0, 1, \ldots$. Hence, it will be shown how these coefficients can be calculated with the help of a series of recursive formulae. Before proceeding, we shall need some further notation and preliminary (relatively standard) results that we introduce below, without proofs. The reader is referred to references in the bibliographic notes and problem sections for further information about these results.

To be consistent with the previous formulation of the perturbed MC we now assume that the generator of the perturbed chain is an analytic function and hence that it can be expanded as a power series at $\varepsilon = 0$,

$$G(\varepsilon) = G_0 + \varepsilon G_1 + \varepsilon^2 G_2 + \dots, \quad 0 \le \varepsilon \le \varepsilon_{max}, \tag{6.17}$$

where G_0 is a generator of the unperturbed MC. Recall that since we make no assumptions about the ergodic structure, both the perturbed and unperturbed models may have several ergodic classes and sets of transient states.

Hence, the starting point of our analysis is to rearrange the states of the perturbed MC, as before, in such a way that the generator $G(\varepsilon)$ can be written in the canonical form

$$G(\varepsilon) = \begin{bmatrix} A_1(\varepsilon) & \cdots & 0 & 0 \\ \vdots & \ddots & \vdots & \vdots \\ 0 & \cdots & A_n(\varepsilon) & 0 \\ R_1(\varepsilon) & \cdots & R_n(\varepsilon) & T(\varepsilon) \end{bmatrix} \begin{matrix} \}\,\Omega_1 \\ \vdots \\ \}\,\Omega_n \\ \}\,\Omega_T. \end{matrix} \tag{6.18}$$

Now note that all invariant measures $m_i(\varepsilon)$ of the perturbed MC can be immediately constructed from the invariant measures of the ergodic classes associated with stochastic subgenerators $A_i(\varepsilon), i = 1,\dots,n$. Namely, $m_i(\varepsilon) = [0\cdots 0 \ \mu_i(\varepsilon) \ 0\cdots 0]$, where $\mu_i(\varepsilon)$ is uniquely determined by the system

$$\begin{cases} \mu_i(\varepsilon) A_i(\varepsilon) = 0, \\ \mu_i(\varepsilon)\mathbf{1} = 1. \end{cases} \tag{6.19}$$

The above is exactly the perturbation problem under the irreducibility assumption.

Note that our task of calculating the expansion of $\Pi(\varepsilon)$ will be complete once we calculate the following:

1. The expansion of $\mu_i(\varepsilon)$ for each $i = 1,\dots,n$, which determines the expansion of the matrix $M(\varepsilon)$ defined as in (6.5).

2. The expansion of the right eigenvectors $q_i(\varepsilon)$ for each $i = 1,\dots,n$, containing the probabilities to be absorbed in one of ergodic classes after perturbation (see (6.7)–(6.8)). This determines the expansion of the matrix $Q(\varepsilon)$ defined as in (6.6).

3. The product $\Pi(\varepsilon) = Q(\varepsilon)M(\varepsilon)$, which yields the desired power series.

These tasks may be accomplished in more or less complex ways depending on the availability of suitable special structure. The remaining subsections present many of the available results.

6.2.1 ▪ The irreducible perturbation

To make the present section self-contained, let us briefly outline the calculation of the asymptotic expansion for the perturbed invariant measure in the case of an *irreducible perturbed chain*. That is, throughout this subsection, we assume the following.

Definition 6.3. *For any $\varepsilon > 0$ and sufficiently small, $P(\varepsilon)$ is an irreducible probability transition matrix and $G(\varepsilon) := P(\varepsilon) - I$ is the generator of the corresponding irreducible, perturbed MC. Such a perturbation is called irreducible.*

Remark 6.1. *Note that this case includes both the case when the unperturbed transition matrix $P(0)$ is irreducible and the case when it is multichain, sometimes called the* regular *and the* singular *cases, respectively.*

One may consider this problem as the perturbation of the left null space of the generator matrix. Therefore, the results of Chapter 3 are immediately applicable.

Let us substitute the power series $\mu(\varepsilon)=\mu_0+\varepsilon\mu_1+\varepsilon^2\mu_2+\ldots$ and $G(\varepsilon)=G_0+\varepsilon G_1+\varepsilon^2 G_2+\ldots$ into the system

$$\begin{cases} \mu(\varepsilon)G(\varepsilon)=0, \\ \mu(\varepsilon)\mathbf{1}=1 \end{cases}$$

and equate coefficients with the same powers of ε. The latter results in the system of *fundamental equations*

$$\begin{aligned} \mu_0 G_0 &= 0 \quad (MF0), \\ \mu_1 G_0 + \mu_0 G_1 &= 0 \quad (MF1), \\ \mu_2 G_0 + \mu_1 G_1 + \mu_0 G_2 &= 0 \quad (MF2), \\ &\;\;\vdots \qquad\quad \vdots \\ \mu_k G_0 + \mu_{k-1} G_1 + \cdots + \mu_1 G_{k-1} + \mu_0 G_k &= 0 \quad (MFk), \\ &\;\;\vdots \qquad\quad \vdots \end{aligned}$$

and the system of *normalization conditions*

$$\begin{aligned} \mu_0\mathbf{1} &= 1 \quad (MN0), \\ \mu_1\mathbf{1} &= 0 \quad (MN1), \\ &\;\;\vdots \qquad \vdots \\ \mu_k\mathbf{1} &= 0 \quad (MNk). \\ &\;\;\vdots \qquad \vdots \end{aligned}$$

Now we may reduce the above system to another equivalent system with matrix coefficients of smaller dimensions. Roughly speaking, the reduction replaces each ergodic class by a single state.

Proposition 6.1. *A solution of the fundamental equations (MF) together with the normalization conditions (MN) is given by the recursive formulae*

$$\mu_0 = \mu_0^{(1)} M, \tag{6.20}$$

$$\mu_k = \mu_k^{(1)} M + \sum_{j=1}^{k} \mu_{k-j} G_j H, \quad k \geq 1, \tag{6.21}$$

where the auxiliary sequence $\mu_k^{(1)}$, $k \geq 0$, is a unique solution to the following system of reduced fundamental equations (RMF)

$$\begin{aligned} \mu_0^{(1)} G_0^{(1)} &= 0 \quad (RMF0), \\ \mu_1^{(1)} G_0^{(1)} + \mu_0^{(1)} G_1^{(1)} &= 0 \quad (RMF1), \\ &\;\;\vdots \qquad\quad \vdots \\ \mu_k^{(1)} G_0^{(1)} + \mu_{k-1}^{(1)} G_1^{(1)} + \cdots + \mu_1^{(1)} G_{k-1}^{(1)} + \mu_0^{(1)} G_k^{(1)} &= 0 \quad (RMFk), \\ &\;\;\vdots \qquad\quad \vdots \end{aligned}$$

coupled with new reduced normalization conditions (RMN)

$$\begin{aligned}
\mu_0^{(1)}\mathbf{1} &= 1 \quad (RMN0),\\
\mu_1^{(1)}\mathbf{1} &= 0 \quad (RMN1),\\
&\vdots \qquad \vdots\\
\mu_k^{(1)}\mathbf{1} &= 0 \quad (RMNk),\\
&\vdots \qquad \vdots
\end{aligned}$$

where the coefficient matrices $G_k^{(1)} \in R^{n\times n}$, $k \geq 0$, are given by the formula

$$G_k^{(1)} = M\left(G_k + \sum_{p=1}^{k}\left(\sum_{r=1}^{p}\left(\sum_{v_1+v_2+\cdots+v_r=p}\prod_{j=1}^{r}\left(G_{v_j}H\right)\right)G_{k-p}\right)\right)Q \tag{6.22}$$

and $G_0^{(1)} = MG_0Q$. In (6.22) $M \in R^{n\times N}$ is a matrix whose rows are invariant measures of the unperturbed MC, $Q \in R^{N\times n}$ is a matrix of right eigenvectors corresponding to the zero eigenvalue of the unperturbed generator, and $H = [\Pi - G]^{-1} - \Pi$ is a deviation matrix of the unperturbed chain.

We refer the reader to Problem 6.2 for the verification of the validity of equation (6.22). Note that the dimension of the coefficients $G_j^{(1)}$, $j \geq 0$, is equal to n, the number of ergodic classes of the unperturbed MC, which is usually much smaller than N, the number of states in the original MC. Moreover, matrix $G_0^{(1)}$ can be considered as a generator of the aggregated MC whose states represent the ergodic classes of the original chain. Next, we illustrate this result with a simple example.

Example 6.2. *Consider an MC with a linearly perturbed transition matrix $P(\varepsilon) = P(0) + \varepsilon C$:*

$$P(0) = \begin{pmatrix} 1 & 0 & 0\\ 0 & 1/2 & 1/2\\ 0 & 1/2 & 1/2 \end{pmatrix} \quad \text{and} \quad C = \begin{pmatrix} -2 & 1 & 1\\ 1 & -1 & 0\\ 0 & 1 & -1 \end{pmatrix}.$$

Note that the unperturbed chain $P(0)$ has two ergodic classes and the perturbed chain $P(\varepsilon)$ has only one and, indeed, is irreducible. Our goal is to find $\mu(\varepsilon) = \mu_0 + \varepsilon\mu_1 + \varepsilon^2\mu_2 + \ldots$. After calculating the stationary distributions of the two ergodic classes in $P(0)$, one may check that

$$M = \begin{pmatrix} 1 & 0 & 0\\ 0 & 1/2 & 1/2 \end{pmatrix}, \quad Q = \begin{pmatrix} 1 & 0\\ 0 & 1\\ 0 & 1 \end{pmatrix}.$$

In order to derive the deviation matrix H of the unperturbed chain, we may compute deviation matrices for each ergodic class i in $P(0)$ separately using $H_i = [\Pi_i - A_i]^{-1} - \Pi_i$. Now the matrix H is given by

$$H = \begin{bmatrix} H_1 & \cdots & 0\\ \vdots & \ddots & \vdots\\ 0 & \cdots & H_n \end{bmatrix}.$$

One may verify that, in our example, $\Pi(0) = P(0)$, and hence H is given by

$$H = \begin{bmatrix} 0 & 0 & 0\\ 0 & 1/2 & -1/2\\ 0 & -1/2 & 1/2 \end{bmatrix}.$$

Now we use (6.22) *to calculate the matrices* $G_k^{(1)}$. *Note that* $G_0 = P(0) - I$, $G_1 = C$, *and* $G_2 = G_3 = \cdots = 0$.

$$G_0^{(1)} = MG_0Q = \begin{bmatrix} 0 & 0 \\ 0 & 0 \end{bmatrix},$$

$$G_1^{(1)} = M(G_1HG_0 + G_1)Q = \begin{bmatrix} -2 & 2 \\ 1/2 & -1/2 \end{bmatrix},$$

$$G_2^{(1)} = M((G_1H)^2G_0 + G_1HG_1)Q = \begin{bmatrix} 0 & 0 \\ 1/4 & -1/4 \end{bmatrix},$$

$$G_3^{(1)} = M((G_1H)^3G_0 + (G_1H)^2G_1)Q = \begin{bmatrix} 0 & 0 \\ -3/8 & 3/8 \end{bmatrix}.$$

$$\ldots$$

In this case, the matrix $G_0^{(1)}$ *in* (6.22) *is a zero matrix, as there are no transient states (nearly completely decomposable case). Now, we may calculate the reduced vectors* $\mu_k^{(1)}$ *by solving the system of reduced fundamental equations. The result is shown below:*

$$\mu_0^{(1)} = \frac{1}{5}[1\ \ 4], \quad \mu_1^{(1)} = \frac{1}{5^2}[2\ \ -2],$$

$$\mu_2^{(1)} = \frac{1}{5^3}[-16\ \ 16], \quad \mu_3^{(1)} = \frac{1}{5^4}[128\ \ -128] \cdots .$$

Our final step is the calculation of the perturbed stationary distribution coefficients using formulae (6.20) *and* (6.21):

$$\mu_0 = \mu_0^{(1)}M = \frac{1}{5}[1\ \ 2\ \ 2],$$

$$\mu_1 = \mu_1^{(1)}M + \mu_0G_1H = \frac{1}{5^2}[2\ \ 4\ \ -6],$$

$$\mu_2 = \mu_2^{(1)}M + \mu_1G_1H = \frac{1}{5^3}[-16\ \ -32\ \ 48].$$

$$\ldots$$

Finally, we conclude that the stationary probabilities that the perturbed system is in state 1, 2, *or* 3, *respectively, are now obtainable from the expansion of* $\mu(\varepsilon)$, *which in this case has the form*

$$\mu(\varepsilon) = \left[\tfrac{1}{5}\ \ \tfrac{2}{5}\ \ \tfrac{2}{5}\right] + \left[\tfrac{2}{25}\ \ \tfrac{4}{25}\ \ -\tfrac{6}{25}\right]\varepsilon + \left[-\tfrac{16}{125}\ \ -\tfrac{32}{125}\ \ \tfrac{48}{125}\right]\varepsilon^2 + \cdots .$$

Since the reduced system (RMF) has essentialy the same structure as the original fundamental system (MF), we may perform a sequence of reduction steps. We terminate the reduction process, say, after s steps, when the system

$$\begin{cases} \mu_0^{(s)}G_0^{(s)} = 0, \\ \mu_0^{(s)}\xi_{n_s} = 1 \end{cases}$$

has a unique solution. In particular, we obtain the following representation for the *limiting* invariant measure:

$$\mu_0 = \mu_0^{(s)}M^{(s-1)}\cdots M^{(1)}M,$$

where $M^{(k)}$ is a matrix of invariant measures for the aggregated chain at the kth reduction step. And the solution to the final step reduced system is given by the recursive formula

$$\mu_k^{(s)} = \sum_{j=1}^{k} \mu_{k-j}^{(s)} G_j^{(s)} H, \quad k \geq 1.$$

See Problem 6.3 for an alternative approach based on the generalized inverses and augmented matrices.

6.2.2 ▪ The multichain perturbation

We are now in a position to analyze the general case of analytic perturbation when the perturbed chain has a multichain structure. As shown earlier, after we have obtained the invariant measures $\mu_1(\varepsilon), \dots, \mu_n(\varepsilon)$ for the ergodic classes $\Omega_1, \dots, \Omega_n$ of the perturbed MC, we can immediately construct the invariant measures of the entire perturbed chain. They are simply of the form $m_i(\varepsilon) = [0 \cdots 0 \ \mu_i(\varepsilon)\ 0 \cdots 0]$, where $\mu_i(\varepsilon)$ is put in the place of states that constitute the ergodic class Ω_i.

Now we demonstrate how to calculate the asymptotic expansions for the right 0-eigenvectors of the perturbed MC generator. The elements of the ith eigenvector are probabilities of absorption in the ergodic class Ω_i starting from some particular state. According to (6.7), the right 0-eigenvectors of the perturbed chain can be written in the form

$$q_i(\varepsilon) = \begin{bmatrix} 0 \\ \mathbf{1} \\ 0 \\ \varphi_i(\varepsilon) \end{bmatrix} \begin{matrix} \\ \}\,\Omega_i \\ \\ \}\,\Omega_T, \end{matrix} \tag{6.23}$$

where the subvector $\varphi_i(\varepsilon)$ is given by (see (6.8))

$$\varphi_i(\varepsilon) = -T^{-1}(\varepsilon) R_i(\varepsilon) \mathbf{1}. \tag{6.24}$$

Note that if some ergodic classes become transient sets after the perturbation, then the matrix-valued function $T^{-1}(\varepsilon)$ has a singularity at $\varepsilon = 0$. To explain this phenomenon, let us consider the first term in the perturbation series $T(\varepsilon) = T_0 + \varepsilon T_1 + \varepsilon^2 T_2 + \dots$, the bottom right block of (6.18). In turn, the first term T_0 has the following canonical structure:

$$T_0 = \begin{bmatrix} \tilde{A}_1 & \cdots & 0 & 0 \\ \vdots & \ddots & \vdots & \vdots \\ 0 & \cdots & \tilde{A}_m & 0 \\ \tilde{R}_1 & \cdots & \tilde{R}_n & \tilde{T} \end{bmatrix}.$$

Blocks $\tilde{A}_1, \dots, \tilde{A}_m$ represent the ergodic classes of the original MC that merged with the transient set after the perturbation. Since each of $\tilde{A}_1, \dots, \tilde{A}_m$ is an MC generator, we conclude that the matrix T_0 has at least m zero eigenvalues and, of course, is not invertible. However, the matrix $T(\varepsilon)$ is invertible for $\varepsilon \neq 0$ and sufficiently small. From the discussion of Sections 2.2 and 3.3, it follows that one can expand $T^{-1}(\varepsilon)$ as a Laurent series at $\varepsilon = 0$:

$$T^{-1}(\varepsilon) = \frac{1}{\varepsilon^s} U_{-s} + \dots + \frac{1}{\varepsilon} U_{-1} + U_0 + \varepsilon U_1 + \dots. \tag{6.25}$$

One may also use the methods of Sections 2.2 and 3.3 to calculate the coefficients of the above series. Substituting the power series $R_i(\varepsilon) = R_{i0} + \varepsilon R_{i1} + \varepsilon^2 R_{i2} + \dots$ and the Laurent series (6.25) into (6.24), we obtain the asymptotic expansion for $\varphi_i(\varepsilon)$. Since the elements of $\varphi_i(\varepsilon)$ are probabilities, the function $\varphi_i(\varepsilon)$ is bounded, and hence the singular terms of (6.25) satisfy the conditions

$$\sum_{j=-s}^{k} U_j R_{k-j} \mathbf{1} = 0, \quad k = -s, \ldots, -1, \tag{6.26}$$

and $\varphi_i(\varepsilon)$ is expanded as a series with nonnegative powers of ε,

$$\varphi_i(\varepsilon) = \varphi_{i0} + \varepsilon \varphi_{i1} + \varepsilon^2 \varphi_{i2} + \ldots, \tag{6.27}$$

where

$$\varphi_{ik} = -\sum_{j=-s}^{k} U_j R_{k-j} \mathbf{1}, \quad k \geq 0. \tag{6.28}$$

The above formulae are valid in the general setting. Now we would like to discuss several important particular cases. First we discuss the situation when no ergodic classes merge with the transient set. In other words, $T_0 = \tilde{T}$, where $\tilde{T}$ is a proper substochastic matrix. The latter implies that T_0 has an inverse and the asymptotic expansion for $\varphi_i(\varepsilon)$ can be immediately constructed, using Neumann expansion for $T^{-1}(\varepsilon)$, that is,

$$\varphi_{i0} = -T_0^{-1} R_{i0} \mathbf{1}, \tag{6.29}$$

$$\varphi_{ik} = -T_0^{-1} \left[R_{ik} \mathbf{1} + \sum_{j=1}^{k} T_j \varphi_{ik-j} \right]. \tag{6.30}$$

This case is interesting, since, even if the perturbation were singular, the calculation of the asymptotic expansions for the right 0-eigenvectors is quite simple.

Example 6.3. *Consider a* (5×5) *MC with two transient states and two ergodic classes before and after the perturbation.*

$$P(\varepsilon) = P(0) + \varepsilon C = \begin{pmatrix} 1 & 0 & 0 & 0 & 0 \\ 0 & 1/2 & 1/2 & 0 & 0 \\ 0 & 1/2 & 1/2 & 0 & 0 \\ 1/2 & 0 & 0 & 0 & 1/2 \\ 1/4 & 1/4 & 0 & 1/4 & 1/4 \end{pmatrix} + \varepsilon \begin{pmatrix} 0 & 0 & 0 & 0 & 0 \\ 0 & 0 & 0 & 0 & 0 \\ 0 & 0 & 0 & 0 & 0 \\ 1 & 0 & 1 & 0 & -2 \\ -1 & 1 & 1 & -1 & 0 \end{pmatrix}.$$

In order to calculate $\varphi_i(\varepsilon)$, *the following matrices are relevant:*

$$T_0 = S_0 - I = \begin{pmatrix} -1 & 1/2 \\ 1/4 & -3/4 \end{pmatrix}, T_1 = \begin{pmatrix} 0 & -2 \\ -1 & 0 \end{pmatrix},$$

$$R_{10} = \begin{pmatrix} 1/2 \\ 1/4 \end{pmatrix}, R_{11} = \begin{pmatrix} 1 \\ -1 \end{pmatrix}, R_{20} = \begin{pmatrix} 0 & 0 \\ 1/4 & 0 \end{pmatrix}, R_{21} = \begin{pmatrix} 0 & 1 \\ 1 & 1 \end{pmatrix}.$$

One may verify that

$$T_0^{-1} = \frac{2}{5} \begin{pmatrix} -3 & -2 \\ -1 & -4 \end{pmatrix}.$$

Using formulae (6.29) *and* (6.30), *we can easily compute* $\varphi_i(\varepsilon)$.

$$\varphi_{10} = \frac{2}{5}\begin{pmatrix} -3 & -2 \\ -1 & -4 \end{pmatrix}\begin{pmatrix} 1/2 \\ 1/4 \end{pmatrix} = \begin{pmatrix} 4/5 \\ 3/5 \end{pmatrix},$$

$$\varphi_{20} = \frac{2}{5}\begin{pmatrix} -3 & -2 \\ -1 & -4 \end{pmatrix}\begin{pmatrix} 0 & 0 \\ 1/4 & 0 \end{pmatrix}\begin{pmatrix} 1 \\ 1 \end{pmatrix} = \begin{pmatrix} 1/5 \\ 2/5 \end{pmatrix},$$

$$\varphi_{11} = -T_0^{-1}[R_{11} + T_1\varphi_{10}] = \begin{pmatrix} -42/25 \\ -74/25 \end{pmatrix},$$

$$\varphi_{21} = -T_0^{-1}[R_{21}\mathbf{1} + T_1\varphi_{20}] = \begin{pmatrix} 42/25 \\ 74/25 \end{pmatrix}.$$

The reader may check that

$$\varphi_1(\varepsilon) = \frac{1}{5}\begin{pmatrix} 4 \\ 3 \end{pmatrix} - \frac{\varepsilon}{25}\begin{pmatrix} 42 \\ 74 \end{pmatrix} + \frac{\varepsilon^2}{125}\begin{pmatrix} 1056 \\ 632 \end{pmatrix} + \cdots,$$

$$\varphi_2(\varepsilon) = \frac{1}{5}\begin{pmatrix} 1 \\ 2 \end{pmatrix} + \frac{\varepsilon}{25}\begin{pmatrix} 42 \\ 74 \end{pmatrix} - \frac{\varepsilon^2}{125}\begin{pmatrix} 1056 \\ 632 \end{pmatrix} + \cdots.$$

Now let us consider an important particular case of a *linear* perturbation, that is, $T(\varepsilon) = T_0 + \varepsilon T_1$. We can now construct an elegant updating formula for the perturbed subvector $\varphi_i(\varepsilon)$. Toward this end, we consider the *singular* and *regular* parts of the Laurent series (6.25)

$$U^S(\varepsilon) = \frac{1}{\varepsilon^s}U_{-s} + \cdots + \frac{1}{\varepsilon}U_{-1}$$

and

$$U^R(\varepsilon) = U_0 + \varepsilon U_1 + \ldots,$$

respectively. In Problem 6.4 we ask the reader to verify that we can conclude that the regular part $U^R(\varepsilon)$ can be written in the closed analytic form

$$U^R(\varepsilon) = (I + \varepsilon U_0 T_1)^{-1} U_0. \tag{6.31}$$

Then, $\varphi_i(\varepsilon) = -U_{-1}R_{i1}\mathbf{1} - U^R(\varepsilon)R_i(\varepsilon)\mathbf{1}$ can be calculated by the updating formula

$$\varphi_i(\varepsilon) = -U_{-1}R_{i1}\mathbf{1} - (I + \varepsilon U_0 T_1)^{-1} U_0 R_i(\varepsilon)\mathbf{1}$$

or in terms of the limiting value φ_{i0}

$$\varphi_i(\varepsilon) = \varphi_{i0} - \varepsilon[U_0 R_{i1} - (I + \varepsilon U_0 T_1)^{-1} U_0 T_1 U_0 R_i(\varepsilon)]\mathbf{1}.$$

We would like to emphasize that the above updating formulae are computationally stable for small values of ε in contrast to the original formula (6.24), where $T^{-1}(\varepsilon)$ is ill-conditioned when ε is close to zero.

Next we consider the case of first order singular perturbations. By this we mean that the Laurent series (6.25) has a simple pole. According to our experience, in general it is quite unlikely that the Laurent series (6.25) has negative powers of ε smaller than -1. In other words, the case of a simple pole is generic. In particular, this setting permits us to derive a nice expression for the limiting value of $\varphi_i(\varepsilon)$ as ε goes to zero. Recall that we have deduced conditions (6.26) from the probabilistic interpretation. In the case of first order singularity it is easy to demonstrate by algebraic methods that the asymptotic expansion for $\varphi_i(\varepsilon)$ does not have a singular part. Toward this end, we write $\varphi_i(\varepsilon)$ as

$$\varphi_i(\varepsilon) = \frac{1}{\varepsilon}\varphi_{i,-1} + \varphi_{i0} + \varepsilon\varphi_{i1} + \dots$$

and show that $\varphi_{i,-1} = 0$. Upon substitution of the above series and the series for $T(\varepsilon)$ and $R_i(\varepsilon)$ into the equation (also see (6.24))

$$T(\varepsilon)\varphi_i(\varepsilon) = -R_i(\varepsilon)\mathbf{1}, \tag{6.32}$$

we obtain the following system of equations:

$$T_0\varphi_{i,-1} = 0, \tag{6.33}$$

$$T_0\varphi_{i0} + T_1\varphi_{i,-1} = -R_{i0}\mathbf{1}. \tag{6.34}$$

$$\dots$$

From equation (6.33) we conclude that

$$\varphi_{i,-1} = \tilde{Q}c_{-1}, \tag{6.35}$$

where $\tilde{Q}$ is a matrix of right 0-eigenvectors of the matrix T_0 (which can be calculated by (6.7)) and $c_{-1} \in R^{m\times 1}$ is a vector of coefficients that can be determined from the feasibility condition of equation (6.34). The feasibility condition for equation (6.34) can be written in the form

$$\tilde{M}T_1\varphi_{i,-1} = -\tilde{M}R_{i0}\mathbf{1}, \tag{6.36}$$

where $\tilde{M}$ is a matrix whose rows $\tilde{m}_k, k = 1,\dots,m$, are "invariant measures" of the substochastic matrix T_0. Namely, $\tilde{m}_k = [0\cdots 0\ \tilde{\mu}_k\ 0\cdots 0]$, where $\tilde{\mu}_k$ is an invariant measure associated with stochastic matrix generator $\tilde{A}_k$. Next we substitute (6.35) into (6.36) to obtain

$$\tilde{M}T_1\tilde{Q}c_{-1} = -\tilde{M}R_{i0}\mathbf{1}.$$

Note that in the case of first order singularity the matrix $\tilde{M}T_1\tilde{Q}$ is invertible. Therefore $c_{-1} = 0$ (and thereby $\varphi_{i,-1} = 0$) provided that $\tilde{M}R_{i0} = 0$. However, the latter holds, since $\tilde{\Omega}_1,\dots,\tilde{\Omega}_m$ correspond to ergodic classes when $\varepsilon = 0$ and the submatrix R_{i0} must have zeros in places corresponding to those ergodic classes. Namely, R_{i0} has the following structure:

$$R_{i0} = \begin{bmatrix} 0 \\ \vdots \\ 0 \\ \rho_{i0} \end{bmatrix} \begin{matrix} \}\tilde{\Omega}_1 \\ \vdots \\ \}\tilde{\Omega}_m \\ \}\tilde{\Omega}_{\tilde{T}}. \end{matrix} \tag{6.37}$$

Now we can write the asymptotic expansion for $\varphi_i(\varepsilon)$ in the form

$$\varphi_i(\varepsilon) = \varphi_{i0} + \varepsilon\varphi_{i1} + \varepsilon^2\varphi_{i2} + \dots.$$

Again from (6.32), we obtain the system of fundamental equations for the coefficients $\varphi_{ik}, k \geq 0$.

$$T_0\varphi_{i0} = -R_{i0}\mathbf{1}, \tag{6.38}$$

$$T_0\varphi_{i1} + T_1\varphi_{i0} = -R_{i1}\mathbf{1}. \tag{6.39}$$

$$\dots$$

The general solution of equation (6.38) can be written in the form

$$\varphi_{i0} = \tilde{Q}c_{i0} + \varphi_{i,pt}, \tag{6.40}$$

where $\varphi_{i,pt}$ is any particular solution of (6.38). For instance, as a particular solution $\varphi_{i,pt}$ we may take

$$\varphi_{i,pt} = \begin{bmatrix} 0 \\ \vdots \\ 0 \\ -\tilde{T}^{-1}\rho_{i0}\mathbf{1} \end{bmatrix} \begin{matrix} \}\tilde{\Omega}_1, \\ \vdots \\ \}\tilde{\Omega}_m, \\ \}\tilde{\Omega}_{\tilde{T}}. \end{matrix} \tag{6.41}$$

Next we substitute the general solution (6.40) into the feasibility condition for the next fundamental equation (6.39).

$$\tilde{M}T_1\varphi_{i0} = -\tilde{M}R_{i1}\mathbf{1},$$

$$\tilde{M}T_1\tilde{Q}c_{i0} + \tilde{M}T_1\varphi_{i,pt} = -\tilde{M}R_{i1}\mathbf{1}.$$

Because of the first order singularity assumption, the matrix $\tilde{M}T_1\tilde{Q}$ is invertible, and we obtain

$$c_{i0} = -(\tilde{M}T_1\tilde{Q})^{-1}\tilde{M}(T_1\varphi_{i,pt} + R_{i1}\mathbf{1}), \tag{6.42}$$

and finally,

$$\varphi_{i0} = -\tilde{Q}(\tilde{M}T_1\tilde{Q})^{-1}\tilde{M}(T_1\varphi_{i,pt} + R_{i1}\mathbf{1}) + \varphi_{i,pt}\,. \tag{6.43}$$

Now let us illustrate the above theoretical development with the help of the following examples.

Example 6.4. *Consider an MC with the perturbed transition matrix*

$$P(\varepsilon) = \begin{bmatrix} 1 & 0 & 0 & 0 \\ 0 & 1 & 0 & 0 \\ 0 & \varepsilon & 1-2\varepsilon & \varepsilon \\ \varepsilon & 0 & 0 & 1-\varepsilon \end{bmatrix} \begin{matrix} \}\,\Omega_1, \\ \}\,\Omega_2, \\ \}\,\tilde{\Omega}_1, \\ \}\,\tilde{\Omega}_2. \end{matrix}$$

In this example we have

$$T_0 = \begin{bmatrix} 0 & 0 \\ 0 & 0 \end{bmatrix}, \quad T_1 = \begin{bmatrix} -2 & 1 \\ 0 & -1 \end{bmatrix},$$

$$R_{10} = \begin{bmatrix} 0 \\ 0 \end{bmatrix}, \quad R_{11} = \begin{bmatrix} 0 \\ 1 \end{bmatrix}, \quad R_{20} = \begin{bmatrix} 0 \\ 0 \end{bmatrix}, \quad R_{21} = \begin{bmatrix} 1 \\ 0 \end{bmatrix}.$$

Next we conclude that

$$\tilde{M} = \begin{bmatrix} 1 & 0 \\ 0 & 1 \end{bmatrix}, \quad \tilde{Q} = \begin{bmatrix} 1 & 0 \\ 0 & 1 \end{bmatrix},$$

and $\varphi_{pt} = 0$ *for both ergodic classes* Ω_1 *and* Ω_2, *since there is no submatrix* $\tilde{T}$, *which represents the states that are transient in the perturbed chain as well as in the unperturbed chain. Then, using the formula* (6.43), *we obtain*

$$\varphi_{10} = -\tilde{Q}(\tilde{M}T_1\tilde{Q})^{-1}\tilde{M}R_{11}\mathbf{1} = -\frac{1}{2}\begin{bmatrix} -1 & -1 \\ 0 & -2 \end{bmatrix}\begin{bmatrix} 0 \\ 1 \end{bmatrix} = \begin{bmatrix} 0.5 \\ 1 \end{bmatrix}$$

and

$$\varphi_{20} = -\tilde{Q}(\tilde{M}T_1\tilde{Q})^{-1}\tilde{M}R_{21}\mathbf{1} = -\frac{1}{2}\begin{bmatrix} -1 & -1 \\ 0 & -2 \end{bmatrix}\begin{bmatrix} 1 \\ 0 \end{bmatrix} = \begin{bmatrix} 0.5 \\ 0 \end{bmatrix}.$$

The above result is rather interesting. Of course, it is apparent that if the process were initiated in the second transient state $\tilde{\Omega}_2$, then it will be absorbed in the first ergodic class Ω_1 with probability one. However, it is a bit surprising that if the process is initiated in the first transient state $\tilde{\Omega}_1$, then it will enter the two ergodic states with equal probabilities. Since to enter the first ergodic class Ω_1 from the first transient state takes two steps and to enter the second ergodic class Ω_2 from the same transient state takes only one step, one might have expected the probabilities of absorption in these two ergodic classes to be different. Nevertheless, the above analysis shows that this is not the case.

Example 6.5. *Consider a (6×6) MC with one transient state and three ergodic classes before the perturbation. After the perturbation, there are three transient states and two ergodic classes, as one ergodic class becomes transient after the perturbation.*

$$P(\varepsilon) = \begin{pmatrix} 1 & 0 & 0 & 0 & 0 & 0 \\ 0 & 1/2 & 1/2 & 0 & 0 & 0 \\ 0 & 1/2 & 1/2 & 0 & 0 & 0 \\ 0 & 0 & 0 & 1/4 & 3/4 & 0 \\ 0 & 0 & 0 & 3/5 & 2/5 & 0 \\ 1/5 & 1/5 & 1/5 & 1/10 & 1/10 & 1/5 \end{pmatrix} + \varepsilon \begin{pmatrix} 0 & 0 & 0 & 0 & 0 & 0 \\ 0 & 0 & 0 & 0 & 0 & 0 \\ 0 & 0 & 0 & 0 & 0 & 0 \\ 2 & 0 & 0 & 0 & -3 & 1 \\ 0 & 1 & 1 & 1 & -4 & 1 \\ 0 & 0 & 0 & 0 & 0 & 0 \end{pmatrix}.$$

In order to calculate $\varphi_i(\varepsilon)$ the following matrices are relevant:

$$T_0 = S_0 - I = \begin{pmatrix} -3/4 & 3/4 & 0 \\ 3/5 & -3/5 & 0 \\ 1/10 & 1/10 & -4/5 \end{pmatrix}, \quad T_1 = \begin{pmatrix} 0 & -3 & 1 \\ 1 & -4 & 1 \\ 0 & 0 & 0 \end{pmatrix},$$

$$R_{10} = \begin{pmatrix} 0 \\ 0 \\ 1/5 \end{pmatrix}, \quad R_{11} = \begin{pmatrix} 2 \\ 0 \\ 0 \end{pmatrix}, \quad R_{20} = \begin{pmatrix} 0 & 0 \\ 0 & 0 \\ 1/5 & 1/5 \end{pmatrix}, \quad R_{21} = \begin{pmatrix} 0 & 0 \\ 1 & 1 \\ 0 & 0 \end{pmatrix}.$$

First, we need to construct $\tilde{Q}$ and $\tilde{M}$. Recall that $\tilde{Q}$ is the matrix of right 0-eigenvectors of T_0, and $\tilde{M}$ is the matrix containing the invariant measure of T_0. That is,

$$\tilde{Q} = \begin{pmatrix} 0 \\ 0 \\ 1/4 \end{pmatrix} \text{ and } \tilde{M} = \begin{pmatrix} 4/9 & 5/9 & 0 \end{pmatrix}.$$

Note that we have to calculate $\varphi_{i,pt}$ for each ergodic class i. In this case there are two such classes, and

$$\varphi_{1,pt} = \begin{pmatrix} 0 \\ 0 \\ 1/4 \end{pmatrix} \text{ and } \varphi_{2,pt} = \begin{pmatrix} 0 \\ 0 \\ 1/2 \end{pmatrix}.$$

Using formula (6.42), one can check that $c_{10} = \frac{41}{99}$ and $c_{20} = \frac{58}{99}$.

Finally, we are in the position the calculate φ_i using formula (6.41) or (6.43). This results in

$$\varphi_{10} = \frac{1}{99}\begin{pmatrix} 41 \\ 41 \\ 35 \end{pmatrix} \text{ and } \varphi_{20} = \frac{1}{99}\begin{pmatrix} 58 \\ 58 \\ 64 \end{pmatrix}.$$

6.2.3 ▪ The nearly completely decomposable case

We conclude this section with the discussion of a specially structured model that is, in fact, a special case of the irreducible perturbation in the sense introduced in the first subsection. We believe that this case deserves separate treatment because of the elegance and the interpretation of the formulae that can be used to compute the expansion of the invariant distribution vector. It is also the case that has received most attention in the literature.

Let $P(0) \in R^{N\times N}$ be a stochastic matrix representing transition probabilities in a completely decomposable MC. By the latter we mean that there exists a partition $\mathscr{S}$ of the state space into n, $n \geq 2$, subsets $\mathscr{S} = \{\Omega_1,\dots,\Omega_n\}$, each of which is an ergodic class. We assume that the order of the rows and of the columns of P is compatible with $\mathscr{S}$, that is, for n stochastic matrices, $P_1,\dots,P_n$,

$$P(0) = \begin{pmatrix} P_1 & 0 & \cdots & 0 \\ 0 & P_2 & \cdots & 0 \\ \vdots & \vdots & \ddots & 0 \\ 0 & 0 & \cdots & P_n \end{pmatrix}. \tag{6.44}$$

Note that we assume above that none of the states is transient. In this subsection we analyze the linear perturbation case. Specifically, let $C \in R^{N\times N}$ be a zero rowsum matrix such that for some $\varepsilon_{\max} > 0$, the matrix $P(\varepsilon) = P(0) + \varepsilon C$ is stochastic for $\varepsilon \in (0, \varepsilon_{\max})$ representing transition probabilities in an irreducible MC. For small values of ε, $P(\varepsilon)$ is called *nearly completely decomposable* (NCD) or sometimes *nearly uncoupled*. Clearly, $c_{ij} \geq 0$ for any pair of states i and j belonging to different subsets Ω_i and Ω_j, as every element in $P(\varepsilon)$ has to be nonnegative.

The highly structured relaxation of the irreducibility of $P(\varepsilon)$ at just the single value of $\varepsilon = 0$ may seem like a very minor change. Nonetheless, it will soon become clear that this small relaxation significantly changes the nature of the series expansions of most of the interesting matrix operators of the perturbed MC by introducing singularities in the expansions of the fundamental, deviation, and mean passage time matrices. Despite the latter, there is sufficient structure remaining in the NCD perturbed MC to permit special analysis that also lends itself to intuitive interpretation.

Next, recall that by irreducibility (for $\varepsilon > 0$), $\Pi(\varepsilon)$ consists of identical rows $\mu(\varepsilon)$, and that $\mu(\varepsilon)$ is analytic in some deleted neighborhood of zero. That is,

$$\mu(\varepsilon) = \sum_{m=0}^{\infty} \varepsilon^m \mu_m,$$

where $\mu_0 = \lim_{\varepsilon\to 0} \mu(\varepsilon)$ and where μ_m, $m \geq 1$, are zerosum vectors. Note that $[\mu_0]_i > 0$ for all i. For any subset $I \in \mathscr{S} = \{\Omega_1,\dots,\Omega_n\}$, let

$$x_I := \Sigma_{i\in I}[\mu_0]_i. \tag{6.45}$$

Note that $x_I > 0$ for any $I \in \mathscr{S}$, and define the probability vector

$$x := (x_1, x_2, \dots, x_n).$$

Also, let γ_I be the subvector of μ_0 corresponding to subset I rescaled so that its entry-sum is now one. Then, γ_I is the unique stationary distribution of A_I. Note that computing γ_I is easy as only the knowledge of A_I is needed.

Next define the matrix $\hat{Q} \in R^{n \times n}$ which is usually referred to as the *aggregated* transition matrix. Each row, and likewise each column, in $\hat{Q}$ corresponds to a subset in Ω. Then, for subsets I and J, $I \neq J$, let

$$\hat{Q}_{IJ} = \sum_{i \in I} (\gamma_I)_i \sum_{j \in J} c_{ij}, \tag{6.46}$$

and let

$$\hat{Q}_{II} = 1 + \sum_{i \in I} (\gamma_I)_i \sum_{j \in I} c_{ij} = 1 - \sum_{J \neq I} Q_{IJ}. \tag{6.47}$$

Note that the matrix C may be divided by any constant and ε may be multiplied by this constant leading to the same $N \times N$ transition matrices. Taking this constant small enough guarantees the stochasticity of $\hat{Q}$, and hence this is assumed without loss of generality. In particular, the stationary distribution of $\hat{Q}$ is invariant with respect to the choice of this constant. Alternatively, one can define $\hat{Q}_{II} := -\Sigma_{J \neq I} \hat{Q}_{IJ}$ and consider $\hat{Q}$ as the generator of the aggregated process, that is, the process among subsets $\Omega_1, \ldots, \Omega_n$ (and hence there is no need to assume anything further with regard to the size of the entries of the matrix C). Moreover, $\hat{Q}$ is irreducible, and the vector $x \in R^n$ (see (6.45)) is easily checked to be its unique stationary distribution.

Often it is convenient to express the aggregated transition matrix $\hat{Q}$ in matrix terms. Specifically, let $M \in R^{n \times N}$ be such that its ith row is full of zeros except for γ_{I_i} at the entries corresponding to subset I_i, and where $Q \in R^{N \times n}$ is such that its jth column is full of zeros except for 1's in the entries corresponding to the subset I_j. Now μ_0 is given by $\mu_0 = xM$. Note that $MQ \in R^{n \times n}$ is the identity matrix. Moreover, M and Q correspond to orthonormal sets of eigenvectors of $P(0)$ belonging to the eigenvalue 1, M made up of left eigenvectors and Q of right eigenvectors. Now, we can write

$$\hat{Q} = I + MCQ.$$

The aggregated stochastic matrix $\hat{Q}$ represents transition probabilities between subsets $\Omega_1, \ldots, \Omega_n$, which, in this context, are sometimes referred to as *macrostates*. However, although the original process among states is Markovian, this is not necessarily the case with the process among macrostates (and, indeed, typically it is not). The process among macrostates is an example of a partially observable Markov process. Yet, as will be seen below, much can be learned about the original process from the analysis of the aggregate process.

Theorem 6.1. *Let the perturbed MC be nearly completely decomposable.*

1. *The stationary distribution $\mu(\varepsilon)$ admits a Maclaurin series expansion in a deleted neighborhood of zero. Specifically, for some vectors $\{\mu_m\}_{m=0}^{\infty}$ with μ_0 being a probability vector positive in all its entries and satisfying $\mu_0 = \mu_0 P(0)$, and for some zerosum vectors μ_m, $m \geq 1$, $\mu(\varepsilon) = \Sigma_{m=0}^{\infty} \mu_m \varepsilon^m$.*

2. *Also, the sequence $\{\mu_m\}_{m=0}^{\infty}$ is geometric; that is, $\mu_m = \mu_0 U^m$ for any $m \geq 0$, where*

$$U = CH(I + CQDM), \tag{6.48}$$

where H is the deviation matrix of the unperturbed MC governed by $P(0)$, and where D is the deviation matrix of the aggregated MC governed by the transition matrix $\hat{Q}$.

Proof: Since $\mu(\varepsilon)$ is the unique solution of the linear system of equations

$$\mu(\varepsilon)P(\varepsilon) = \mu(\varepsilon), \quad \mu(\varepsilon)\mathbf{1} = 1, \tag{6.49}$$

whose coefficients are linear functions of ε, it possesses (at worst) a Laurent series expansion around $\varepsilon = 0$. However, since $\mu(\varepsilon)$ is a probability vector, it is bounded, and hence the latter expansion must constitute a Maclaurin series. Of course, $\mu(\varepsilon)\mathbf{1} = 1$ for all $\varepsilon > 0$ implies that $\mu_0\mathbf{1} = 1$ as well. Hence it follows that $\mu_m\mathbf{1} = 0$ for every positive integer m. Passing to the limit as $\varepsilon \to 0$ in (6.49) yields $\mu_0 = \mu_0 P(0)$.

Concerning the second part, we first note that for $m \geq 2$ all coefficients in the expansion of the generator $G(\varepsilon)$ are 0, $G_0 = P - I$, and $G_1 = C$. Thus the kth and $(k+1)$st fundamental equations (MFk) and (MFk+1) of Subsection 6.2.1 reduce to

$$\mu_k G_0 + \mu_{k-1} G_1 = 0, \quad \mu_{k+1} G_0 + \mu_k G_1 = 0. \tag{6.50}$$

Multiplying the second equation above on the right by Q and using the fact that $G_0 Q = PQ - Q = 0$, we immediately obtain

$$\mu_k G_1 Q = 0. \tag{6.51}$$

Also, equation (6.21) reduces to

$$\mu_k = \mu_k^{(1)} M + \mu_{k-1} G_1 H. \tag{6.52}$$

Multiplying (6.52) on the right by $G_1 Q$ and using (6.51), we obtain

$$\mu_k^{(1)} M G_1 Q + \mu_{k-1} G_1 H G_1 Q = 0, \tag{6.53}$$

together with the normalization condition $\mu_k^{(1)}\mathbf{1} = 0$. Now, the matrix $\bar{G} := M G_1 Q = MCQ$ is the generator of the aggregated MC, and hence the above equation may be thought of as the linear system $\mu_k^{(1)}[-\bar{G}] = b$, where $b = \mu_{k-1} G_1 H G_1 Q$. The solution of the corresponding homogeneous equation $\mu_k^{(1)}[-\bar{G}] = 0$ is a scalar multiple of the unique invariant distribution of the irreducible aggregated chain and can be denoted by $\rho\mu_0^{(1)}$. Furthermore, the deviation matrix, D, of the aggregated chain is also the group inverse of its negative generator, and hence the vector bD constitutes a particular solution of this linear system. Thus, according to Lemma 2.1, we have

$$\mu_k^{(1)} = \rho\mu_0^{(1)} + \mu_{k-1} G_1 H G_1 Q D. \tag{6.54}$$

Multiplying the above on the right by $\mathbf{1}$, using the property $D\mathbf{1} = 0$ of deviation matrices and the preceding normalization condition, we now obtain $\rho = 0$, and hence

$$\mu_k^{(1)} = \mu_{k-1} G_1 H G_1 Q D. \tag{6.55}$$

Substituting the above, and $G_1 = C$, into (6.52) yields

$$\mu_k = \mu_{k-1} G_1 H G_1 Q D M + \mu_{k-1} G_1 H = \mu_{k-1} C H [I + CQDM], \tag{6.56}$$

from which the required geometric nature of our sequence follows by iterating on the index k.

For $\varepsilon, 0 \leq \varepsilon < \varepsilon_{\max}$, let $H(\varepsilon)$ be the deviation matrix of $P(\varepsilon)$. This matrix is uniquely defined, and the case $\varepsilon = 0$ is no exception. Yet, as we will see later, there is a discontinuity in $H(\varepsilon)$ at $\varepsilon = 0$. However, $H(0)$ has the same shape as $P(0)$, namely,

$$H(0) = H = \begin{pmatrix} H_1 & 0 & \cdots & 0 \\ 0 & H_2 & \cdots & 0 \\ \vdots & \vdots & \ddots & \vdots \\ 0 & 0 & \cdots & H_n \end{pmatrix}, \tag{6.57}$$

where H_{I_i} is the deviation matrix of P_i, $1 \leq i \leq n$. □

Example 6.6. *Consider an MC with a linearly perturbed transition matrix* $P(\varepsilon) = P(0) + \varepsilon C$*:*

$$P(0) = \begin{pmatrix} 1 & 0 & 0 \\ 0 & 1/2 & 1/2 \\ 0 & 1/2 & 1/2 \end{pmatrix} \quad \text{and} \quad C = \begin{pmatrix} -2 & 1 & 1 \\ 1 & -1 & 0 \\ 0 & 1 & -1 \end{pmatrix}.$$

The number of ergodic subsets is equal to 2 *with* $\gamma_{I_1} = 1$ *and* $\gamma_{I_2} = (1/2, 1/2)$. *First, we construct the following matrices:*

$$M = \begin{pmatrix} 1 & 0 & 0 \\ 0 & 1/2 & 1/2 \end{pmatrix}, \quad Q = \begin{pmatrix} 1 & 0 \\ 0 & 1 \\ 0 & 1 \end{pmatrix}, \quad H = \begin{pmatrix} 0 & 0 & 0 \\ 0 & 1/2 & -1/2 \\ 0 & -1/2 & 1/2 \end{pmatrix}.$$

The aggregated generator is given by

$$\bar{G} = MCQ = \begin{pmatrix} -2 & 2 \\ 1/2 & -1/2 \end{pmatrix}.$$

Calculating the stationary distribution of matrix $\hat{Q}$ *gives us* $x = (\frac{1}{5}\ \frac{4}{5})$ *and* $\mu_0 = (\frac{1}{5}\ \frac{2}{5}\ \frac{2}{5})$. *Next, we calculate* D*, the deviation matrix of* $\hat{Q}$*,*

$$D = (MCQ + 1x)^{-1} - 1x = \frac{1}{25}\begin{pmatrix} 8 & -8 \\ -2 & 2 \end{pmatrix}.$$

Next, from (6.48) *we obtain*

$$U = CH(I + CQDM) = \frac{1}{5}\begin{pmatrix} 0 & 0 & 0 \\ -1 & -2 & 3 \\ 2 & 4 & -6 \end{pmatrix}.$$

Finally, we conclude that the stationary probabilities that the perturbed system is in state 1, 2, *or* 3 *are now obtainable from the expansions of* $\mu(\varepsilon)$*, which, in this case, has the form*

$$\mu(\varepsilon) = \begin{bmatrix} \frac{1}{5} & \frac{2}{5} & \frac{2}{5} \end{bmatrix} + \begin{bmatrix} \frac{2}{25} & \frac{4}{25} & -\frac{6}{25} \end{bmatrix} \varepsilon + \begin{bmatrix} -\frac{16}{125} & -\frac{32}{125} & \frac{48}{125} \end{bmatrix} \varepsilon^2 + \cdots.$$

Note that these are the same results as in Example 6.2 in the irreducible perturbation case, where we used the same matrices but a different way to find the stationary distribution.

6.3 ▪ Asymptotic Analysis of Deviation, Fundamental, and Mean Passage Time Matrices

First we note that the deviation matrix $H(\varepsilon)$ and the fundamental matrix $Z(\varepsilon)$ may be expressed in terms of matrix inversion (see formulae (6.13), (6.14)). Thus, the direct application of Theorem 2.4 from Section 2.2 implies that the deviation and fundamental matrices of the perturbed MC may be expanded as Laurent power series. Then, from formulae (6.13) and (6.14) we also conclude that the singular parts of the Laurent series for $H(\varepsilon)$ and $Z(\varepsilon)$ coincide. In the next subsection we study the general case of irreducible perturbations. Then, we study in more detail the cases of regular perturbation and NCD MCs. Here we do not provide the asymptotic analysis for the deviation, fundamental, and mean first passage matrices for the case of multichain perturbation. If needed, the interested reader can extend the results of the present section using the approach of Section 3.3.

6.3.1 ▪ The irreducible perturbation

First we present a reduction process for the computation of the coefficients of the Laurent series for the deviation matrix:

$$H(\varepsilon)=\frac{1}{\varepsilon^s}H_{-s}+\frac{1}{\varepsilon^{s-1}}H_{-s+1}+\cdots=\frac{1}{\varepsilon^s}(X_0+\varepsilon X_1+\dots). \tag{6.58}$$

When applying the reduction process, we number the coefficients starting from zero for notational convenience. Even though any number of Laurent series coefficients can be calculated by the reduction process, it is more computationally efficient to calculate by the reduction process only the singular part coefficients and the first regular part coefficient, namely, coefficients $X_k, k=0,\dots,s$. The other coefficients, if needed, may be computed by the recursive formulae provided in the second part of this section.

The reduction process

We shall use the reduction process to compute the first regular term and the singular part of the deviation matrix expansion (6.58). Since the deviation matrix is the negative group inverse of the MC generator, the general approach for the calculation of the Laurent series for perturbed group inverses (see Section 3.3) may be applied to the present problem. In this section we choose to pursue the algebraic reduction technique in the spirit of Section 2.2. Of course, once the matrices $H_{-k}, k=1,\dots,s$, and H_0 are computed, one can immediately recover (see (6.14)) the matrices $Z_{-k}, k=0,\dots,s$, that is

$$Z_{-k}=H_{-k},\ k=1,\dots,s, \quad Z_0=H_0+\Pi_0. \tag{6.59}$$

The reduction process for analytic perturbations has practically the same level of difficulty as for linear perturbations. Therefore, we consider the general case of analytic perturbations (6.15).

Under the assumption of irreducible perturbation, the deviation matrix $H(\varepsilon)$ of the perturbed MC is uniquely defined by the following equations:

$$H(\varepsilon)G(\varepsilon)=\Pi(\varepsilon)-I, \tag{6.60}$$

$$H(\varepsilon)1_N=0. \tag{6.61}$$

To simplify subsequent summations, we now formally introduce a new notation for the coefficients H_k. Let us denote $X_k \stackrel{def}{=} H_{k-s}, k = 0,1,\ldots$, so that $H(\varepsilon) = (X_0 + \varepsilon X_1 + \ldots)/\varepsilon^s = X(\varepsilon)/\varepsilon^s$ and (6.60)–(6.61) become

$$X(\varepsilon)G(\varepsilon) = \varepsilon^s[\Pi(\varepsilon) - I], \tag{6.62}$$

$$X(\varepsilon)\mathbf{1}_N = 0. \tag{6.63}$$

Then, substitute (6.58) and (6.17) into (6.62) and collect terms with the same power of ε to obtain

$$\begin{aligned}
X_0 G_0 &= 0 && (FH.0),\\
X_1 G_0 + X_0 G_1 &= 0 && (FH.1),\\
&\cdots\\
X_{s-1} G_0 + \cdots + X_0 G_{s-1} &= 0 && (FH.s-1),\\
X_s G_0 + \cdots + X_0 G_s &= \Pi_0 - I && (FH.s),\\
X_{s+1} G_0 + \cdots + X_0 G_{s+1} &= \Pi_1 && (FH.s+1),\\
&\cdots
\end{aligned}$$

the system of *fundamental equations.*

It follows from (6.63) that each equation (FH.k) is coupled with the *normalization condition*

$$X_k \mathbf{1}_N = 0, \; k \geq 0. \tag{6.64}$$

Before applying the reduction process, let us formally introduce the nth step reduced system of fundamental equations:

$$\begin{aligned}
X_0^{(n)} G_0^{(n)} &= 0 && (FHn.0),\\
X_1^{(n)} G_0^{(n)} + X_0^{(n)} G_1^{(n)} &= 0 && (FHn.1),\\
&\cdots\\
X_{s-n-1}^{(n)} G_0^{(n)} + \cdots + X_0^{(n)} G_{s-n-1}^{(n)} &= 0 && (FHn.s-n-1),\\
X_{s-n}^{(n)} G_0^{(n)} + \cdots + X_0^{(n)} G_{s-n}^{(n)} &= A_0^{(n)} && (FHn.s-n),\\
X_{s-n+1}^{(n)} G_0^{(n)} + \cdots + X_0^{(n)} G_{s-n+1}^{(n)} &= A_1^{(n)} && (FHn.s-n+1).\\
&\cdots
\end{aligned}$$

With $n = 0$, we retrieve the original system of fundamental equations (FH). Namely, $X_k^{(0)} = X_k, G_k^{(0)} = G_k$, and $A^{(0)} = \Pi_0 - I, A_k^{(0)} = \Pi_k, k \geq 1$. The nth step reduced equations (FHn) are also coupled with the normalization conditions

$$X_k^{(n)} \mathbf{1}_{m_n} = 0, \quad k \geq 0. \tag{6.65}$$

The matrix $G_0^{(n)} \in \mathbb{R}^{m_n \times m_n}$ can be interpreted as the generator of the nth step aggregated MC. Further, m_n is the number of ergodic classes of the $(n-1)$th step aggregated MC, and, in particular, m_1 is the number of ergodic classes in the original chain. The corresponding aggregated ergodic projection $\Pi^{(n)}$ satisfies

$$\Pi^{(n)} G_0^{(n)} = G_0^{(n)} \Pi^{(n)} = 0. \tag{6.66}$$

The aggregated ergodic projection is constructed by the following decomposition:

$$\Pi^{(n)} = Q^{(n)} M^{(n)}, \tag{6.67}$$

where

$$Q^{(n)} \in \mathbb{R}^{m_n \times m_{n+1}} : G_0^{(n)} Q^{(n)} = 0, Q^{(n)} \mathbf{1}_{m_{n+1}} = \mathbf{1}_{m_n}; \tag{6.68}$$

$$M^{(n)} \in \mathbb{R}^{m_{n+1} \times m_n} : M^{(n)} G_0^{(n)} = 0, M^{(n)} \mathbf{1}_{m_n} = \mathbf{1}_{m_{n+1}}. \tag{6.69}$$

In addition,

$$M^{(n)} Q^{(n)} = I_{m_{n+1}}.$$

The nth step deviation matrix $H^{(n)}$ is computed via the formula

$$H^{(n)} = (\Pi^{(n)} - G_0^{(n)})^{-1} - \Pi^{(n)}. \tag{6.70}$$

Of course, this matrix satisfies (6.60)–(6.61), that is,

$$H^{(n)} G_0^{(n)} = G_0^{(n)} H^{(n)} = \Pi^{(n)} - I, \tag{6.71}$$

$$H^{(n)} \Pi^{(n)} = \Pi^{(n)} H^{(n)} = 0. \tag{6.72}$$

We may now formulate the main result of this section, which allows us to solve the system (FH) step by step.

Theorem 6.2. *Let the nth step* ($n < s$) *reduced fundamental system* (FHn) *with the normalization conditions* (6.65) *be given. Then the unknown matrices* $X_k^{(n)}$ *satisfy*

$$X_k^{(n)} = X_k^{(n+1)} M^{(n)} + \begin{cases} \sum_{i=0}^{k-1} X_i^{(n)} G_{k-i}^{(n)} H^{(n)}, & k < s-n, \\ \sum_{i=0}^{k-1} X_i^{(n)} G_{k-i}^{(n)} H^{(n)} - A_{k+n-s}^{(n)} H^{(n)}, & k \geq s-n, \end{cases} \tag{6.73}$$

where $X_k^{(n+1)} \in \mathbb{R}^{m_n \times m_{n+1}}, k = 0, 1, \ldots,$ *are solutions of the next* $(n+1)$*st step reduced fundamental equations*

$$X_0^{(n+1)} G_0^{(n+1)} = 0 \qquad (FHn+1.0),$$

$$X_1^{(n+1)} G_0^{(n+1)} + X_0^{(n+1)} G_1^{(n+1)} = 0 \qquad (FHn+1.1),$$

$$\cdots$$

$$X_{s-n-2}^{(n+1)} G_0^{(n+1)} + \cdots + X_0^{(n+1)} G_{s-n-2}^{(n+1)} = 0 \quad (FHn+1.s-n-2),$$

$$X_{s-n-1}^{(n+1)} G_0^{(n+1)} + \cdots + X_0^{(n+1)} G_{s-n-1}^{(n+1)} = A_0^{(n+1)} (FHn+1.s-n-1),$$

$$X_{s-n}^{(n+1)} G_0^{(n+1)} + \cdots + X_0^{(n+1)} G_{s-n}^{(n+1)} = A_1^{(n+1)} \ (FHn+1.s-n).$$

$$\cdots$$

The matrices $G_k^{(n+1)}$ *and the right-hand-sides* $A_k^{(n+1)}$ *are given by*

$$G_k^{(n+1)} = M^{(n)} \sum_{p=1}^{k+1} \sum_{\nu_1+\cdots+\nu_p=k+1} G_{\nu_1}^{(n)} H^{(n)} G_{\nu_2}^{(n)} \cdots H^{(n)} G_{\nu_p}^{(n)} Q^{(n)}, \quad k = 0, 1, \ldots, \tag{6.74}$$

$$A_k^{(n+1)} = \sum_{i=0}^{k} A_i^{(n)} \sum_{p=1}^{k-i} \sum_{\nu_1+\cdots+\nu_p=k-i} G_{\nu_1}^{(n)} H^{(n)} G_{\nu_2}^{(n)} \cdots H^{(n)} G_{\nu_p}^{(n)} Q^{(n)}, \quad k = 0, 1, \ldots. \tag{6.75}$$

The new reduced equations (FHn+1) *are also coupled with the normalization conditions*

$$X_k^{(n+1)}\mathbf{1}_{m_{n+1}} = 0, \quad k = 0, 1, \ldots. \tag{6.76}$$

Proof: Define the new unknowns

$$X_k^{(n+1)} \stackrel{def}{=} X_k^{(n)} Q^{(n)}. \tag{6.77}$$

The first fundamental equation (FHn.0) implies

$$X_0^{(n)} = X_0^{(n)} \Pi^{(n)}. \tag{6.78}$$

Substituting the decomposition of the ergodic projection (6.67) into (6.78), we obtain

$$X_0^{(n)} = X_0^{(n)} Q^{(n)} M^{(n)} = X_0^{(n+1)} M^{(n)}. \tag{6.79}$$

Multiply (FHn.1) by $Q^{(n)}$ from the right and use (6.68) to obtain

$$X_0^{(n)} G_1^{(n)} Q^{(n)} = 0. \tag{6.80}$$

Then, substitute (6.79) into (6.80) to obtain

$$X_0^{(n+1)} M^{(n)} G_1^{(n)} Q^{(n)} = 0$$

or

$$X_0^{(n+1)} G_0^{(n+1)} = 0,$$

where $G_0^{(n+1)} \stackrel{def}{=} M^{(n)} G_1^{(n)} Q^{(n)}$. The above equation is the first required equation (FHn+1.0). The nth step deviation matrix $H^{(n)}$ plays a crucial role in obtaining the subsequent reduced equations. Indeed, consider the following decomposition of $X_1^{(n)}$:

$$\begin{aligned} X_1^{(n)} &= X_1^{(n)} \Pi^{(n)} + X_1^{(n)} (I - \Pi^{(n)}) \\ &= X_1^{(n)} Q^{(n)} M^{(n)} - X_1^{(n)} G_0^{(n)} H^{(n)} \\ &= X_1^{(n+1)} M^{(n)} - X_1^{(n)} G_0^{(n)} H^{(n)}. \end{aligned} \tag{6.81}$$

In the above the definition of $X_1^{(n+1)}$ (6.77) and the property (6.71) of the deviation matrix $H^{(n)}$ have been used. Now, using (FHn.1) $X_1^{(n)} G_0^{(n)} = -X_0^{(n)} G_1^{(n)}$ and substituting it into (6.81) yields

$$\begin{aligned} X_1^{(n)} &= X_1^{(n+1)} M^{(n)} + X_0^{(n)} G_1^{(n)} H^{(n)} \\ &= X_1^{(n+1)} M^{(n)} + X_0^{(n+1)} M^{(n)} G_1^{(n)} H^{(n)}, \end{aligned} \tag{6.82}$$

where the last equality follows from (6.79). Note that we have expressed the nth step unknown $X_1^{(n)}$ in terms of new unknowns $X_0^{(n+1)}, X_1^{(n+1)}$. Similar expressions are obtained for $X_k^{(n)}, k \geq 2$.

Now substitute (6.79) and (6.82) into (FHn.2) to obtain

$$[X_1^{(n+1)} M^{(n)} + X_0^{(n+1)} M^{(n)} G_1^{(n)} H^{(n)}] G_1^{(n)} + X_0^{(n+1)} M^{(n)} G_2^{(n)} = 0.$$

Multiplying (FHn.2) by $Q^{(n)}$ from the right and using (6.68) yields

$$X_1^{(n+1)}M^{(n)}G_1^{(n)}Q^{(n)}+X_0^{(n+1)}M^{(n)}(G_1^{(n)}H^{(n)}G_1^{(n)}+G_2^{(n)})Q^{(n)}=0,$$

or, equivalently,

$$X_1^{(n+1)}G_0^{(n+1)}+X_0^{(n+1)}G_1^{(n+1)}=0,$$

with $G_1^{(n+1)} \stackrel{def}{=} M^{(n)}(G_1^{(n)}H^{(n)}G_1^{(n)}+G_2^{(n)})Q^{(n)}$. Thus, we have the second step reduced equation (FHn+1.1) and an expression for $G_1^{(n+1)}$. The subsequent next step reduced equations are obtained with similar arguments. The general formulae (6.74) and (6.75) can be proved by induction (Problem 6.5).

Note that if $X_k^{(n+1)}, k \geq 0$, are known, then the coefficients $X_k^{(n)}, k \geq 0$, are easily calculated by the recursive formula (6.73). Indeed,

$$\begin{aligned}
X_k^{(n)} &= X_k^{(n)}\Pi^{(n)}+X_k^{(n)}(I-\Pi^{(n)}) \\
&= X_k^{(n+1)}M^{(n)}-X_k^{(n)}G_0^{(n)}H^{(n)} \\
&= X_k^{(n+1)}M^{(n)}+\begin{cases} \sum_{i=0}^{k-1} X_i^{(n)}G_{k-i}^{(n)}H^{(n)}, & k<s-n, \\ \sum_{i=0}^{k-1} X_i^{(n)}G_{k-i}^{(n)}H^{(n)}-A_{k+n-s}^{(n)}H^{(n)}, & k\geq s-n. \end{cases}
\end{aligned}$$

Finally, we show that the normalization condition (6.76) holds. To prove this, we need the second identity in (6.69) and the property of the ergodic projection $\Pi^{(n)}\mathbf{1}_{m_n}=\mathbf{1}_{m_n}$. For example, consider the case $k<s-n$.

$$\begin{aligned}
0=X_k^{(n)}\mathbf{1}_{m_n} &= X_k^{(n+1)}M^{(n)}\mathbf{1}_{m_n}+\sum_{i=0}^{k-1}X_i^{(n)}G_{k-i}^{(n)}H^{(n)}\mathbf{1}_{m_n} \\
&= X_k^{(n+1)}\mathbf{1}_{m_{n+1}}+\sum_{i=0}^{k-1}X_i^{(n)}G_{k-i}^{(n)}H^{(n)}\Pi^{(n)}\mathbf{1}_{m_n} \\
&= X_k^{(n+1)}\mathbf{1}_{m_{n+1}}
\end{aligned}$$

since $H^{(n)}\Pi^{(n)}=0$. □

Note that the $(n+1)$st step reduced system (FHn+1) has a structure very similar to that of the nth step reduced system (FHn). The only, but important, difference between the structures of these two systems is that the system (FHn+1) has fewer equations with null right-hand sides. Thus, after s reduced steps, the system (below) of reduced equations has nonzero right-hand sides.

$$\begin{aligned}
X_0^{(s)}G_0^{(s)} &= A_0^{(s)} \qquad (FHs.0), \\
X_1^{(s)}G_0^{(s)}+X_0^{(s)}G_1^{(s)} &= A_1^{(s)} \qquad (FHs.1), \\
&\cdots \\
X_k^{(s)}G_0^{(s)}+\cdots+X_0^{(s)}G_k^{(s)} &= A_k^{(s)} (FHs.k). \\
&\cdots
\end{aligned}$$

The next proposition gives simple recursive formulae for the solution of the (final step) reduced system (FHs).

Proposition 6.2. *The solutions* $X_k^{(s)}, k=0,1,\ldots,$ *of the system (FHs) are given by*

$$X_0^{(s)} = -A_0^{(s)} H^{(s)}; \; X_k^{(s)} = \left[\sum_{i=0}^{k-1} X_i^{(s)} G_{k-i}^{(s)} - A_k^{(s)} \right] H^{(s)}, \quad k \geq 1. \tag{6.83}$$

Proof: The final-step aggregated generator $G_0^{(s)}$ has the same number of ergodic classes as the perturbed chain described by the generator $G(\varepsilon), \varepsilon > 0$. Hence, in view of the irreducible perturbation assumption, the aggregated generator $G_0^{(s)}$ is a unichain generator, and the corresponding ergodic projection is just $\Pi^{(s)} = \mathbf{1}_{m_s} \mu_0^{(s)}$, where $\mu_0^{(s)} \in \mathbb{R}^{1 \times m_s}$ is a unique stationary distribution vector of the aggregated generator $G_0^{(s)}$.

Of course, the final-step reduced system (FHs) is coupled with the normalization conditions $X_k^{(s)} \mathbf{1}_{m_s} = 0, k = 0,1,\ldots.$ Multiplying by $\mu_0^{(s)}$, we obtain

$$X_k^{(s)} \mathbf{1}_{m_s} \mu_0^{(s)} = X_k^{(s)} \Pi^{(s)} = 0, \quad k = 0,1,\ldots. \tag{6.84}$$

Now, using the modified normalization conditions (6.84) and the decomposition of $X_k^{(s)}$ into subspaces $\mathscr{R}(\Pi^{(s)})$ and $\mathscr{R}(I - \Pi^{(s)})$, we obtain the recursive formulae (6.83):

$$\begin{aligned} X_k^{(s)} &= X_k^{(s)} P^{*(s)} + X_k^{(s)} (I - \Pi^{(s)}) \\ &= X_k^{(s)} (I - \Pi^{(s)}) = -X_k^{(s)} G_0^{(s)} H^{(s)} \\ &= \left[\sum_{i=0}^{k-1} X_i^{(s)} G_{k-i}^{(s)} - A_k^{(s)} \right] H^{(s)}. \end{aligned}$$ □

Using Theorem 6.2 and Proposition 6.2, we are now able to outline a practical algorithm for the computation of matrices H_{-k} and Z_{-k} for $k = 0,\ldots,s$.

Computational algorithm for the series coefficients of fundamental and deviation matrices

1. Set $s = 1$.
2. Carry out a reduction step. If $G_0^{(s)}$ has rank $m_s - 1$, the pole has order s. One can now proceed to the next step. If $G_0^{(s)}$ has rank smaller than $m_s - 1$, one should increment s and carry out another reduction step.
3. By using the formulae in Theorem 6.2, successively calculate the matrices $G_k^{(n)}, k = 0,\ldots,2s-n$, and the right-hand sides $A_k^{(n)}, k = 0,\ldots,s$, for $n = 1,\ldots,s$. As a result, one obtains the final system of reduced fundamental equations (FHs).
4. Calculate $X_k^{(s)}, k = 0,\ldots,s$, in (FHs) using the recurrent formulae (6.83).
5. Using (6.73), reconstruct successively all the $X_k^{(n)}, k = 0,\ldots,s$, from $n = s-1$ down to $n = 0$. In particular,

$$H_k = X_{k+s} = X_{k+s}^{(0)}, \quad k = -s,\ldots,0.$$

6. Finally, via (6.59), compute the matrices of the fundamental matrix expansion

$$Z_{-k} = H_{-k}, k = 1,\dots,s; \quad Z_0 = H_0 + \Pi_0.$$

Remark 6.2. *To calculate the matrices* $G_k^{(n+1)}$ *and* $A_k^{(n+1)}$*, instead of using* (6.74) *and* (6.75)*, one may also use recursive formulae that are more efficient and simpler. Define*

$$V_k^{(n)} = \sum_{p=1}^{k+1} \sum_{\nu_1+\cdots+\nu_p=k+1} G_{\nu_1}^{(n)} H^{(n)} G_{\nu_2}^{(n)} \cdots H^{(n)} G_{\nu_p}^{(n)},$$

which can be calculated by the simple recursive formula

$$V_k^{(n)} = G_{k+1}^{(n)} + \sum_{i=1}^{k} G_i^{(n)} H^{(n)} V_{k-i}^{(n)}, \quad k = 0, 1, \dots. \tag{6.85}$$

We then have

$$G_k^{(n+1)} = M^{(n)} V_k^{(n)} Q^{(n)} \quad and \quad A_k^{(n+1)} = \sum_{i=0}^{k} A_i^{(n)} V_{k-i-1}^{(n)} Q^{(n)}, \tag{6.86}$$

which definitely appear considerably simpler than (6.74)–(6.75).

One may obtain *every* matrix $Z_k, k = -s, -s+1, \dots,$ by the reduction process. At first sight, it seems that if we already have $Z_{-s}, \dots, Z_k$ and want to obtain the next coefficient Z_{k+1}, we need to calculate a lot of additional reduced coefficients $G_k^{(n)}$. Fortunately, we can avoid this. From Section 2.2, we know that once $X_0, \dots, X_k$ are computed, the next coefficient X_{k+1} is uniquely determined by the next finite subsystem of fundamental equations (FH),

$$X_k G_0 = R_k - \sum_{i=1}^{k} X_{k-i} G_i,$$

$$X_{k+1} G_0 + X_k G_1 = R_{k+1} - \sum_{i=1}^{k} X_{k-i} G_{i+1},$$

$$\dots$$

$$X_{k+s} G_0 + \cdots + X_k G_s = R_{k+s} - \sum_{i=1}^{k} X_{k-i} G_{i+s},$$

where

$$R_k = \begin{cases} 0, & k < s, \\ \Pi_0 - I, & k = s, \\ \Pi_{k-s}, & k > s, \end{cases}$$

plus the corresponding normalization conditions. Note that the above system can be efficiently solved by the same reduction process as before. Moreover, we need only recompute the right-hand sides $A_k^{(n)}$. The coefficient matrices $G_k^{(n)}, k = 0, \dots, s-n, n = 1, \dots, s,$ computed before can be used again. By doing this, one can even accelerate the computational procedure for $Z_k, k = -s, \dots, 0$, outlined above.

However, despite the above elegant modification of the reduction process, we still recommend calculating the regular part of the Laurent series by using an even simpler recursive formula described below.

Since the deviation matrix $H(\varepsilon)$ is a negative group inverse of the perturbed MC generator $G(\varepsilon)$, we may use the recursive formula (3.49) from Section 3.3. In particular, this formula allows us to deal with analytic perturbations. Here (3.49) takes the form

$$H_{k+1} = \sum_{i=0}^{k+s}\left(\sum_{j=0}^{s} H_{-j}G_{i+j+1}\right)H_{k-i} - \sum_{i=1}^{m}\Pi_{k+1-i}H_i$$
$$-(\Pi_{m+1}H_0 + \cdots + \Pi_{m+1+s}H_{-s}), \tag{6.87}$$

where $m \geq 0$ and $\Pi_k, k \geq 0$, are coefficients of the Taylor series for the ergodic projection $\Pi(\varepsilon)$ of the perturbed MC. Note that the term $(H_{-s}\Pi_{m+1+s} + \cdots + H_0\Pi_{m+1})$ in (3.49) vanishes, since according to the irreducible perturbation assumption $H_k\mathbf{1} = 0$ and $\Pi_k = \mathbf{1}\mu_k$.

Finally, we discuss the computational complexity of the above algorithm. Obviously, Steps 2 and 4 have the highest computational burden. In fact, Step 2 is computationally the most demanding. Therefore, it suffices to estimate the number of arithmetic operations in Step 2 to obtain the computational complexity of the reduction process.

Step 2 consists of s reduction steps. Note that the first reduction step is the most demanding from a computational point of view, since it reduces the determining system from the full state space into the aggregated chain subspace with dimension m_1 equal to the number of ergodic classes in the original unperturbed chain. It is not difficult to see that the number of operations in this procedure is $O(s^2N^3)$. Indeed, multiplying two $N \times N$ matrices requires $O(N^3)$ operations, and the recursive formulae (6.85), (6.86) for $k = 0,\ldots,2s-1$ require $O(s^2)$ such multiplications. After this crucial first step, we deal only with matrices whose dimension does not exceed m_1. The complexity of the other reduction steps can be estimated as $O(s^3m_1^3)$. Thus, Step 2 requires $O(s^2N^3 + s^3m_1^3)$ operations.

Let us now discuss this evaluation. In most practical applications, $m_1 \ll N$ and $s \ll N$; that is, the number of ergodic classes and the order of singularity are much less than N, the number of states of the original chain. Therefore, the complexity of the algorithm is in fact not much worse than $O(N^3)$ (or $O(N^4)$; see the remark below about the determination of s). However, if $m_1 \sim N$, then the complexity of the algorithm is $O(s^3N^3)$. The latter may increase significantly (even up to $O(N^6)$) if s is of the same order of magnitude as N. However, based on our experience, we believe that the cases of large s are quite rare.

One may choose not to determine the order of the pole before proceeding with the reduction process. In such a case the reduction process algorithm needs to be run with $s := 1,2,\ldots$ until $G_0^{(s)}$ has rank $m_s - 1$, in which case s is the order of the pole. Therefore, assuming that $m_1 \ll N$, the computational complexity to determine both s and the singular part Z^s is just $O(s^3N^3)$. When $s \ll N$ (as one may expect in practice), compare the above with $O(N^4)$ for just obtaining s in the Hassin and Haviv combinatorial algorithm outlined in the following subsection.

6.3.2 ▪ The linear perturbation

In this subsection, in addition to the assumption that the perturbed MC is irreducible, we assume that the perturbation is linear, that is,

$$P(\varepsilon) = P + \varepsilon C,$$

and, consequently,

$$G(\varepsilon) = G_0 + \varepsilon G_1,$$

with $G_0 = P - I$ and $G_1 = C$.

In the case of linear perturbation there exists a combinatorial algorithm for the determination of the order of the pole of the Laurent series for deviation and mean first passage time matrices. Before presenting the algorithm, let us introduce some necessary notations.

We say that $f(\varepsilon)$ is of order of magnitude k and denote it by $f(\varepsilon) = \Theta(\varepsilon^k)$ if there exist positive real numbers m and M such that, for all $\varepsilon > 0$ small enough,

$$m\varepsilon^k \le |f(\varepsilon)| \le M\varepsilon^k.$$

Let us associate a graph $G = (V, E)$ with the transition matrix $P(\varepsilon)$. Each node in V corresponds to an MC state, and each edge in E corresponds to a positive transition probability $p_{ij}(\varepsilon)$. Furthermore, we divide the edge set as $E = E_r \cup E_e$ on the basis of the order of magnitude of the transition probabilities $p_{ij}(\varepsilon)$. Namely, if $p_{ij}(\varepsilon) = \Theta(1)$, we classify the edge as $(i, j) \in E_r$, and if $p_{ij}(\varepsilon) = \Theta(\varepsilon)$, we classify it as $(i, j) \in E_e$. The edges of E_r are called *r-edges* (regular edges) and the edges of E_e are called *e-edges* (epsilon edges). A path (cycle) in G is called an *r-path* (r-cycle, resp.) if it consists only of r-edges. For a subset of vertices C, denote $\delta(C)$ the set of its outward-oriented boundary edges. Namely, $\delta(C) = \{(i, j) \in E | i \in C, j \notin C\}$.

Let us fix a state $s \in V$ and denote by $\mathcal{M}_i(\varepsilon)$ the expected time of the first passage to state i when the process starts from state s. Since $\mathcal{M}_i(\varepsilon)$ may be found from a solution of a linear system, we have that $\mathcal{M}_i(\varepsilon) = \Theta(\varepsilon^{-u(i)})$ for some integer $u(i)$ which is zero or positive. The following algorithm determines $u(i)$ for all $i \in V$.

Combinatorial algorithm for the determination of the order of the pole for the expected mean passage times:

Input: $G = (V, E_r, E_e)$ and node s.

Output: $u(i)$ for all $i \in V$.

Step 1 (initialization): Construct a graph $G' = (V', E'_r, E'_e)$ from G by deleting all loops $(i, i) \in E_e$ and all edges emanating out of s. Set $u(i) = 0$ and $S(i) = \{i\}$ for all $i \in V$.

Step 2 (condensation of cycles): If G' does not contain directed r-cycles, go to Step 3. Otherwise, let C be such a cycle. Condense C into a single node c, and set the value of $u(c)$ according to the following two cases:
Case (i) $\delta(C) \cap E'_r \neq \emptyset$. Set $u(c) = \max\{u(i) | i \in C\}$.
Case (ii) $\delta(C) \subset E'_e$. Set $u(c) = 1 + \max\{u(i) | i \in C\}$.
Change E'_r to $E'_r \cup \delta(C)$, and change E'_e to $E'_e \backslash \delta(C)$. Set $S(c) = \cup_{i \in C} S(i)$. Repeat Step 2.

Step 3 (solution of the problem for r-acyclic graphs): Set $T = V'$. Let $u(j) = \max\{u(i) | i \in T\}$, breaking ties arbitrarily. Delete j from T. For r-edges (i, j) with $i \in T$, set $u(i) = u(j)$. For e-edges (i, j) with $i \in T$, set $u(i)$ as $\max\{u(i), u(j) - 1\}$. If $T = \emptyset$, go to Step 4; else repeat Step 3.

Step 4 (determination of $u(i), i \in V \backslash \{s\}$): The collection of sets $\{S(v') | v' \in V'\}$ is a partition of V. For each $v \in V$, find $v' \in V'$ such that $v \in S(v')$, and set $u(v) = u(v')$.

Step 5 (determination of $u(s)$): Set $u(s) = \max\{\max\{u(i) | (s, i) \in E_r\}, \max\{u(i) - 1 | (s, i) \in E_e\}\}$.

For ease of understanding the above algorithm, we recommend executing the algorithm on the example given in Problem 6.6.

Now denote by u_{kl} the order of the pole for the expected mean passage time $\mathcal{M}_{kl}$ from state k to state l. Then, thanks to the formula (6.11),

$$\frac{\mathcal{M}_{kl}}{\mathcal{M}_{ll}} = \delta_{kl} + H_{ll} - H_{kl},$$

and the fact that $H_{ll} \geq H_{kl}$ (see Problem 6.7), we can immediately retrieve the order of the pole of the deviation matrix in (6.58):

$$s = \max_{k,l}\{u_{kl} - u_{ll}\}.$$

Once the order of the pole is determined, the reduction process for the computation of the singular part coefficients becomes straightforward.

The refined reduction process computational algorithm:

1. Determine the order of singularity s using the combinatorial Hassin and Haviv algorithm.

2. By using the formulae in Theorem 6.2, carry out s reduction steps (i.e., successively calculate the matrices $G_k^{(n)}, k = 0,\dots,2s-n$, and the right-hand sides $A_k^{(n)}, k = 0,\dots,s$, for $n = 1,\dots,s$). As a result, obtain the final system of reduced fundamental equations (FHs).

3. Calculate $X_k^{(s)}, k = 0,\dots,s$, in (FHs) using the recursive formulae (6.83).

4. Using (6.73), reconstruct successively all the $X_k^{(n)}, k = 0,\dots,s$, from $n = s-1$ down to $n = 0$. In particular,

$$H_k = X_{k+s} = X_{k+s}^{(0)}, \quad k = -s,\dots,0.$$

5. Finally, via (6.59), calculate the matrices of the fundamental matrix expansion

$$Z_{-k} = H_{-k}, k = 1,\dots,s; \quad Z_0 = H_0 + \Pi_0.$$

The regular part of the Laurent series for the fundamental matrix $Z^R(\varepsilon) = Z_0 + \varepsilon Z_1 + \dots$ may now be expressed by an updating formula given in the next theorem.

Theorem 6.3. *Let $P(\varepsilon) = P + \varepsilon C$ be the transition matrix of a linearly perturbed MC, and let the perturbation be irreducible. Then the regular part $Z^R(\epsilon)$ of the fundamental matrix $Z(\epsilon)$ is given by*

$$Z^R(\varepsilon) = \{[I - \Pi(\varepsilon)]Z_0 + \Pi_0\}[I - \varepsilon C Z_0]^{-1} - \Pi(\varepsilon)Z^S(\varepsilon). \tag{6.88}$$

Proof: For arbitrary $0 < \epsilon_1, \epsilon_2$, we have the following identity (see Problem 6.8):

$$Z(\varepsilon_1) - Z(\varepsilon_2) = (\varepsilon_1 - \varepsilon_2)Z(\varepsilon_1)CZ(\varepsilon_2) + Z(\varepsilon_1)\Pi(\varepsilon_2) - \Pi(\varepsilon_1)Z(\varepsilon_2).$$

Under the assumption of the irreducible perturbation, and using $Z(\varepsilon_1)\mathbf{1} = \mathbf{1}$, we have $Z(\varepsilon_1)\Pi(\varepsilon_2) = Z(\varepsilon_1)\mathbf{1}\mu(\varepsilon_2) = \mathbf{1}\mu(\varepsilon_2) = \Pi(\varepsilon_2)$. Hence,

$$Z(\varepsilon_1) - Z(\varepsilon_2) = (\varepsilon_1 - \varepsilon_2)Z(\varepsilon_1)CZ(\varepsilon_2) + \Pi(\varepsilon_2) - \Pi(\varepsilon_1)Z(\varepsilon_2). \tag{6.89}$$

With ε_2 fixed in (6.89), the regular parts with respect to ε_1 satisfy

$$Z^R(\varepsilon_1) - Z(\varepsilon_2) = (\varepsilon_1 - \varepsilon_2)Z^R(\varepsilon_1)CZ(\varepsilon_2) + Z_{-1}CZ(\varepsilon_2) + \Pi(\varepsilon_2) - \Pi(\varepsilon_1)Z(\varepsilon_2). \tag{6.90}$$

Letting $\varepsilon_1 = \varepsilon_2$ in (6.90) yields

$$Z^R(\varepsilon_2) - Z(\varepsilon_2) = Z_{-1}CZ(\varepsilon_2) + \Pi(\varepsilon_2) - \Pi(\varepsilon_2)Z(\varepsilon_2), \tag{6.91}$$

and since $\Pi(\varepsilon) = \Pi(\varepsilon)Z(\varepsilon)$, we have

$$Z^R(\varepsilon_2) - Z(\varepsilon_2) = Z_{-1}CZ(\varepsilon_2)$$

so that

$$Z^S(\varepsilon) = -Z_{-1}CZ^S(\varepsilon) \tag{6.92}$$

and

$$0 = Z_{-1}CZ^R(\varepsilon). \tag{6.93}$$

If, instead, we fix ε_1 in (6.89) and consider the regular parts with respect to ε_2, we obtain

$$Z^S(\varepsilon) = -Z^S(\varepsilon)CZ_{-1} \tag{6.94}$$

and

$$Z^R(\varepsilon)CZ_{-1} = \Pi(\varepsilon) - \Pi(\varepsilon)Z^R(\varepsilon) = \Pi(\varepsilon)Z^S(\varepsilon). \tag{6.95}$$

Taking the regular parts in (6.90) with respect to ε_2 (with ε_1 fixed) yields

$$\begin{aligned} Z^R(\varepsilon_1) - Z^R(\varepsilon_2) = (\varepsilon_1 - \varepsilon_2)Z^R(\varepsilon_1)CZ^R(\varepsilon_2) - Z^R(\varepsilon_1)CZ_{-1} + Z_{-1}CZ^R(\varepsilon_2) + \Pi(\varepsilon_2) \\ - \Pi(\varepsilon_1)Z^R(\varepsilon_2). \end{aligned}$$

The term $Z_{-1}CZ^R(\varepsilon_2)$ vanishes in view of (6.93). Then with $\varepsilon_1 := \varepsilon$ and letting $\varepsilon_2 \to 0$, one obtains

$$Z^R(\varepsilon) - Z_0 = \varepsilon Z^R(\varepsilon)CZ_0 - Z^R(\varepsilon)CZ_{-1} + \Pi_0 - \Pi(\varepsilon)Z_0,$$

$$Z^R(\varepsilon)[I - \varepsilon CZ_0] = [I - \Pi(\varepsilon)]Z_0 + \Pi_0 - Z^R(\varepsilon)CZ_{-1}.$$

Using (6.95), we obtain

$$Z^R(\varepsilon)[I - \varepsilon CZ_0] = [I - \Pi(\varepsilon)]Z_0 + \Pi_0 - \Pi(\varepsilon)Z^S(\varepsilon). \tag{6.96}$$

We now show that

$$\Pi(\varepsilon)Z^S(\varepsilon) = \Pi(\varepsilon)Z^S(\varepsilon)[I - \varepsilon CZ_0]. \tag{6.97}$$

It suffices to prove that $Z^S(\varepsilon)CZ_0 = 0$. Indeed, from (6.93) and (6.94),

$$Z^S(\varepsilon)CZ_0 = -Z^S(\varepsilon)CZ_{-1}CZ_0 = 0.$$

Finally, substituting (6.97) into (6.96) and multiplying (6.96) by $[I-\varepsilon CZ_0]^{-1}$ from the right-hand side, we obtain the required formula (6.88). □

Two useful corollaries follow directly from Theorem 6.3. First, consider the term $\Pi(\varepsilon)Z^S(\varepsilon)$. This term is regular, despite the fact that it is the product of the perturbed ergodic projection and the *singular* part of the fundamental matrix. The next corollary shows the explicit regular structure of this product.

Corollary 6.1. *The formula* (6.88) *is equivalent to*

$$Z^R(\varepsilon)=\{[I-\Pi(\varepsilon)]Z_0+\Pi_0\}[I-\varepsilon CZ_0]^{-1}-\Pi(\varepsilon)\sum_{i=1}^{s}(CZ_0)^iZ_{-i}. \tag{6.98}$$

Proof: Since in the case of the irreducible perturbation we have

$$\Pi(\varepsilon)=\mathbf{1}\mu(\varepsilon)=\mathbf{1}\mu_0[I-\varepsilon CZ_0]^{-1}=\Pi_0[I-\varepsilon CZ_0]^{-1},$$

we conclude that $\Pi_k=\Pi_0(CZ_0)^k, k=0,1,\dots$, in the power series (6.16).

Note that *only* the regular part of $\Pi(\varepsilon)Z^S(\varepsilon)$ contributes to equation (6.88), since $Z^R(\varepsilon)$ and $\{[I-\Pi(\varepsilon)]Z_0+\Pi_0\}[I-\varepsilon CZ_0]^{-1}$ are both regular. Then,

$$[\Pi(\varepsilon)Z^S(\varepsilon)]^R=\left[\sum_{k=1}^{\infty}\Pi_0(\varepsilon CZ_0)^k\sum_{i=1}^{s}\varepsilon^{-i}Z_{-i}\right]^R=\sum_{i=1}^{s}\sum_{k=i}^{\infty}\Pi_0(CZ_0)^kZ_{-i}\varepsilon^{k-i}$$

$$=\sum_{i=1}^{s}\sum_{k=0}^{\infty}\Pi_0(\varepsilon CZ_0)^k(CZ_0)^iZ_{-i}=\sum_{k=0}^{\infty}\Pi_0(\varepsilon CZ_0)^k\sum_{i=1}^{s}(CZ_0)^iZ_{-i}=\Pi(\varepsilon)\sum_{i=1}^{s}(CZ_0)^iZ_{-i}.$$

This yields formula (6.98). □

By using (6.13) and (6.14), one easily obtains the counterpart of (6.88) for the deviation matrix $H(\varepsilon)$.

Corollary 6.2. *The regular part of the deviation matrix $H(\varepsilon)$ is given by*

$$H^R(\varepsilon)=[I-\Pi(\varepsilon)]H_0[I-\varepsilon CZ_0]^{-1}-\Pi(\varepsilon)H^S(\varepsilon).$$

Remark 6.3. *The well-known formula (see Bibliographic Notes) for regular perturbations*

$$Z(\varepsilon)=\{[I-\Pi(\varepsilon)]Z(0)+\Pi\}[I-\varepsilon CZ(0)]^{-1}$$

is a particular case of (6.88) (*since in this case* $Z^S(\epsilon)=0$).

Remark 6.4. *The matrices Z_k, $k\geq 1$, are easily obtained from* (6.98), *with*

$$Z_k=Z_0(CZ_0)^k-\Pi_0\sum_{j=0}^{k}(CZ_0)^jZ_0(CZ_0)^{k-j}+\Pi_0(CZ_0)^k\left\{I-\sum_{i=1}^{s}(CZ_0)^iZ_{-i}\right\}. \tag{6.99}$$

Of course, in practical applications, a more efficient (and numerically stable) computational scheme for the above matrices $Z_k, k\geq 1$, is via recurrent formulae. For instance, if we define the two k-dependent expressions

$$U_k=(CZ_0)^k \quad \textit{and} \quad W_k=\sum_{j=0}^{k}(CZ_0)^jZ_0(CZ_0)^{k-j},\ k=1,\dots,$$

then (6.99) *is equivalent to*

$$Z_k = Z_0 U_k - P_0^* \left[W_k - U_k \left\{ I - \sum_{i=1}^{s} (CZ_0)^i Z_{-i} \right\} \right], \quad k = 1, \ldots, \tag{6.100}$$

with U_k *and* W_k *computed recursively in* (6.101) *below:*

$$U_{k+1} = U_k(CZ_0), \quad W_{k+1} = W_k(CZ_0) + U_{k+1}Z_0, \quad k = 1, \ldots. \tag{6.101}$$

6.3.3 ▪ The regular perturbation

In this subsection we assume that the unperturbed MC has a probability transition matrix $P = P(0)$ that is irreducible. This leads to the case of regular perturbations. In this case, all of the previously mentioned important matrices corresponding to the perturbed MC possess well-behaved Maclaurin series expansions. The following theorem summarizes the salient properties of these expansions.

Theorem 6.4. *Assume that the unperturbed MC is irreducible. Then the following hold:*

(i) *The matrix functions* $\Pi(\varepsilon)$, $H(\varepsilon)$, *and* $M(\varepsilon)$, *representing the Cesaro limit matrix, the deviation matrix, and the matrix of mean first passage times, respectively, are analytic in some (undeleted) neighborhood of zero. In particular, they all admit Maclaurin series expansions:*

$$\Pi(\varepsilon) = \sum_{m=0}^{\infty} \varepsilon^m \Pi^{(m)}, \; H(\varepsilon) = \sum_{m=0}^{\infty} \varepsilon^m H^{(m)}, \; \text{and} \; \mathcal{M}(\varepsilon) = \sum_{m=0}^{\infty} \varepsilon^m \mathcal{M}^{(m)},$$

with coefficient sequences $\{\Pi^{(m)}\}_{m=0}^{\infty}$, $\{H^{(m)}\}_{m=0}^{\infty}$, *and* $\{\mathcal{M}^{(m)}\}_{m=0}^{\infty}$.

(ii) *The Cesaro limit matrix* $\Pi(\varepsilon)$ *and the deviation matrix of the perturbed MC admit the updating formulae*

$$\Pi(\varepsilon) = \Pi(0)[I - \varepsilon U]^{-1}, \tag{6.102}$$

and

$$\begin{aligned} H(\varepsilon) &= [I - \Pi(\varepsilon)]H(0)[I - \varepsilon U]^{-1} \\ &= H(0)[I - \varepsilon U]^{-1} - \Pi(0)[I - \varepsilon U]^{-1} H(0)[I - \varepsilon U]^{-1}, \end{aligned}$$

where $U = CH(0)$.

(iii) *These updating formulae yield the following expressions for the power series coefficients:*

$$\Pi^{(m)} = \Pi^{(0)} U^m, \quad m \geq 0,$$

$$H^{(m)} = H(0)U^m - \Pi(0) \sum_{j=1}^{m} U^j H(0) U^{m-j}, \quad m \geq 0,$$

$$\mathcal{M}_{ij}^{(m)} = \frac{1}{\Pi_j^{(0)}} (H_{jj}^{(m)} - H_{ij}^{(m)}) - \frac{1}{\Pi_j^{(0)}} \sum_{l=1}^{m} \Pi_j^{(l)} \mathcal{M}_{ij}^{(m-l)}, \quad m \geq 0.$$

(iv) *The validity of any of the above series expansion holds for any* ε, $0 \leq \varepsilon < \min\{\varepsilon_{\max}, \rho^{-1}(U)\}$, *where* $\rho(U)$ *is the spectral radius of* U.

We do not prove this theorem in full, as the algebraic technique used to prove (6.102) contains the "flavor" of the required analysis. We refer the reader to Problems 6.9–6.11 to reconstruct the proofs of these results.

Next we just show the validity of only the statement (6.102):

$$\Pi(\varepsilon)-\Pi(0)=\varepsilon\Pi(0)U(I-\varepsilon U)^{-1}.$$

The latter follows from the observation that

$$\begin{aligned}\Pi(\varepsilon)-\Pi(0)&=\Pi(\varepsilon)P(\varepsilon)-\Pi(0)P(0)=\Pi(\varepsilon)(P(0)+\varepsilon C)-\Pi(0)P(0)\\&=(\Pi(\varepsilon)-\Pi(0))P(0)+\Pi(\varepsilon)\varepsilon C\end{aligned}$$

or

$$(\Pi(\varepsilon)-\Pi(0))(I-P(0))=\Pi(\varepsilon)\varepsilon C.$$

Postmultiply the last equation by $H(0)$ and use (6.14) in order to obtain $(\Pi(\varepsilon)-\Pi(0))(I-\Pi(0))=\varepsilon\Pi(\varepsilon)U$. But $(\Pi(\varepsilon)-\Pi(0))\Pi(0)=0$ (as we multiply a zero row sum matrix by a matrix with identical rows). Hence, $\Pi(\varepsilon)-\Pi(0)=\varepsilon\Pi(\varepsilon)U$. Replace $\Pi(\varepsilon)$ in the right-hand side with $[\Pi(\varepsilon)-\Pi(0)]+\Pi(0)$, and move the product due to the term in brackets to the left-hand side to obtain $(\Pi(\varepsilon)-\Pi(0))(I-\varepsilon U)=\varepsilon\Pi(0)U$. Postmultiplication of both sides by $(I-\varepsilon U)^{-1}$ yields $\Pi(\varepsilon)-\Pi(0)=\varepsilon\Pi(0)U(I-\varepsilon U)^{-1}$, as required. Naturally, the latter implies that

$$\Pi(\varepsilon)=\Pi(0)[I+\varepsilon U(I-\varepsilon U)^{-1}]=\Pi(0)(I-\varepsilon U)^{-1}.$$

Example 6.7. *For* $0\le\varepsilon<1/4$, *let*

$$P(\varepsilon)=P(0)+\varepsilon C=\begin{pmatrix}1/2&1/2\\1/2&1/2\end{pmatrix}+\varepsilon\begin{pmatrix}2&-2\\-1&1\end{pmatrix}.$$

Clearly,

$$\Pi(0)=\Pi^{(0)}=\begin{pmatrix}1/2&1/2\\1/2&1/2\end{pmatrix}.$$

Also,

$$H(0)=H^{(0)}=\begin{pmatrix}1/2&-1/2\\-1/2&1/2\end{pmatrix}\ \text{and}\ \mathcal{M}(0)=\mathcal{M}^{(0)}=\begin{pmatrix}2&2\\2&2\end{pmatrix}.$$

Hence,

$$U=CH(0)=\begin{pmatrix}2&-2\\-1&1\end{pmatrix}.$$

It is easy to see that for $m\ge1$, $U^m=3^{m-1}U$, *and hence for* $m\ge1$,

$$\Pi^{(m)}=\Pi(0)U^m=3^{m-1}\begin{pmatrix}1/2&-1/2\\1/2&-1/2\end{pmatrix}.$$

Also, for $m\ge1$,

$$H^{(m)}=3^{m-1}H(0)U-3^{m-2}(m-1)\Pi(0)UH(0)U-3^{m-1}\Pi(0)UH(0)$$

$$=3^{m-1}\begin{pmatrix}3/2&-3/2\\-3/2&3/2\end{pmatrix}-3^{m-2}(m-1)\begin{pmatrix}3/2&-3/2\\3/2&-3/2\end{pmatrix}-3^{m-1}\begin{pmatrix}1/2&-1/2\\1/2&-1/2\end{pmatrix}.$$

Finally,

$$\mathcal{M}_{12}(\varepsilon)=2+8\varepsilon+\dots\quad\text{and}\quad\mathcal{M}_{21}(\varepsilon)=2+4\varepsilon+\dots.$$

6.3.4 ▪ The nearly completely decomposable Markov chains

We now return to the nearly completely decomposable (NCD) case discussed in Subsection 6.2.3. We use the notation introduced in that subsection. Recall that the transition matrix of the perturbed MC had the form

$$P(\varepsilon) = P(0) + \varepsilon C, \quad \varepsilon \in (0, \varepsilon_{max}),$$

where $P(0)$ had the NCD block structure (6.44). We now address in more detail expansions of the corresponding perturbed deviation and mean first passage time matrices.

For $\varepsilon \in (0, \varepsilon_{max})$, let $H(\varepsilon)$ be the deviation matrix of $P(\varepsilon)$. This matrix is uniquely defined, and the case $\varepsilon = 0$ is no exception. Yet, there is no continuity of $H(\varepsilon)$ at $\varepsilon = 0$. In particular, $H(0)$ has the same shape P has, namely,

$$H(0) = \begin{pmatrix} H_1 & 0 & \cdots & 0 \\ 0 & H_2 & \cdots & 0 \\ \vdots & \vdots & \ddots & \vdots \\ 0 & 0 & \cdots & H_n \end{pmatrix}, \tag{6.103}$$

where H_i is the deviation matrix of P_i, $1 \le i \le n$.

Theorem 6.5. *In the case of NCD MCs, the matrix $H(\varepsilon)$ admits a Laurent series expansion in a deleted neighborhood of zero with the order of the pole being exactly one. Specifically, for some matrices $\{H^{(m)}\}_{m=-1}^{\infty}$ with $H^{(-1)} \neq 0$,*

$$H(\varepsilon) = \frac{1}{\varepsilon} H^{(-1)} + H^{(0)} + \varepsilon H^{(1)} + \varepsilon^2 H^{(2)} + \cdots \tag{6.104}$$

for $0 < \varepsilon < \varepsilon_{\max}$. In particular,

$$H^{(-1)} = QDM,$$

or, in a component form,

$$H_{ij}^{(-1)} = D_{IJ}(\gamma_J)_j, \quad i \in I,\ j \in J, \tag{6.105}$$

where D is the deviation matrix of the aggregated transition matrix $\hat{Q}$. In addition, the matrix U in (6.48) may alternatively be expressed as

$$U = CH^{(0)}. \tag{6.106}$$

We now focus our attention on $\mathcal{M}(\varepsilon)$, the mean first passage time matrix of the perturbed MC. Note that, as opposed to $H(0)$, $\mathcal{M}(0)$ is not well defined as the corresponding mean value (when $\varepsilon = 0$ and states i and j belong to two different ergodic classes) does not exist. Let $E \in R^{p \times p}$ be the mean passage time matrix associated with the aggregated process. That is, for any pair of subsets I and J ($I = J$ included), E_{IJ} is the mean passage time from the macrostate I into the macrostate J when transition probabilities are governed by the stochastic matrix $\hat{Q}$.

Theorem 6.6. *The matrix $\mathcal{M}(\varepsilon)$ admits a Laurent series expansion in a deleted neighborhood of zero with the order of the pole being exactly one. Specifically, for some matrices $\{\mathcal{M}^{(m)}\}_{m=-1}^{\infty}$ with $\mathcal{M}^{(-1)} \neq 0$,*

$$\mathcal{M}(\varepsilon) = \frac{1}{\varepsilon} \mathcal{M}^{(-1)} + \mathcal{M}^{(0)} + \varepsilon \mathcal{M}^{(1)} + \varepsilon^2 \mathcal{M}^{(2)} + \cdots \tag{6.107}$$

for $0 < \varepsilon < \varepsilon_{\max}$. *Moreover, for* $i \in I$ *and* $j \in J$,

$$\mathcal{M}_{ij}^{(-1)} = \begin{cases} 0 & \text{if } J = I, \\ E_{IJ} & \text{if } J \neq I, \end{cases} \tag{6.108}$$

$$\mathcal{M}_{ij}^{(m)} = \frac{1}{\pi_j^{(0)}}(H_{jj}^{(m)} - H_{ij}^{(m)}) - \frac{1}{\pi_j^{(0)}} \sum_{l=1}^{m+1} \pi_j^{(l)} \mathcal{M}_{ij}^{(m-l)}, \quad m \geq -1.$$

Proof: From (6.11) coupled with the fact that the MC is ergodic when $0 < \varepsilon < \varepsilon_{\max}$,

$$\mathcal{M}_{ij}(\varepsilon) = \frac{\delta_{ij} + H_{jj}(\varepsilon) - H_{ij}(\varepsilon)}{\pi_j(\varepsilon)}, \quad 0 < \varepsilon < \varepsilon_{\max}. \tag{6.109}$$

Hence, by (6.104),

$$\mathcal{M}_{ij}^{(-1)} = \frac{H_{jj}^{(-1)} - H_{ij}^{(-1)}}{\pi_j^{(0)}}. \tag{6.110}$$

By (6.105), $H_{jj}^{(-1)} = H_{ij}^{(-1)}$ whenever states i and j are in the same subset; hence $\mathcal{M}_{ij}^{(-1)} = 0$ in this case. Using (6.105) again for the case where $J \neq I$, (6.110) has a numerator which is equal to $(D_{JJ} - D_{IJ})(\gamma_J)_j$. By (6.45) and the definition of γ_J, the denominator is equal to $x_J(\gamma_J)_j$. Thus for this case, $\mathcal{M}_{ij}^{(-1)}$ is equal to $(D_{JJ} - D_{JI})/x_J$. Using (6.11) for the aggregated MC, we conclude that

$$\mathcal{M}_{ij}^{(-1)} = \frac{D_{JJ} - D_{JI}}{x_J} = E_{IJ}$$

whenever $i \in I$, $j \in J$, and $J \neq I$. □

Example 6.8. *Let*

$$P(0) = \begin{pmatrix} 1 & 0 & 0 \\ 0 & 1/2 & 1/2 \\ 0 & 1/2 & 1/2 \end{pmatrix} \quad \text{and} \quad C = \frac{2}{7}\begin{pmatrix} -2 & 1 & 1 \\ 3 & -1 & -2 \\ 4 & -3 & -1 \end{pmatrix}.$$

The number of subsets is equal to 2 with $\gamma_{I_1} = 1$ *and* $\gamma_{I_2} = (1/2, 1/2)$. *First, we construct the following matrices:*

$$M = \begin{pmatrix} 1 & 0 & 0 \\ 0 & 1/2 & 1/2 \end{pmatrix}, \quad Q = \begin{pmatrix} 1 & 0 \\ 0 & 1 \\ 0 & 1 \end{pmatrix}, \quad H(0) = \begin{pmatrix} 0 & 0 & 0 \\ 0 & 1/2 & -1/2 \\ 0 & -1/2 & 1/2 \end{pmatrix}.$$

The aggregated transition matrix is given by

$$\hat{Q} = I + MCQ = \begin{pmatrix} 3/7 & 4/7 \\ 1 & 0 \end{pmatrix}.$$

Hence, $x = (7/11, 4/11)$ *and* $\mu^{(0)} = (7/11, 2/11, 2/11)$. *Next, we calculate D, the deviation matrix of Q,*

$$D = (I - Q + 1x)^{-1} - 1x = \frac{1}{121}\begin{pmatrix} 28 & -28 \\ -49 & 49 \end{pmatrix},$$

and hence, using (6.105),

$$H^{(-1)} = QDM = \frac{1}{242}\begin{pmatrix} 56 & -28 & -28 \\ -98 & 49 & 49 \\ -98 & 49 & 49 \end{pmatrix}.$$

The matrix E, which is the mean passage time matrix for the aggregated process, is equal to

$$E = \begin{pmatrix} 11/7 & 7/4 \\ 1 & 11/4 \end{pmatrix},$$

and hence, using (6.108),

$$M^{(-1)} = \begin{pmatrix} 0 & 7/4 & 7/4 \\ 1 & 0 & 0 \\ 1 & 0 & 0 \end{pmatrix}.$$

Finally, from (6.48) *we obtain*

$$U = CH(0)(I + CQDM) = \frac{1}{77}\begin{pmatrix} 0 & 0 & 0 \\ -2 & 12 & -10 \\ 4 & -24 & 20 \end{pmatrix}.$$

6.3.5 ▪ The general case: Absorbing states communicate via transient states

First we recall that $H(\varepsilon)$ and $\mathcal{M}(\varepsilon)$ always possess Laurent series expansions around zero. This is the case since these functions can be obtained as solutions to linear systems, and hence, they are rational functions. Namely, elements of $H(\varepsilon)$ and of $\mathcal{M}(\varepsilon)$ can be expressed as ratios of polynomials.

The next important issue is the order of the poles of $H(\varepsilon)$ and $\mathcal{M}(\varepsilon)$ at $\varepsilon = 0$. Assuming the perturbed process to be irreducible, as we have done throughout this section, the first question to address here is if some of the results of the preceding subsections still hold in the general case. For example, are the orders of the poles of $H(\varepsilon)$ and of $M(\varepsilon)$ always smaller than or equal to one? Or, do these orders always coincide?

In Sections 6.3.3 and 6.3.4 we have assumed that no transient states (under $P(\varepsilon)$) exist and this was a sufficient condition for the order of the poles at zero to coincide and to be smaller than or equal to one. Thus, the existence of transient states is a necessary condition for a higher order singularity. Yet, as examples show, this is not a sufficient condition, and some additional structure (besides the presence of transient states) is needed in order to encounter higher order singularities.

Indeed, suppose (as is done in Problem 6.12) that in a perturbed MC $P(\varepsilon)$, a recurrent (under $P(0)$) state j can be reached from another recurrent (under $P(0)$) state i, where i and j belong to different ergodic classes (under $P(0)$). Then, this can be achieved only through a path which contains transient states (under $P(0)$). Also, in such a case the deviation and mean passage time matrices may contain poles of order greater than 1. The following perturbed transition matrix illustrates these phenomena.

Example 6.9.

$$P(\varepsilon) = P(0) + \varepsilon C = \begin{pmatrix} 0 & 1 & 0 & 0 \\ 0 & 1 & 0 & 0 \\ 0 & 0 & 0 & 1 \\ 0 & 0 & 0 & 1 \end{pmatrix} + \varepsilon \begin{pmatrix} 0 & -1 & 0 & 1 \\ 1 & -1 & 0 & 0 \\ 0 & 1 & 0 & -1 \\ 0 & 0 & 1 & -1 \end{pmatrix}.$$

In this example the unperturbed chain contains two ergodic classes (states 2 and 4) and two transient states (states 1 and 3). They all are coupled in a single ergodic class when $\varepsilon > 0$. Moreover, states 2 and 4 (i.e., the ergodic chains in the unperturbed process) communicate under the perturbation only via states 1 and 3 (i.e., transient states in the unperturbed case). This, in particular, implies that the expected time it takes to reach state 3 for a process which starts in state 1 is of the order of magnitude of $O(\varepsilon^{-2})$. In other words, the order of the pole of $M_{13}(\varepsilon)$ at zero is two.

In the general case a stochastic matrix has the form

$$P(0) = P = \begin{pmatrix} P_1 & 0 & \cdots & 0 & 0 \\ 0 & P_2 & \cdots & 0 & 0 \\ \vdots & \vdots & \ddots & 0 & 0 \\ 0 & 0 & \cdots & P_n & 0 \\ R_1 & R_2 & \cdots & R_n & S \end{pmatrix}.$$

Corresponding to the above, let Ω_0 denote the (possibly empty) set of transient states (i.e., $\lim_{t\to\infty} S^t = 0$) and where the rest of the states are as before with $n \geq 1$. Here we limit ourselves to the case of linear perturbation, that is, $P(\varepsilon) = P(0) + \varepsilon C$ for some matrix C. Yet, for the reduction process defined below, we need to consider analytic perturbations (of lower dimensions) which are not necessarily linear. Thus, although it seems redundant at this stage, we assume that

$$G(\varepsilon) = P(\varepsilon) - I = \sum_{k=0}^{\infty} \varepsilon^k G_k \tag{6.111}$$

for some matrix sequence $\{G_k\}_{k=0}^{\infty}$. Of course, $G_0 = P(0) - I$, $G_1 = C$, and $G_k = 0$ for $k \geq 2$ in the present setting.

Since the deviation matrix $H(\varepsilon)$ of the perturbed Markov chain involves a matrix inverse, it is clear by results of Section 2.2 that it possesses a Laurent series expansion

$$H(\varepsilon) = \frac{1}{\varepsilon^s}(H_0 + \varepsilon H_1 + \cdots) \tag{6.112}$$

for some integer $s \geq 0$ and with $H_0 \neq 0$. Note that, in the present case, s (the order of the singularity) can vary with the particular structure of the chain. In the previously studied regular case this order was 0 in all instances, and in the NCD case it was always 1.

Consequently, we adopt a slightly different notation here. Specifically, H_0 is the leading coefficient, and subscripts, rather than superscripts, denote the order of the coefficients. The superscript index is preserved for other purposes.

The value of s can be determined by the algorithm suggested in Section 2.2 or by other methods (see, e.g., Problem 6.13). Note that in order to apply the numerical procedure outlined below it is necessary to first determine the value of s.

Recall that the deviation matrix $H(\varepsilon)$ is the unique solution of the system

$$H(\varepsilon)G(\varepsilon) = \Pi(\varepsilon) - I, \quad H(\varepsilon)\mathbf{1} = 0. \tag{6.113}$$

By the results of Section 6.2 we know that $\Pi(\varepsilon)$ can be expanded as a power series

$$\Pi(\varepsilon) = \Pi_0 + \varepsilon \Pi_1 + \varepsilon^2 \Pi_2 + \cdots$$

with $\Pi_0 \neq \Pi(0)$ in the singular perturbation case. Upon substitution of the above series for $\Pi(\varepsilon)$, (6.111), and (6.112) into (6.113), and then collecting the terms with the same power of ε, we obtain the following system of fundamental equations for H_i, $i \geq 0$:

$$H_0 G_0 = 0, \qquad (F0)$$

$$H_1 G_0 + H_0 G_1 = 0, \qquad (F1)$$

$$\vdots$$

$$H_s G_0 + H_{s-1} G_1 + \cdots + H_1 G_{s-1} + H_0 G_s = \Pi_0 - I. \qquad (Fs)$$

Note that the above system contains only $s+1$ fundamental equations, even though equating coefficients yields a system of infinitely many such equations. In Problem 6.14 we leave it to the reader to verify that it is, indeed, sufficient to solve only these $s+1$ equations. Now, we outline how the reduction process of Section 2.2 is used to solve the fundamental equations.

Extending the definition of the matrices M and Q given in Subsection 6.2.3, let $M \in R^{n\times N}$ be such that its Ith row is full of zeros excepts for γ_I at the entries corresponding to subset Ω_I that is exactly the same definition for M as given in the NCD case. Now, let $Q \in R^{N\times n}$ be such that Q_{iJ} is equal to the probability (under the unperturbed transition matrix) that a process which initializes in state i is eventually absorbed into the ergodic subset Ω_J. Of course, if i is recurrent and $i \in \Omega_J$, then $Q_{iJ} = 1$. If i is recurrent and $i \notin \Omega_J$, then $Q_{iJ} = 0$. Finally, if i is transient, $Q_{iJ} = [(I-S)^{-1} R_J \underline{1}]_i$. Let

$$G_k^{(1)} = M \sum_{p=1}^{k+1} \sum_{\nu_1+\cdots+\nu_p=k+1} G_{\nu_1} H(0) G_{\nu_2} \cdots H(0) G_{\nu_p} Q.$$

Note that in the case where $G_k = 0$ for $k \geq 2$, we have $G_k^{(1)} = M G_1 (H(0) G_1)^k Q$.

It is straightforward to check (see Problem 6.15) that the system $(F0)$–(Fs) is equivalent to the following reduced system with variables $H_i^{(1)}$:

$$H_0^{(1)} G_0^{(1)} = 0, \qquad (RF0)$$

$$H_1^{(1)} G_0^{(1)} + H_0^{(1)} G_1^{(1)} = 0, \qquad (RF1)$$

$$\vdots$$

$$H_{s-1}^{(1)} G_0^{(1)} + H_{s-2}^{(1)} G_1^{(1)} + \cdots + H_1^{(1)} G_{s-2}^{(1)} + H_0^{(1)} G_{s-1}^{(1)} = (\Pi_0 - I) Q. \qquad (RFs-1)$$

The superscript (1) corresponds to the fact that only the first reduction step is done here and there will be more steps to come. Note that $H_k^{(1)} \in R^{n\times n}$, $k \geq 0$. The matrix $H_0^{(1)}$ is uniquely determined by the above equations and the normalization condition $H_0^{(1)} 1 = 0$. Once $H_0^{(1)}$ is obtained, H_0 can be calculated by

$$H_0 = H_0^{(1)} M.$$

Note that the system (RF) has s matrix equations in comparison to $s+1$ matrix equations in (F). The dimension of aggregated matrices $G_k^{(1)}$ is equal to the number of ergodic sets in the unperturbed MC. As in the NCD case, we refer to $G_0^{(1)} = I + MP(0)Q$ as a generator of the aggregated MC.

We can apply the reduction technique again but now to the reduced system (RF). After the second reduction step the number of matrix equations is reduced to $s-1$. Similarly, one can perform s reduction steps. Specifically, define in a recursive manner, for $j = 1,\dots,s$,

$$G_k^{(j)} = M^{(j-1)} \sum_{p=1}^{k+1} \sum_{\nu_1+\cdots+\nu_p=k+1} G_{\nu_1}^{(j-1)} H^{(j-1)} G_{\nu_2}^{(j-1)} \cdots H^{(j-1)} G_{\nu_p}^{(j-1)} Q^{(j-1)},$$

where $H^{(j-1)}$ is the deviation matrix corresponding to the generator $G_0^{(j-1)}$. As $G_0^{(j)}$ is an MC generator, let the matrices $M^{(j)}$ and $Q^{(j)}$ be defined similarly to the matrices M and Q for the original MC. By convention, let $M^{(0)} = M$ and $Q^{(0)} = Q$. Note that by the nature of the final reduction step, $M^{(s)}$ is a row vector, while $Q^{(s)}$ is a column vector, the latter being full of ones. Then, the jth step reduces the fundamental system into the form

$$H_0^{(j)} G_0^{(j)} = 0, \qquad (RjF0)$$

$$H_1^{(j)} G_0^{(j)} + H_0^{(j)} G_1^{(j)} = 0, \qquad (RjF1)$$

$$\vdots \quad \vdots \quad \vdots$$

$$H_{s-1}^{(j)} G_0^{(j)} + H_{s-2}^{(j)} G_1^{(j)} + \cdots + H_1^{(j)} G_{s-2}^{(j)} + H_0^{(j)} G_{s-1}^{(j)} = (\Pi_0 - I) Q Q^{(1)} \cdots Q^{(j-1)}. \qquad (RjFs-1)$$

The limiting stationary distribution μ_0 can be given by the following formula (see Problem 6.16):

$$\mu^{(0)} = M^{(s)} M^{(s-1)} \cdots M^{(1)} M. \qquad (6.114)$$

To specify the above formula for each element $\mu_i^{(0)}$, $1 \le i \le n$, we introduce the integer-valued function $I^{(k)}(i)$, $k = 0,\dots,s-1$. Specifically, let $I^{(k)}(i)$ be the index of the ergodic set in the kth reduction step to which state i belongs. Then, formula (6.114) can be rewritten in the component form

$$\mu_i^{(0)} = M_{I^{(s-1)}(i)}^{(s)} M_{I^{(s-1)}(i),I^{(s-2)}(i)}^{(s-1)} \cdots M_{I^{(1)}(i),I^{(0)}(i)}^{(1)} M_{I^{(0)}(i),i}. \qquad (6.115)$$

From (6.115) one can learn whether a state i is transient at some level of the aggregation, since the corresponding element $\mu_i^{(0)}$ is equal to zero.

Mean first passage times

We continue by studying the relation between the Laurent expansion for the mean first passage time matrix and the Laurent expansion for the deviation matrix. Before proceeding to the next result, we introduce a useful notion of the degree of transience.

Definition 6.4. *For a state i, define its* degree of transience, *denoted by $t(i)$, as follows:*

$$t(i) = \min\{\, m \mid \mu_i^{(m)} > 0;\ m = 0,1,\dots\}.$$

Since $\mu_i(\varepsilon) = 1/\mathcal{M}_{ii}(\varepsilon)$, it is clear that $t(i)$ is equal to the order of the pole of $\mathcal{M}_{ii}(\varepsilon)$ at zero. Furthermore, there always exists at least one state i such that $t(i) = 0$; otherwise the elements of $\mu^{(0)}$ would not sum to one.

Theorem 6.7. *The most singular coefficient of the Laurent series for the deviation matrix of the perturbed MC is given by*

$$H_0 = QQ^{(1)} \cdots Q^{(s-1)} H^{(s)} M^{(s-1)} \cdots M^{(1)} M, \tag{6.116}$$

where $H^{(s)} = [-G_0^{(s)}]^{\#}$ is the deviation matrix for the sth level aggregated MC. Furthermore, let state i belong to some ergodic set of the $(s-1)$st level aggregated process, and let state j have zero degree of transience, that is, $t(j) = 0$. Then, the most singular coefficient of the Laurent series for $\mathcal{M}_{ij}(\varepsilon)$ is given by

$$\mathcal{M}_{ij}^{(-s)} = \begin{cases} \dfrac{H^{(s)}_{I^{(s-1)}(j),I^{(s-1)}(j)} - H^{(s)}_{I^{(s-1)}(i),I^{(s-1)}(j)}}{M^{(s)}_{I^{(s-1)}(j)}} & \text{if } I^{(s-1)}(i) \neq I^{(s-1)}(j), \\ 0 & \text{if } I^{(s-1)}(i) = I^{(s-1)}(j). \end{cases} \tag{6.117}$$

Proof: After s reduction steps we obtain the following equation (RsF0):

$$H_0^{(s)} G_0^{(s)} = (\Pi_0 - I) Q \cdots Q^{(s-1)}.$$

Since $\Pi_0 = Q \cdots Q^{(s-1)} \mathbf{1} M^{(s)} M^{(s-1)} \cdots M$ and $M^{(k)} Q^{(k)} = I$, the right-hand side of the above equation can be transformed as follows:

$$(\Pi_0 - I) Q \cdots Q^{(s-1)} = (Q \cdots Q^{(s-1)} \mathbf{1} M^{(s)} M^{(s-1)} \cdots M - I) Q \cdots Q^{(s-1)}$$

$$= Q \cdots Q^{(s-1)} \mathbf{1} M^{(s)} - Q \cdots Q^{(s-1)} = Q \cdots Q^{(s-1)} (\mathbf{1} M^{(s)} - I).$$

Next we recall that if $G_0^{(s)}$ has a simple zero eigenvalue, the equation

$$H_0^{(s)} G_0^{(s)} = W \cdots W^{(s-1)} (\mathbf{1} V^{(s)} - I)$$

coupled with the normalization condition $H_0^{(s)} \mathbf{1} = 0$ yields a unique solution for $H_0^{(s)}$. Hence, applying the group generalized inverse (see Section 2.1), we obtain

$$H_0^{(s)} = Q \cdots Q^{(s-1)} (\mathbf{1} M^{(s)} - I)(G_0^{(s)})^{\#} = Q \cdots Q^{(s-1)} (-G_0^{(s)})^{\#} = Q \cdots Q^{(s-1)} H^{(s)}.$$

Finally, we have

$$H_0 = H_0^{(1)} M = H_0^{(s)} M^{(s-1)} \cdots M = Q \cdots Q^{(s-1)} H^{(s)} M^{(s-1)} \cdots M,$$

which is the required expression (6.116). Similarly, using (6.109) and (6.115), we derive (6.117). □

Note that an immediate corollary from the above theorem is as follows.

Corollary 6.3. *Let t_{ij} be the order of the pole of $\mathcal{M}_{ij}(\varepsilon)$ at zero. Let j be such that $t_{jj} = 0$ (or, equivalently, $t(j) = 0$). Then,*

$$s = \max_{1 \le i,j \le n} \{t_{ij}\}.$$

Example 6.10. *Consider the following perturbed transition matrix:*

$$P(\varepsilon)=P(0)+\varepsilon C=\begin{pmatrix} 1 & 0 & 0 & 0 \\ 0 & 1 & 0 & 0 \\ 0 & 1 & 0 & 0 \\ 0 & 0 & 1 & 0 \end{pmatrix}+\varepsilon\begin{pmatrix} -1 & 1 & 0 & 0 \\ 0 & -1 & 1 & 0 \\ 0 & -1 & 0 & 1 \\ 1 & 0 & -1 & 0 \end{pmatrix}.$$

As before, we denote by t_{ij} the order of the pole of $\mathcal{M}_{ij}(\varepsilon)$ at $\varepsilon=0$. By the algorithm discussed in Problem 6.17 we calculate all the t_{ij}'s to obtain

$$(t_{ij})=\begin{pmatrix} 2 & 1 & 1 & 2 \\ 3 & 0 & 1 & 2 \\ 3 & 0 & 1 & 2 \\ 3 & 0 & 0 & 2 \end{pmatrix}.$$

Hence, the order of singularity of $\mathcal{M}(\varepsilon)$, given by $\max_{ij}\{t_{ij}\}$, is three. The order of singularity of the deviation matrix $H(\varepsilon)$, denoted above by s and given by $\max_{ij}\{t_{ij}-t_{jj}\}$, is then equal to one. In particular, denoting by t the order of singularity of $\mathcal{M}(\varepsilon)$ at zero, we have constructed an example where $s<t$. Also, $t(j)=t_{jj}$, the degree of transience, can be read from the diagonal of the matrix (t_{ij}). Alternatively, one may apply Corollary 6.3 to determine that $s=1$.

Next let us apply Theorem 6.7 to this example. Here we have

$$M=\begin{pmatrix} 1 & 0 & 0 & 0 \\ 0 & 1 & 0 & 0 \end{pmatrix}$$

and

$$Q=\begin{pmatrix} 1 & 0 \\ 0 & 1 \\ 0 & 1 \\ 0 & 1 \end{pmatrix}.$$

Hence,

$$G_0^{(1)}=MCQ=\begin{pmatrix} -1 & 1 \\ 0 & 0 \end{pmatrix},\quad H^{(1)}=[-G_0^{(1)}]^{\#}=\begin{pmatrix} 1 & -1 \\ 0 & 0 \end{pmatrix}.$$

As zero is a simple eigenvalue of $G_0^{(1)}$, only one reduction step is required here. This is, of course, an alternative way to verify that $s=1$. Next, we calculate $\mu^{(0)}$ and H_0.

$$\mu^{(0)}=M^{(1)}M=\begin{pmatrix} 0 & 1 \end{pmatrix}\begin{pmatrix} 1 & 0 & 0 & 0 \\ 0 & 1 & 0 & 0 \end{pmatrix}=\begin{pmatrix} 0 & 1 & 0 & 0 \end{pmatrix},$$

$$H_0=QH^{(1)}M=\begin{pmatrix} 1 & 0 \\ 0 & 1 \\ 0 & 1 \\ 0 & 1 \end{pmatrix}\begin{pmatrix} 1 & -1 \\ 0 & 0 \end{pmatrix}\begin{pmatrix} 1 & 0 & 0 & 0 \\ 0 & 1 & 0 & 0 \end{pmatrix}=\begin{pmatrix} 1 & -1 & 0 & 0 \\ 0 & 0 & 0 & 0 \\ 0 & 0 & 0 & 0 \\ 0 & 0 & 0 & 0 \end{pmatrix}.$$

Inspecting the entries of $\mu^{(0)}$ we see that all transient states i in the unperturbed process are, as always, with $\mu_i^{(0)}=0$. A phenomenon we observe here is that state 1, although being recurrent in the unperturbed system, also has $\mu_1^{(0)}=0$. This is of course a priori possible (yet not all recurrent states can, simultaneously, have this property). In particular, here the recurrent state 1, as opposed to state 2, possesses some degree of transience in the perturbed MC.

Furthermore, the degrees of transience (see Definition 6.4) for the states in this example are $t(1)=2$, $t(2)=0$, $t(3)=1$, *and* $t(4)=2$. *Applying formula* (6.117) *of Theorem 6.7, we obtain*

$$\mathcal{M}_{12}(\varepsilon)=\frac{H_{22}^{(1)}-H_{12}^{(1)}}{M_2^{(1)}}\varepsilon^{-1}+o(\varepsilon^{-1})=\frac{0-(-1)}{1}\varepsilon^{-1}+o(\varepsilon^{-1})=\varepsilon^{-1}+o(\varepsilon^{-1}).$$

Note that if a fifth state were added so that this state would be related to the fourth as currently the fourth is related to the third, $t(5)$ *would be equal to* 3, *but the value of* s *would still be preserved at* $s=1$. *Also, the values of* $t(2)$, $t(3)$, *and* $t(4)$ *would stay unchanged. Finally, in the modified example we would have* $t(1)=t(5)=3$.

6.4 ▪ Google PageRank as a Perturbed Markov Chain

Surfers on the Internet frequently use search engines to find pages satisfying their query. However, there are typically hundreds or thousands of relevant pages available on the web. Thus, listing them in a proper order is a crucial and difficult task. One can use several criteria to sort relevant answers. It turns out that the link-based criteria that capture the importance of web pages provide rankings that appear to be very satisfactory to internet users. Examples of link-based criteria are PageRank used by the search engine Google, HITS used by search engines Teoma and Ask, and SALSA. In the link-based ranking criteria a hyperlink pointing to a web page is interpreted as a recommendation for this page. In this section we describe in detail the PageRank and show that it is an example of a singularly perturbed MC. The singular perturbation approach allows us to tune the main PageRank parameter, the so-called damping factor.

A page is called *dangling* if it does not have outgoing hyperlinks. Denote by n the total number of pages on the web, and define the $n\times n$ *hyperlink matrix* W as follows:

$$w_{ij}=\begin{cases}1/d_i & \text{if page } i \text{ links to } j,\\ 1/n & \text{if page } i \text{ is dangling,}\\ 0 & \text{otherwise,}\end{cases} \tag{6.118}$$

for $i,j=1,\ldots,n$, where d_i is the number of outgoing hyperlinks from page i. Then, PageRank is defined as a stationary distribution of an MC whose state space is the set of all web pages, and the transition matrix is

$$G=cW+(1-c)(1/n)\mathbf{1}^T\mathbf{1}. \tag{6.119}$$

We refer to the matrix G as *Google matrix*. Recall that we use the symbol $\mathbf{1}$ to denote a column vector of ones having by default an appropriate dimension. In (6.119), $\mathbf{1}^T\mathbf{1}$ is a matrix whose entries are all equal to one, and $c\in(0,1)$ is the parameter known as a *damping factor*. Let π be the PageRank vector. Then by definition, $\pi G=\pi$, and $||\pi||=\pi\mathbf{1}=1$, where we write $||\mathbf{x}||$ for the L_1-norm of the vector $\mathbf{x}$.

The damping factor c is a crucial parameter in the PageRank definition. It regulates the level of the uniform noise introduced to the system. Based on the publicly available information Google originally used, $c=0.85$, which appears to be a reasonable compromise between the true reflection of the web structure and numerical efficiency. As we demonstrate below, when $c=1$ there are several absorbing sets for the random walk defined by matrix W. However, if c is less than one but greater than zero, the MC induced by matrix G is ergodic. Thus, PageRank is a stationary distribution of the singularly perturbed MC with $\varepsilon=1-c$.

6.4.1 ▪ Illustrative datasets

We illustrate all theoretical results of the present section on two samples of the web graph, which we denote by INRIA and FMI. The web graph INRIA was taken from the site of INRIA, the French Research Institute of Informatics and Automatics. The seed for the INRIA collection was web page `http://www.inria.fr`. It is a typical large web site with around 300,000 pages and 2,000,000 hyperlinks. We collected all pages belonging to INRIA. The web graph FMI was crawled with the initial seeds of 50 French Mathematics and Informatics (FMI) laboratories, taken from Google Directory. The crawl was executed by breadth first search of depth 6. The FMI web graph contains around 700,000 pages and 8,000,000 hyperlinks. Because of the fractal-like structure of the web we expect our datasets to be sufficiently representative.

6.4.2 ▪ The structure of the web graph

The web graph can be divided into three principal components. The giant strongly connected component (SCC) contains a large group of pages all having a hyperlink path to each other. The pages in the IN (OUT) component have a path to (from) the SCC but not back. Furthermore, the SCC component is larger than the second largest strongly connected component by several orders of magnitude. With this bow-tie web structure in mind, we would like to analyze a stationary distribution of a Markov random walk governed by the hyperlink transition matrix W given by (6.118). Such a random walk follows an outgoing link chosen uniformly at random, and dangling nodes are assumed to have links to all pages in the web.

Obviously, the graph induced by W has a much higher connectivity than the original web graph. In particular, if the random walk can move from a dangling node to an arbitrary node with the uniform distribution, then the giant SCC component increases further in size. We refer to this new strongly connected component as the extended strongly connected component (ESCC). Due to the artificial links from the dangling nodes, the SCC and IN components are now interconnected and are parts of the ESCC. Furthermore, if there are dangling nodes in the OUT component, then these nodes together with all their predecessors become a part of the ESCC.

In the miniexample in Figure 6.1, node 0 represents the IN component, nodes from 1 to 3 form the SCC component, and the rest of the nodes (4 to 11) are in the OUT component. Node 5 is a dangling node; thus, artificial links go from the dangling node 5 to all other nodes. After addition of the artificial links, all nodes from 0 to 5 form the ESCC.

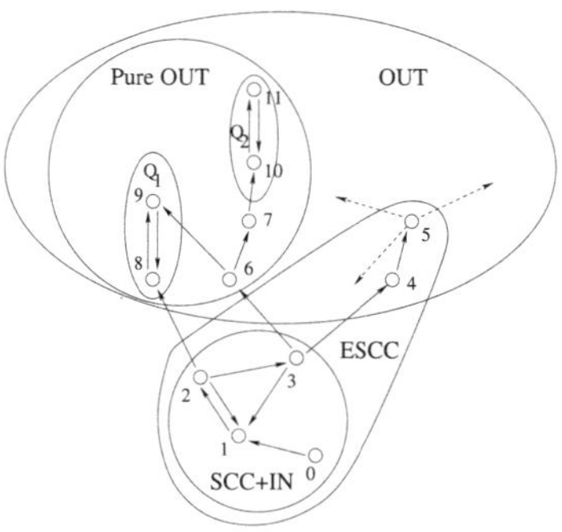

Figure 6.1. *Miniexample of a web graph [16]*

In the MC induced by the matrix W, all states in ESCC are *transient*; that is, with probability 1, the MC eventually leaves this set of states and never returns. The stationary probability of all these states is zero. The part of the OUT component without dangling nodes and their predecessors forms a block that we refer to as a Pure OUT component. In Figure 6.1 the Pure OUT component consists of nodes from 6 to 11. Typically, the Pure OUT component is much smaller than the ESCC. However, this is the set where the total stationary probability mass is concentrated in the long run. The sizes of all components for our two datasets are displayed in Table 6.1. Our algorithms for discovering the structures of the web graph are based on breadth first search and depth first search methods, which are linear in the sum of number of nodes and links. Here the size of the IN components is zero because in the web crawl we used the breadth first search method and we started from important pages in the giant SCC. For the purposes of the present analysis it does not make any difference since we always consider IN and SCC together.

Table 6.1. *Component sizes in INRIA and FMI datasets [16]*

	INRIA	*FMI*
Total size	318585	764119
Number of nodes in SCC	154142	333175
Number of nodes in IN	0	0
Number of nodes in OUT	164443	430944
Number of nodes in ESCC	300682	760016
Number of nodes in Pure OUT	17903	4103
Number of SCCs in OUT	1148	1382
Number of SCCs in Pure OUT	631	379

Let us now analyze the structure of the Pure OUT component in more detail. It turns out that inside Pure OUT there are many disjoint strongly connected components. All states in these sub-SCCs (or, absorbing sets) are *recurrent*. There are many absorbing sets of size two and three.

The Pure OUT component also contains transient states that eventually bring the random walk into one of the absorbing sets. For simplicity, we add these states to the giant transient ESCC component.

Now, by appropriate renumbering of the states, we can refine the hyperlink matrix W by subdividing all states into one giant transient block and a number of small recurrent blocks as follows:

$$W = \begin{bmatrix} Q_1 & & 0 & 0 \\ & \ddots & & \\ 0 & & Q_m & 0 \\ R_1 & \cdots & R_m & T \end{bmatrix} \begin{array}{l} \text{absorbing set (recurrent)} \\ \cdots \\ \text{absorbing set (recurrent)} \\ \text{ESCC+[transient states in Pure OUT] (transient).} \end{array} \tag{6.120}$$

Here for $i = 1, \ldots, m$, a block Q_i corresponds to transitions inside the ith recurrent block, and a block R_i contains transition probabilities from transient states to the ith recurrent block. Block T corresponds to transitions between the transient states. For instance, in the example of the graph from Figure 6.1, the nodes 8 and 9 correspond to block Q_1, nodes 10 and 11 correspond to block Q_2, and all other nodes belong to block T.

We would like to emphasize that the recurrent blocks here are really small, constituting altogether about 5% for INRIA and about 0.5% for FMI. We believe that for larger datasets, this percentage will be even less. By far the most important part of the web is contained in ESCC, which constitutes the major part of the giant transient block.

Next, we note that if $c < 1$, then all states in the MC induced by the Google matrix G are recurrent, which automatically implies that they all have positive stationary probabilities. However, if $c = 1$, the majority of pages turn into transient states with stationary probability zero. Hence, the random walk governed by the Google matrix (6.119) is in fact a singularly perturbed MC with $\varepsilon = 1 - c$. Using our general results on the singular perturbation of MCs, in the next proposition we characterize explicitly the limiting PageRank vector as $c \to 1$ or, equivalently, $\varepsilon \to 0$.

Proposition 6.3. *Let $\bar{\pi}_{O,i}$ be a stationary distribution of the MC governed by Q_i ($\bar{\pi}_{O,i} Q_i = \bar{\pi}_{O,i}$), $i = 1, \ldots, m$. Then, we have*

$$\lim_{c \to 1} \pi(c) = \left[\pi_{O,1} \; \cdots \; \pi_{O,m} \; 0 \right],$$

where

$$\pi_{O,i} = \left(\frac{|Q_i|}{n} + \frac{1}{n} \mathbf{1}^T [I - T]^{-1} R_i \mathbf{1} \right) \bar{\pi}_{O,i} \tag{6.121}$$

for $i = 1, \ldots, m$, and where $|Q_i|$ is the number of states in block Q_i, I is the identity matrix, and 0 is a row vector of zeros that correspond to stationary probabilities of the states in the transient block.

As the proof is rather straightforward, in Problem 6.18 we invite the reader to verify this statement.

The second term inside the brackets in formula (6.121) corresponds to the PageRank mass (the sum of corresponding elements of the PageRank vector) received by an absorbing set from the ESCC. If c is close to one, then this contribution can by far outweigh the fair share of the PageRank, whereas the PageRank mass of the giant transient block decreases to zero. How large is the neighborhood of one where the ranking is skewed toward the Pure OUT? Is the value $c = 0.85$ already too large? We address these questions in the remainder of this section. In the next subsection we analyze the PageRank mass of the IN+SCC component, which is an important part of the transient block.

6.4.3 ▪ PageRank mass of IN+SCC

In Figure 6.2 for the FMI dataset we depict the PageRank mass of the giant SCC as a function of the damping factor. Here we see a typical behavior of PageRank for important pages: the PageRank first grows with c and then decreases to zero. In our case, the PageRank mass of SCC drops drastically starting from some value c close to one. We can explain this phenomenon by highlighting the role of the dangling nodes.

We start the analysis by subdividing the web graph sample into three subsets of nodes: IN+SCC, OUT, and the set of dangling nodes DN. To simplify the algebra, we assume that no dangling node originates from OUT. A more general situation is left to the reader as an exercise (see Problem 6.19). Then the web hyperlink matrix W in (6.118) can be written in the form

$$W = \begin{bmatrix} Q & 0 & 0 \\ R & P & S \\ \frac{1}{n}\mathbf{1}\mathbf{1}^T & \frac{1}{n}\mathbf{1}\mathbf{1}^T & \frac{1}{n}\mathbf{1}\mathbf{1}^T \end{bmatrix} \begin{matrix} OUT \\ IN+SCC \\ DN \end{matrix} \tag{6.122}$$

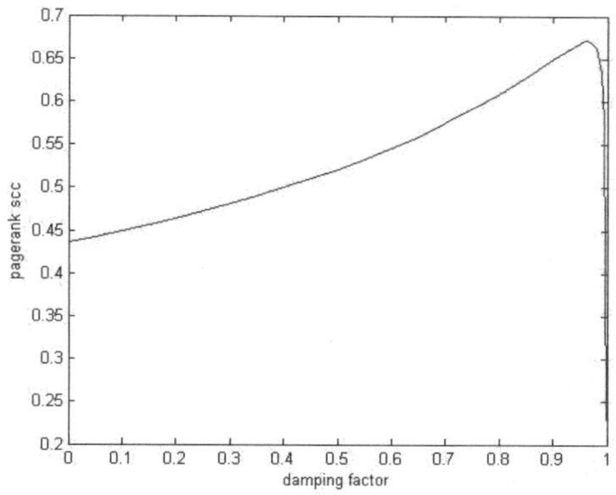

Figure 6.2. *PageRank mass of SCC as a function of c [16]*

where the block Q corresponds to the hyperlinks inside the OUT component, the block R corresponds to the hyperlinks from IN+SCC to OUT, the block P corresponds to the hyperlinks inside the IN+SCC component, and the block S corresponds to the hyperlinks from SCC to dangling nodes. Recall that n is the total number of pages in the web graph sample, and the blocks $\mathbf{1}\mathbf{1}^T$ are the matrices of ones adjusted to appropriate dimensions.

We note (see Problem 6.20) that the PageRank vector can be written with the explicit formula

$$\pi = \frac{1-c}{n}\mathbf{1}^T[I - cW]^{-1}. \tag{6.123}$$

Next, dividing the PageRank vector into segments corresponding to the blocks OUT, IN+SCC, and DN,

$$\pi = [\pi_{\mathrm{O}}\ \pi_{\mathrm{I+S}}\ \pi_{\mathrm{D}}],$$

we can rewrite (6.123) as a system of three linear equations:

$$\pi_{\mathrm{O}}[I - cQ] - \pi_{\mathrm{I+S}}cR - \frac{c}{n}\pi_{\mathrm{D}}\mathbf{1}\mathbf{1}^T = \frac{1-c}{n}\mathbf{1}^T, \tag{6.124}$$

$$\pi_{\mathrm{I+S}}[I - cP] - \frac{c}{n}\pi_{\mathrm{D}}\mathbf{1}\mathbf{1}^T = \frac{1-c}{n}\mathbf{1}^T, \tag{6.125}$$

$$-\pi_{\mathrm{I+S}}cS + \pi_{\mathrm{D}} - \frac{c}{n}\pi_{\mathrm{D}}\mathbf{1}\mathbf{1}^T = \frac{1-c}{n}\mathbf{1}^T. \tag{6.126}$$

First, we observe that if $\pi_{\mathrm{I+S}}$ and $\pi_{\mathrm{D}}\mathbf{1}$ are known, then it is straightforward to calculate π_{O}. Namely, we have

$$\pi_{\mathrm{O}} = \pi_{\mathrm{I+S}}cR[I - cQ]^{-1} + \left(\frac{1-c}{n} + \pi_{\mathrm{D}}\mathbf{1}\frac{c}{n}\right)\mathbf{1}^T[I - cQ]^{-1}.$$

Therefore, let us first solve the equations (6.125) and (6.126). Toward this goal, we sum the elements of the vector equation (6.126), which corresponds to the postmultiplication of equation (6.126) by vector $\mathbf{1}$.

$$-\pi_{\mathrm{I+S}}cS\mathbf{1} + \pi_{\mathrm{D}}\mathbf{1} - \frac{c}{n}\pi_{\mathrm{D}}\mathbf{1}\mathbf{1}^T\mathbf{1} = \frac{1-c}{n}\mathbf{1}^T\mathbf{1}.$$

Now, denote by n_I, n_O, n_S, and n_D the number of pages in the IN component, OUT component, and SCC component and the number of dangling nodes. Since $\mathbf{1}^T\mathbf{1} = n_D$ with the dimension of $\mathbf{1}$ as in (6.122), we have

$$\pi_{\mathrm{D}}\mathbf{1} = \frac{n}{n-cn_D}\left(\pi_{\mathrm{I+S}}cS\mathbf{1} + \frac{1-c}{n}n_D\right).$$

Substituting the above expression for $\pi_{\mathrm{D}}\mathbf{1}$ into (6.125), we obtain

$$\pi_{\mathrm{I+S}}\left[I - cP - \frac{c^2}{n-cn_D}S\mathbf{1}\mathbf{1}^T\right] = \frac{c}{n-cn_D}\frac{1-c}{n}n_D\mathbf{1}^T + \frac{1-c}{n}\mathbf{1}^T,$$

which implies

$$\pi_{\mathrm{I+S}}(c) = \frac{(1-c)\alpha}{1-c\beta}\mathbf{u}_{\mathrm{I+S}}\left[I - cP - \frac{c^2\alpha}{1-c\beta}S\mathbf{1}\mathbf{u}_{\mathrm{I+S}}\right]^{-1}, \tag{6.127}$$

where

$$\alpha = \frac{n_I + n_S}{n} \quad \text{and} \quad \beta = \frac{n_D}{n}$$

are the fractions of nodes in IN+SCC and DN, respectively, and $\mathbf{u}_{\mathrm{I+S}} = [n_I + n_S]^{-1}\mathbf{1}^T$ is a uniform probability row-vector of dimension $n_I + n_S$.

Now, define

$$k(c) = \frac{(1-c)\alpha}{1-c\beta} \quad \text{and} \quad U(c) = P + \frac{c\alpha}{1-c\beta}S\mathbf{1}\mathbf{u}_{\mathrm{I+S}}. \tag{6.128}$$

Then the derivative of $\pi_{\mathrm{I+S}}(c)$ with respect to c is given by

$$\pi'_{\mathrm{I+S}}(c) = \mathbf{u}_{\mathrm{I+S}}\left\{k'(c)I + k(c)[I - cU(c)]^{-1}(cU(c))'\right\}[I - cU(c)]^{-1}, \tag{6.129}$$

where, using (6.128), after simple calculations, we obtain

$$k'(c) = -\frac{(1-\beta)\alpha}{(1-c\beta)^2}, \quad (cU(c))' = U(c) + \frac{c\alpha}{(1-c\beta)^2}S\mathbf{1}\mathbf{u}_{\mathrm{I+S}}.$$

Let us consider the point $c = 0$. Using (6.129), we obtain

$$\pi'_{\mathrm{I+S}}(0) = -\alpha(1-\beta)\mathbf{u}_{\mathrm{I+S}} + \alpha\mathbf{u}_{\mathrm{I+S}}P. \tag{6.130}$$

One can see from the above equation that the PageRank mass of pages in IN+SCC with many incoming links will increase as c increases from zero.

Next, let us analyze the total mass of the IN+SCC component. From (6.130) we obtain

$$||\pi'_{\mathrm{I+S}}(0)|| = -\alpha(1-\beta)\mathbf{u}_{\mathrm{I+S}} + \alpha\mathbf{u}_{\mathrm{I+S}}P\mathbf{1} = \alpha(-1+\beta+p_1),$$

where $p_1 = \mathbf{u}_{\mathrm{I+S}}P\mathbf{1}$ is the probability that a random walk on the hyperlink matrix stays in IN+SCC for one step if the initial distribution is uniform over IN+SCC. If $1-\beta < p_1$, then the derivative at 0 is positive. Since dangling nodes typically constitute more than 25% of the web graph, and p_1 is usually close to one, the condition $1-\beta < p_1$ seems to be comfortably satisfied in typical web graph samples. Thus, the total PageRank mass of IN+SCC increases in c when c is small. Note that if $\beta = 0$, then $||\pi_{\mathrm{I+S}}(c)||$ is strictly decreasing in c. Hence, surprisingly, the presence of dangling nodes qualitatively changes the behavior of the IN+SCC PageRank mass.

Now let us consider the point $c=1$. Again using (6.129), we obtain

$$\pi'_{\mathrm{I+S}}(1) = -\frac{\alpha}{1-\beta}\mathbf{u}_{\mathrm{I+S}}[I-P-\frac{\alpha}{1-\beta}S\mathbf{1}\mathbf{u}_{\mathrm{I+S}}]^{-1}. \tag{6.131}$$

Note that the matrix in the square braces is close to singular. Let us state an auxiliary result which is a particular case of Theorem 2.9 when the perturbation matrix is rank one (Problem 6.21).

Lemma 6.8. *Let $A(\varepsilon) = A - \varepsilon C$ be a perturbation of irreducible stochastic matrix A such that $A(\varepsilon)$ is substochastic. Then, for sufficiently small and positive ε the following Laurent series expansion holds:*

$$[I-A(\varepsilon)]^{-1} = \frac{1}{\varepsilon}X_{-1} + X_0 + \varepsilon X_1 + \dots,$$

with

$$X_{-1} = \frac{1}{\mu C\mathbf{1}}\mathbf{1}\mu,$$

where μ is the stationary distribution of A. It follows that

$$[I-A(\varepsilon)]^{-1} = \frac{1}{\mu\varepsilon C\mathbf{1}}\mathbf{1}\mu + O(1) \quad \textit{as } \varepsilon \to 0. \tag{6.132}$$

Denote by $\bar{P}$ the hyperlink matrix of IN+SCC when the outer links are neglected. Then, $\bar{P}$ is an irreducible stochastic matrix. Denote its stationary distribution by $\bar{\pi}_{\mathrm{I+S}}$. Then we can apply Lemma 6.8 to (6.131) by taking

$$A = \bar{P}, \quad \varepsilon C = \bar{P} - P - \frac{\alpha}{1-\beta}S\mathbf{1}\mathbf{u}_{\mathrm{I+S}}$$

and noting that $\varepsilon C\mathbf{1} = R\mathbf{1} + (1-\alpha-\beta)(1-\beta)^{-1}S\mathbf{1}$. Combining all terms together and using $\bar{\pi}_{\mathrm{I+S}}\mathbf{1} = ||\bar{\pi}_{\mathrm{I+S}}|| = 1$ and $\mathbf{u}_{\mathrm{I+S}}\mathbf{1} = ||\mathbf{u}_{\mathrm{I+S}}|| = 1$, from (6.132) we obtain

$$||\pi'_{\mathrm{I+S}}(1)|| \approx -\frac{\alpha}{1-\beta}\frac{1}{\bar{\pi}_{\mathrm{I+S}}R\mathbf{1} + \frac{1-\beta-\alpha}{1-\beta}\bar{\pi}_{\mathrm{I+S}}S\mathbf{1}}.$$

Typically for the web graph the value of $\bar{\pi}_{\mathrm{I+S}}R\mathbf{1} + \frac{1-\beta-\alpha}{1-\beta}\bar{\pi}_{\mathrm{I+S}}S\mathbf{1}$ is small, and hence the mass $||\pi_{\mathrm{I+S}}(c)||$ decreases very quickly as c approaches one.

Having described the behavior of the PageRank mass $||\pi_{\mathrm{I+S}}(c)||$ at the boundary points $c=0$ and $c=1$, we would now like to show that there is at most one extremum on $(0,1)$. It is sufficient to prove that if $||\pi'_{\mathrm{I+S}}(c_0)|| \le 0$ for some $c_0 \in (0,1)$, then $||\pi'_{\mathrm{I+S}}(c)|| \le 0$ for all $c > c_0$. To this end, we apply the Sherman–Morrison formula to (6.127), which yields

$$\pi_{\mathrm{I+S}}(c) = \tilde{\pi}_{\mathrm{I+S}}(c) + \frac{\frac{c^2\alpha}{1-c\beta}\mathbf{u}_{\mathrm{I+S}}[I-cP]^{-1}S\mathbf{1}}{1+\frac{c^2\alpha}{1-c\beta}\mathbf{u}_{\mathrm{I+S}}[I-cP]^{-1}S\mathbf{1}}\tilde{\pi}_{\mathrm{I+S}}(c), \tag{6.133}$$

where

$$\tilde{\pi}_{\mathrm{I+S}}(c) = \frac{(1-c)\alpha}{1-c\beta}\mathbf{u}_{\mathrm{I+S}}[I-cP]^{-1} \tag{6.134}$$

represents the most significant order term in the right-hand side of (6.133). Now the behavior of $\pi_{\mathrm{I+S}}(c)$ in Figure 6.2 can be explained by the next proposition.

Proposition 6.4. *The function $||\tilde{\pi}_{I+S}(c)||$ associated with (6.134) has exactly one local maximum at some $c_0 \in [0,1]$. Moreover, $||\tilde{\pi}''_{I+S}(c)|| < 0$ for $c \in (c_0, 1]$.*

Proof: Multiplying both sides of (6.134) by **1** and taking the derivatives, after some tedious algebra, we obtain

$$||\tilde{\pi}'_{I+S}(c)|| = -a(c) + \frac{\beta}{1-c\beta}||\tilde{\pi}_{I+S}(c)||, \tag{6.135}$$

where the real-valued function $a(c)$ is given by

$$a(c) = \frac{\alpha}{1-c\beta}\mathbf{u}_{I+S}[I-cP]^{-1}[I-P][I-cP]^{-1}\mathbf{1}.$$

Differentiating (6.135) and substituting $\frac{\beta}{1-c\beta}||\tilde{\pi}_{SCC}(c)||$ from (6.135) into the resulting expression, we get

$$||\tilde{\pi}''_{I+S}(c)|| = \left\{-a'(c) + \frac{\beta}{1-c\beta}a(c)\right\} + \frac{2\beta}{1-c\beta}||\tilde{\pi}'_{SCC}(c)||.$$

Note that the term in the curly brackets is negative by definition of $a(c)$. Hence, if $||\tilde{\pi}'_{I+S}(c)|| \le 0$ for some $c \in [0,1]$, then $||\tilde{\pi}''_{I+S}(c)|| < 0$ for this value of c. □

We conclude that $||\tilde{\pi}_{I+S}(c)||$ is decreasing and concave for $c \in [c_0, 1]$, where $||\tilde{\pi}'_{I+S}(c_0)|| = 0$. This is exactly the behavior we observe in the experiments. The analysis and experiments suggest that c_0 is definitely larger than 0.85 and actually is quite close to one. Thus, one may want to choose large c in order to maximize the PageRank mass of IN+SCC. However, in the next section we will indicate important drawbacks of this choice.

6.4.4 ▪ PageRank mass of ESCC

Let us now consider the PageRank mass of the extended strongly connected component (ESCC) described in Section 6.4.2, as a function of $c \in [0,1]$. Subdividing the PageRank vector into the blocks $\pi = [\pi_P, \pi_E]$, according to Pure OUT and ESCC components, and using formula (6.123), we obtain

$$\pi_E(c) = (1-c)\gamma\mathbf{u}_E[I-cT]^{-1} = (1-c)\gamma\mathbf{u}_E\sum_{k=1}^{\infty}c^kT^k, \tag{6.136}$$

where T represents the transition probabilities inside the ESCC block, $\gamma = |ESCC|/n$ is the fraction of pages contained in the ESCC, and $\mathbf{u}_E$ is a uniform probability row-vector over ESCC. Clearly, we have $||\pi_E(0)|| = \gamma$ and $||\pi_E(1)|| = 0$. Furthermore, it is easy to see that $||\pi_E(c)||$ is a concave decreasing function, since

$$\frac{d}{dc}||\pi_E(c)|| = -\gamma\mathbf{u}_E[I-cT]^{-2}[I-T]\mathbf{1} < 0$$

and

$$\frac{d^2}{dc^2}||\pi_E(c)|| = -2\gamma\mathbf{u}_E[I-cT]^{-3}T[I-T]\mathbf{1} < 0.$$

The next proposition establishes the upper and lower bounds for $||\pi_E(c)||$.

Proposition 6.5. *Let λ_1 be the Perron–Frobenius eigenvalue of T, and let $p_1 = \mathbf{u}_E T\mathbf{1}$ be the probability that the random walk started from a randomly chosen state in the ESCC stays in the ESCC for one step. If $p_1 \le \lambda_1$ and*

$$p_1 \le \frac{\mathbf{u}_E T^k \mathbf{1}}{\mathbf{u}_E T^{k-1}\mathbf{1}} \le \lambda_1 \quad \forall k \ge 1, \tag{6.137}$$

then

$$\frac{\gamma(1-c)}{1-cp_1} < ||\pi_E(c)|| < \frac{\gamma(1-c)}{1-c\lambda_1}, \quad c \in (0,1). \tag{6.138}$$

Proof: From condition (6.137) it follows by induction that

$$p_1^k \le \mathbf{u}_E T^k \mathbf{1} \le \lambda_1^k, \quad k \ge 1,$$

and thus the statement of the proposition is obtained directly from the series expansion of $\pi_E(c)$ in (6.136). □

The conditions of Proposition 6.5 have a natural probabilistic interpretation. The value p_1 is the probability that the Markov random walk on the web sample stays in the block T for one step, starting from the uniform distribution over T. Furthermore, $p_k = \mathbf{u}_E T^k \mathbf{1}/(\mathbf{u}_E T^{k-1}\mathbf{1})$ is the probability that the random walk stays in T for one step provided that it has stayed there for the first $k-1$ steps. It is a well-known fact that, as $k \to \infty$, p_k converges to λ_1, the Perron–Frobenius eigenvalue of T. Let $\hat{\pi}_E$ be the probability-normed left Perron–Frobenius eigenvector of T. Then $\hat{\pi}_E$, also known as a *quasi-stationary* distribution of T, is the limiting probability distribution of the MC given that the random walk never leaves the block T. Since $\hat{\pi}_E T = \lambda_1 \hat{\pi}_E$, the condition $p_1 < \lambda_1$ means that the probability of staying in the ESCC for one step in the quasi-stationary regime is higher than that of starting from the uniform distribution $\mathbf{u}_E$. This is quite natural since the quasi-stationary distribution tends to avoid the states from which the random walk is likely to leave the block T. Furthermore, the condition in (6.137) says that if the random walk is about to make its kth step in T, then it leaves T most easily at step $k = 1$, and it is least likely to leave T after an infinite number of steps. Both conditions of Proposition 6.5 are satisfied in our experiments on both datasets. Moreover, we noticed that the sequence $(p_k, k \ge 1)$ was increasing from p_1 to λ_1.

With the help of the derived bounds we conclude that $||\pi_E(c)||$ decreases very slowly for small and moderate values of c, and it decreases extremely fast when c becomes close to 1. This typical behavior is clearly seen in Figure 6.3, where $||\pi_E(c)||$ is plotted with a solid line. The bounds are plotted in Figure 6.3 with dashed lines. For the INRIA dataset we have $p_1 = 0.97557$ and $\lambda_1 = 0.99954$, and for the FMI dataset we have $p_1 = 0.99659$ and $\lambda_1 = 0.99937$.

From the above we conclude that the PageRank mass of the ESCC is smaller than γ for any value $c > 0$. On the contrary, the PageRank mass of Pure OUT increases in c beyond its "fair share" $\delta = |PureOUT|/n$. With $c = 0.85$, the PageRank mass of the Pure OUT component in the INRIA dataset is equal to 1.95δ. In the FMI dataset, the unfairness is even more pronounced: the PageRank mass of the Pure OUT component is equal to 3.44δ. This gives users an incentive to create dead-ends: groups of pages that link only to each other. Clearly, this can be mitigated by choosing a smaller damping factor. Below we propose one way to determine an "optimal" value of c.

Since the PageRank mass of the ESCC is always smaller than γ, we would like to choose the damping factor in such a way that the ESCC receives a "fair" fraction of γ.

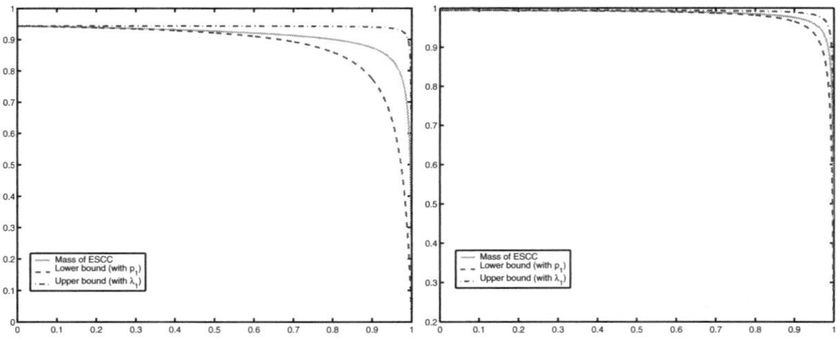

Figure 6.3. *PageRank mass of the ESCC and bounds; INRIA (left) and FMI (right) [16]*

Formally, we would like to define a number $\rho \in (0,1)$ such that a desirable PageRank mass of the ESCC could be written as $\rho\gamma$, and then find the value c^* that satisfies

$$\|\pi_{\mathrm{E}}(c^*)\| = \rho\gamma. \tag{6.139}$$

Then $c \leq c^*$ will ensure that $\|\pi_{\mathrm{E}}(c)\| \geq \rho\gamma$. Naturally, ρ should somehow reflect the properties of the substochastic block T. For instance, as T becomes closer to being a stochastic matrix, ρ should also increase. One possibility to do it is to define

$$\rho = \mathbf{v}T\mathbf{1},$$

where $\mathbf{v}$ is a row vector representing some probability distribution on the ESCC. Then the damping factor c should satisfy

$$c \leq c^*,$$

where c^* is given by

$$\|\pi_{\mathrm{E}}(c^*)\| = \gamma\mathbf{v}T\mathbf{1}. \tag{6.140}$$

In this setting, ρ is a probability of staying in the ESCC for one step if initial distribution is $\mathbf{v}$. For given $\mathbf{v}$, this number increases as T becomes closer to a stochastic matrix. Now, the problem of choosing ρ comes down to the problem of choosing $\mathbf{v}$. The advantage of this approach is twofold. First, we still have all the flexibility because, depending on $\mathbf{v}$, the value of ρ may vary considerably, except it cannot become too small if T is really close to a stochastic matrix. Second, we can use a probabilistic interpretation of $\mathbf{v}$ to make a reasonable choice. One can think, for instance, of the following three intuitive choices of $\mathbf{v}$: (1) $\hat{\pi}_{\mathrm{E}}$, the quasi-stationary distribution of T, (2) the uniform vector $\mathbf{u}_{\mathrm{E}}$, and (3) the normalized PageRank vector $\pi_{\mathrm{E}}(c)/\|\pi_{\mathrm{E}}(c)\|$. The first choice reflects the proximity of T to a stochastic matrix. The second choice is inspired by definition of PageRank (restart from uniform distribution), and the third choice combines both these features.

If the conditions of Proposition 6.5 are satisfied, then (6.138) holds, and thus the value of c^* satisfying (6.140) must be in the interval (c_1, c_2), where

$$(1-c_1)/(1-p_1c_1) = \|\mathbf{v}T\|, \quad (1-c_2)/(1-\lambda_1c_2) = \|\mathbf{v}T\|.$$

Numerical results for all three choices of $\mathbf{v}$ are presented in Table 6.2.

If $\mathbf{v} = \hat{\pi}_{\mathrm{E}}$, then we have $\|\mathbf{v}T\| = \lambda_1$, which implies $c_1 = (1-\lambda_1)/(1-\lambda_1 p_1)$ and $c_2 = 1/(\lambda_1+1)$. In this case, the upper bound c_2 is only slightly larger than $1/2$ and c^* is close to zero in our datasets (see Table 6.2). Such small c, however, leads to ranking that takes into account only local information about the web graph. The choice $\mathbf{v} = \hat{\pi}_{\mathrm{E}}$

Table 6.2. *Values of c^* with bounds [16]*

v	c	INRIA	FMI
$\hat{\pi}_E$	c_1	0.0184	0.1956
	c_2	0.5001	0.5002
	c^*	.02	.16
$\mathbf{u}_E$	c_1	0.5062	0.5009
	c_2	0.9820	0.8051
	c^*	.604	.535
$\pi_E/\|\|\pi_E\|\|$	$1/(1+\lambda_1)$	0.5001	0.5002
	$1/(1+p_1)$	0.5062	0.5009

does not seem to represent the dynamics of the system, probably because the "easily bored surfer" random walk that is used in PageRank computations never follows a quasi-stationary distribution since it often restarts itself from the uniform probability vector.

For the uniform vector $\mathbf{v} = \mathbf{u}_E$, we have $||\mathbf{v}T|| = p_1$, which gives c_1, c_2, c^* presented in the second row of Table 6.2. We have obtained a higher upper bound, but the values of c^* are still much smaller than 0.85.

Finally, consider the normalized PageRank vector $\mathbf{v}(c) = \pi_E(c)/||\pi_E(c)||$. This choice of $\mathbf{v}$ can also be justified as follows. Consider the derivative of the total PageRank mass of the ESCC. Since $[I - cT]^{-1}$ and $[I - T]$ commute, we can write

$$\frac{d}{dc}||\pi_E(c)|| = -\gamma \mathbf{u}_E[I - cT]^{-1}[I - T][I - cT]^{-1}\mathbf{1},$$

or, equivalently,

$$\begin{aligned}\frac{d}{dc}||\pi_E(c)|| &= -\frac{1}{1-c}\pi_E[I - T][I - cT]^{-1}\mathbf{1}\\ &= -\frac{1}{1-c}\left(\pi_E - ||\pi_E||\frac{\pi_E}{||\pi_E||}T\right)[I - cT]^{-1}\mathbf{1}\\ &= -\frac{1}{1-c}(\pi_E - ||\pi_E||\mathbf{v}(c)T)[I - cT]^{-1}\mathbf{1},\end{aligned}$$

with $\mathbf{v}(c) = \pi_E/||\pi_E||$. It is easy to see that

$$||\pi_E(c)|| = \gamma - \gamma(1 - \mathbf{u}_E T\mathbf{1})c + o(c).$$

Consequently, we obtain

$$\frac{d}{dc}||\pi_E(c)|| = -\frac{1}{1-c}(\pi_E - \gamma\mathbf{v}(c)T + \gamma(1 - \mathbf{u}_E T\mathbf{1})c\mathbf{v}(c)T + o(c))[I - cT]^{-1}\mathbf{1}.$$

Since in practice T is very close to being stochastic, we have

$$1 - \mathbf{u}_E T\mathbf{1} \approx 0 \quad \text{and} \quad [I - cT]^{-1}\mathbf{1} \approx \frac{1}{1-c}\mathbf{1}.$$

The latter approximation follows from Lemma 6.8. Thus, satisfying condition (6.140) means keeping the value of the derivative small.

Let us now solve (6.140) for $\mathbf{v}(c) = \pi_E(c)/||\pi_E(c)||$. Using (6.136), we rewrite (6.140) as

$$||\pi_E(c)|| = \frac{\gamma}{||\pi_E(c)||}\pi_E(c)T\mathbf{1} = \frac{\gamma^2(1-c)}{||\pi_E(c)||}\mathbf{u}_{I+S}[I-cT]^{-1}T\mathbf{1}.$$

Multiplying by $||\pi_E(c)||$, after some algebra, we obtain

$$||\pi_E(c)||^2 = \frac{\gamma}{c}\,||\pi_E(c)|| - \frac{(1-c)\gamma^2}{c}.$$

Solving the quadratic equation for $||\pi_E(c)||$, we obtain

$$||\pi_E(c)|| = r(c) = \begin{cases} \gamma & \text{if } c \le 1/2, \\ \frac{\gamma(1-c)}{c} & \text{if } c > 1/2. \end{cases}$$

Hence, the value c^* solving (6.140) corresponds to the point where the graphs of $||\pi_E(c)||$ and $r(c)$ cross each other. There is only one such point on (0,1), and since $||\pi_E(c)||$ decreases very slowly unless c is close to one, whereas $r(c)$ decreases relatively quickly for $c > 1/2$, we expect that c^* is only slightly larger than 1/2. Under the conditions of Proposition 6.5, $r(c)$ first crosses the line $\gamma(1-c)/(1-\lambda_1 c)$, then $||\pi_E(c)||_1$, and then $\gamma(1-c)/(1-p_1 c)$. This yields $(1+\lambda_1)^{-1} < c^* < (1+p_1)^{-1}$. Since both λ_1 and p_1 are large, this suggests that c should be chosen around 1/2. This is also reflected in Table 6.2.

Last but not least, to support our theoretical argument about the undeserved high ranking of pages from Pure OUT, we carry out the following experiment. In the INRIA dataset we have chosen an absorbing component in Pure OUT consisting of just two nodes. We have added an artificial link from one of these nodes to a node in the giant SCC and recomputed the PageRank. In Table 6.3 in the column "PR rank w/o link" we give a ranking of a page according to the PageRank value computed before the addition of the artificial link, and in the column "PR rank with link" we give a ranking of a page according to the PageRank value computed after the addition of the artificial link. We have also analyzed the log file of the site INRIA Sophia Antipolis (`http://www-sop.inria.fr`) and ranked the pages according to the number of clicks for the period of one year up to May 2007. We note that since we have access only to the log file of the INRIA Sophia Antipolis site, we use the PageRank ranking also only for the pages from the INRIA Sophia Antipolis site. For instance, for $c = 0.85$, the ranking of Page A without an artificial link is 731 (this means that 730 pages are ranked better than Page A among the pages of INRIA Sophia Antipolis). However, its ranking according to the number of clicks is much lower—2588. This confirms our conjecture that the nodes in Pure OUT obtain unjustifiably high ranking. Next we note that the addition of an artificial link significantly diminishes the ranking. In fact, it brings it close to the ranking provided by the number of clicks. Finally, we draw the attention of the reader to the fact that choosing $c = 1/2$ also significantly reduces the gap between the ranking by PageRank and the ranking by the number of clicks.

To summarize, our results indicate that with $c = 0.85$, the Pure OUT component receives an unfairly large share of the PageRank mass. Remarkably, in order to satisfy any of the three intuitive criteria of fairness presented above, the value of c should be drastically reduced. The experiment with the log files confirms the same. Of course, a drastic reduction of c also considerably accelerates the computation of PageRank by numerical methods.

Even though our statement that c should be 1/2 might be received with healthy skepticism, we hope to have convinced the reader that the study of the perturbed MC structure on the web graph helps in understanding and improving link-based ranking criteria.

Table 6.3. *Comparison between PR and click based rankings [16]*

c	PR rank w/o link	PR rank with link	rank by no. of clicks
Node A			
0.5	1648	2307	2588
0.85	731	2101	2588
0.95	226	2116	2588
Node B			
0.5	1648	4009	3649
0.85	731	3279	3649
0.95	226	3563	3649

6.5 ▪ Problems

Problem 6.1. Prove the following well-known formulae for the fundamental matrix Z and the deviation matrix H of an MC:

$$Z=[\Pi+I-P]^{-1}=[\Pi-G]^{-1},$$

$$H=Z-\Pi=[\Pi+I-P]^{-1}-\Pi=[\Pi-G]^{-1}-\Pi.$$

Hint: Proofs of these identities can either be derived or be found in many sources including [101] and [26].

Problem 6.2. Exploit the structure of the systems of equations (MF) and (RMF) in Section 6.2.1 to verify the validity of the recursive formula (6.22). Recall that the dimension of the coefficients $G_j^{(1)}$, $j \geq 0$, is equal to n, the number of ergodic classes of the unperturbed MC, and that the matrix $G_0^{(1)}$ can be considered as a generator of the aggregated MC whose states represent the ergodic classes of the original chain. *Hint: See the proofs of Theorems 2.14 and 3.7.*

Problem 6.3. In the discussion at the end of Section 6.2.1 show that we can stop after the first reduction step, and then solve the system (RMF) with the help of generalized inverses and augmented matrices using the results of Sections 2.2–3.3. Of course, one can make any number of reduction steps between 1 and s and then apply the approach based on the generalized inverses and augmented matrices. *Hint: This approach is in line with the work of Haviv and Ritov [76, 77].*

Problem 6.4. Use the results of Schweitzer and Stewart [141] to show that in (6.31) the regular part $U^R(\varepsilon)$ can be written in the closed analytic form

$$U^R(\varepsilon)=(I+\varepsilon U_0T_1)^{-1}U_0. \tag{6.141}$$

Then, verify that $\varphi_i(\varepsilon)=-U_{-1}R_{i1}\mathbf{1}-U^R(\varepsilon)R_i(\varepsilon)\mathbf{1}$ can be calculated by the updating formula

$$\varphi_i(\varepsilon)=-U_{-1}R_{i1}\mathbf{1}-(I+\varepsilon U_0T_1)^{-1}U_0R_i(\varepsilon)\mathbf{1}$$

or in terms of the limiting value φ_{i0},

$$\varphi_i(\varepsilon)=\varphi_{i0}-\varepsilon[U_0R_{i1}-(I+\varepsilon U_0T_1)^{-1}U_0T_1U_0R_i(\varepsilon)]\mathbf{1}.$$

Problem 6.5. Using the induction argument, prove formulae (6.74) and (6.75). *Hint: Similar formulae can be found in [104].*

Problem 6.6. Execute the combinatorial algorithm of Hassin and Haviv [73] for the perturbed MC with the transition matrix

$$P(\varepsilon)=P(0)+\varepsilon C=\begin{pmatrix} 0 & 1 & 0 & 0 \\ 0 & 1 & 0 & 0 \\ 0 & 0 & 0 & 1 \\ 0 & 0 & 0 & 1 \end{pmatrix}+\varepsilon\begin{pmatrix} 0 & -1 & 0 & 1 \\ 1 & -1 & 0 & 0 \\ 0 & 1 & 0 & -1 \\ 0 & 0 & 1 & -1 \end{pmatrix},$$

and hence find the degree of the pole of the expansion of the deviation matrix.

Problem 6.7. Let H be a deviation matrix. Show that the diagonal elements dominate all the other elements, that is, $H_{ll} \geq H_{kl}$ for all k and l.

Problem 6.8. Let $P(\varepsilon) = P + \varepsilon C$. Prove the following resolvent-type identity for the perturbed fundamental matrix $Z(\varepsilon)=[I-P(\varepsilon)+\Pi(\varepsilon)]^{-1}$:

$$Z(\varepsilon_1)-Z(\varepsilon_2)=(\varepsilon_1-\varepsilon_2)Z(\varepsilon_1)CZ(\varepsilon_2)+Z(\varepsilon_1)\Pi(\varepsilon_2)-\Pi(\varepsilon_1)Z(\varepsilon_2).$$

Hint: The proof is similar in spirit to the proof of the more general identity (3.47) *(also see [112]).*

Problem 6.9. Consider the deviation matrix $H(\varepsilon)$ of the *regularly* perturbed MC and its Taylor series, as specified in part (i) of Theorem 6.4. Verify that

$$H(\varepsilon)=H(0)[I-\varepsilon U]^{-1}-\Pi(0)[I-\varepsilon U]^{-1}H(0)[I-\varepsilon U]^{-1},$$

where $U = CH(0)$, as specified in part (ii) of Theorem 6.4. *Hint: This is based on Section 3 of [15]. See also [11, 138]. It might be convenient to first derive an analogous expression for the perturbed fundamental matrix* $Z(\varepsilon)$.

Problem 6.10. Under the assumptions of Theorem 6.4 derive the updating formulae stated in part (iii) of that theorem. *Hint: See Remarks 6.3 and 6.4 and Section 5 in [15]. See also [11].*

Problem 6.11. Under the assumptions of Theorem 6.4 establish the validity of part (iv) of that theorem. *Hint: See [11].*

Problem 6.12. Assume that in a perturbed MC $P(\varepsilon)$, a recurrent (under $P(0)$) state j can be reached from another recurrent (under $P(0)$) state i, where i and j belong to different ergodic classes (under $P(0)$). Show that this can be achieved only through a path which contains transient under P states and that, in such a case the deviation and mean passage time matrices may contain poles of order greater than 1. In particular, consider Example 6.9 in Section 6.3.5. *Hint: See [77], [73], and [11].*

Problem 6.13. Consider the deviation matrix $Y(\varepsilon)$ of the perturbed MC and its Laurent (6.112). Verify that the algorithm in [73] can be used to determine s, the order of the singularity.

Problem 6.14. Consider the system of infinitely many equations obtained upon substitution of series expansions for $\Pi(\varepsilon)$, (6.111), and (6.112) into (6.113) and then collect the terms with the same power of ε. Show that it suffices to solve the system of $s+1$ fundamental equations $(F0)$–(Fs), as given in Section 6.3.5. *Hint: Use the requirement that* $Y(\varepsilon)\mathbf{1}=0$ *leads to a unique solution for* Y_0 *(but not for the other coefficients). See also [77] and [14].*

Problem 6.15. Verify that the system $(F0)$–(Fs) is equivalent to the reduced system of equations $(RF0)$–$(RFs-1)$.

Problem 6.16. Prove formula (6.114). Namely, prove that the limiting stationary distribution π_0 can be given by the following formula:

$$\pi^{(0)} = V^{(s)}V^{(s-1)}\dots V^{(1)}V,$$

where $V^{(k)}$ is defined in Subsection 6.3.5.

Problem 6.17. In Example 6.10 of Section 6.3.5, use the algorithm discussed in Problem 6.13 (also see [73]) to verify that $t'_{ij}s$ is the order of poles of $\mathcal{M}_{ij}(\varepsilon)$ at $\varepsilon=0$ stated in that example.

Problem 6.18. Prove Proposition 6.3. *Hint: Use the results of Section 6.2.2.*

Problem 6.19. Extend the calculation of Subsection 6.4.3 to the case when some dangling nodes originate from the OUT component. *Hint: To model such a, more general, situation, distinguish between absorbing sets with dangling nodes and absorbing sets without dangling nodes.*

Problem 6.20. Prove formula (6.123).

Problem 6.21. Prove Lemma 6.8.

6.6 ▪ Bibliographic Notes

Many authors contributed to the topics covered in this chapter. To the best of our knowledge, the study of perturbed MCs was pioneered by Schweitzer in [137]. In that fundamental paper Schweitzer rigorously analyzed the regular perturbation of MCs.

The first motivation to study the singular perturbed MCs was given in the paper by Simon and Ando [144]. They demonstrated that several problems in econometrics lead to the mathematical model based on singularly perturbed MCs. Perhaps the first rigorous theoretical developments of the singularly perturbed MCs were carried out by Pervozvanskii and Smirnov [127] and Gaitsgori and Pervozvanskii [62]. They used the so-called *aggregation* approach. Similar ideas were developed in the works of Courtois and his co-authors [45, 46, 47]. As in [127, 62], they have also assumed what is now called the nearly completely decomposable (NCD) structure. In the early 1980s Schweitzer (see technical report [138] and the subsequent publication [140]) generalized the updating formula (stated in Remark 6.3) to an extended NCD case that allows for transient states after perturbation.

Delebecque and Quadrat [49] were the first to investigate a more general case, when the original MC has a set of transient states in addition to the ergodic classes. Additional theoretical and computational developments in this more general setting were given in [25, 104, 134, 152]. In particular, Korolyuk and Turbin [104] and later Bielecki and

Stettner [25] have analyzed the perturbation of MCs with transient states in the context of general Borel state space. In the above authors considered a two time scale model. However, in the presence of transient states, the perturbed MC exhibits multi-time scale behavior. This phenomenon was thoroughly investigated in the fundamental paper of Delebecque [48] that also made a link with Kato's approach [99] based on spectral theory. Coderch et al. [40, 41] carried out similar development for continuous-time Markov processes.

The study of *continuous-time* singularly perturbed MCs was pioneered by Phillips and Kokotovic [128], and then it proceeded pretty much in parallel with the developments of the discrete-time model. The reader interested in the analysis of singular perturbations for continuous-time MCs is referred to the comprehensive book by Yin and Zhang [162].

In this literature review we also would like to mention the papers by Hunter [91, 92, 93] and Seneta [142], where the authors investigate the rank one perturbation of MCs and derive several updating formulae. Probably the most general updating formula was obtained by Lasserre [112]. As was shown in the paper of Abbad and Filar [1], there is no limit for the ergodic projection in the case of general additive perturbation. We also refer the reader to the surveys by Abbad and Filar [2] and Avrachenkov, Filar, and Haviv [11].

Next we include some bibliographic notes on specific sections.

The treatment in Section 6.2 is based primarily on Chapter 2, which stems from the results in the 1999 PhD thesis of Avrachenkov [8].

Results in Section 6.3 were significantly influenced by the works of Delebecque and Quadrat [49, 48, 131], Latouche and Louchard [114, 116], Latouche [113], Haviv and Ritov [76, 77], Hassin and Haviv [73], and, of course, Schweitzer's key papers [138] and [137]. Indeed, we named the combinatorial algorithm for finding the order of the pole after Hassin and Haviv [73]. Again, we note the works of Courtois and his co-authors [45, 46, 47], Haviv and his co-authors [74, 75, 76, 77], and others [139, 141, 152].

Finally, the Internet search application discussed in Section 6.4 is based on Avrachenkov, Litvak, and Pham [17]. To the best of our knowledge, this was the first formulation of the Google PageRank as a manifestation of a singularly perturbed MC technique and constitutes perhaps the largest dimensional instance of such a chain discussed in the literature hitherto.

Chapter 7

Applications to Markov Decision Processes

7.1 ▪ Markov Decision Processes: Concepts and Introduction

Whereas Markov chains (MCs) form a good description of some discrete event stochastic processes, they are not automatically equipped with a capability to model situations where there may be a "decision-maker" or a "controller" who—by a judicious choice of actions—can influence the trajectory of the process. Hence, in this chapter, we consider discrete-time *Markov decision processes* (MDPs) with finite state and action spaces and study the dependence of optimal policies/controls of these decision processes on certain important parameters. In this context our usual ε-perturbation is seen as simply an instance of a more generic parametric dependence.

7.1.1 ▪ Preliminaries and notation for MDP

Consider a discrete-time MDP with a finite state space $\mathbb{X} = \{1,\dots,N\}$ and a finite action space $\mathbb{A}(i) = \{1,\dots,m_i\}$ for each state $i \in \mathbb{X}$. At any time point t the system is in one of the states $i \in \mathbb{X}$ and the "decision-maker" chooses an action $a \in \mathbb{A}(i)$; as a result the following occur: (a) the decision-maker gains an immediate reward $r(i,a)$, and (b) the process moves to a state $j \in \mathbb{X}$ with transition probability $p(j|i,a)$, where $p(j|i,a) \geq 0$ and $\sum_{j\in\mathbb{X}} p(j|i,a) = 1$.

A *decision rule* π_t at time t is a function which assigns a probability to the event that any particular action a is taken at time t. In general, π_t may depend on all history $h_t = (i_0,a_0,i_1,a_1,\dots,a_{t-1},i_t)$ up to time t. The distribution $\pi_t(a_t|h_t)$ defines the probability of selecting the action a_t at time t given the history h_t.

A *policy* (or *control*) is a sequence of decision rules $\pi = (\pi_0,\pi_1,\dots,\pi_t,\dots)$. A policy π is called *Markov* if $\pi_t(\cdot|h_t) = \pi_t(\cdot|i_t)$. If $\pi_t(\cdot|i) = \pi_{t'}(\cdot|i)$ for all $t,t' \in \mathbb{N}$, then the Markov policy π is called *stationary*. Furthermore, a *deterministic* policy π is a stationary policy whose single decision rule is nonrandomized. It can be defined by the function $f(i) = a, a \in \mathbb{A}(i)$.

Let $\mathcal{U}$, $\mathcal{U}^{\mathcal{M}}$, $\mathcal{U}^{\mathcal{S}}$, and $\mathcal{U}^{\mathcal{D}}$ denote the sets of all policies, all Markov policies, all stationary policies, and all deterministic policies, respectively. It is known that, in many contexts, there is no loss of generality in restricting consideration to stationary or even deterministic policies. Indeed, the latter are also most useful for the purpose of asymptotic analyses carried out in this chapter.

For any stationary policy $\pi \in \mathcal{U}^{\mathcal{S}}$ we define the corresponding transition matrix $P(\pi)=\{p_{ij}(\pi)\}_{i,j=1}^N$ and the reward vector $r(\pi)=\{r_i(\pi)\}_{i=1}^N$,

$$p_{ij}(\pi):=\sum_{a\in\mathbb{A}(i)} p(j|i,a)\pi_{ia}, \quad r_i(\pi):=\sum_{a\in\mathbb{A}(i)} r(i,a)\pi_{ia},$$

where π_{ia} denotes the probability of choosing action a in state i, whenever that state is visited. Of course, $\pi \in \mathcal{U}^{\mathcal{S}}$ uniquely defines all possible π_{ia}'s. The *expected average reward* $g_i(\pi)$ and the *expected discounted reward* $v_i^\lambda(\pi)$ can be defined as follows for any $\pi \in \mathcal{U}^{\mathcal{S}}$:

$$g_i(\pi):=\lim_{T\to\infty}\frac{1}{T}\sum_{t=1}^{T}\left[P^{t-1}(\pi)r(\pi)\right]_i=[\Pi(\pi)r(\pi)]_i \tag{7.1}$$

and

$$v_i^\lambda(\pi):=\sum_{t=1}^{\infty}\lambda^{t-1}\left[P^{t-1}(\pi)r(\pi)\right]_i=\left[(I-\lambda P(\pi))^{-1}r(\pi)\right]_i, \tag{7.2}$$

where $i\in\mathbb{X}$ is an initial state and $\lambda\in(0,1)$ is the so-called *discount factor*. It is important to note that frequently it is natural to relate the latter parameter to an *interest rate* denoted by $\rho\in[0,\infty)$. In such a case it is customary to make the substitution $\lambda:=\frac{1}{1+\rho}$ and replace $v_i^\lambda(\pi)$ by $v_i^\rho(\pi)$.

We now introduce three commonly used optimality criteria. Two of these, the discount optimality and the average optimality, are basic criteria in MDP models.

Definition 7.1. *A stationary policy π_* is called discount optimal for fixed $\lambda\in(0,1)$ if $v_i^\lambda(\pi_*)\geq v_i^\lambda(\pi)$ for each $i\in\mathbb{X}$ and all $\pi\in\mathcal{U}^{\mathcal{S}}$.*

Definition 7.2. *A stationary policy π_* is called the average optimal if $g_i(\pi_*)\geq g_i(\pi)$ for each $i\in\mathbb{X}$ and all $\pi\in\mathcal{U}^{\mathcal{S}}$.*

Definition 7.3. *We say that a policy π_* is Blackwell optimal if there exists some $\rho_0>0$ such that $v^\rho(\pi^*)\geq v^\rho(\pi)$ for all $\rho\in(0,\rho_0]$ and for all $\pi\in\mathcal{U}^{\mathcal{S}}$. Equivalently, $v^\lambda(\pi^*)\geq v^\lambda(\pi)$ for all $\lambda\in(\lambda_0,1]$ and for all $\pi\in\mathcal{U}^{\mathcal{S}}$.*

In other words, a Blackwell optimal policy is the policy which is discount optimal for any discount factor sufficiently close to one. Furthermore, the dependence of a discount optimal policy on the discount factor (or interest rate) naturally raises the issue of general parametric analysis of an MDP and of particular dependence of optimal policies and rewards as the value of the parameter of interest tends to some "critical value" such as a discount factor equal to 1 (or an interest rate equal to 0). The latter opens the possibility of applying results of analytic perturbation theory to MDPs.

In the example below we introduce a perturbation parameter ε in the transition probabilities and consider the behavior of solutions as $\varepsilon\downarrow 0$. The example shows that policies that are optimal for the unperturbed MDP ($\varepsilon=0$) may not coincide with optimal policies for the perturbed MDP.

Example 7.1. *Let us consider a long-run average MDP model with $\mathbb{X}=\{1,2\}$, $\mathbb{A}(1)=\{a_1,b_1\}$, $\mathbb{A}(2)=\{a_2\}$, and*

$$p^\varepsilon(1|1,a_1)=1, \quad p^\varepsilon(2|1,a_1)=0;$$

$$p^\varepsilon(1|1,b_1)=1-\varepsilon,\quad p^\varepsilon(2|1,b_1)=\varepsilon;$$
$$p^\varepsilon(1|2,a_2)=\varepsilon,\quad p^\varepsilon(2|2,a_2)=1-\varepsilon;$$
$$r(1,a_1)=1,\quad r(1,b_1)=1.5,\quad r(2,a_2)=0.$$

There are only two deterministic policies: $u=[u(1),u(2)]=[a_1,a_2]$ *and* $v=[v(1),v(2)]=[b_1,a_2]$*. These induce MCs with perturbed probability transition matrices*

$$P^\varepsilon(u)=\begin{bmatrix} 1 & 0 \\ \varepsilon & 1-\varepsilon \end{bmatrix},\quad P^\varepsilon(v)=\begin{bmatrix} 1-\varepsilon & \varepsilon \\ \varepsilon & 1-\varepsilon \end{bmatrix}.$$

This is a case of singular perturbations since

$$\Pi^\varepsilon(u)=\begin{bmatrix} 1 & 0 \\ 1 & 0 \end{bmatrix},\quad \Pi^\varepsilon(v)=\begin{bmatrix} 1/2 & 1/2 \\ 1/2 & 1/2 \end{bmatrix}$$

for $\varepsilon>0$*, but at* $\varepsilon=0$

$$\Pi^0(u)=\Pi^0(v)=\begin{bmatrix} 1 & 0 \\ 0 & 1 \end{bmatrix}.$$

For the policies u and v one may now directly use the definition (7.1) *to calculate the average reward vectors* $g^\varepsilon(u):=(g_1^\varepsilon(u),g_2^\varepsilon(u))^T$ *and* $g^\varepsilon(v):=(g_1^\varepsilon(v),g_2^\varepsilon(v))^T$*. Namely,*

$$g^\varepsilon(u)=\begin{cases} \begin{bmatrix} 1 \\ 0 \end{bmatrix} & \varepsilon=0, \\ \begin{bmatrix} 1 \\ 1 \end{bmatrix} & \varepsilon>0, \end{cases} \quad \text{and} \quad g^\varepsilon(v)=\begin{cases} \begin{bmatrix} 3/2 \\ 0 \end{bmatrix} & \varepsilon=0, \\ \begin{bmatrix} 3/4 \\ 3/4 \end{bmatrix} & \varepsilon>0. \end{cases}$$

Thus, we can see that for $\varepsilon=0$ *the average optimal policy is* v*, whereas for* $\varepsilon>0$ *the average optimal policy is* u*.*

More generally, the average reward optimization problem for the perturbed MDP can be written in the form

$$g_i^{op,\varepsilon}=\max_{\pi\in\mathcal{U}^\mathcal{S}}[\Pi^\varepsilon(\pi)r^\varepsilon(\pi)]_i\ \ \forall\ i\in\mathbb{X}\quad (L^\varepsilon),$$

where $\Pi^\varepsilon(\pi)$ is the perturbed stationary distribution matrix and $r^\varepsilon(\pi)$ is the perturbed immediate reward vector induced by a policy $\pi\in\mathcal{U}^\mathcal{S}$. Of course, in the generic case, the original unperturbed problem is merely the case when $\varepsilon=0$, namely, (L^0).

Since often we do not know the exact value of the perturbation parameter ε, we are interested in finding the policy which is "close" to the optimal one for ε small but different from zero. Of course, if it were possible to find a policy optimal for all values of ε near 0, that would be even better.

Definition 7.4. *We say that a policy* π_* *is uniform optimal (in* ε*) if there exists some* $\varepsilon_0>0$ *such that* $g_i^\varepsilon(\pi_*)\geq g_i^\varepsilon(\pi)$, $i\in\mathbb{X}$*, for all* $\varepsilon\in[0,\varepsilon_0]$ *and for all* $\pi\in\mathcal{U}^\mathcal{S}$*.*

Remarkably, it will be seen in what follows that under rather general assumptions, in cases of most interest, there exists a uniform optimal (often deterministic) policy. We will be especially interested in the case of *singular* perturbations, that is, when the perturbation changes the ergodic structure of the underlying MCs.

7.2 ▪ Nearly Completely Decomposable Markov Decision Processes

One particular, specially structured, perturbed MDP has attracted the most attention by researchers in the field. As in the uncontrolled case, it is the so-called *nearly completely decomposable* (NCD) process. Recall that a critical structural assumption in this process is that the corresponding unperturbed process consists of a collection of uncoupled ergodic processes and that interdependence of these processes arises as a result of a small perturbation that is, in fact, a form of "weak coupling."

This structure has an obvious appeal in the context of MDPs because one may readily imagine a collection of independent decision-makers, in charge of certain parts/components of the system, who already know optimal subpolicies for these separate components. However, their decision-making is complicated by the presence of a central decision-maker responsible for the performance of the system as a whole. It is desirable for the "interference" of the central decision-maker to be small so as not to interfere unduly with the operation of the components. However, this interference cannot be fully eliminated if the system needs to perform well as a whole. Hence a very natural question arises: Can the entire system be optimally (or nearly optimally) controlled and yet permit a large degree of "autonomy" to its constituent subsystems?

To make the above more precise, we now introduce the following four assumptions:

(A1) State space partition: $\mathbb{X} = \cup_{k=1}^{n} \mathbb{X}_k$, where $\mathbb{X}_k \cap \mathbb{X}_l = \emptyset$ if $k \neq l$, $n > 1$, $\sum_{k=1}^{n} n_k = N$, and $n_k := card(\mathbb{X}_k)$.

(A2) Uncoupled components: $p(j|i,a) = 0$ whenever $i \in \mathbb{X}_k, j \in \mathbb{X}_l$, and $k \neq l$.

(A3) For every $i = 1, \ldots, n$ the unperturbed MDP associated with the subspace $\mathbb{X}_k$ is ergodic.

(A4) Transition probabilities have the linearly perturbed structure $p^{\varepsilon}(j|i,a) = p(j|i,a) + \varepsilon d(j|i,a)$, for all $i, j \in \mathbb{X}$ and for all $a \in \mathbb{A}(i)$, where $\sum_j d(j|i,a) = 0$. The transition matrix $P^{\varepsilon}(\pi)$ is irreducible for any $\pi \in \mathcal{U}^{\mathcal{S}}$ and any ε sufficiently small but different from zero; that is, the perturbed MDP is ergodic.

Hence, as intended, the perturbed MDP model can be viewed as a complex system consisting of n "weakly interacting" subsystems associated with $\mathbb{X}_k, k = 1, \ldots, n$. Note that perturbation $\varepsilon d(j|i,a)$, where i and j are the states of different subsystems $\mathbb{X}_k$ and $\mathbb{X}_l$, respectively, represents the probability of rare transitions between the subsystems, which are independent in the unperturbed process.

If the value of the perturbation parameter ε were known, it is clear that the solution of the average MDP problem (L^{ε}), $\max_{\pi \in \mathcal{U}^{\mathcal{S}}} [\Pi^{\varepsilon}(\pi) r(\pi)]_i$ for all $i \in \mathbb{X}$, would provide an optimal policy for that particular value ε. However, since ε will frequently be unknown, it is desirable to find—if possible—a policy that is at least approximately optimal for all values of $\varepsilon > 0$ and small. From now on we shall denote the perturbed MDP by Γ^{ε} and the unperturbed MDP by Γ^0.

The so-called *limit control principle* provides a formal setting for the concept of *suboptimal policies*. First, we note that, by the results of Section 6.2, for any stationary policy $\pi \in \mathcal{U}^{\mathcal{S}}$ there exists a *limiting stationary distribution matrix*

$$\bar{\Pi}(\pi) := \lim_{\varepsilon \to 0} \Pi^{\varepsilon}(\pi).$$

The *limit control principle* states that instead of the singular optimization problem (L^{ε}) one may consider a well-defined *limit Markov control problem*:

$$\bar{g}_i^{opt} = \max_{\pi \in \mathcal{U}^{\mathcal{S}}} [\bar{\Pi}(\pi) r(\pi)]_i \quad (L).$$

It is natural to expect that an optimal strategy, if it exists, for (L) could be approximately optimal for the perturbed MDP Γ^{ε}, when the perturbation parameter is small. Namely, if π_* is any maximizer in (L), then

$$\lim_{\varepsilon \to 0} \max_{i \in \mathbb{X}} |g_i^{\varepsilon}(\pi_*) - g^{opt,\varepsilon}| = 0.$$

However, a policy that solves (L) will, in general, be only suboptimal in Γ^{ε}. Of course, if a uniform optimal policy introduced at the end of the preceding section could be easily found, then such a policy would also be limit control optimal (suboptimal). The next example shows that a suboptimal policy need not be uniform optimal.

Example 7.2. *Consider* $\mathbb{X} = \{1,2\}$, $\mathbb{A}(1) = \{a_1, b_1\}$, *and* $\mathbb{A}(2) = \{a_2\}$; *let*

$$\begin{array}{ll} p^{\varepsilon}(1|1,a_1) = 1, & r(1,a_1) = 10, \\ p^{\varepsilon}(1|1,b_1) = 1-\varepsilon, & r(1,b_1) = 10, \\ p^{\varepsilon}(2|1,b_1) = \varepsilon, & \\ p^{\varepsilon}(1|2,a_2) = 1, & r(2,a_2) = 0. \end{array}$$

Again, let u *be the deterministic policy that chooses* a_1 *in state* 1 *and* v *be the one that chooses* b_1 *in state* 1 *(the choice in state* 2 *is, of course,* a_2*). Clearly, for* $\varepsilon \geq 0$, u *and* v *induce MCs with probability transition matrices*

$$P^{\varepsilon}(u) = \begin{bmatrix} 1 & 0 \\ 1 & 0 \end{bmatrix}, \quad P^{\varepsilon}(v) = \begin{bmatrix} 1-\varepsilon & \varepsilon \\ 1 & 0 \end{bmatrix}$$

and stationary distribution matrices

$$\Pi^{\varepsilon}(u) = \begin{bmatrix} 1 & 0 \\ 1 & 0 \end{bmatrix}, \quad \Pi^{\varepsilon}(v) = \frac{1}{1+\varepsilon} \begin{bmatrix} 1 & \varepsilon \\ 1 & \varepsilon \end{bmatrix}.$$

Then the stationary policy $u(1) = a_1, u(2) = a_2$ *is uniformly optimal with expected average reward* $g_i^{\varepsilon}(u) \equiv 10$. *The stationary policy* $v(1) = b_1, v(2) = a_2$ *is limit control optimal as* $\lim_{\varepsilon \to 0} g_i^{\varepsilon}(v) = 10$, *but for every* $\varepsilon > 0$,

$$g_i^{\varepsilon}(v) = \frac{10}{1+\varepsilon} < g_i^{\varepsilon}(u).$$

The main rationale for focusing on suboptimal policies stems from the fact that they are much easier to calculate than uniform optimal policies and, for practical purposes, may perform nearly as well. Indeed, we will demonstrate that under assumptions (A1)–(A4) the limit Markov control problem (L) can be solved by the following linear programming problem (LP):

$$\max \textstyle\sum_{k=1}^{n} \sum_{i \in \mathbb{X}_k} \sum_{a \in A(i)} r(i,a) z_{ia}^k$$

subject to

$$\text{(i)} \sum_{i\in\mathbb{X}_k}\sum_{a\in A(i)} (\delta_{ij} - p(j|i,a))z^k_{ia} = 0, \quad j \in \mathbb{X}_k,\ k = 1,\dots,n,$$

$$\text{(ii)} \sum_{k=1}^{n}\sum_{j\in\mathbb{X}_\ell}\sum_{i\in\mathbb{X}_k}\sum_{a\in A(i)} d(j|i,a)z^k_{ia} = 0, \quad \ell = 1,\dots,n,$$

$$\text{(iii)} \sum_{k=1}^{n}\sum_{i\in\mathbb{X}_k}\sum_{a\in A(i)} z^k_{ia} = 1,$$

$$\text{(iv)}\ z^k_{ia} \ge 0, \quad k = 1,\dots,n, \quad i \in \mathbb{X}_k,\ a \in A(i).$$

Subsections 7.2.1–7.2.2 are devoted to proving that an optimal policy in the limit Markov control problem (L) can be constructed as follows.

Theorem 7.1. *Let $\{z^k_{ia} | k = 1,\dots,n;\ i \in \mathbb{X}_k;\ a \in A(i)\}$ be an optimal extreme solution to the linear program (LP); then the deterministic strategy defined by*

$$f_*(i) = a, \quad i \in \mathbb{X}_k,\ k = 1,\dots,n \Longleftrightarrow z^k_{ia} > 0$$

is optimal in the limit Markov control problem (L).

Before proceeding to the proof of the above theorem, we provide in the next subsection a series of auxiliary results.

Remark 7.1. *An important feature of the above linear program is that it possesses the so-called staircase structure. Namely, constraints* (i) *for $k = 1,2,\dots,n$ define decoupled diagonal blocks of the coefficient matrix of (LP) and together will typically contain the great bulk of all the constraints. These blocks are coupled by the, typically few in number, constraints* (ii)–(iii). *Of course, this special structure is inherited from the NCD structure of the underlying MDP. A classical linear programming technique known as "Wolf–Dantzig decomposition" shows that it is possible to exploit this structure algorithmically.*

7.2.1 ▪ Aggregated process $\hat{\Gamma}$ and the intermediate nonlinear program

We shall exploit the special structure implied by assumptions (A1)–(A4) to construct an "aggregated decision process" $\hat{\Gamma}$ in a manner analogous to that used in the analysis of the uncontrolled NCD case in Subsection 6.2.3. However, because of the presence of policies we need to develop a somewhat more detailed notation.

In particular, we shall regard the unperturbed MDP Γ^0 as a collection of n uncoupled unperturbed subprocesses Γ^0_k, $k = 1,2,\dots,n$, with respective state spaces $\mathbb{X}_k$ and unchanged action spaces and rewards. Let $\mathcal{U}^S_k$ be the space of stationary policies of Γ^0_k for each $k = 1,2,\dots,n$. It is now clear that every stationary policy π in the perturbed (unperturbed) process Γ^ε (Γ^0) can be viewed as being composed of stationary subpolicies π^k in Γ^0_k for each $k = 1,2,\dots,n$. More formally, we shall write for every $\pi \in \mathcal{U}^{\mathcal{S}}$

$$\pi = (\pi^1,\pi^2,\dots,\pi^n), \ \pi^k \in \mathcal{U}^S_k, \ k = 1,2,\dots,n.$$

Clearly, each π^k induces a probability transition matrix $P^0_k(\pi^k)$ in the corresponding unperturbed subprocess Γ^0_k, while in the composite unperturbed MDP Γ^0, π induces the block-diagonal structured probability transition matrix

$$P^0(\pi) = \begin{pmatrix} P_1^0(\pi^1) & 0 & \cdots & 0 \\ 0 & P_2^0(\pi^2) & \cdots & 0 \\ \vdots & \vdots & \ddots & 0 \\ 0 & 0 & \cdots & P_n^0(\pi^n) \end{pmatrix}.$$

Similarly, with the same $\pi \in \mathcal{U}^{\mathcal{S}}$ we associate the induced $N \times N$ perturbation matrix $D(\pi) := [d_{ij}(\pi)]_{i=1,j=1}^N$, where

$$d_{ij}(\pi) := \sum_{a \in \mathbb{A}(i)} d(j|i,a)\pi_{ia} \;\; \forall \; i,j \in \mathbb{X}.$$

In the perturbed MDP Γ^ε (and for $\varepsilon > 0$ and sufficiently small) the same stationary policy now induces the perturbed probability transition matrix and the associated MC generator

$$P^\varepsilon(\pi) = P^0(\pi) + \varepsilon D(\pi), \quad G^\varepsilon(\pi) := P^\varepsilon(\pi) - I = G^0(\pi) + \varepsilon D(\pi), \tag{7.3}$$

where $G^0(\pi)$ is the corresponding unperturbed generator. Of course, the block-diagonal structure is lost in the above as $P^\varepsilon(\pi)$ is irreducible by assumption (A4).

Next, let $\mu_k(\pi^k)$ be the unique invariant distribution of $P_k^0(\pi)$ for each $k = 1,2,\ldots,n$. Note that $\mu_k(\pi^k)$ is a $1 \times n_k$ vector whose entries will be denoted by $[\mu_k(\pi^k)]_i$, $i \in \mathbb{X}_k$. Hence, $\pi \in \mathcal{U}^{\mathcal{S}}$ also identifies an auxiliary $n \times N$ matrix $M(\pi)$ whose entries are defined by

$$m_{ki}(\pi) = \begin{cases} [\mu_k(\pi^k)]_i & \text{if } \sum_{r=1}^{k-1} n_r < i \leq \sum_{r=1}^{k} n_r, \\ 0 & \text{otherwise} \end{cases}$$

for $k = 1,2,\ldots,n$, $i = 1,2,\ldots,N$, and $\sum_{r=1}^0 n_r := 0$. Hence, the kth row of $M(\pi)$ consists of blocks of zeros with the exception of the invariant distribution $\mu_k(\pi^k)$ residing in the kth block.

Another useful auxiliary matrix is the $N \times n$ matrix Q whose entries are defined by

$$q_{ik} = \begin{cases} 1 & \text{if } \sum_{r=1}^{k-1} n_r < i \leq \sum_{r=1}^{k} n_r, \\ 0 & \text{otherwise} \end{cases}$$

for $i = 1,2,\ldots,N$, $k = 1,2,\ldots,n$, and $\sum_{r=1}^0 n_r := 0$, as before. Hence, the kth column of Q consists of blocks of zeros with the exception of n_k 1's residing in the kth block. The following useful identities follow immediately from the definitions of Q and $M(\pi)$:

$$\Pi^0(\pi) = QM(\pi) \quad \text{and} \quad M(\pi)Q = I_n \tag{7.4}$$

for I_n an $n \times n$ identity matrix and any $\pi \in \mathcal{U}^{\mathcal{S}}$.

In what follows it is useful to follow an approach analogous to that used in the construction of an "aggregated process" in (6.46)–(6.47) in Subsection 6.2.3. Essentially, we consider a process $\hat{\Gamma}$ on the much smaller state space $S := \{1,2,\ldots,n\}$, and with each $\pi \in \mathcal{U}^{\mathcal{S}}$ we associate aggregated MC generator and transition matrices

$$\hat{G}(\pi) := M(\pi)D(\pi)Q \quad \text{and} \quad \hat{Q}(\pi) = I_n + M(\pi)D(\pi)Q. \tag{7.5}$$

The unique 0-eigenvector $x(\pi)$ (scaled to be a probability vector) of $\hat{G}(\pi)$ captures the long-run frequencies of the "macrostates" of the process $\hat{\Gamma}$ when the policy π is used in the original process. Of course, the macrostate k corresponds to the set of states $\mathbb{X}_k$ in Γ^ε for each $k=1,2,\dots,n$. Now, the ergodic projection at infinity corresponding to $\hat{G}(\pi)$ is an $n\times n$ matrix $\hat{\Pi}(\pi)$ with $x(\pi)$ in every row. It now follows from the above and Theorem 6.1 that $\bar{\Pi}(\pi):=\lim_{\varepsilon\to 0}\Pi^\varepsilon(\pi)$, from the limit control problem (L), can be calculated by the simple formula

$$\bar{\Pi}(\pi)=Q\hat{\Pi}(\pi)M(\pi). \tag{7.6}$$

Note that the above formula has a natural intuitive interpretation: the product $\hat{\Pi}(\pi)M(\pi)$ simply weights the stationary distribution vectors $\mu_k(\pi^k)$ from the decoupled subprocesses Γ^0_k by the long-run frequency of the corresponding macrostate $k=1,2,\dots,n$. The first factor Q merely arranges the resulting component vectors in the correct places.

We are now in a position to start deriving constraints of an intermediate nonlinear program, the solution of which will also provide a solution to the limit control problem (L).

The key step is the well-known correspondence (see Problem 7.1 and references in the bibliographic notes) between stationary policies of an irreducible MDP and its space of long-run state-action frequencies. In particular, in our context, consider the irreducible MDP Γ^0_k on the state space $\mathbb{X}_k$ and its set of stationary policies $\mathcal{U}^S_k$. Every subpolicy π^k defines a vector $x^k(\pi^k)$ of long-run state-action frequencies whose entries are defined by

$$x^k_{ia}=x^k_{ia}(\pi^k):=[\mu_k(\pi^k)]_i\pi^k_{ia}\ \forall\ i\in\mathbb{X}_k,\ a\in\mathbb{A}(i), \tag{7.7}$$

where the dependence on the policy will be suppressed when it is clear from the context. Now, for each $k=1,2,\dots,n$ define a polyhedral set

$$L^k:=\left\{x^k \,\middle|\, \sum_{i\in\mathbb{X}_k}\sum_{a\in\mathbb{A}(i)}(\delta_{ij}-p(j|i,a))x^k_{ia}=0\ \forall\, j\in\mathbb{X}_k,\right.$$

$$\left.\sum_{i\in\mathbb{X}_k}\sum_{a\in\mathbb{A}(i)}x^k_{ia}=1;\ \&\ x^k_{ia}\ge 0\ \forall\ i\in\mathbb{X}_k,\ a\in\mathbb{A}(i)\right\}.$$

It is left as an exercise (see Problem 7.2) to check that x^k defined by (7.7) satisfies the constraints of L^k. Thus equation (7.7) actually defines a map $T\colon \mathcal{U}^S_k\to L^k$. The irreducibility of Γ^0_k can also be exploited to prove (again, see Problem 7.1) that the inverse map $T^{-1}:L^k\to\mathcal{U}^S_k$ is well defined by the equation

$$\pi^k_{ia}=\pi^k_{ia}(x^k)=\frac{x^k_{ia}}{\sum_{a\in\mathbb{A}(i)}x^k_{ia}}\ \forall\ i\in\mathbb{X}_k,\ a\in\mathbb{A}(i). \tag{7.8}$$

Now, let $x=(x^1,x^2,\dots,x^n)$, where $x^k\in L^k$ for each k. Define an $n\times n$ matrix $\hat{D}(x)$ whose (i,j)th entry is defined by

$$\hat{d}_{ij}(x):=\sum_{\ell\in\mathbb{X}_j}\sum_{m\in\mathbb{X}_i}\sum_{a\in\mathbb{A}(m)}d(\ell|m,a)x^k_{ma}.$$

Next, consider the following intermediate nonlinear programming problem (NL):

maximize $\sum_{k=1}^n\sum_{i\in\mathbb{X}_k}\sum_{a\in A(i)}r(i,a)\mu_k x^k_{ia}$

subject to

$$\text{(i) } x^k \in L^k,\ k=1,2,\ldots,n,$$

$$\text{(ii) } \mu_k \geq 0,\ k=1,2,\ldots,n,\ \sum_{k=1}^{n} \mu_k = 1,$$

$$\text{(iii) } \sum_{i=1}^{n} \mu_i \hat{d}_{ij}(x) = \sum_{i=1}^{n} \sum_{\ell \in \mathbb{X}_j} \sum_{m \in \mathbb{X}_i} \sum_{a \in \mathbb{A}(m)} d(\ell|m,a)\mu_i x^i_{ma} = 0,\ j=1,2,\ldots,n.$$

We will show that an optimal solution of (NL) yields an optimal policy in the limit control problem (L) in the following sense.

Proposition 7.2. *Let $(\bar{x},\bar{\mu}) = (\bar{x}^1,\bar{x}^2,\ldots,\bar{x}^n,\bar{\mu})$ be an optimal solution of the nonlinear program (NL). For each k, construct $\bar{\pi}^k = T^{-1}(\bar{x}^k)$. Then, $\bar{\pi} = (\bar{\pi}^1,\bar{\pi}^2,\ldots,\bar{\pi}^n)$ is an optimal policy in the limit control problem (L).*

Proof: First we shall show that every feasible policy $\pi \in \mathcal{U}^S$ induces a point (x,μ) feasible in the nonlinear program (NL) in such a way that the objective function in (NL) evaluated at (x,μ) coincides with the objective function of the limit control problem (L), evaluated at π.

Let $\bar{g}_\ell(\pi)$ denote the objective function of the limit control problem for the starting state $\ell \in \mathbb{X}$. We shall exploit the one-to-one correspondence between subpolicies $\pi^k \in \mathcal{U}^S_k$ and points $x^k \in L^k$, namely,

$$T(\pi^k) = x^k \text{ and } T^{-1}(x^k) = \pi^k,\ k=1,2,\ldots,n. \tag{7.9}$$

In particular, note that for any $\ell \in \mathbb{X}_i$

$$[\mu_i(\pi^i)]_\ell = [\mu_i(T^{-1}(x^i))]_\ell = \sum_{a\in\mathbb{A}(\ell)} x^i_{\ell a},\quad \ell \in \mathbb{X}_i,\ i=1,2,\ldots,n, \tag{7.10}$$

where the last equality above follows from (7.7). Now we use (7.6) to obtain

$$\bar{g}_\ell(\pi) = [\bar{\Pi}(\pi)r(\pi)]_\ell = [Q\hat{\Pi}(\pi)M(\pi)r(\pi)]_\ell = \sum_{i=1}^{n}[x(\pi)]_i[\mu_i(\pi^i)\cdot r^i(\pi)]_\ell, \tag{7.11}$$

where $r^i(\pi)$ is an n_i-vector whose entries are $\sum_{a\in\mathbb{A}(\ell)} r(\ell,a)\pi^i_{\ell a}$ and $\mu_i(\pi^i)\cdot r^i(\pi)$ is the inner product of these two vectors. Note that in the above, the dependence on ℓ vanishes on the right-hand side because $\bar{\Pi}(\pi)$ has identical rows. Furthermore, from (7.10) we have that

$$[\mu_i(\pi^i).r(\pi)]_\ell = \sum_{\ell\in\mathbb{X}_i}[\mu_i(\pi^i)]_\ell \sum_{a\in\mathbb{A}(\ell)} r(\ell,a)\frac{x^i_{\ell a}}{\sum_{a\in\mathbb{A}(\ell)} x^i_{\ell a}} = \sum_{\ell\in\mathbb{X}_i}\sum_{a\in\mathbb{A}(\ell)} r(\ell,a)x^i_{\ell a}.$$

The above together with (7.11) now allow us to express the objective function of the limit control problem in terms of variables of the nonlinear program (NL), namely,

$$\bar{g}_\ell(\pi) = [\bar{\Pi}(\pi)r(\pi)]_\ell = \sum_{i=1}^{n}[x(\pi)]_i \sum_{\ell\in\mathbb{X}_i}\sum_{a\in\mathbb{A}(\ell)} r(\ell,a)x^i_{\ell a} = \sum_{i=1}^{n}\sum_{\ell\in\mathbb{X}_i}\sum_{a\in\mathbb{A}(\ell)} r(\ell,a)\mu_i x^i_{\ell a}, \tag{7.12}$$

where in the last equality we merely substituted $\mu_i := [x(\pi)]_i$. While it is clear from the construction that the (x, μ) variables separately satisfy constraints (i) and (ii) of (NL), it is not immediately obvious that together they satisfy constraints (iii).

However, once we recall that the vector $x(\pi)$ is the unique invariant distribution of the aggregated chain induced by π, we have that $\mu := (\mu_1, \mu_2, \dots, \mu_n)$ is the unique solution of

$$\mu\hat{Q}(\pi) = \mu \quad \text{and} \quad \sum_{i=1}^{n} \mu_i = 1,$$

or, equivalently, by formulae (7.4) and (7.5) we have

$$\mu[I_n + M(\pi)D(\pi)Q] = \mu.$$

This immediately leads to

$$\mu M(\pi)D(\pi)Q = 0.$$

Now, substituting $\pi = (T^{-1}(x^1), T^{-1}(x^2), \dots, T^{-1}(x^n))$ into the above and manipulating the resulting equations, we obtain

$$\sum_{i=1}^{n} \mu_i \hat{d}_{ij}(x) = 0, \quad i = 1,2,\dots,n.$$

Thus constraints (iii) are also satisfied and (x, μ) is a feasible point of (NL).

Finally, from the optimality of $(\bar{x}, \bar{\mu})$ in this nonlinear program and equation (7.12) we immediately conclude that

$$\bar{g}_\ell(\pi) = \sum_{i=1}^{n} \sum_{\ell \in \mathbb{X}_i} \sum_{a \in \mathbb{A}(\ell)} r(\ell, a)\mu_i x^i_{\ell a} \le \sum_{i=1}^{n} \sum_{\ell \in \mathbb{X}_i} \sum_{a \in \mathbb{A}(\ell)} r(\ell, a)\bar{\mu}_i \bar{x}^i_{\ell a} = \bar{g}_\ell(\bar{\pi}). \tag{7.13}$$

This completes the proof. □

7.2.2 ▪ Linear programming solution of the limit control problem

It is well known in optimization that bilinear terms $\mu_k x^k_{ia}$ appearing in both the objective and constraint functions of (NL) make that problem algorithmically difficult. Of course, the naive way to try to avoid this difficulty is by linearizing the problem by merely substituting

$$z^k_{ia} := \mu_k x^k_{ia} \quad \forall\ k = 1,2,\dots,n,\ i \in \mathbb{X}_k,\ a \in \mathbb{A}(i). \tag{7.14}$$

Clearly, in general, the above linearization "trick" will not work. Remarkably, in our setting, (7.14) leads to the linear program (*LP*), which, as will be shown below, does yield an optimal limit control policy.

The fact that the substitution of (7.14) into (NL) yields the linear program (*LP*) is immediate. However, if the variables z^k_{ia} are to have the required interpretation we must be able to use them to construct a strictly positive invariant distribution of an appropriate aggregate MC. To achieve the latter we require the following result.

Lemma 7.3. *Let z be a feasible point of the linear program (LP); then*

$$\sum_{i \in \mathbb{X}_k} \sum_{a \in \mathbb{A}(i)} z^k_{ia} > 0 \ \forall\ k = 1,2,\dots,n.$$

Proof: The feasible point z of (LP) consists of entries z^k_{ia}. Partition the states of $S = \{1,2,\dots,n\}$ into $F(z)$ and its complement $F^c(z) := S\backslash F(z)$, where

$$F(z) := \left\{ k \in S \;\middle|\; \sum_{i\in\mathbb{X}_k}\sum_{a\in\mathbb{A}(i)} z^k_{ia} > 0 \right\} \text{ and } F^c(z) := \left\{ k \in S \;\middle|\; \sum_{i\in\mathbb{X}_k}\sum_{a\in\mathbb{A}(i)} z^k_{ia} = 0 \right\}.$$

We shall show that $F^c(z) = \emptyset$. Suppose that $F^c(z) \neq \emptyset$. Next define a policy π^k on the components $\mathbb{X}_k$ depending on whether $k \in F^c(z)$ or otherwise. In particular, if $k \in F^c(z)$, set $\mu_k := 0$ and choose and fix an arbitrary stationary strategy in each state $i \in \mathbb{X}_k$. Denote this strategy by π^k. If $k \in F(z)$, define

$$\mu_k := \sum_{i\in\mathbb{X}_k}\sum_{a\in\mathbb{A}(i)} z^k_{ia} \tag{7.15}$$

and

$$x^k_{ia} := \frac{z^k_{ia}}{\mu_k} \quad \forall\ i \in \mathbb{X}_k,\ a \in \mathbb{A}(i). \tag{7.16}$$

It immediately follows that $\sum_{i\in\mathbb{X}_k}\sum_{a\in\mathbb{A}(i)} x^k_{ia} = 1$ for each $k \in F(z)$. Note also that, by construction, we now have

$$z^k_{ia} := \mu_k x^k_{ia} \quad \forall\ k = 1,2,\dots,n,\ i \in \mathbb{X}_k,\ a \in \mathbb{A}(i).$$

That is, (x,μ) has been constructed from z so that (7.14) holds.

Furthermore, for $k \in F(z)$ it follows from constraints (i) of (LP) that for all $j \in \mathbb{X}_k$, $\sum_{i\in\mathbb{X}_k}\sum_{a\in A(i)}(\delta_{ij} - p(j|i,a))\mu_k x^k_{ia} = 0$, which upon dividing by $\mu_k > 0$ yields

$$\sum_{i\in\mathbb{X}_k}\sum_{a\in A(i)}(\delta_{ij} - p(j|i,a))x^k_{ia} = 0, \quad j \in \mathbb{X}_k.$$

Thus, x^k made up of x^k_{ia}'s so constructed lies in L^k for $k \in F(z)$. However, since the map $T^{-1} : L^k \to \mathcal{U}^S_k$ is a bijection, there exists a stationary policy $\pi^k \in \mathcal{U}^S_k$ such that $x^k = T(\pi^k)$. Together with the previously fixed subpolicies $\pi^k \in F^c(z)$ we now have a complete policy $\pi = (\pi^1,\pi^2,\dots,\pi^n)$ that induces (x,μ) satisfying (7.14). Now, since z also satisfies constraints (ii) of (LP), it now follows that for each $\ell = 1,2,\dots,n$

$$\begin{aligned}
0 &= \sum_{k=1}^{n}\sum_{j\in\mathbb{X}_\ell}\sum_{i\in\mathbb{X}_k}\sum_{a\in A(i)} d(j|i,a) z^k_{ia} \\
&= \sum_{k=1}^{n}\sum_{j\in\mathbb{X}_\ell}\sum_{i\in\mathbb{X}_k}\sum_{a\in A(i)} d(j|i,a)\mu_k x^k_{ia} \\
&= \sum_{k=1}^{n}\mu_k\left[\sum_{j\in\mathbb{X}_\ell}\sum_{i\in\mathbb{X}_k}\sum_{a\in A(i)} d(j|i,a)[\mu_k(\pi^k)]_i \pi^k_{ia}\right] \\
&= \sum_{k=1}^{n}\mu_k\left[\sum_{j\in\mathbb{X}_\ell}\sum_{i\in\mathbb{X}_k}[\mu_k(\pi^k)]_i d_{ij}(\pi^k)\right] \\
&= [\mu M(\pi) D(\pi) Q]_\ell .
\end{aligned}$$

Hence, as a vector, $\mu M(\pi)D(\pi)Q = 0$, and because $M(\pi)I_N Q = I_n$ we have from (7.5) that

$$\mu M(\pi)[D(\pi)+I_N]Q = \mu\hat{Q}(\pi) = \mu, \tag{7.17}$$

where μ is nonnegative and by (7.15) and constraint (iii) of (LP) it satisfies

$$\sum_{k=1}^{n}\mu_k = \sum_{k=1}^{n}\sum_{i\in\mathbb{X}_k}\sum_{a\in A(i)} z_{ia}^k = 1.$$

Hence, μ is the unique invariant distribution of the irreducible aggregated chain $\hat{Q}(\pi)$, and so we must have $\mu > 0$, thereby contradicting $F^c(z) \neq \emptyset$. □

Lemma 7.4. *Let $\bar{z}$ be an optimal solution of the linear program (LP) and define $(\bar{x},\bar{\mu})$ by*

$$\bar{\mu}_k := \sum_{i\in\mathbb{X}_k}\sum_{a\in\mathbb{A}(i)} \bar{z}_{ia}^k \;\; \forall\; k=1,2,\ldots,n \;\; and \;\; \bar{x}_{ia}^k := \frac{\bar{z}_{ia}^k}{\mu_k} \;\; \forall\; i\in\mathbb{X}_k,\; a\in\mathbb{A}(i).$$

Then $(\bar{x},\bar{\mu})$ is an optimal solution of the nonlinear program (NL).

Proof: It follows from the proof of Lemma 7.3 that $(\bar{x},\bar{\mu})$ is well defined and feasible in the nonlinear program (NL). To establish its optimality, consider any other (x,μ) feasible in (NL) and define a vector z with entries

$$z_{ia}^k := \mu_k x_{ia}^k, \;\; k=1,2,\ldots,n, \;\; \forall\; i\in\mathbb{X}_k,\;\; a\in\mathbb{A}(i).$$

It is clear from the constraints of (NL) that z is also feasible in the linear program (LP). Comparing the objective function values at (x,μ) and $(\bar{x},\bar{\mu})$ and exploiting the optimality of $\bar{z}$ in (LP), we see that

$$\begin{aligned}
&\sum_{k=1}^{n}\sum_{i\in\mathbb{X}_k}\sum_{a\in A(i)} r(i,a)\mu_k x_{ia}^k \\
&= \sum_{k=1}^{n}\sum_{i\in\mathbb{X}_k}\sum_{a\in A(i)} r(i,a) z_{ia}^k \leq \sum_{k=1}^{n}\sum_{i\in\mathbb{X}_k}\sum_{a\in A(i)} r(i,a)\bar{z}_{ia}^k \\
&= \sum_{k=1}^{n}\sum_{i\in\mathbb{X}_k}\sum_{a\in A(i)} r(i,a)\bar{\mu}_k\bar{x}_{ia}^k .
\end{aligned}$$

Thus $(\bar{x},\bar{\mu})$ is optimal in (NL). □

It can now be shown that there exists a deterministic optimal policy in the limit Markov control problem (L). Toward this goal we shall need the following technical result.

Lemma 7.5. *Let z be any extreme (basic) feasible solution of the linear program (LP). Then for any $k\in S$ and any $i\in\mathbb{X}_k$ there exists a unique $a\in\mathbb{A}(i)$ such that $z_{ia}^k > 0$.*

Proof: It follows from the proof of Lemma 7.3 that for any $k\in S$ there exists a policy π^k such that for all $i\in\mathbb{X}_k$, $a\in\mathbb{A}(i)$

$$x_{ia}^k = [T(\pi^k)]_{ia} = [\mu_k(\pi^k)]_i \pi_{ia}^k = \frac{z_{ia}^k}{\sum_{i\in\mathbb{X}_k}\sum_{a\in\mathbb{A}(i)} z_{ia}^k}. \tag{7.18}$$

Since $[\mu_k(\pi^k)]_i > 0$ and $\sum_{a\in\mathbb{A}(i)} \pi^k_{ia} = 1$, there must exist at least one $a \in \mathbb{A}(i)$ such that x^k_{ia}, and hence z^k_{ia} is strictly positive. Hence, the number of positive entries of z must be at least $\sum_{k=1}^n n_k = N$.

However, since z is a basic feasible solution of (LP), the number of its positive entries is less than or equal to r, the rank of its coefficient matrix (determined by the constraints (i)–(iii)). In Problem 7.3 the reader is invited to verify that for each $k \in S$, summing over $j \in \mathbb{X}_k$, the block of constraints (i) corresponding to that k yield 0. Thus that block cannot have more than $(n_k - 1)$ linearly independent rows. Similarly, summing over $\ell \in S$, the block of constraints (ii) also yields zero. Thus, this block can have at most $(n-1)$ linearly independent rows. Consequently, the upper bound for the number of linearly independent rows contributed by constraints (i)–(iii) and hence on the rank is

$$r \leq (n_1 - 1) + (n_2 - 1) + \cdots + (n_n - 1) + (n-1) + 1 = N - n + (n-1) + 1 = N.$$

Hence, the number of positive entries in z must be exactly N. Thus we conclude that there is exactly one $a \in \mathbb{A}(i)$ such that $z^k_{ia} > 0$ for every $i \in \mathbb{X}_k$ and $k \in S$. □

Proposition 7.6. *There exists a deterministic stationary policy $\bar{\pi}$ that is optimal for the limit Markov control problem (L). That is,*

$$\bar{g}_i^{opt} = [\bar{\Pi}(\bar{\pi})r(\bar{\pi})]_i = \max_{\pi \in \mathcal{U}^{\mathcal{S}}} [\bar{\Pi}(\pi)r(\pi)]_i \quad \forall\ i \in \mathbb{X}.$$

Proof: In the proof of Proposition 7.2 it was shown that every feasible policy $\pi \in \mathcal{U}^S$ induces a point (x, μ) feasible in the nonlinear program (NL). Furthermore, we have seen that z constructed by $z^k_{ia} := \mu_k x^k_{ia}$ is feasible for the linear program (LP). Since the constraints of the latter define a bounded polyhedron, an optimal solution must exist. Hence, by fundamental theorem of linear programming, there must also exist an extreme optimal solution $\bar{z}$. From the latter, by Lemma 7.4, we may construct $(\bar{x}, \bar{\mu})$ optimal in the intermediate nonlinear program (NL). Now, by Proposition 7.2 an optimal policy in the limit control problem (L) can be constructed by setting $\bar{\pi}^k = T^{-1}(\bar{x}^k)$ for each $k \in S$. However, it is clear from (7.18) and the definition of the T^{-1} map that the policy so constructed from an extreme feasible point of (L) is deterministic. □

The proof of the main result of this section is now merely a consequence of the preceding sequence of lemmas and propositions.

Proof of Theorem 7.1: By the proof of Proposition 7.6 we have that there exists an optimal deterministic policy $\bar{\pi}$ that can be constructed from an extreme optimal solution $\bar{z}$ of (L) with entries $\{\bar{z}^k_{ia} | k = 1, \ldots, n;\ i \in \mathbb{X}_k;\ a \in A(i)\}$. According to that construction

$$\bar{\pi}^k_{ia} = \begin{cases} 1 & \text{if } \bar{z}^k_{ia} > 0, \\ 0 & \text{otherwise.} \end{cases}$$

This is equivalent to a policy $f_* \in \mathcal{U}^D$ defined by

$$f_*(i) = a, \quad i \in \mathbb{X}_k,\ k = 1, \ldots, n, \quad \Longleftrightarrow \quad \bar{z}^k_{ia} > 0.$$ □

7.3 ▪ Parametric Analysis of Markov Decision Processes

In this section we draw on a well-known correspondence between MDPs and suitably constructed linear programs. The decision variables of the latter are merely certain "frequencies" defined with the help of key matrices (e.g., stationary distribution and deviation matrices) of the associated MCs.

The generic approach of this section will be to consider for each MDP model the corresponding parametric linear program (LP^θ) of the generic form

$$\max c(\theta)x$$

subject to

$$A(\theta)x = b(\theta), \quad x \geq 0,$$

where the elements of $A(\theta), b(\theta)$, and $c(\theta)$ are polynomial functions of θ. Indeed, in accordance with the theory developed in Chapter 5, rational functions or their Laurent series expansions are also permissible here.

The unifying methodology for solving these linear programs, in all the models considered below, will be via an application of the "asymptotic simplex method" discussed in detail in the preceding chapter.

It will be seen that discount and Blackwell optimality, branching, and singularly perturbed MDPs with killing interest rate can all be considered in a unified framework based on the asymptotic simplex method. In one way or another many of the connections between these optimality criteria stem from the following "Blackwell expansion" of the resolvent-like matrix operator that underlies the discounted MDP model:

$$[I - \lambda P(\pi)]^{-1} = \frac{1}{1-\lambda}\Pi(\pi) + H(\pi) + o(1-\lambda) \quad \forall\ \pi \in \mathcal{U}^{\mathcal{S}}. \tag{7.19}$$

Note that the above expansion has a simple pole at $\lambda = 1$.

7.3.1 ▪ Discount and Blackwell optimality criteria and their generalizations

Suppose that $\nu_j > 0$ is the probability that j is the initial state and $\sum_j \nu_j = 1$, and let ν denote a column vector whose entries are ν_j. A standard linear programming formulation of the discounted MDP (with the discount factor $\lambda \in [0,1)$ fixed) relies on defining "discounted state-action frequencies" induced by a policy $\pi \in \mathcal{U}^{\mathcal{S}}$ in accordance with

$$x_{ia}(\pi) := \{\nu^T[I - \lambda P(\pi)]^{-1}\}_i \pi_{ia} \quad \forall\ i \in \mathbb{X};\ a \in \mathbb{A}(i), \tag{7.20}$$

where the set of all such $x_{ia}(\pi)$'s enumerated in the natural fashion makes up the *discounted frequency vector* $x(\pi)$ induced by the policy π.

In Problem 7.4 the reader is invited to verify that the set of all frequency vectors induced by policies in $\mathcal{U}^{\mathcal{S}}$ is precisely the linear polytope X_λ defined by the constraints

$$\sum_{i,a}[\delta_{ij} - \lambda p(j|i,a)]x_{ia} = \nu_j \ \forall\ j \in \mathbb{X}, \quad x_{ia} \geq 0 \ \forall\ i \in \mathbb{X},\ a \in \mathbb{A}(i). \tag{7.21}$$

Note that equation (7.20) can be viewed as defining a map $M : \mathcal{U}^{\mathcal{S}} \to X_\lambda$ that is invertible, with the inverse map $M^{-1} : X_\lambda \to \mathcal{U}^{\mathcal{S}}$ defined by

$$\pi_{ia}(x) := \frac{x_{ia}}{\sum_{a\in\mathbb{A}(i)} x_{ia}} \quad \forall\ i \in \mathbb{X},\ a \in \mathbb{A}(i), \tag{7.22}$$

for every $x \in X_\lambda$. The above immediately leads to the linear program (see Problem 7.4)

$$\max \sum_{i,a} r(i,a)x_{ia}$$

$$\sum_{i,a} [\delta_{ij} - \lambda p(j|i,a)]x_{ia} = \nu_j \;\; \forall \; j \in \mathbb{X}, \quad x_{ia} \geq 0 \; \forall \; i \in \mathbb{X}, \; a \in \mathbb{A}(i), \tag{7.23}$$

which solves the discounted MDP in the sense that if x_* is an optimal solution of the above linear program, then $\pi_* = M^{-1}(x_*)$ is a discount optimal policy in this MDP. Summing constraints (7.23) over the index j shows that for every feasible x, $\sum_{i,a} x_{ia} = \frac{1}{1-\lambda}$, indicating that the norm of these points tends to infinity as the discount factor tends to 1, as does the objective function value. To avoid this unboundedness (in the limit) frequently these constraints are multiplied by $1-\lambda$, and the variables x_{ia} are replaced by new variables $(1-\lambda)x_{ia}$. For notational simplicity the latter are also denoted by x_{ia}. Constraints (7.23) so modified will be called the normalized constraints.

Model I: Blackwell optimality criterion

In the above classical development the discount factor λ is fixed at a particular value. However, a Blackwell optimal policy is a discount optimal policy for all discount factors sufficiently close to one or, equivalently, the policy which is optimal for all interest rates sufficiently close to zero. This suggests that the problem of finding a Blackwell optimal policy might be expressible as a perturbed mathematical program in the sense studied in the preceding chapter. Indeed, the relationship between the discount factor and the interest rate $\lambda = \frac{1}{1+\rho}$ and $1-\lambda = \frac{\rho}{1+\rho}$ immediately suggests the natural transformation: merely substitute the latter for λ in the normalized constraints (7.23), and then multiply by $1+\rho$ to obtain

$$\sum_{i,a} [(1+\rho)\delta_{ij} - p(j|i,a)]x_{ia} = \rho\nu_j$$

for each state j. Now, coefficients of the variables and the right-hand side values in the above can be rewritten in the linearly perturbed form

$$(1-p(j|i,a)) + \rho\delta_{ij} \;\; \& \;\; 0 + \rho\nu_j$$

for each state j and state-action pair (i,a).

Hence, by results from the preceding chapter, a Blackwell optimal policy can be determined by applying the asymptotic simplex method to the (linearly) perturbed linear program:

$$\max \sum_{i,a} (1+\rho)r(i,a)x_{ia}$$

$$\sum_{i,a} [(1-p(j|i,a)) + \rho\delta_{ij}]x_{ia} = 0 + \rho\nu_j \;\; \forall j \in \mathbb{X}, \quad x_{ia} \geq 0 \quad \forall \; i \in \mathbb{X}, \; a \in \mathbb{A}(i). \tag{7.24}$$

Note that the above linear program can be immediately written in the familiar form (5.4), (5.5) with $\varepsilon = \rho$.

Of course, an application of the asymptotic simplex method to the above will yield an optimal solution x_* that is optimal for all ρ in some interval $[0,\rho_0) \subset [0,1)$, and hence $\pi_* := M^{-1}(x_*)$ is a Blackwell optimal policy.

Model II: Markov branching decision chains

Markov branching decision chains are MDPs where the immediate rewards are dependent on the interest rate. Namely, it is assumed that $r(i,a) = r^\rho(i,a)$ is a known polynomial function in the interest rate ρ. To find a policy which is optimal for all sufficiently small ρ we simply need to apply the asymptotic simplex method to only a slightly modified version of (7.24), that is,

$$\max \sum_{i,a}(1+\rho)r^{\rho}(i,a)x_{ia}$$

$$\sum_{i,a}[(1-p(j|i,a))+\rho\delta_{ij}]x_{ia}=\rho\nu_j \ \ \forall j\in\mathbb{X}, \quad x_{ia}\geq 0 \ \ \forall\ i\in\mathbb{X},\ a\in\mathbb{A}(i). \tag{7.25}$$

Model III: Singularly perturbed chains with killing interest rate

A related model also considered in the literature is that of a singularly perturbed MDP with "killing interest rate" $\rho(\varepsilon)=\mu\varepsilon^{\ell}$, where ℓ is the order of a time scale. In addition, it is assumed that the transition probabilities have the linearly perturbed structure

$$p^{\varepsilon}(j|i,a)=p(j|i,a)+\varepsilon d(j|i,a) \ \ \forall\ i,j\in\mathbb{X},\ \ \forall a\in\mathbb{A}(i),$$

where $p(j|i,a)$ are transition probabilities of the original unperturbed chain, ε is a "small" perturbation parameter, and $\sum_j d(j|i,a)=0$. This model exhibits the necessity of different control regimes for different time scales. Once again, the extension of our asymptotic simplex method for the polynomial perturbation can be used to solve this problem. Here the parametric linear program takes the form

$$\max \sum_{i,a}(1+\mu\varepsilon^{\ell})r(i,a)x_{ia}$$

$$\sum_{i,a}[(1+\mu\varepsilon^{\ell})\delta_{ij}-p(j|i,a)-\varepsilon d(j|i,a)]x_{ia}=\nu_j \ \ \forall j\in\mathbb{X}, \quad x_{ia}\geq 0\ \forall\ i\in\mathbb{X},\ a\in\mathbb{A}(i). \tag{7.26}$$

Generalized model

Finally, we would like to note that Models I, II, and III can all be viewed as particular cases of a unified scheme. In particular, consider a parametric MDP model where the transition probabilities $p^{\varepsilon}(j|i,a)$, immediate rewards $r^{\varepsilon}(i,a)$, and the interest rate $\rho(\varepsilon)$ are all given polynomials of the parameter ε. Then a policy which is optimal for all sufficiently small values of parameter ε can be found, by the asymptotic simplex method, from the following perturbed linear program:

$$\max \sum_{i,a}(1+\rho(\varepsilon))r^{\varepsilon}(i,a)x_{ia}$$

$$\sum_{i,a}[(1+\rho(\varepsilon))\delta_{ij}-p^{\varepsilon}(j|i,a)]x_{ia}=\rho(\varepsilon)\nu_j \ \ \forall j\in\mathbb{X}, \quad x_{ia}\geq 0\ \forall\ i\in\mathbb{X},\ a\in\mathbb{A}(i). \tag{7.27}$$

Note that we retrieve

1. Model I with $\rho(\varepsilon)=\varepsilon$, $r^{\varepsilon}(i,a)=r(i,a)$, $p^{\varepsilon}(j|i,a)=p(j|i,a)$;
2. Model II with $\rho(\varepsilon)=\varepsilon$, $r^{\varepsilon}(i,a)=\sum_{k=0}^{p}\varepsilon^k r_k(i,a)$, $p^{\varepsilon}(j|i,a)=p(j|i,a)$; and
3. Model III with $\rho(\varepsilon)=\mu\varepsilon^{\ell}$, $r^{\varepsilon}(i,a)=r(i,a)$, $p^{\varepsilon}(j|i,a)=p(j|i,a)+\varepsilon d(j|i,a)$.

7.3.2 ▪ Singularly perturbed average MDP and parametric linear programming

Whereas discounted MDPs have an immediate economic interpretation, in many engineering problems the so-called steady-state or long-run average method of aggregating rewards (or costs/outputs) is preferred. The corresponding long-run average, linearly perturbed MDP is merely the optimization problem

$$\min_{\pi\in\mathscr{U}}[g_i^{\varepsilon}(\pi)] = \min_{\pi\in\mathscr{U}^{\mathscr{S}}}[\Pi^{\varepsilon}(\pi)r(\pi)]_i \quad \forall i\in\mathbb{X},$$

where it is assumed that the transition probabilities have the familiar linearly perturbed structure

$$p^{\varepsilon}(j|i,a) = p(j|i,a) + \varepsilon d(j|i,a) \ \forall\, i,j\in\mathbb{X}, \ \forall a\in\mathbb{A}(i), \tag{7.28}$$

where $p(j|i,a)$ are transition probabilities of the original unperturbed chain, ε is a "small" perturbation parameter, and $\sum_j d(j|i,a) = 0$. The reader is referred to the bibliographical notes at the end of the chapter for a reference which shows that there is no loss of generality in considering this maximization problem over $\mathscr{U}^{\mathscr{S}}$ instead of $\mathscr{U}$.

Note that the above perturbed average MDP can also be seen as corresponding to the "boundary case" $\ell = 0$ and $\mu = 0$ of Model III introduced above. The zero interest rate case has an intrinsic interest of its own, but a mere substitution of the boundary parameter values in the linear program (7.26) does not lead to useful results because it changes the rank of the coefficient matrix of that program. Indeed, we are especially interested in the case of *singular* perturbations, that is, the case when the perturbation changes the ergodic structure of the underlying MC and these changes manifest themselves as changes of rank of certain matrices of interest.

A standard linear programming formulation of the long-run average MDP relies on defining "long-run state-action frequencies" induced by a policy $\pi\in\mathscr{U}^{\mathscr{S}}$ in accordance with

$$z_{ia}(\pi) := [\Pi^{\varepsilon}(\pi)]_i \pi_{ia} \quad \forall\ i\in\mathbb{X},\ a\in\mathbb{A}(i), \tag{7.29}$$

where the set of all such $z_{ia}(\pi)$'s enumerated in the natural fashion makes up the *long-run frequency vector* $z(\pi)$ induced by the policy π.

For the long-run average MDP the construction of a linear program whose solution yields an average optimal policy is well known but more involved than in the case of the discounted MDP. Below, we merely describe this construction and refer the reader to Problem 7.5 and references in the bibliographic notes for a verification of its validity.

Let $K = \{(i,a)\mid i\in\mathbb{X};\ a\in\mathbb{A}(i)\}$ be the set of all state-action pairs, and let $|K|$ denote its cardinality. Given the initial distribution ν over $\mathbb{X}$, define X_ν^{ε} to be the set of $\{(z,\zeta)\}$, $z,\zeta\in\mathbb{R}^{|\mathbb{K}|}$, that satisfy

$$\sum_{i\in\mathbb{X}}\sum_{a\in\mathbb{A}(i)}(\delta_{ij} - p(j|i,a) - \varepsilon d(j|i,a))z_{ia} = 0 \qquad \forall j\in\mathbb{X}, \tag{7.30}$$

$$\sum_{a\in\mathbb{A}_j} z_{ja} + \sum_{i\in\mathbb{X}}\sum_{a\in\mathbb{A}(i)}(\delta_{ij} - p(j|i,a) - \varepsilon d(j|i,a))\zeta_{ia} = \nu(j) \qquad \forall j\in\mathbb{X}, \tag{7.31}$$

$$z\ge 0, \qquad \zeta\ge 0. \tag{7.32}$$

Remark 7.2. (i) *Every* $z(\cdot,\cdot) \in X_\nu^\varepsilon$ *satisfies* $\sum_{i,a} z_{ia} = 1$. *This can be seen by summing equation* (7.31) *over all* $j \in \mathbb{X}$.

(ii) *We may delete one of the constraints among* (7.30). *This follows from the fact that coefficients of* z_{ia} *variables in* (7.30) *sum to* 0.

Next we consider the perturbed linear program with fixed ε: Find $z, \zeta \in \mathbb{R}^{|\mathbb{K}|}$ such that

$$\max\{r \cdot z\} \qquad (LP^\varepsilon)$$

subject to

$$(z,\zeta) \in X_\nu^\varepsilon.$$

This linear program (LP^ε) is related to the long-run average perturbed MDP in the following way. Given any $(z,\zeta) \in X_\nu^\varepsilon$, define the stationary policy $\pi \in \mathcal{U}^{\mathcal{S}}$ by

$$\pi_{ia} = \begin{cases} \dfrac{z_{ia}}{\sum_{a' \in \mathbb{A}(i)} z_{ia'}} & \text{if } \sum_{a' \in \mathbb{A}(i)} z_{ia'} > 0, \\ \dfrac{\zeta_{ia}}{\sum_{a' \in \mathbb{A}(i)} \zeta_{ia'}} & \text{if } \sum_{a' \in \mathbb{A}(i)} z_{ia'} = 0 \text{ and } \sum_{a' \in \mathbb{A}(i)} \zeta_{ia'} > 0, \\ \text{arbitrary} & \text{otherwise.} \end{cases} \tag{7.33}$$

Lemma 7.7. *Fix* $\varepsilon > 0$. *Suppose that* $(z^*(\varepsilon), \zeta^*(\varepsilon))$ *is an optimal solution of* (LP^ε) *with an associated policy* π_* *constructed via* (7.33); *then* π_* *is an average optimal policy in the perturbed long-run average MDP.*

The above lemma is an immediate corollary of known results (see Problem 7.5 and references cited therein). However, prior results do not permit us to find a uniform (in ε) average optimal deterministic policy. The latter is a more difficult problem both from a theoretical point of view and due to the fact that the rank of the coefficient matrix of (LP^ε) can change at $\varepsilon = 0$ (the case of the singular perturbation). This can also create numerical problems when $\varepsilon > 0$ is small. Nonetheless, the asymptotic simplex method of the preceding chapter still applies to this problem.

7.3.3 ▪ Numerical example

Consider the following very simple, but illustrative, example.

Example 7.3. *Consider an MDP with* $\mathbb{X} = \{1,2\}$, $\mathbb{A}(1) = \{a_1, b_1\}$, *and* $\mathbb{A}(2) = \{a_2, b_2\}$; *let*

$$\begin{array}{ll} p^\varepsilon(1|1,a_1) = 1, & r(1,a_1) = 10, \\ p^\varepsilon(1|1,b_1) = 1 - \varepsilon, & r(1,b_1) = 10, \\ p^\varepsilon(2|1,b_1) = \varepsilon, & \\ p^\varepsilon(1|2,a_2) = 1, & r(2,a_2) = 0, \\ p^\varepsilon(1|2,b_2) = p^\varepsilon(2|2,b_2) = 1/2, & r(2,b_2) = 5. \end{array}$$

We take $\nu(1) = \nu(2) = 1/2$.

By adding the artificial variables, the linear program (LP^ε) *becomes*

$$\max\left\{10z_{1a_1} + 10z_{1b_1} + 5z_{2b_2} - 100\xi_1 - 100\xi_2 - 100\xi_3 - 100\xi_4\right\}$$

Table 7.1. *First tableau of Example 7.3 [6]*

Basis	Artif. variables ξ_1	ξ_3	ξ_4	z variables $1a_1$	$1b_1$	$2a_2$	$2b_2$	ζ variables $1a_1$	$1b_1$	$2a_2$	$2b_2$	r.h.s.
ξ_1	1	0	0	0	ε	-1	-1/2	0	0	0	0	0
ξ_3	0	1	0	$\underline{1}$	1	0	0	0	ε	-1	-1/2	1/2
ξ_4	0	0	1	0	0	1	1	0	$-\varepsilon$	1	1/2	1/2
Red. cost	0	0	0	110	$110+100\varepsilon$	0	55	0	0	0	0	

Table 7.2. *Second tableau of Example 7.3 [6]*

Basis	Artif. variables ξ_1	ξ_3	ξ_4	z variables $1a_1$	$1b_1$	$2a_2$	$2b_2$	ζ variables $1a_1$	$1b_1$	$2a_2$	$2b_2$	r.h.s.
ξ_1	1	0	0	0	ε	-1	-1/2	0	0	0	0	0
z_{1a_1}	0	1	0	1	1	0	0	0	ε	-1	-1/2	1/2
ξ_4	0	0	1	0	0	1	1	0	$-\varepsilon$	$\underline{1}$	1/2	1/2
Red. cost	0	-110	0	0	100ε	0	55	0	-110ε	110	55	

Table 7.3. *Third tableau of Example 7.3 [6]*

Basis	Artif. variables ξ_1	ξ_3	ξ_4	z variables $1a_1$	$1b_1$	$2a_2$	$2b_2$	ζ variables $1a_1$	$1b_1$	$2a_2$	$2b_2$	r.h.s.
ξ_1	1	0	0	0	$\underline{\varepsilon}$	-1	-1/2	0	0	0	0	0
z_{1a_1}	0	1	1	1	1	1	1	0	0	0	0	1
z_{2a_2}	0	0	1	0	0	1	1	0	$-\varepsilon$	1	1/2	1/2
Red. cost	0	-110	-110	0	100ε	-110	-55	0	0	0	0	

subject to the constraints

$$\begin{array}{llllll}
+\varepsilon z_{1b_1} - z_{2a_2} - 0.5 z_{2b_2} & & +\xi_1 & & = & 0,\\
-\varepsilon z_{1b_1} + z_{2a_2} + 0.5 z_{2b_2} & & +\xi_2 & & = & 0,\\
+z_{1a_1} + z_{1b_1} & +\varepsilon\zeta_{1b_1} - \zeta_{2a_2} - 0.5\zeta_{2b_2} & & +\xi_3 & = & 0.5,\\
\quad +z_{2a_2} + z_{2b_2} & -\varepsilon\zeta_{1b_1} + \zeta_{2a_2} + 0.5\zeta_{2b_2} & & +\xi_4 & = & 0.5,
\end{array}$$

$$z_{1a_1}, z_{1b_1}, z_{2a_2}, z_{2b_2}, \zeta_{1a_1}, \zeta_{1b_1}, \zeta_{2a_2}, \zeta_{2b_2}, \xi_1, \xi_2, \xi_3, \xi_4 \geq 0.$$

A reader familiar with MDPs will note that this example is of the so-called unichain model (for $\varepsilon > 0$). Consequently, a simpler version of (LP^{ε}) could have been used (see, e.g., Problem 7.1). However, the present version of (LP^{ε}) applies generally and hence is better suited for demonstrating the technique.

We added a penalty term for the artificial variables to ensure that they exit the basis. We shall delete the second constraint as it is redundant (and will thus not use ξ_2).

The first simplex tableau is given in Table 7.1. We then choose the first column z_{1a_1} to enter. The row/variable to exit is the second one, ξ_3. In all the tableaux the pivoting element is underlined.

The second simplex tableau is given in Table 7.2. The column that enters the basis is ζ_{1a_1} for which the reduced cost 110 is the largest. The column to exit is ξ_4.

The third and fourth simplex tableaux are given in Tables 7.3 and 7.4.

At this stage we have obtained an optimal solution over the field of Laurent series with real coefficients (see Section 5.2). A uniformly optimal policy uses actions a_1 and a_2 in states 1

Table 7.4. *Fourth tableau of Example 7.3 [6]*

	Artif. variables			z variables				ζ variables				
Basis	ξ_1	ξ_3	ξ_4	$1a_1$	$1b_1$	$2a_2$	$2b_2$	$1a_1$	$1b_1$	$2a_2$	$2b_2$	r.h.s.
z_{1b_1}	1	0	0	0	ε	-1	-1/2	0	0	0	0	0
z_{1a_1}	-1	ε	ε	ε	0	$1+\varepsilon$	$1/2+\varepsilon$	0	0	0	0	ε
z_{2a_2}	0	0	ε	0	0	ε	ε	0	$-\varepsilon^2$	ε	$1/2\varepsilon$	$1/2\varepsilon$
Red. cost	-100ε	$-110\,\varepsilon$	-110ε	0	0	-10ε	-5ε	0	0	0	0	

and 2, respectively, as follows from (7.33). Note that it is uniformly optimal for all $\varepsilon > 0$ and sufficiently small. The value of this MDP is 10, independently of the initial state and ε, in this simple example. The stationary deterministic policies that choose action b_1 in state 1 are optimal for the limit problem but are not optimal for any positive ε.

7.4 ▪ Perturbed Markov Chains and the Hamiltonian Cycle Problem

In this section we discuss an application that may be seen as belonging to a line of research which aims to exploit the tools of perturbed MDPs to study the properties of a famous problem of combinatorial optimization: the Hamiltonian cycle problem (HCP).

We consider the following version of the HPC: *given a directed graph, find a simple cycle that contains all vertices of the graph (Hamiltonian cycle) or prove that the Hamiltonian cycle does not exist.* With respect to this property—*Hamiltonicity*—graphs possessing the Hamiltonian cycle are called Hamiltonian. The above are so named due to the fact that Sir William Hamilton investigated the existence of such cycles on the dodecahedron graph.

In particular, we shall show that the HCP can be embedded in a discounted MDP in the sense that a search for Hamiltonian cycles reduces to a search for stationary policies that trace out such cycles in the graph. The latter will be seen to induce Hamiltonian feasible solutions in singularly perturbed polytopes that arise from the underlying so-called frequency or occupational measure space that is usually associated with a discounted MDP.

7.4.1 ▪ Formulation and preliminaries

Our dynamic stochastic approach to the HCP considers a moving object tracing out a directed path on the graph $\mathcal{G}$ with its movement "controlled" by a function f mapping the set of nodes $\mathcal{V} = \mathcal{V}(\mathcal{G}) = \{1,2,\dots,N\}$ of $\mathcal{G}$ into the set of arcs $\mathcal{A} = \mathcal{A}(\mathcal{G})$ of $\mathcal{G}$. We think of this set of nodes as the *state space* of an MDP $\Gamma = \Gamma(\mathcal{G})$, where for each state/node i, the *action space* $\mathbb{A}(i) := \{a | (i,a) \in \mathcal{A}\}$ is in one-to-one correspondence with the set of arcs emanating from that node or, equivalently, with the set of endpoints of those arcs.

Throughout this section it will be more convenient to use the symbol f to denote a policy of our MDP rather than the more mnemonic π used in earlier sections. This is because the kind of policies we are ultimately searching for are best thought of as maps $f : \mathcal{V} \to \mathcal{V}$, as will be made clear in the illustrative example discussed below. In addition, our hypothetical "decision-maker" will usually be called "controller" in the current context.

Illustration: Consider the complete graph $\mathcal{G}_5$ on five nodes (with no self-loops), and think of the nodes as the states of an MDP, denoted by Γ, and of the arcs emanating from a given node as actions available at that state. In a natural way the Hamiltonian cycle

$c_1 : 1 \to 2 \to 3 \to 4 \to 5 \to 1$ corresponds to the "deterministic policy" $f_1 : \{1,2,3,4,5\} \to \{2,3,4,5,1\}$, where $f_1(2) = 3$ corresponds to the controller choosing arc (2,3) in state 2 with probability 1. The MC induced by f_1 is given by the "zero-one" transition matrix $P(f_1)$ which, clearly, is irreducible. On the other hand, the union of two subcycles $1 \to 2 \to 3 \to 1$ and $4 \to 5 \to 4$ corresponds to the policy $f_2 : \{1,2,3,4,5\} \to \{2,3,1,5,4\}$ which identifies the MC transition matrix $P(f_2)$ (see below) containing two distinct ergodic classes. This leads to a natural embedding of the HCP in a Markov decision problem Γ. The latter MDP has a multichain ergodic structure. For instance, the probability transition matrix induced by policy f_2 has the form

$$P(f_2) = \begin{pmatrix} 0 & 1 & 0 & 0 & 0 \\ 0 & 0 & 1 & 0 & 0 \\ 1 & 0 & 0 & 0 & 0 \\ 0 & 0 & 0 & 0 & 1 \\ 0 & 0 & 0 & 1 & 0 \end{pmatrix}.$$

Next, we consider an MC induced by a stationary *randomized policy*. As usual, the latter can be defined by an $N \times N$ *stochastic* matrix $\mathbf{f}$ with entries representing probabilities $f(i,a)$ of choosing a possible action a at a particular state i whenever this state is visited. Of course, $f(i,a) = 0$ whenever $a \notin \mathbb{A}(i)$. Randomized policies compose the *strategy space* $\mathcal{U}^S$. The discrete nature of the HCP focuses our attention on special paths which our moving object can trace out in $\mathcal{G}$. These paths correspond to the subspace $\mathcal{U}^D \subset \mathcal{U}^S$ of *deterministic* policies arising when the controller at every fixed state chooses some particular action with probability 1 whenever this state is visited (f_1 and f_2 above are instances of the latter). To illustrate these definitions consider the simple case where f_ν is obtained from the strictly deterministic policy f_2 by the "controller" deciding to randomize at node 4 by choosing the arcs $(4,5)$ and $(4,3)$ with probabilities $f(4,5) = 1-\nu$ and $f(4,3) = \nu$, respectively. The transition probability matrix of the resulting policy f_ν is given by

$$P(f_\nu) = \begin{pmatrix} 0 & 1 & 0 & 0 & 0 \\ 0 & 0 & 1 & 0 & 0 \\ 1 & 0 & 0 & 0 & 0 \\ 0 & 0 & \nu & 0 & 1-\nu \\ 0 & 0 & 0 & 1 & 0 \end{pmatrix}.$$

As ν ranges from 0 to 1 the MC ranges from the one induced by f_2 to the one induced by another deterministic policy.

7.4.2 ▪ Embedding in a discounted Markov decision process

Clearly, the formulation and definition of the preceding subsection suggest that the HCP may be embedded in any one of the MDPs discussed earlier in this chapter. This is because the search for a Hamiltonian cycle in a graph $\mathcal{G}$ reduces to a search for a stationary policy f_h in the MDP $\Gamma(\mathcal{G})$ that induces a probability transition matrix $P(f_h)$ that is an incidence matrix of a Hamiltonian cycle.

The reward structure of such an MDP plays a somewhat secondary role because we are primarily interested in determining the existence of such a policy rather than in evaluating its "performance." Indeed, in what follows we postulate that all reward vectors $r(f)$ for $f \in \mathcal{U}^S$ will be equal to e_1, the first vector of the unit basis. This corresponds to the notion that the controller receives a reward of one unit whenever the home node is visited and no reward otherwise. Nonetheless, it will be seen below that the manner of aggregation of the infinite stream of such awards can be usefully exploited.

In particular, below we focus on the previously introduced discounted MDP, with a discount factor $\lambda \in [0,1)$. We recall that in this model, a key object of interest associated with any given $f \in \mathcal{U}^S$ is the resolvent-like matrix

$$R_f(\lambda) := [I - \lambda P(f)]^{-1} = \sum_{t=0}^{\infty} \lambda^t P(f)^t, \tag{7.34}$$

where $\lambda \in [0,1)$ and $P(f)^0 := I$. Note that $\lambda = 1$ is excluded as it is an eigenvalue of $P(f)$.

7.4.3 ▪ MDP inspired characterizations of Hamiltonian cycles

Let us define $X(\lambda)$ to be the polytope defined by the system of linear constraints

(i) $\sum_i \sum_a \left(\delta_{ij} - \lambda p(j|i,a)\right) x_{ia} = \left(1 - \lambda^N\right) \delta_{1j}$ for $j = 1, \ldots, N$,

(ii) $\sum_{a \in \mathbb{A}(1)} x_{1a} = 1$,

(iii) $x_{ia} \geq 0$ for $i = 1, \ldots, N$, $a \in \mathbb{A}(i)$.

Note that in this application $p(j|i,a) = 1$ whenever $a \in \mathbb{A}(i)$ and is equal to 0 otherwise. Let $\mathbf{x} \in X(\lambda)$; then the entries of $\mathbf{x}$ are in one-to-one correspondence with node-arc pairs (i,a) so that $x_{ia} = [\mathbf{x}]_{ia}$ for all $i = 1, \ldots, N$, $a \in \mathbb{A}(i)$. We say that $\mathbf{x}_h \in X(\lambda)$ is Hamiltonian if there exists a Hamiltonian cycle h in the graph $\mathcal{G}$ such that for every $\lambda \in (0,1)$

(a) $[x_h]_{ia} = 0$ whenever $(i,a) \notin h$, and

(b) $[x_h]_{ia} > 0$ otherwise.

Note that all vectors x satisfying (i) must satisfy the matrix equation

$$W(\lambda)x = \left(1 - \lambda^N\right) e_1, \tag{7.35}$$

where $e_1 = (1, 0, \ldots, 0)^T \in \mathbb{R}^N$ and $W(\lambda)$ is an $N \times m$ matrix (with m denoting the total number of arcs) whose rows will be subscripted by j and whose columns will be subscripted by the pairs ia. That is, a typical (j, ia)-entry of $W(\lambda)$ is given by

$$w_{j,ia} := [W(\lambda)]_{j,ia} = \delta_{ij} - \lambda p(j|i,a), \; i,j = 1, \ldots, N, \; a \in \mathbb{A}(i). \tag{7.36}$$

Example 7.4. *Consider the four node graph given in Figure 7.1. It is clear that* $\mathbb{A}(1) = \{2,3,4\}$, $\mathbb{A}(2) = \{1,3\}$, $\mathbb{A}(3) = \{2,4\}$, $\mathbb{A}(4) = \{1,2,3\}$. *Hence any* $x \in X(\lambda)$ *must be of the form* $x^T = (x_{12}, x_{13}, x_{14}, x_{21}, x_{23}, x_{32}, x_{34}, x_{41}, x_{42}, x_{43})$.

Furthermore, $W(\lambda)$ *is a* 4×10 *matrix and equation* (7.35) *becomes*

$$\left[\begin{array}{ccc:cc:cc:ccc} 1 & 1 & 1 & -\lambda & 0 & 0 & 0 & -\lambda & 0 & 0 \\ -\lambda & 0 & 0 & 1 & 1 & -\lambda & 0 & 0 & -\lambda & 0 \\ 0 & -\lambda & 0 & 0 & -\lambda & 1 & 1 & 0 & 0 & -\lambda \\ 0 & 0 & -\lambda & 0 & 0 & 0 & -\lambda & 1 & 1 & 1 \end{array}\right] \begin{bmatrix} x_{12} \\ x_{13} \\ x_{14} \\ x_{21} \\ x_{23} \\ x_{32} \\ x_{34} \\ x_{41} \\ x_{42} \\ x_{43} \end{bmatrix} = \left(1 - \lambda^4\right) \begin{bmatrix} 1 \\ 0 \\ 0 \\ 0 \end{bmatrix}.$$

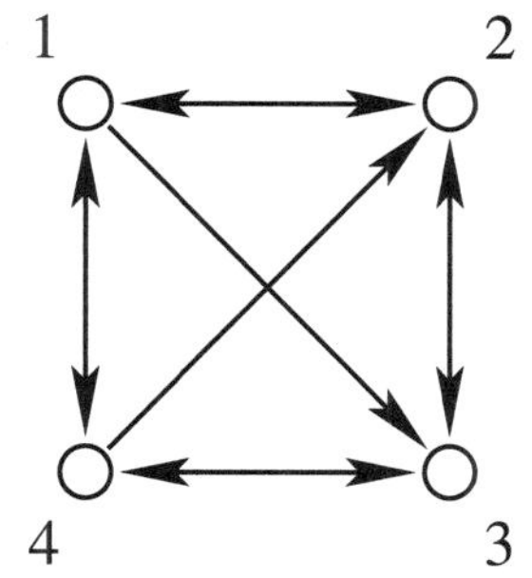

Figure 7.1. *A four-node graph with 10 arcs and two Hamiltonian cycles*

Let h_1 be the Hamiltonian cycle $1 \to 2 \to 3 \to 4 \to 1$ and h_2 be the reverse cycle $1 \to 4 \to 3 \to 2 \to 1$. Clearly, both h_1 and h_2 belong to Γ. Also, they may be viewed as collections of arcs, namely, $h_1 = \{(1,2),(2,3),(3,4),(4,1)\}$ and $h_2 = \{(1,4),(4,3),(3,2),(2,1)\}$. Now, let x_1^T and x_2^T be two 10-dimensional vectors

$$x_1^T = \begin{pmatrix} 1 & 0 & 0 & 0 & \lambda & 0 & \lambda^2 & \lambda^3 & 0 & 0 \end{pmatrix} \text{ and}$$
$$x_2^T = \begin{pmatrix} 0 & 0 & 1 & \lambda^3 & 0 & \lambda^2 & 0 & 0 & 0 & \lambda \end{pmatrix}.$$

It is now easy to check that both x_1 and x_2 satisfy the above stated version of (7.35). Indeed, they also satisfy (ii) and (iii) and their positive entries correspond to linearly independent columns of $W(\lambda)$, respectively. It follows that x_1 and x_2 are extreme points of $X(\lambda)$. They also happen to be the Hamiltonian points in $X(\lambda)$.

The characteristics of Hamiltonian points x_1 and x_2 demonstrated in Example 7.4 suggest that these may hold, generally, for all Hamiltonian solutions.

In what follows we consider the Hamiltonian cycle

$$h : j_0 = 1 \to j_1 \to j_2 \to \cdots \to j_{N-2} \to j_{N-1} \to 1 = j_N \tag{7.37}$$

consisting of the selection of arcs

$$(1, j_1), (j_1, j_2), \ldots, (j_{N-2}, j_{N-1}), (j_{N-1}, 1).$$

Thus j_k is the kth node on h following the home node $j_0 = 1$ for each $k = 1, 2, \ldots, N$. Motivated by Example 7.4, we construct a vector $x_h = x_h(\lambda)$ (with $\lambda \in [0, 1)$) according to

$$[x_h]_{ia} = \begin{cases} 0 & \text{if} \quad (i,a) \notin h, \\ \lambda^k & \text{if} \quad (i,a) = (j_k, j_{k+1}),\ k = 0, 1, 2, \ldots, N-1. \end{cases} \tag{7.38}$$

In Problem 7.6 the reader is asked to verify the following, now natural, property.

Lemma 7.8. *Let $X(\lambda)$ be defined by (i)–(iii), as above, let h be any Hamiltonian cycle, and let x_h be constructed by (7.38). It follows that x_h is an extreme point of $X(\lambda)$.*

Our previous assumption concerning the reward structure of the discounted MDP implies that

$$r(i,a) = \begin{cases} 1 & \text{if } i = 1,\ a \in \mathbb{A}(1), \\ 0 & \text{otherwise.} \end{cases}$$

This helps simplify the expression for the expected discounted reward $v_i^\lambda(f)$ corresponding to any $f \in \mathcal{U}^S$. In particular, we observe that if we let i_m denote the state/node visited at stage m, then an alternative probabilistic expression for the discounted reward starting from node 1 is

$$v_1^\lambda(f) = \sum_{m=0}^{\infty} \lambda^m P_1^f(i_m = 1), \tag{7.39}$$

where $P_1^f(\cdot)$ denotes the probability measure induced by f and the initial state $i_0 = 1$. It now immediately follows that

$$P_1^f(i_m = 1) = \frac{1}{m!}\left[\frac{\partial^m}{\partial \lambda^m}(v_1^\lambda(f))\right]_{\lambda=0}. \tag{7.40}$$

Next, we observe from (7.39) that if a policy f traces out a Hamiltonian cycle, then the home node is visited periodically after N steps, and this results in a deterministic sequence of discounted rewards

$$1, \lambda^N, \lambda^{2N}, \ldots, \lambda^{mN}, \ldots$$

that sums to $(1-\lambda^N)^{-1}$.

The above observations lead to some interesting characterizations of Hamiltonian cycles that are summarized in the result stated below.

Theorem 7.9. *With the embedding in Γ described above the following statements are equivalent:*

(i) *A policy f is deterministic and a Hamiltonian cycle in $\mathcal{G}$.*

(ii) *A policy f is stationary and a Hamiltonian cycle in $\mathcal{G}$.*

(iii) *A policy f is deterministic and $v_1^\lambda(f) = (1-\lambda^N)^{-1}$ for at least one $\lambda \in (0,1)$.*

(iv) *A policy f is stationary and $v_1^\lambda(f) = (1-\lambda^N)^{-1}$ for $2N-1$ distinct discount factors $\lambda_k \in (0,1)$, $k = 1,2,\ldots,2N-1$.*

In Problem 7.7, the interested reader is invited to reconstruct the proof of the above theorem (see also the bibliographic notes for the original source).

The above characterizations can be used to derive a number of alternative mathematical programming and feasibility formulations of both HCP and the traveling salesman problem (TSP). One of these is based on the following refinement of the $X(\lambda)$ polytope. Consider the polyhedral set $\mathcal{H}_\lambda$ defined by the linear constraints

$$\sum_{i=1}^{N} \sum_{a \in \mathbb{A}(i)} \left(\delta_{ij} - \lambda p(j|i,a)\right) x_{ia} = \delta_{1j}(1-\lambda^N) \;\forall\, j \in S, \tag{7.41}$$

$$\sum_{a \in \mathbb{A}(1)} x_{1a} = 1, \tag{7.42}$$

$$\sum_{a \in \mathbb{A}(i)} x_{ia} \ge \lambda^{N-1} \;\forall\, i \ne 1, \tag{7.43}$$

$$\sum_{a \in \mathbb{A}(i)} x_{ia} \le \lambda \;\forall\, i \ne 1, \tag{7.44}$$

$$x_{ia} \ge 0 \;\forall\, i \in S, a \in \mathbb{A}(i). \tag{7.45}$$

Note that by Lemma 7.8, all Hamiltonian solutions lie in $\mathcal{H}_\lambda$ and that the "wedge constraints" (7.43)–(7.44) can be made extremely narrow by choosing λ sufficiently near 1.

Furthermore, suppose that $x \in \mathcal{H}_\lambda$ satisfies the additional "complementarity constraint"

$$x_{ia}x_{ib} = 0 \ \forall \ i \in S, \ a \neq b \ \in \mathbb{A}(i). \tag{7.46}$$

Note that by (7.42)–(7.42), $\sum_{a\in\mathbb{A}(i)} x_{ia} > 0$ for each $i \in S$. Hence, whenever (7.46) holds, this means that $x_{ia} > 0$ for exactly one $a \in \mathbb{A}(i)$, for each $i \in S$. Now, if we map that $x \in \mathcal{H}_\lambda$ onto a policy by the usual transformation $f_x = T^{-1}(x)$ such that

$$f_x(i,a) = \frac{x_{ia}}{\sum_{a\in\mathbb{A}(i)} x_{ia}} \quad \forall \ i \in S, a \in \mathbb{A}(i),$$

then f_x is clearly deterministic and hence a Hamiltonian cycle by Theorem 7.9 (iii) (see also Problem 7.4).

The above leads to a quadratic programming formulation of HCP that requires the following notation. Let m_i be the cardinality of $\mathbb{A}(i)$ for each i, and let $m := \sum_{i\in S} m_i$, the total number of arcs in the original graph. Let J_i denote the $m_i \times m_i$ matrix of ones, and let I_i be the identity matrix of the same dimension. Define $Q_i := J_i - I_i$ for each $i \in S$, and the $m \times m$ block-diagonal matrix $Q := diag(Q_1, Q_2, \ldots, Q_N)$. It should now be clear that for any $x \in \mathcal{H}_\lambda$ we can define a quadratic function

$$\theta(x) := \sum_{i=1}^{N} \sum_{a\neq b} x_{ia}x_{ib} = \frac{1}{2} x^T Q x.$$

Proposition 7.10. *With the embedding in Γ described earlier and the above notation assume that $\mathcal{H}_\lambda \neq \emptyset$ and consider the quadratic programming problem*

$$\min\left\{\frac{1}{2} x^T Q x \mid x \in \mathcal{H}_\lambda\right\},$$

where $\lambda \in (0,1)$, and let x^* denote any one of its global minima. Then the following statements hold:

(i) *The above quadratic program is indefinite and possesses a global minimum x^* such that $\theta(x^*) \geq 0$.*

(ii) *If the graph $\mathcal{G}$ is Hamiltonian, then there exists a global optimum $x^* \in \mathcal{H}_\lambda$ such that $\theta(x^*) = 0$. Furthermore, the policy $f_{x^*} = T^{-1}(x^*)$ is deterministic and identifies a Hamiltonian cycle in $\mathcal{G}$.*

(iii) *If the graph $\mathcal{G}$ is non-Hamiltonian, then $\theta(x^*) > 0$.*

Proof: First note that at least one global minimum of the continuous function $\theta(x) = \frac{1}{2}x^T Q x$ must exist in $\mathcal{H}_\lambda$ as the latter is a compact set in m-dimensional Euclidean space. Then $\theta(x^*) \geq 0$ follows immediately from constraints (7.45). It is easy to check that, by construction, each Q_i (and hence also Q) possesses both positive and negative eigenvalues. Thus $\theta(x)$ is indefinite and part (i) holds. The same constraints (7.45) and the condition $\theta(x^*) = 0$ immediately imply that $x_{ia}x_{ib} = 0$ for all $i \in S$, $a \neq b \ \in \mathbb{A}(i)$, and hence $f_{x^*} = T^{-1}(x^*)$ is a deterministic policy which defines a Hamiltonian cycle by Theorem 7.9 (iii). Hence part (ii) holds. Finally, we claim that if $\mathcal{G}$ is non-Hamiltonian, then $\theta(x^*) > 0$.

Otherwise, by part (i) we must have that there exists $x^* \in \mathcal{H}_\lambda$ such that $\theta(x^*) = 0$, which by part (ii) allows us to construct a Hamiltonian cycle f_{x^*}, contradicting the hypothesis of non-Hamiltonicity. □

Another way to model the difficult "either-or" constraints (7.46) is with the help of auxiliary binary variables. For instance, define a set $\mathcal{B}$ of vectors u whose binary entries are indexed both by vertices of the graphs and by distinct pairs of arcs emanating from these vertices. More formally

$$\mathcal{B} := \{u \mid u_{iab} := [u]_{iab} \in \{0,1\} \ \forall \ i \in S, \ a \neq b \ \in \mathbb{A}(i)\}.$$

Now the following result shows that a whole family of mixed integer linear programming programs can be used to solve the HCP.

Proposition 7.11. *With the above notation consider the set*

$$\mathcal{H}\mathcal{B}_\lambda := \{(x,u) \in \mathcal{H}_\lambda \times \mathcal{B} \mid x_{ia} \le u_{iab};\ x_{ib} \le (1 - u_{iab}) \ \forall \ i \in S, \ a \neq b \ \in \mathbb{A}(i)\}.$$

Then the following statements hold:

(i) *The graph $\mathcal{G}$ is Hamiltonian if and only if $\mathcal{H}\mathcal{B}_\lambda \neq \emptyset$.*

(ii) *If $\ell(x,u)$ is any linear objective function made up of variables of (x,u), then the mixed linear integer mathematical program*

$$\min\{\ell(x,u) \mid (x,u) \in \mathcal{H}\mathcal{B}_\lambda\}$$

solves the HCP.

Proof: Suppose the graph $\mathcal{G}$ is Hamiltonian. Then by Proposition 7.10 there exists $x^* \in \mathcal{H}\mathcal{B}_\lambda$ such that $f_{x^*} = T^{-1}(x^*)$ is a deterministic policy which defines a Hamiltonian cycle. Hence, for each $i \in S$ there exists exactly one positive $a_i^* \in \mathbb{A}(i)$. Define u^*_{iab} to be 1 if $a = a_i^*$ and to be 0 otherwise. Clearly, $x^*_{ia^*} \le 1 = u^*_{ia^*b}$ for any $b \neq a^* \in \mathbb{A}(i)$ and $x^*_{ib} = 0 = 1 - u^*_{ia^*b}$ for any $b \neq a^* \in \mathbb{A}(i)$. Hence, $\mathcal{H}\mathcal{B}_\lambda \neq \emptyset$. On the other hand, if there exists $(\tilde{x},\tilde{u}) \in \mathcal{H}\mathcal{B}_\lambda$, then $\tilde{x}$ satisfies constraints (7.46) and $f_{\tilde{x}} = T^{-1}(\tilde{x})$ is a deterministic policy which defines a Hamiltonian cycle. This proves part (i).

For part (ii) note that a mixed linear integer program with an arbitrary linear objective function $\ell(x,u)$ either will yield infeasibility, which implies non-Hamiltonicity of HCP by part (i), or will supply at least one $(\tilde{x},\tilde{u}) \in \mathcal{H}\mathcal{B}_\lambda$, from which the Hamiltonian cycle $f_{\tilde{x}} = T^{-1}(\tilde{x})$ can be constructed, as above. □

We remark that, since $\mathcal{H}_\lambda$ becomes smaller as λ approaches 1 from below, it is natural to conjecture that identifying the non-Hamiltonicity of a graph will be easier when the discount factor is a neighborhood on unity. However, we note that $\lambda = 1$ is also the value where $I - \lambda P$ becomes singular. The latter in turn suggests that it might be worthwhile to regard this problem from the perspective of singular perturbations, which are the subject of this book. This perspective is outlined below.

7.4.4 ▪ Perspective of singular perturbations

In order to apply the singular perturbation techniques developed in this book to the preceding embedding of the HCP in a discounted MDP, it is convenient to make the standard substitutions

$$\lambda = \frac{1}{1+\rho}, \quad \rho \in (0,\infty), \quad \text{and} \quad \rho = \frac{1-\lambda}{\lambda}, \quad \lambda \in [0,1). \tag{7.47}$$

With the above definitions it is common to refer to the parameter ρ as an *interest rate* and to study the asymptotic behavior of a given problem as $\rho \to 0$ from above. Thus in the remainder of this section ρ will play the role that the perturbation parameter ε has played throughout most of this book.

Now, as before, with any given $f \in \mathcal{U}^S$ we can rewrite the resolvent-like matrix

$$\begin{aligned} R_f(\lambda) &:= [I - \lambda P(f)]^{-1} \\ &= \left((1+\rho)^{-1}[(1+\rho)I - P(f)]\right)^{-1} \\ &= ((1+\rho)[(I-P(f)) + \rho I])^{-1} \\ &= (1+\rho)[A(f) + \rho B]^{-1} = (1+\rho)R_f(\rho), \end{aligned} \tag{7.48}$$

where $A(f) = (I - P(f))$ and $B = I$. Note that now $R_f(\rho)$ is equivalent to the classical resolvent of the negative generator matrix $(I - P(f))$ of the MC induced by the policy f.

In the spirit of this book, we wish to analyze the problem as $\rho \to 0$. Thus the first question to answer concerns the expansion of the resolvent $R_f(\rho)$ as a Laurent series in the powers of ρ.

Proposition 7.12. *Let f be an arbitrary stationary policy in Γ_ρ, $P(f)$ be the probability transition matrix of the corresponding MC, and $R_f(\rho)$ be the resolvent of its negative generator as derived in* (7.48). *Then*

$$R_f(\rho) = \sum_{k=-1}^{\infty} \rho^k Y_k(f) = \frac{1}{\rho}\Pi(f) + H(f) - \rho H^2(f) + \rho^2 H^3(f) - \rho^3 H^4(f) + \dots. \tag{7.49}$$

That is, $Y_{-1}(f) = \Pi(f)$, $Y_0(f) = H(f)$, and $Y_k(f) = (-H(f))^k H(f)$, $k = 1,2,\dots$, where $\Pi(f)$ and $H(f)$ are the stationary distribution matrix and the deviation matrix, respectively.

Proof: Of course, the above expansion can be formally derived using the techniques of Chapter 2. However, in this special application it is possible to conjecture (on the basis of the classical Blackwell expansion) that the order of the pole at $\rho = 0$ is one and that the coefficients of ρ^{-1} and ρ^0 are the stationary distribution matrix $\Pi(f)$ and the deviation matrix $H(f)$, respectively. In such a case the form $Y_k(f) = (-H(f))^k H(f)$ for $k = 1,2,\dots$ follows immediately from equation (2.38) in Chapter 2 and the fact that $B = I$ in (7.48). Thus it is sufficient to verify that $[(I - P(f)) + \rho I]^{-1}[\sum_{k=-1}^{\infty} \rho^k Y_k(f)] = I$. However, we see that

$$\begin{aligned} &[R_f(\rho)]^{-1}\left[\sum_{k=-1}^{\infty} \rho^k Y_k(f)\right] \\ &= [(I-P(f)) + \rho I]\left[\frac{1}{\rho}\Pi(f) + H(f) - \rho H^2(f) + \rho^2 H^3(f) - \rho^3 H^4(f) + \dots\right]. \end{aligned}$$

Now, the right side of the above can be rearranged as

$$\begin{aligned} &(I-P(f))\frac{1}{\rho}\Pi(f) + (I-P(f))H(f)\left[I - \rho H(f) + \rho^2 H^2(f) - \rho^3 H^3(f) + \dots\right] \\ &\quad + \Pi(f) + \rho H(f)\left[I - \rho H(f) + \rho^2 H^2(f) - \rho^3 H^3(f) + \dots\right]. \end{aligned}$$

However, it is now possible to exploit the identities $P(f)\Pi(f)=\Pi(f)P(f)=\Pi(f)\Pi(f)=\Pi(f)$, $H(f)\Pi(f)=\Pi(f)H(f)=0$, and $(I-P(f))H(f)=I-\Pi(f)$ to verify that

$$\begin{aligned}[R_f(\rho)]^{-1}\left[\sum_{k=-1}^{\infty}\rho^k Y_k(f)\right] &= (I-\Pi(f))[I+\rho H(f)]^{-1}+\Pi(f)+\rho H(f)[I+\rho H(f)]^{-1}\\ &= \Pi(f)+[I+\rho H(f)-\Pi(f)][I+\rho H(f)]^{-1}\\ &= \Pi(f)-\Pi(f)[I+\rho H(f)]^{-1}+I\\ &= \Pi(f)\big[I-[I+\rho H(f)]^{-1}\big]+I=I. \end{aligned} \tag{7.50}$$

□

Now, the essential constraints

$$\sum_i\sum_a\big(\delta_{ij}-\lambda p(j|i,a)\big)y_{ia}=\delta_{1j},\quad j=1,\dots,N, \tag{7.51}$$

normally used in the linear programming formulations of the discounted MDP are satisfied by the vector $y_\lambda(f)$ variables constructed from any given $f\in\mathcal{U}^S$ according to

$$[y_\lambda(f)]_{ia}:=[\mathbf{e}_1^T R_f(\lambda)]_i f(i,a)\ \forall\ i\in S,\ a\in\mathbb{A}(i). \tag{7.52}$$

Hence, similarly to (7.48), the preceding equations (7.51) can be rewritten as

$$(1+\rho)^{-1}\sum_i\sum_a\big((1+\rho)\delta_{ij}-p(j|i,a)\big)y_{ia}=\delta_{1j},\quad j=1,\dots,N.$$

However, using (7.48), we note that $(1+\rho)^{-1}[\mathbf{e}_1^T R_f(\lambda)]_i f(i,a)=[\mathbf{e}_1^T R_f(\rho)]_i f(i,a)$ for all $i\in S,\ a\in\mathbb{A}(i)$, and so the above can be replaced by

$$\sum_i\sum_a\big((1+\rho)\delta_{ij}-p(j|i,a)\big)y_{ia}=\delta_{1j},\quad j=1,\dots,N, \tag{7.53}$$

where we search for a vector $y_\rho(f)$ of variables constructed from any given $f\in\mathcal{U}^S$ according to

$$[y_\rho(f)]_{ia}:=[\mathbf{e}_1^T R_f(\rho)]_i f(i,a)\ \forall\ i\in S,\ a\in\mathbb{A}(i). \tag{7.54}$$

Of course, the above system of equations can be viewed as a linearly perturbed system of the form

$$U(\rho)y=[U_0+\rho U_1]y=b, \tag{7.55}$$

where $b=\mathbf{e}_1$ and matrices U_0 and U_1 have entries defined by

$$[U_0]_{j,ia}=\delta_{ij}-p(j|i,a)\ \text{and}\ [U_1]_{j,ia}=\delta_{ij},\ i,j=1,\dots,N,\ a\in\mathbb{A}(i). \tag{7.56}$$

Since for each fixed pair (i,a) the sum $\sum_j(\delta_{ij}-p(j|i,a))=0$, it is clear that the row rank of U_0 is strictly less than N. Thus the above system of equations is, indeed, singularly perturbed. Of course, the singularly perturbed system (7.55) has solutions that depend on the parameter ρ, including solution vectors $y_\rho(f)$ of variables constructed from any given $f\in\mathcal{U}^S$ according to (7.54).

Now, in view of Proposition 7.12 it is reasonable to expect that, in a neighborhood of $\rho = 0$, the solution $y(\rho)$ of the above system will be a Laurent series of the form

$$y_\rho(f) := \sum_{k=-1}^{\infty} \rho^k y_k(f), \tag{7.57}$$

where the sequence of vectors $\{y_k(f)\}_{k=-1}^{\infty}$, whose entries are in one-to-one correspondence with the arcs of the underlying graph $\mathcal{G}$, is defined by

$$[y_k(f)]_{ia} := [\mathbf{e}_1^T Y_k(f)]_i f(i,a), \quad k = -1, 0, 1, 2, \ldots, \tag{7.58}$$

with $Y_k(f)$'s as given in Proposition 7.12.

By substituting the series $\sum_{k=-1}^{\infty} \rho^k y_k(f)$ into (7.55) and collecting coefficients of the same powers of ρ, one obtains the following system, which, as before, we refer to as the *fundamental equations* (FE) for $k = -1, 0, \ldots$:

$$\begin{aligned} U_0 y_{-1} &= 0, \qquad \text{(FE)} \\ U_0 y_0 + U_1 y_{-1} &= b, \\ U_0 y_1 + U_1 y_0 &= 0, \\ &\vdots \\ U_0 y_{k+1} + U_1 y_k &= 0. \\ &\vdots \end{aligned}$$

Of course, any solution $\{y_k\}_{k=-1}^{\infty}$ of the fundamental equations (FE) defines a ρ-dependent solution of (7.55) that is obtained by setting $y_\rho := \sum_{k=-1}^{\infty} \rho^k y_k$.

7.4.5 ▪ HCP from the perspective of singular perturbations

At first sight the system of fundamental equations (FE) obtained above by the singular perturbation approach appears rather cumbersome as it involves determining $\{y_k\}_{k=-1}^{\infty}$, a set of m-dimensional vectors of variables. However, in this subsection, it will be shown that when searching for a Hamiltonian cycle, a significant reduction in complexity can be achieved. This reduction stems from the simple observation that if $f \in \mathcal{U}^S$ is Hamiltonian, then the probability transition matrix $P(f)$ of the induced MC is periodic with period N. This property leads to the following representation of fundamental and deviation matrices induced by Hamiltonian policies.

Proposition 7.13. *Let f be a policy in Γ_ρ corresponding to an arbitrary Hamiltonian cycle in $\mathcal{G}$. Suppressing the notational dependence on f, let $P := P(f)$ be the probability transition matrix of the corresponding MC, $\Pi := \Pi(f)$ be its stationary matrix, $Z := Z(f)$ be its fundamental matrix, and $H := H(f)$ be its deviation matrix. Then there exist scalars π_r, z_r, h_r, $r = 0, 1, \ldots, N-1$, independent of the particular Hamiltonian cycle considered and such that the following representations hold:*

$$\Pi = \sum_{r=0}^{N-1} \pi_r P^r, \quad \pi_r = \frac{1}{N}, \qquad r = 0, 1, \ldots, N-1, \tag{7.59}$$

$$Z = \sum_{r=0}^{N-1} z_r P^r, \quad z_r = \frac{N+1-2r}{2N}, \; r = 0, 1, \ldots, N-1, \tag{7.60}$$

$$H=\sum_{r=0}^{N-1} h_r P^r, \quad h_r=\frac{N-1-2r}{2N}, \ r=0,1,\ldots,N-1. \tag{7.61}$$

Proof: The critical observations are that since f is Hamiltonian, it induces an MC with period N, and it follows that $I=P^0=P^N$. Furthermore, since P is doubly stochastic and irreducible, $\mathbf{1}^T P=\mathbf{1}^T$ and $\frac{1}{N}\mathbf{1}^T$ constitutes its unique invariant distribution, irrespective of which Hamiltonian cycle is specified by f. Thus we have

$$\Pi=\frac{1}{N}J=\frac{1}{N}[I+P+P^2+\cdots+P^{N-1}],$$

where J is an $N\times N$ matrix with all entries equal to 1. This proves (7.59).

To establish (7.60) we exploit the identities

$$Z(I-P+\Pi)=I \quad \text{and} \quad Z\Pi=\Pi.$$

By the uniqueness of the matrix inverse, if we can show that, with appropriately constructed scalars, z_r, $r=0,1,\ldots,N-1$, the sum $\sum_{r=0}^{N-1} z_r P^r$ satisfies the first of these identities, then the validity of (7.60) will be proved. Hence we formally substitute into that identity the above, desired, form of Z to obtain

$$\left(\sum_{r=0}^{N-1} z_r P^r\right)(I-P+\Pi)=\sum_{r=0}^{N-1} z_r P^r-\sum_{r=0}^{N-1} z_r P^{r+1}+\sum_{r=0}^{N-1}\frac{1}{N}P^r=I,$$

where the second equality follows from $Z\Pi=\Pi$ and (7.59). Now, equating coefficients of like powers of P^r, $r=0,1,\ldots,N-1$, on both sides of the above, we obtain the set of difference equations

$$z_0-z_{N-1}+\frac{1}{N}=1 \quad \text{and} \quad z_r-z_{r-1}+\frac{1}{N}=0, \ r=1,\ldots,N-1.$$

In addition, $Z\Pi=(\sum_{r=0}^{N-1} z_r P^r)[(1/N)J]=(1/N)J=\Pi$ implies that

$$\sum_{r=0}^{N-1} z_r=1.$$

The above equations can be easily manipulated to obtain the unique explicit solution for the z_r coefficients, namely,

$$z_r=z_0-\frac{r}{N}=\frac{N+1}{2N}-\frac{r}{N}=\frac{N+1-2r}{2N}, \ r=0,1,\ldots,N-1.$$

This proves (7.60). Now (7.61) follows immediately from the fact that $H=Z-\Pi$ and (7.59). □

Corollary 7.14. *Let f be a policy in Γ_ρ corresponding to an arbitrary Hamiltonian cycle in $\mathcal{G}$. Using the same notation as in Proposition 7.13, consider all powers of the corresponding fundamental and deviation matrices and Z^k and H^k for $k=1,2,\ldots$. For the case $k=1$ define scalars*

$$z_r^1:=z_r=\frac{N+1-2r}{2N} \quad \text{and} \quad h_r^1:=h_r=\frac{N-1-2r}{2N}, \ r=0,1,2,\ldots,N-1.$$

Then there exist scalars z_r^k, h_r^k, $r = 0, 1, \ldots, N-1$, $k = 2, 3, \ldots$, *independent of the particular Hamiltonian cycle considered and such that the following representations hold for each* $k = 1, 2, \ldots$:

$$Z^k = \sum_{r=0}^{N-1} z_r^k P^r, \tag{7.62}$$

where

$$z_0^{k+1} = 1 + \frac{1}{N}\sum_{s=0}^{N-1} s z_s^k \quad \text{and} \quad z_r^{k+1} = z_0^{k+1} + \sum_{\ell=1}^{r} z_\ell^k - \frac{r}{N}, \; r = 1, \ldots, N-1, \tag{7.63}$$

and

$$H^k = \sum_{r=0}^{N-1} h_r^k P^r, \tag{7.64}$$

where

$$h_0^{k+1} = \frac{3(N-1)}{2N} + \frac{1}{N}\sum_{s=0}^{N-1} s h_s^k \quad \text{and} \quad h_r^{k+1} = h_0^{k+1} + \sum_{\ell=1}^{r} h_\ell^k, \; r = 1, \ldots, N-1. \tag{7.65}$$

Proof: Of course, (7.60) and (7.61) show that the case $k = 1$ holds. Hence, for any $k = 1, 2, \ldots$, we have

$$Z^{k+1} = Z^k Z = \left(\sum_{r=0}^{N-1} z_r^k P^r\right)\left(\sum_{r=0}^{N-1} z_r^1 P^r\right).$$

Now, the fact that for each $\ell = 0, 1, 2, \ldots, N-1$ we have $P^{N+\ell} = P^\ell$ implies that some expansion of the form (7.62) must hold. The precise recursion for the coefficients of that expansion may be derived by grouping coefficients corresponding to powers of P^r, $r = 0, 1, \ldots, N-1$, in the above equation. The corresponding statements for the powers of the deviation matrix are derived analogously (see Problem 7.8). □

An interesting consequence of the above results is that whenever f is Hamiltonian the (difficult to interpret) solutions of the fundamental equations (FE) of the form $\{y_k(f)\}_{k=-1}^{\infty}$ can be expressed with the help of a finite collection of variable vectors $\{x_r(f)\}_{r=0}^{N-1}$ that are easy to interpret. The latter are constructed as follows.

Recalling that the entries of $y_k(f)$ are defined by $[y_k(f)]_{ia} := [\mathbf{e}_1^T Y_k(f)]_i f(i,a)$, we now define the entries of vectors $x_r(f)$, for each $r = 0, 1, \ldots, N-1$, by

$$[x_r(f)]_{ia} := x_{r,ia}(f) = [\mathbf{e}_1^T P^r(f)]_i f(i,a) \; \forall\, i \in S, \; a \in \mathbb{A}(i). \tag{7.66}$$

Thus we see that $x_{r,ia}(f) = 1$ if and only if arc (i,a) is the rth arc on the Hamiltonian cycle defined by f, starting at vertex 1. Otherwise, $x_{r,ia}(f) = 0$. Note that here it is convenient to regard the initial position (and arc) emanating from 1 as the 0th position (and arc), as it corresponds to $P^0(f) = I$.

In view of the above and recalling the definition of $Y_k(f)$ in Proposition 7.12, for $k = -1, 0, 1, 2, \ldots$, we can easily check that for any Hamiltonian $f \in \mathcal{U}^S$

$$Y_{-1}(f) = \sum_{r=0}^{N-1} \frac{1}{N} P^r \quad \text{and} \quad Y_k(f) = (-1)^k \sum_{r=0}^{N-1} h_r^{k+1} P^r, \; k = 0, 1, 2, \ldots.$$

The above, together with (7.66), now immediately implies that

$$y_{-1}(f) = \sum_{r=0}^{N-1} \frac{1}{N} x_r(f) \quad \text{and} \quad y_k(f) = (-1)^k \sum_{r=0}^{N-1} h_r^{k+1} x_r(f), \; k = 0, 1, 2, \ldots. \tag{7.67}$$

Thus we see that when searching for a Hamiltonian cycle in the original graph $\mathcal{G}$, instead of considering the previously derived fundamental equations (FE), we might as well consider the reduced system in the finite set of vectors of variables $\{x_r\}_{r=0}^{N-1}$ obtained by substitution of (7.67) into (FE). Note that argument f is suppressed because, at this stage, we do not even know whether the graph possesses a Hamiltonian cycle. However, we do know that if it does, then the following reduced system of equations (RE) possesses a solution:

$$\sum_{r=0}^{N-1} \left[\frac{1}{N} U_0 \right] x_r(f) = 0, \tag{RE}$$

$$\sum_{r=0}^{N-1} \left[h^1 U_0 + \frac{1}{N} U_1 \right] x_r(f) = b,$$

$$\sum_{r=0}^{N-1} \left[h^2 U_0 + h^1 U_1 \right] x_r(f) = 0,$$

$$\vdots$$

$$\sum_{r=0}^{N-1} [h^{k+1} U_0 + h^k U_1] x_r(f) = 0.$$

$$\vdots$$

Essentially, the above system (RE) is an equivalent parameter-free representation of the constraints (7.41) that are part of the definition of the polytope $\mathcal{H}_\lambda$. In Problem 7.9 we invite the reader to verify that the remaining constraints (7.42)–(7.45) can also be represented by countably many "layers" of parameter-free constraints in the $\{x_r\}_{r=0}^{N-1}$ that must be feasible whenever the underlying graph $\mathcal{G}$ is Hamiltonian.

We conclude this section with the following, currently still open, problem. Recall that in Proposition 7.11 the HCP was shown to be equivalent to the feasibility of a certain finite set of mixed linear integer constraints. In view of the preceding discussion, it is natural to ask whether there exists a finite number of the above-mentioned layers of linear constraints in the variables $\{x_r\}_{r=0}^{N-1}$ whose feasibility would also determine that the graph $\mathcal{G}$ is Hamiltonian. Note that such constraints would drop the integrality requirement.

7.5 ▪ Problems

Problem 7.1. For any $\pi \in \mathcal{U}^S$ in the NCD MDP Γ^ε consider the auxiliary matrices $M(\pi)$ and Q, as defined in the discussion preceding equation (7.4). Use the structure of Γ^ε to verify the correctness of that equation, namely, of the identities

$$\Pi^0(\pi) = QM(\pi) \quad \text{and} \quad M(\pi)Q = I_n.$$

Problem 7.2. Consider the frequency space of an irreducible MDP Γ on the state space $\mathbb{X}$ as characterized by the polyhedron

$$L := \left\{ x \,\middle|\, \sum_{i\in\mathbb{X}} \sum_{a\in\mathbb{A}(i)} (\delta_{ij} - p(j|i,a))x_{ia} = 0 \ \forall\, j \in \mathbb{X}, \right.$$

$$\left. \sum_{i\in\mathbb{X}} \sum_{a\in\mathbb{A}(i)} x_{ia} = 1, \quad \text{and} \quad x_{ia} \geq 0 \ \forall\, i \in \mathbb{X},\ a \in \mathbb{A}(i) \right\}.$$

Take any policy $\pi \in \mathcal{U}^S$ and its stationary distribution matrix $\Pi(\pi)$ consisting of identical rows $\mu(\pi)$. Let $x(\pi)$ be its associated vector of long-run state-action frequencies whose entries are defined by

$$x_{ia}(\pi) := [\mu(\pi)]_i \pi_{ia} \ \forall\, i \in \mathbb{X},\ a \in \mathbb{A}(i),$$

which defines a map $T:\ \mathcal{U}^S \to L$.

1. Prove that $x(\pi) \in L$.

2. Now define the map $T^{-1} : L \to \mathcal{U}^S$ by

$$\pi_{ia}(x) = \frac{x_{ia}}{\sum_{a\in\mathbb{A}(i)} x_{ia}} \ \forall\, i \in \mathbb{X},\ a \in \mathbb{A}(i).$$

   Prove that T^{-1} is well defined and, indeed, constitutes the inverse map of T.

Problem 7.3. Consider the feasible region of the linear program discussed in Section 7.2, namely, the region characterized by the constraints

$$\text{(i)} \ \sum_{i\in\mathbb{X}_k} \sum_{a\in A(i)} (\delta_{ij} - p(j|i,a))z^k_{ia} = 0, \quad j \in \mathbb{X}_k,\ k = 1,\ldots,n,$$

$$\text{(ii)} \ \sum_{k=1}^{n} \sum_{j\in\mathbb{X}_\ell} \sum_{i\in\mathbb{X}_k} \sum_{a\in A(i)} d(j|i,a)z^k_{ia} = 0, \quad \ell = 1,\ldots,n,$$

$$\text{(iii)} \ \sum_{k=1}^{n} \sum_{i\in\mathbb{X}_k} \sum_{a\in A(i)} z^k_{ia} = 1,$$

$$\text{(iv)} \ z^k_{ia} \geq 0, \quad k = 1,\ldots,n, \quad i \in \mathbb{X}_k, a \in A(i).$$

1. Verify that for each $k \in S$, summing over $j \in \mathbb{X}_k$, the block of constraints (i) corresponding to that k yields 0.

2. Verify that summing over $\ell \in S$, the block of constraints (ii) also yields zero.

3. Hence, or otherwise, prove that the rank of the coefficient matrix defined by the constraints (i)–(iii) is at most N.

4. Use the above and equation (7.18) to prove that the rank of the coefficient matrix defined by the constraints (i)–(iii) is equal to N.

Problem 7.4. Consider the frequency space of a discounted MDP Γ on the state space $\mathbb{X}$ as characterized by the polyhedron

$$L := \left\{ x \,\middle|\, \sum_{i\in\mathbb{X}} \sum_{a\in\mathbb{A}(i)} (\delta_{ij} - \lambda p(j|i,a))x_{ia} = \nu_j \ \forall\, j \in \mathbb{X}, \right. \\ \left. x_{ia} \geq 0 \ \forall\ i \in \mathbb{X},\ a \in \mathbb{A}(i) \right\},$$

where $\nu_j > 0$ denotes the probability that j is the initial stat and $\sum_j \nu_j = 1$. Take any policy $\pi \in \mathcal{U}^S$, and let $x(\pi)$ be its associated vector of discounted state-action frequencies whose entries are defined by (7.20), which defines a map $M:\ \mathcal{U}^S \to L$.

1. Prove that $x(\pi) \in L$.
2. Now define the map $M^{-1} : L \to \mathcal{U}^S$ by

$$\pi_{ia}(x) = \frac{x_{ia}}{\sum_{a\in\mathbb{A}(i)} x_{ia}} \quad \forall\ i \in \mathbb{X},\ a \in \mathbb{A}(i).$$

Prove that M^{-1} is well defined and, indeed, constitutes the inverse map of T.

Problem 7.5. Consider the perturbed linear program (LP^ε) introduced in Section 7.3.2. Verify the validity of Lemma 7.7, which shows that the policy constructed in (7.33) is indeed an average optimal policy in the general perturbed long-run average MDP. *Hint: Consider a pair of optimal solutions to both* (LP^ε) *and its dual* (DLP^ε)*, and invoke the complementary slackness theorem. This problem is based on analysis that can be found in [82] and [97].*

Problem 7.6. Let $f_h \in \mathcal{U}^{\mathcal{D}}$ be a Hamiltonian policy tracing out the standard Hamiltonian cycle, and let x_h be defined as in Lemma 7.8. Let $\overline{x}_h$ be an N-component vector consisting of only the positive entries of x_h. Show that

$$(I - \lambda P(f_h))^T \overline{x}_h = (1-\lambda^N)e_1.$$

Hence, or otherwise, prove that $\overline{x}_h$ is an extreme point of $X(\lambda)$.

Problem 7.7. Prove the validity of the four equivalent characterizations of Hamiltonian cycles given in Theorem 7.9. *Hint: See [54].*

Problem 7.8. Consider the finite series expansions of Corollary 7.14,

$$Z^k = \sum_{r=0}^{N-1} z_r^k P^r \quad \text{and} \quad H^k = \sum_{r=0}^{N-1} h_r^k P^r,$$

with the powers of the fundamental and deviation matrices when the probability transition matrix P corresponds to Hamiltonian cycle. Prove that the coefficients of these expansions satisfy the recursive equations (7.63) and (7.65), respectively. *Hint: It may be advisable to begin with the trivial equation* $Z^k = Z^{k+1}Z^{-1} = Z^{k+1}(I - P + \Pi)$*. Now, we can expand the latter as*

$$\sum_{r=0}^{n-1} z_r^k P^r = \sum_{r=0}^{n-1} z_r^{k+1} P^r - \sum_{r=0}^{n-1} z_r^{k+1} P^{r+1} + \sum_{r=0}^{n-1} z_r^{k+1} \Pi$$

and proceed to equate coefficients of the same powers of P *in the above.*

Problem 7.9. Consider constraints (7.41)–(7.45) defining the $X(\lambda)$ polytope. Show the following:

1. For any Hamiltonian policy $f \in \mathcal{U}^S$ variables $x_{ia} := (1-\lambda^N)[\mathbf{e}_1^T R_f(\lambda)]_i f(i,a)$ for all $i \in S,\ a \in \mathbb{A}(i)$ satisfy these constraints.

2. Hence show that when we use variables $[y_\rho(f)]_{ia} := [\mathbf{e}_1^T R_f(\rho)]_i f(i,a)$ for all $i \in S,\ a \in \mathbb{A}(i)$ that satisfy (7.53), constraint (7.42) becomes
$$\left(1-\left(\frac{1}{1+\rho}\right)^N\right)(1+\rho)\sum_{a\in\mathbb{A}(1)}[y_\rho(f)]_{1a}=1,$$
or, equivalently, in matrix form
$$[(1+\rho)^{N+1}-1-\rho]Vy_\rho(f)=(1+\rho)^N,$$
where V is a $1-0$ row vector that simply aggregates the entries in $\mathbb{A}(1)$.

3. As in Section 7.4.4, use the substitution $y_\rho(f) := \sum_{k=-1}^{\infty}\rho^k y_k(f)$ in the above to obtain
$$[(1+\rho)^{N+1}-1-\rho]\sum_{k=-1}^{\infty}\rho^k V y_k(f)=(1+\rho)^N.$$

4. Equate coefficients of powers of ρ in the above to derive the parameter-free system of layered linear constraints extending (FE) by incorporating the constraint (7.42).

5. Use the change of variables (7.67) in the preceding system to derive the parameter-free system of layered linear constraints extending (RE), obtained in Section 7.4.5, by incorporating the constraint (7.42) and expressing all the constraints in terms of a finite collection of variable vectors $\{x_r(f)\}_{r=0}^{N-1}$.

6. Devise a simple way of expressing the wedge constraints (7.43)–(7.44) in terms of a finite collection of variable vectors $\{x_r(f)\}_{r=0}^{N-1}$, as above.

7.6 ▪ Bibliographic Notes

Perhaps inadvertently, Blackwell [26] launched investigations of perturbed MDPs by deriving the so-called Blackwell expansion (see (7.19)). This initiated many studies of the properties of discounted MDPs in the neighborhood of the singularity at zero interest rate (discount factor of one). Notable among the latter are the results of Veinott and his co-workers (see, e.g., [122, 154]), many of which are summarized in Puterman's comprehensive book [130].

For an introduction and more detailed study of general MDPs we refer the reader to the books and surveys [50, 78, 83, 97, 98, 130, 61, 155] and references therein. The latter provide ample background reading for the material discussed in Sections 7.1–7.3.

The analysis of the singularly perturbed NCD MDP reported in Section 7.2 was inspired by the important paper of Delebecque and Quadrat [49]. Bielecki and Filar [24] and Abbad and Filar [1] proved that there exists a deterministic policy that solves the limit control problem, and the linear programming treatment given in Section 7.2 is based on Abbad et al. [3] and is also related to a method developed in Pervozvanski and Gaitsgori [126]. Bielecki and Stettner [25] generalized the limit control principle to MDPs with general space. Altman and Gaitsgori [7] analyzed singularly perturbed MDPs with constraints.

The results of Section 7.3 follow from Altman et al. [6] and Filar et al. [58]. However, asymptotic linear programming was first introduced by Jeroslow [95, 96] and later refined by Hordijk, Dekker, and Kallenberg [81]. Huang and Veinott [90] studied a similar problem in the context of Markov branching decision chains.

The approach to the HCP via singularly perturbed MDPs discussed in Section 7.4 was initiated in Filar and Krass [60]. The results of Section 7.4.3 are based on Feinberg [54], whose embedding of a graph in the discounted (rather than long-run average) MDP offers a number of advantages. For a survey of the MDP based approach to the HCP, see Filar [57]. A comprehensive research monograph, Borkar et al. [30], on MCs and the HCP contains details of many results obtained by this line of investigation.

Part III

Infinite Dimensional Perturbations

In mathematics you don't understand things. You just get used to them.

—John von Neumann (1903–1957)

Chapter 8

Analytic Perturbation of Linear Operators

8.1 ▪ Introduction

In this chapter we consider systems defined by linear operators on Hilbert or Banach space where the perturbation parameter is a single complex number. Let H and K be Hilbert or Banach spaces, and let

$$A : U \to \mathscr{L}(H,K)$$

be an analytic function where

$$U = \{z \mid |z| < \delta\} \subseteq \mathbb{C}$$

is a neighborhood of the origin in the complex plane. When

$$A(0) \in \mathscr{L}(H,K)$$

is singular, we wish to find conditions under which the inverse operator

$$A^{-1} : V \subseteq C \to \mathscr{L}(K,H)$$

is a well-defined analytic function for some deleted neighborhood of the origin

$$V = \{z \mid 0 < |z| < \epsilon\}.$$

We will begin by discussing the basic principles using matrix operators on finite dimensional spaces. Although this topic was considered in detail in Chapter 2, the treatment here is different. We will illustrate the main ideas with appropriate examples, particularly those that offer an easy comparison of results from the finite and infinite dimensional theories. Subsequently, we move on to consider the general theory which will be introduced with some more difficult examples and applications involving integral and differential operators.

8.2 ▪ Preliminaries from Finite Dimensional Theory

Suppose that a linear system is defined by a matrix $A \in \mathbb{C}^{m\times m}$. Imagine that the elements of

$$A = \left[a_{ij}\right]$$

are determined by an experimental process. We could think of the elements as functions that depend systematically on some inherent experimental error and write such perturbation as a Maclaurin series

$$A(z) = A_0 + A_1 z + A_2 z^2 + \cdots$$

valid in some neighborhood $|z| < r$ of the origin in the complex plane, with coefficients $A_i \in \mathbb{C}^{m \times m}$, and a supposed inverse Maclaurin series

$$X(z) = X_0 + X_1 z + X_2 z^2 + \cdots$$

valid in the same neighborhood, with coefficients $X_j \in \mathbb{C}^{m \times m}$. Then by equating coefficients of the various powers of z in the intuitive identities

$$A(z)X(z) = X(z)A(z) = I, \tag{8.1}$$

we obtain a system of fundamental equations. The fundamental equations

$$\begin{array}{rcl} A_0X_0 & = & I, \\ A_1X_0 + A_0X_1 & = & 0, \\ A_2X_0 + A_1X_1 + A_0X_2 & = & 0, \\ \vdots & & \vdots \end{array} \quad \text{and} \quad \begin{array}{rcl} X_0A_0 & = & I, \\ X_0A_1 + X_1A_0 & = & 0, \\ X_0A_2 + X_1A_1 + X_2A_0 & = & 0 \\ \vdots & & \vdots \end{array} \tag{8.2}$$

have a solution if and only if A_0 is nonsingular, in which case the solution is unique. When A_0 is singular, it is somewhat less obvious that, in the generic case, we may have an inverse Laurent series

$$X(z) = \frac{1}{z}\left[X_0 + X_1 z + X_2 z^2 + \cdots\right]$$

valid in some punctured neighborhood $0 < |z| < s$ and that by equating coefficients in the identities

$$A(z)X(z) = X(z)A(z) = zI \tag{8.3}$$

we can obtain a modified system of fundamental equations. The modified fundamental equations

$$\begin{array}{rcl} A_0X_0 & = & 0, \\ A_1X_0 + A_0X_1 & = & I, \\ A_2X_0 + A_1X_1 + A_0X_2 & = & 0, \\ \vdots & & \vdots \end{array} \quad \text{and} \quad \begin{array}{rcl} X_0A_0 & = & 0, \\ X_0A_1 + X_1A_0 & = & I, \\ X_0A_2 + X_1A_1 + X_2A_0 & = & 0 \\ \vdots & & \vdots \end{array} \tag{8.4}$$

have a solution if and only if we can find nonsingular matrices $F \in \mathbb{C}^{m \times m}$ and $G \in \mathbb{C}^{m \times m}$ such that

$$A_0' = F^{-1}A_0G = \begin{bmatrix} I_{m_1} & 0 \\ 0 & 0 \end{bmatrix} \quad \text{and} \quad A_1' = F^{-1}A_1G = \begin{bmatrix} A_{111}' & A_{112}' \\ A_{121}' & I_{m_2} \end{bmatrix}, \tag{8.5}$$

where $m_1 > 0$, $m_2 > 0$, and $m_1 + m_2 = m$. In the special case where

$$A_0 = \begin{bmatrix} I_{m_1} & 0 \\ 0 & 0 \end{bmatrix} \quad \text{and} \quad A_1 = \begin{bmatrix} A_{111} & A_{112} \\ A_{121} & I_{m_2} \end{bmatrix}$$

we can use elementary linear algebra to show that the modified fundamental equations have a unique solution. In this special case we can also see that

$$\operatorname{rank}\begin{bmatrix} A_0 & 0 \\ A_1 & A_0 \end{bmatrix} - \operatorname{rank} A_0 = m. \tag{8.6}$$

When A_0 is singular the rank condition (8.6) is equivalent to the earlier condition (8.5) on the existence of suitable nonsingular matrices F and G. Hence the rank condition is also necessary and sufficient for a unique solution. Similar ideas can be applied to analyze higher order singularities. More details about the rank condition can be found in the problems at the end of the chapter.

Let us define

$$||A|| = \sup_{x\in\mathbb{C}^m,\ x\neq 0} \frac{||Ax||}{||x||}. \tag{8.7}$$

The following theorem summarizes the results about the inversion of a regularly perturbed matrix.

Theorem 8.1. *Let $\{A_j\}_{j=0}^{\infty} \subseteq \mathbb{C}^{m\times m}$ be a sequence of square matrices such that A_0 is nonsingular. If $||A_j|| < r^{j+1}$ for some $r > 0$, then the series $A(z) = \sum_{j=0}^{\infty} A_j z^j$ is absolutely convergent for $|z| < 1/r$. Furthermore, we can find a real number $s > 0$ and a uniquely defined sequence $\{X_j\}_{j=0}^{\infty} \subseteq \mathbb{C}^{m\times m}$ of square matrices such that the series $X(z) = \sum_{j=0}^{\infty} X_j z^j$ is well defined and absolutely convergent for $|z| < 1/s$ and such that $A(z)X(z) = X(z)A(z) = I$ for $|z| < \max\{1/r, 1/s\}$. We write $X(z) = [A(z)]^{-1}$.*

In the next theorem we present the results of Subsection 2.2 in a convenient form involving matrix inverses for the generic case of a singularity of order one.

Theorem 8.2. *Let $\{A_j\}_{j=0}^{\infty} \subseteq \mathbb{C}^{m\times m}$ be a sequence of square matrices such that A_0 is singular, and suppose we can find nonsingular matrices $F \in \mathbb{C}^{m\times m}$ and $G \in \mathbb{C}^{m\times m}$ such that*

$$A_0' = F^{-1}A_0G = \begin{bmatrix} I_{m_1} & 0 \\ 0 & 0 \end{bmatrix} \quad \text{and} \quad A_1' = F^{-1}A_1G = \begin{bmatrix} A_{111}' & A_{112}' \\ A_{121}' & I_{m_2} \end{bmatrix}, \tag{8.8}$$

where $m_1 > 0$, $m_2 > 0$, and $m_1 + m_2 = m$. If $||A_j|| < r^{j+1}$ for some $r > 0$, then we can find a real number $s > 0$ and a uniquely determined sequence $\{X_j\}_{j=0}^{\infty} \subseteq \mathbb{C}^{m\times m}$ of square matrices such that the series $X(z) = \sum_{j=0}^{\infty} X_j z^j$ is well defined and absolutely convergent for $|z| < 1/s$ and such that $A(z)X(z) = X(z)A(z) = zI$ for $0 < |z| < \max\{1/r, 1/s\}$. We write $[A(z)]^{-1} = X(z)/z$.

Example 8.1. *Consider $(A_0 + A_1 z)^{-1}$, where*

$$A_0 = \begin{bmatrix} 1 & 1 & 0 \\ 0 & 1 & 1 \\ 1 & 2 & 1 \end{bmatrix} \quad \text{and} \quad A_1 = \begin{bmatrix} 1 & 0 & 0 \\ 0 & 1 & 0 \\ 0 & 0 & 1 \end{bmatrix}.$$

If we define

$$F = \begin{bmatrix} 1 & 0 & 0 \\ 0 & 1 & 0 \\ 1 & 2 & 1 \end{bmatrix} \quad \text{and} \quad G = \begin{bmatrix} 1 & -1 & 1 \\ 0 & 1 & -1 \\ 0 & 0 & 1 \end{bmatrix},$$

then

$$A_0' = F^{-1}A_0G = \begin{bmatrix} 1 & 0 & 0 \\ 0 & 1 & 0 \\ 0 & 0 & 0 \end{bmatrix} \quad \text{and} \quad A_1' = F^{-1}A_1G = \begin{bmatrix} 1 & -1 & 1 \\ 0 & 1 & -1 \\ -1 & 0 & 1 \end{bmatrix}.$$

The first equation

$$A_0'X_0' = 0 \quad \Leftrightarrow \quad \begin{bmatrix} I_2 & 0 \\ 0 & 0 \end{bmatrix}\begin{bmatrix} X_{011}' & X_{012}' \\ X_{021}' & X_{022}' \end{bmatrix} = \begin{bmatrix} 0 & 0 \\ 0 & 0 \end{bmatrix}$$

gives

$$X_{011}' = \begin{bmatrix} 0 & 0 \\ 0 & 0 \end{bmatrix} \quad \text{and} \quad X_{012}' = \begin{bmatrix} 0 \\ 0 \end{bmatrix},$$

and the second equation

$$A_1'X_0' + A_0'X_1' = I$$

$$\Leftrightarrow \begin{bmatrix} A_{111}' & A_{112}' \\ A_{121}' & 1 \end{bmatrix}\begin{bmatrix} 0 & 0 \\ X_{021}' & X_{022}' \end{bmatrix} + \begin{bmatrix} I_2 & 0 \\ 0 & 0 \end{bmatrix}\begin{bmatrix} X_{111}' & X_{112}' \\ X_{121}' & X_{122}' \end{bmatrix} = \begin{bmatrix} I_2 & 0 \\ 0 & 1 \end{bmatrix}$$

gives

$$X_{021}' = \begin{bmatrix} 0 & 0 \end{bmatrix} \quad \text{and} \quad X_{022}' = [\,1\,].$$

Thus X_0' is completely determined. The second equation also gives

$$X_{111}' = -A_{112}'X_{021}' = \begin{bmatrix} 0 & 0 \\ 0 & 0 \end{bmatrix} \quad \text{and} \quad X_{112}' = -A_{112}'X_{022}' = -A_{112}' = \begin{bmatrix} -1 \\ 1 \end{bmatrix}.$$

The third equation

$$A_1'X_1' + A_0'X_2' = 0$$

$$\Leftrightarrow \begin{bmatrix} A_{111}' & A_{112}' \\ A_{121}' & 1 \end{bmatrix}\begin{bmatrix} 0 & -A_{112}' \\ X_{121}' & X_{122}' \end{bmatrix} + \begin{bmatrix} I_2 & 0 \\ 0 & 0 \end{bmatrix}\begin{bmatrix} X_{211}' & X_{212}' \\ X_{221}' & X_{222}' \end{bmatrix} = \begin{bmatrix} 0 & 0 \\ 0 & 0 \end{bmatrix}$$

gives

$$X_{121}' = \begin{bmatrix} 0 & 0 \end{bmatrix} \quad \text{and} \quad X_{122}' = A_{121}'A_{112}' = \begin{bmatrix} -1 & 0 \end{bmatrix}\begin{bmatrix} 1 \\ -1 \end{bmatrix} = [\,-1\,],$$

and hence X_1' is completely determined. The third equation also allows us to determine X_{211}' and X_{212}'. By continuing in this way we can determine as many of the terms of the sequence $\{X_j'\}_{j=0}^{\infty}$ as we please. The sequence $\{X_j\}_{j=0}^{\infty}$ can now be reconstructed using the formula $X_j = GX_j'F^{-1}$.

Let $\{A_j\}_{j=0}^{\infty} \subseteq \mathbb{C}^{m\times m}$ be a sequence of square matrices. For each $k = 1, 2, \ldots$ define a corresponding sequence $\{\mathscr{A}_j^{(k)}\}_{j=0}^{\infty} \subseteq \mathbb{C}^{km\times km}$ of square matrices by the formulae

$$\mathscr{A}_0^{(k)} = \begin{bmatrix} A_0 & 0 & 0 & \cdots & 0 \\ A_1 & A_0 & 0 & \cdots & 0 \\ A_2 & A_1 & A_0 & \cdots & 0 \\ \vdots & \vdots & \vdots & \ddots & 0 \\ A_{k-1} & A_{k-2} & A_{k-3} & \cdots & A_0 \end{bmatrix} \tag{8.9}$$

and

$$\mathscr{A}_j^{(k)} = \begin{bmatrix} A_{jk} & A_{jk-1} & \cdots & A_{(j-1)k+1} \\ A_{jk+1} & A_{jk} & \cdots & A_{(j-1)k+2} \\ \vdots & \vdots & \ddots & \vdots \\ A_{(j+1)k} & A_{(j+1)k-1} & \cdots & A_{jk} \end{bmatrix} \tag{8.10}$$

for each $j > 0$. Then we obtain a generalization of Theorem 8.2 for cases when higher order singularities arise.

Theorem 8.3. *Let $\{A_j\}_{j=0}^{\infty} \subseteq \mathbb{C}^{m\times m}$ be a sequence of square matrices such that A_0 is singular but such that condition* (8.8) *is not satisfied. Let p be the smallest positive integer for which we can find nonsingular matrices $\mathscr{F} \in \mathbb{C}^{pm\times pm}$ and $\mathscr{G} \in \mathbb{C}^{pm\times pm}$ such that*

$$\mathscr{A}_0{}' = \mathscr{F}^{-1}\mathscr{A}_0^{(p)}\mathscr{G} = \begin{bmatrix} I_{m_1} & 0 \\ 0 & 0 \end{bmatrix} \text{ and } \mathscr{A}_1{}' = \mathscr{F}^{-1}\mathscr{A}_1^{(p)}\mathscr{G} = \begin{bmatrix} \mathscr{A}_{111}' & \mathscr{A}_{112}' \\ \mathscr{A}_{121}' & I_{m_2} \end{bmatrix}, \tag{8.11}$$

where $m_1 > 0$, $m_2 > 0$, and $m_1 + m_2 = pm$. If $\|A_j\| < r^{j+1}$ for some real number $r > 0$, then we can find a real number $s > 0$ and a uniquely determined sequence $\{X_j\}_{j=0}^{\infty} \subseteq \mathbb{C}^{m\times m}$ of square matrices such that the series $X(z) = \sum_{j=0}^{\infty} X_j z^j$ is well defined and absolutely convergent for $|z| < 1/s$ and such that $A(z)X(z) = X(z)A(z) = z^p I$ for $|z| < \min\{1/r, 1/s\}$. We write $[A(z)]^{-1} = X(z)/z^p$.

Example 8.2. *Consider $(A_0 + A_1 z + A_2 z^2)^{-1}$, where*

$$A_0 = \begin{bmatrix} 1 & 0 \\ 0 & 0 \end{bmatrix}, \quad A_1 = \begin{bmatrix} 1 & 1 \\ 0 & 0 \end{bmatrix}, \quad \text{and} \quad A_2 = \begin{bmatrix} 1 & 0 \\ 0 & 1 \end{bmatrix}.$$

If we define

$$\mathscr{A}_0^{(2)} = \begin{bmatrix} A_0 & 0 \\ A_1 & A_0 \end{bmatrix} = \begin{bmatrix} 1 & 0 & 0 & 0 \\ 0 & 0 & 0 & 0 \\ 1 & 1 & 1 & 0 \\ 0 & 0 & 0 & 0 \end{bmatrix},$$

$$\mathscr{A}_1^{(2)} = \begin{bmatrix} A_2 & A_1 \\ 0 & A_2 \end{bmatrix} = \begin{bmatrix} 1 & 0 & 1 & 1 \\ 0 & 1 & 0 & 0 \\ 0 & 0 & 1 & 0 \\ 0 & 0 & 0 & 1 \end{bmatrix},$$

and

$$\mathscr{X}_0^{(2)} = \begin{bmatrix} X_0 & 0 \\ X_1 & X_0 \end{bmatrix}, \quad \mathscr{X}_1^{(2)} = \begin{bmatrix} X_2 & X_1 \\ X_3 & X_2 \end{bmatrix}, \ldots,$$

then the equations

$$A_0X_0 = 0, \quad A_1X_0 + A_0X_1 = 0, \quad A_2X_0 + A_1X_1 + A_0X_2 = I,$$

$$A_2X_1 + A_1X_2 + A_0X_3 = 0, \ldots$$

can be rewritten in the augmented form

$$\mathscr{A}_0^{(2)}\mathscr{X}_0^{(2)} = 0, \quad \mathscr{A}_1^{(2)}\mathscr{X}_0^{(2)} + \mathscr{A}_0^{(2)}\mathscr{X}_1^{(2)} = I, \quad \mathscr{A}_1^{(2)}\mathscr{X}_1^{(2)} + \mathscr{A}_0^{(2)}\mathscr{X}_2^{(2)} = 0, \ldots,$$

which can be solved by applying Theorem 8.2. Indeed, if we define

$$\mathscr{F} = \begin{bmatrix} 1 & 0 & 0 & 0 \\ 0 & 0 & 1 & 0 \\ 0 & 1 & 0 & 0 \\ 0 & 0 & 0 & 1 \end{bmatrix} \quad \text{and} \quad \mathscr{G} = \begin{bmatrix} 1 & 0 & 0 & 0 \\ -1 & 1 & 1 & 0 \\ 0 & 0 & -1 & 0 \\ 0 & 0 & 0 & 1 \end{bmatrix},$$

then we obtain

$$\mathscr{A}_0{}' = \mathscr{F}^{-1}\mathscr{A}_0^{(2)}\mathscr{G} = \begin{bmatrix} 1 & 0 & 0 & 0 \\ 0 & 1 & 0 & 0 \\ 0 & 0 & 0 & 0 \\ 0 & 0 & 0 & 0 \end{bmatrix} \text{ and } \mathscr{A}_1{}' = \mathscr{F}^{-1}\mathscr{A}_1^{(2)}\mathscr{G} = \begin{bmatrix} 1 & 0 & -1 & 1 \\ 0 & 0 & -1 & 0 \\ -1 & 1 & 1 & 0 \\ 0 & 0 & 0 & 1 \end{bmatrix},$$

and so the solution to the augmented system can be computed directly as in Example 8.1.

The condition (8.11) is difficult to test directly but it can be reformulated in a more convenient form. If we define $\Delta : \mathbb{Z}^+ \to \mathbb{Z}^+$ by the formula

$$\Delta(k) = \begin{cases} \text{rank } \mathscr{A}_0^{(k+1)} & \text{if } k = 0, \\ \text{rank } \mathscr{A}_0^{(k+1)} - \text{rank } \mathscr{A}_0^{(k)} & \text{if } k = 1, 2, \ldots, \end{cases}$$

then it can be shown (see Problem 8.4) that $\Delta(k+1) \geq \Delta(k)$ for all $k \in \mathbb{Z}^+$.

Theorem 8.4. *Let $\{A_j\}_{j=0}^{\infty} \subseteq \mathbb{C}^{m\times m}$ be a sequence of square matrices, and let $\Delta : \mathbb{Z}^+ \to \mathbb{Z}^+$ be the function defined above. The matrix A_0 is nonsingular if and only if $\Delta(0) = m$. If $\Delta(0) < m$, we can find nonsingular matrices $\mathscr{F} \in \mathbb{C}^{pm\times pm}$ and $\mathscr{G} \in \mathbb{C}^{pm\times pm}$ such that*

$$\mathscr{A}_0{}' = \mathscr{F}^{-1}\mathscr{A}_0^{(p)}\mathscr{G} = \begin{bmatrix} I_{m_1} & 0 \\ 0 & 0 \end{bmatrix} \quad \text{and} \quad \mathscr{A}_1{}' = \mathscr{F}^{-1}\mathscr{A}_1^{(p)}\mathscr{G} = \begin{bmatrix} \mathscr{A}_{111}' & \mathscr{A}_{112}' \\ \mathscr{A}_{121}' & I_{m_2} \end{bmatrix},$$

where m_1, $m_2 > 0$, and $m_1 + m_2 = pm$ if and only if $\Delta(p-1) < m$ and $\Delta(p) = m$.

If we wish to extend the above arguments to linear mappings on Hilbert space, then we need to understand that conditions involving the rank of the augmented matrix $\mathscr{A}_0^{(k)}$ are really conditions to ensure that certain key matrices are invertible. For infinite dimensional Hilbert space we must rewrite these conditions in a more general form to ensure that the corresponding mappings are one-to-one and onto.

8.3 ▪ Key Examples

In infinite dimensional space we are no longer able to use elementary row and column operations to determine whether linear mappings are one-to-one and onto. The following simple example shows that linear independence alone is no longer a satisfactory condition for a set of basis vectors.

Example 8.3. *We consider the space $l_2 = \{x = [x_i]_{i=1,2,\ldots} \mid x_i \in \mathbb{C} \text{ and } \sum_i |x_i|^2 < \infty\}$. The standard basis for l_2 is the well-known set*

$$e_1 = \begin{bmatrix} 1 \\ 0 \\ 0 \\ 0 \\ \vdots \end{bmatrix}, \quad e_2 = \begin{bmatrix} 0 \\ 1 \\ 0 \\ 0 \\ \vdots \end{bmatrix}, \quad e_3 = \begin{bmatrix} 0 \\ 0 \\ 1 \\ 0 \\ \vdots \end{bmatrix}, \ldots$$

of mutually orthogonal unit vectors. We will consider a possible basis of nonorthogonal vectors. The set $\{f_j\}_{j=2,3,\ldots}$ *defined by*

$$f_2 = \begin{bmatrix} 1 \\ -1 \\ 0 \\ 0 \\ 0 \\ \vdots \end{bmatrix}, \quad f_3 = \begin{bmatrix} 1 \\ 0 \\ -1 \\ 0 \\ 0 \\ \vdots \end{bmatrix}, \quad f_4 = \begin{bmatrix} 1 \\ 0 \\ 0 \\ -1 \\ 0 \\ \vdots \end{bmatrix}, \ldots$$

is clearly a linearly independent set because

$$\sum_{j=2}^{n} \alpha_j f_j = 0 \quad \Rightarrow \quad \alpha_j = 0 \quad \text{for each } j = 2,3,\ldots,n,$$

but the set is not a spanning set because the equation

$$\sum_{j=2}^{\infty} \alpha_j f_j = e_1$$

requires $\sum_{j=2}^{\infty} \alpha_j = 1$ *with* $\alpha_j = 0$ *for all* j *and hence has no solution. However, if we define*

$$e_1^{(n)} = \frac{1}{n} \sum_{j=2}^{n+1} f_j,$$

then we can see that

$$||e_1^{(n)} - e_1||^2 = \frac{1}{n} \to 0$$

as $n \to \infty$*. Hence, although* e_1 *lies in the closure of the spanning set* $\mathcal{S}(f_2, f_3, \ldots)$*, there is no satisfactory representation of* e_1 *as a linear combination of the proposed "basis" vectors. Since* $e_j = e_1 - f_j$ *for all* $j = 2,3,\ldots$*, it follows that any vector in* l_2 *can be approximated as accurately as we please by a linear combination of the vectors* f_j*. Hence* $l_2 = \overline{\mathcal{S}(f_2, f_3, \ldots)}$*.*

In finite dimensional problems we showed how elementary row and column operations can be used to simplify representations of the matrices A_j and hence to simplify the system of fundamental equations. These operations can be interpreted as nonsingular linear transformations of the coordinate systems in both the domain space and the range space. For infinite dimensional problems transformations involving row and column operations are no longer suitable. In Hilbert space we will use unitary transformations[7] to

[7]If H is a Hilbert space over the field of complex numbers, the operator $P \in \mathcal{L}(H)$ is said to be a *unitary* operator if $P^*P = PP^* = I \in \mathcal{L}(H)$, where $P^* \in \mathcal{L}(H)$ is the conjugate transpose or *adjoint* operator and I is the identity.

simplify the operator representations. Let H and K be Hilbert spaces over the field of complex numbers $\mathbb{C}$, and let $A_0 : H \to K$ and $A_1 : H \to K$ be bounded linear transformations. For the linear perturbation $A_0 + A_1 z$ the key spaces to consider are the null space $M = A_0^{-1}(\{0\}) \subseteq H$ and the image $N = A_1(M) = A_1 A_0^{-1}(\{0\}) \subseteq K$ of the null space under the operator A_1.

Example 8.4. *Let $H = K = \mathbb{C}^3$. Consider $(A_0 + A_1 z)^{-1}$, where*

$$A_0 = \begin{bmatrix} -2 & -2 & 4 \\ 0 & 0 & 0 \\ -1 & -1 & 2 \end{bmatrix} \quad \text{and} \quad A_1 = \begin{bmatrix} 1 & 1 & 1 \\ 2 & 0 & 1 \\ 1 & -1 & 0 \end{bmatrix}.$$

The null space of A_0 is given by $M = \{x \mid x_1 + x_2 - 2x_3 = 0\}$, and the image of this space under A_1 is defined by $N = A_1(M) = \{y \mid y_1 - y_2 + y_3 = 0\}$. If we transform to orthonormal bases for $M \times M^\perp$ and $N \times N^\perp$ by making the unitary transformations $x = Px'$ and $y = Qy'$, where

$$P = \begin{bmatrix} \frac{\sqrt{3}}{3} & \frac{\sqrt{2}}{2} & \frac{\sqrt{6}}{6} \\ \frac{\sqrt{3}}{3} & \frac{-\sqrt{2}}{2} & \frac{\sqrt{6}}{6} \\ \frac{\sqrt{3}}{3} & 0 & \frac{-\sqrt{6}}{3} \end{bmatrix} \quad \text{and} \quad Q = \begin{bmatrix} \frac{\sqrt{2}}{2} & \frac{\sqrt{6}}{6} & \frac{-\sqrt{3}}{3} \\ \frac{\sqrt{2}}{2} & \frac{-\sqrt{6}}{6} & \frac{\sqrt{3}}{3} \\ 0 & \frac{-\sqrt{6}}{3} & \frac{-\sqrt{3}}{3} \end{bmatrix},$$

then the transformed matrices are

$$A_0' = Q^* A_0 P = \begin{bmatrix} 0 & 0 & -\frac{2\sqrt{3}}{3} \\ 0 & 0 & 0 \\ 0 & 0 & \sqrt{3} \end{bmatrix} \quad \text{and} \quad A_1' = Q^* A_1 P = \begin{bmatrix} \sqrt{6} & 1 & 0 \\ 0 & -\sqrt{3} & 0 \\ 0 & 0 & 0 \end{bmatrix},$$

and the reformulated problem

$$(A_0' + A_1' z)^{-1} = \begin{bmatrix} \sqrt{6} \cdot z & z & \frac{-2\sqrt{3}}{3} \\ 0 & -\sqrt{3} \cdot z & 0 \\ 0 & 0 & \sqrt{3} \end{bmatrix}^{-1}$$

essentially reduces to inversion of the matrices

$$A_{111}' z = \begin{bmatrix} \sqrt{6} & 1 \\ 0 & -\sqrt{3} \end{bmatrix} z \quad \text{and} \quad A_{022}' + A_{122}' z = \begin{bmatrix} \sqrt{3} \end{bmatrix}.$$

By direct calculation we have

$$(A_0' + A_1' z)^{-1} = \begin{bmatrix} \frac{\sqrt{6}}{6} \cdot \frac{1}{z} & \frac{\sqrt{2}}{6} \cdot \frac{1}{z} & \frac{\sqrt{6}}{9} \cdot \frac{1}{z} \\ 0 & \frac{-\sqrt{3}}{3} \cdot \frac{1}{z} & 0 \\ 0 & 0 & \frac{\sqrt{3}}{3} \end{bmatrix}.$$

The formula

$$(A_0 + A_1 z)^{-1} = P (A_0' + A_1' z)^{-1} Q^*$$

allows us to retrieve the desired inverse.

For each $j = 1,2$ let $A_j \in \mathbb{C}^{m\times n}$. If the perturbed matrix $A_0 + A_1 z$ is to be an invertible mapping, then the perturbation matrix A_1 should be one-to-one on the null space $M = A_0^{-1}(\{0\})$ of the unperturbed operator. If we define $N = A_1(M) = A_1 A_0^{-1}(\{0\})$, the transformation to orthonormal bases for $M \times M^{\perp}$ and $N \times N^{\perp}$ makes this obvious by reducing certain key components of the transformation to zero. In the domain space we choose an orthonormal basis $p_1,\ldots,p_m$ for $\mathbb{C}^m$ such that $p_1,\ldots,p_{m-r}$ is an orthonormal basis for M and $p_{m-r+1},\ldots,p_m$ is an orthonormal basis for $M^{\perp}$. In the image space we choose an orthonormal basis $q_1,\ldots,q_n$ for $\mathbb{C}^n$ such that $q_1,\ldots,q_{n-s}$ is an orthonormal basis for N and $q_{n-s+1},\ldots,q_n$ is an orthonormal basis for $N^{\perp}$. We define the matrices $P \in \mathbb{C}^{m\times m}$, $P_1 \in \mathbb{C}^{m\times(m-r)}$, and $P_2 \in \mathbb{C}^{m\times r}$ by

$$P = [P_1, P_2], \quad \text{where} \quad P_1 = [p_1,\ldots,p_{m-r}] \quad \text{and} \quad P_2 = [p_{m-r+1},\ldots,p_m],$$

and $Q \in \mathbb{C}^{n\times n}$, $Q_1 \in \mathbb{C}^{n\times(n-s)}$, and $Q_2 \in \mathbb{C}^{n\times s}$ by

$$Q = [Q_1, Q_2], \quad \text{where} \quad Q_1 = [q_1,\ldots,q_{n-s}] \quad \text{and} \quad Q_2 = [q_{n-s+1},\ldots,q_n],$$

and consider the unitary transformations $x = Px'$ and $y = Qy'$. Now we have

$$A_j' = Q^* A_j P = \begin{bmatrix} Q_1^* \\ Q_2^* \end{bmatrix} A_j \begin{bmatrix} P_1 & P_2 \end{bmatrix} = \begin{bmatrix} Q_1^* A_j P_1 & Q_1^* A_j P_2 \\ Q_2^* A_j P_1 & Q_2^* A_j P_2 \end{bmatrix}.$$

Since $A_0 P_1 = 0$ and $Q_2^* A_1 P_1 = 0$ it follows that

$$A_0' = \begin{bmatrix} 0 & Q_1^* A_0 P_2 \\ 0 & Q_2^* A_0 P_2 \end{bmatrix} \quad \text{and} \quad A_1' = \begin{bmatrix} Q_1^* A_1 P_1 & Q_1^* A_1 P_2 \\ 0 & Q_2^* A_1 P_2 \end{bmatrix},$$

and hence the transformed inversion problem becomes

$$\begin{aligned} (A_0' + A_1' z)^{-1} &= \begin{bmatrix} Q_1^* A_1 P_1 z & Q_1^*(A_0 + A_1 z)P_2 \\ 0 & Q_2^*(A_0 + A_1 z)P_2 \end{bmatrix}^{-1} \\ &= \begin{bmatrix} A_{111}' z & (A_{012}' + A_{112}' z) \\ 0 & (A_{022}' + A_{122}' z) \end{bmatrix}^{-1}. \end{aligned}$$

Thus the matrix $(A_0' + A_1' z)$ is invertible if and only if the matrices $A_{111}' z$ and $(A_{022}' + A_{122}' z)$ are each invertible.

Example 8.5. *Consider $(A_0 + A_1 z)^{-1}$, where*

$$A_0 = \begin{bmatrix} 2 & 0 & 0 & -2 \\ 1 & 1 & -1 & -1 \\ 1 & -1 & 1 & -1 \end{bmatrix} \quad \textit{and} \quad A_1 = \begin{bmatrix} 2 & -1 & 2 & -1 \\ -1 & 1 & -2 & 0 \\ -1 & 2 & -3 & 0 \end{bmatrix}.$$

Using the procedure outlined above we define

$$P_1 = \begin{bmatrix} \frac{\sqrt{2}}{2} & 0 \\ 0 & \frac{\sqrt{2}}{2} \\ 0 & \frac{\sqrt{2}}{2} \\ \frac{\sqrt{2}}{2} & 0 \end{bmatrix} \quad \textit{and} \quad P_2 = \begin{bmatrix} \frac{\sqrt{2}}{2} & 0 \\ 0 & \frac{\sqrt{2}}{2} \\ 0 & \frac{-\sqrt{2}}{2} \\ \frac{-\sqrt{2}}{2} & 0 \end{bmatrix}$$

and

$$Q_1 = \begin{bmatrix} \frac{\sqrt{3}}{3} \\ \frac{-\sqrt{3}}{3} \\ \frac{-\sqrt{3}}{3} \end{bmatrix} \quad \text{and} \quad Q_2 = \begin{bmatrix} 0 & \frac{\sqrt{6}}{3} \\ \frac{\sqrt{2}}{2} & \frac{\sqrt{6}}{6} \\ \frac{-\sqrt{2}}{2} & \frac{\sqrt{6}}{6} \end{bmatrix},$$

from which we obtain

$$A_0' = \begin{bmatrix} 0 & 0 & 0 & 0 \\ 0 & 0 & 0 & 2\sqrt{2} \\ 0 & 0 & 2\sqrt{3} & 0 \end{bmatrix} \quad \text{and} \quad A_1' = \begin{bmatrix} \frac{\sqrt{6}}{2} & \frac{\sqrt{6}}{2} & \frac{5\sqrt{6}}{6} & \frac{-11\sqrt{6}}{6} \\ 0 & 0 & 0 & -1 \\ 0 & 0 & \frac{2\sqrt{3}}{3} & \frac{\sqrt{3}}{3} \end{bmatrix}.$$

Of course in this example it is clear that

$$A_{111}' = \begin{bmatrix} \frac{\sqrt{6}}{2} & \frac{\sqrt{6}}{2} \end{bmatrix}$$

is not invertible because it is not square.

Let us consider the mapping from $\mathbb{C}^m$ to $\mathbb{C}^n$ defined by the matrix $A \in \mathbb{C}^{n\times m}$. We wish to show that the range space $A(\mathbb{C}^m)$ is closed. We consider two related Hermitian matrices $S = A^*A \in \mathbb{C}^{m\times m}$ and $T = AA^* \in \mathbb{C}^{n\times n}$. It is well known that the eigenvalues of a Hermitian matrix are real and that the set of eigenvectors can be chosen to provide an orthonormal basis for the associated vector space. Let us suppose that $p \in \mathbb{C}^m$ is an eigenvector for S corresponding to a nonzero eigenvalue σ. Thus $Sp = \sigma p$. Since

$$||Ap||^2 = p^*A^*Ap = p^*Sp = \sigma p^*p = \sigma ||p||^2,$$

it follows that $\sigma \geq 0$. If we define

$$q = \frac{1}{\sqrt{\sigma}} Ap,$$

then it follows that $Tq = \sigma q$. Hence we can choose corresponding orthonormal sets $p_1, \ldots, p_m$ in $\mathbb{C}^m$ and $q_1, \ldots, q_n$ in $\mathbb{C}^n$ with $Sp_{m-j+1} = \sigma_j p_{m-j+1}$ and $Tq_{n-j+1} = \sigma_j q_{n-j+1}$ and such that

$$q_{n-j+1} = \frac{1}{\sqrt{\sigma_j}} Ap_{m-j+1},$$

where $\sigma_j > 0$ for each $j = 1, \ldots, r$ and with $Sp_j = 0$ for $j = 1, \ldots, m-r$ and $Tq_j = 0$ for $j = 1, \ldots, n-r$. We will assume that $0 < \sigma_1 \leq \sigma_2 \leq \cdots \leq \sigma_r$. We can now show that the null space of A is

$$A^{-1}(\{0\}) = \left\{ x \mid x = \sum_{j=1}^{m-r} \alpha_j p_j \right\}$$

and the range space of A is

$$A(\mathbb{C}^m) = \left\{ y \mid y = \sum_{j=1}^{r} \beta_{n-j+1} q_{n-j+1} \right\}.$$

For $x \in A^{-1}(\{0\})^{\perp}$ we have $\sigma_1||x|| \le ||Ax|| \le \sigma_r||x||$. We say that A is bounded above and below on $A^{-1}(\{0\})^{\perp}$. To show that the range of A is closed, let $y^{(k)} = Ax^{(k)} \in A(\mathbb{C}^m)$, and suppose that $||y^{(k)} - g|| \to 0$ as $k \to \infty$. If we write

$$y^{(k)} = \sum_{j=1}^{r} \beta_{n-j+1}^{(k)} q_{n-j+1} \quad \text{and} \quad g = \sum_{j=1}^{r} \beta_{n-j+1} q_{n-j+1},$$

then we must have $\beta_{n-j+1}^{(k)} \to \beta_{n-j+1}$ as $k \to \infty$ for each $j = 1, \ldots, r$. Since

$$g = \sum_{j=1}^{r} \beta_{n-j+1} \cdot \frac{1}{\sqrt{\sigma_j}} A p_{m-j+1} = Af,$$

where

$$f = \sum_{j=1}^{r} \frac{\beta_{n-j+1}}{\sqrt{\sigma_j}} p_{m-j+1},$$

it follows that $g \in A(\mathbb{C}^m)$. Thus the range of A is closed.

In infinite dimensional problems the range space of a bounded operator need not be closed. In a Banach space it can be shown that the range space is closed if and only if there is some constant $\epsilon > 0$ such that for each y in the range space we can find $x \in A^{-1}(\{y\})$ such that $||y|| \ge \epsilon||x||$. If the inverse mapping A^{-1} is well defined, then the equation $y = Ax$ must have a unique solution for each y in the range space. In this case we must have $||Ax|| \ge \epsilon||x||$ for all x, and so A is bounded below on the entire domain, and the null space of A contains only the zero vector. The next two examples use infinite dimensional spaces. The important general properties of these spaces are reviewed in Section 8.5. Example 8.6 shows that we may be able to modify the topology of an infinite dimensional space to ensure that the range space is closed.

Example 8.6. *Let $\Omega = [0,1]$, and let $H = K = L^2(\Omega)$. For each $x \in H$ define $\mu(x) = \int_\Omega x(s)ds$. Let $A \in \mathcal{L}(H,K)$ be defined by*

$$Ax(t) = \int_{(0,t)} [x(s) - \mu(x)]ds \quad \forall\, x \in H,\ t \in [0,1].$$

The space $A(H) = \mathscr{C}_0^1(\Omega)$ is the space of absolutely continuous functions $y : [0,1] \to \mathbb{C}$ with $y(0) = y(1) = 0$. The space $A(H)$ is not closed in K. In Problem 8.16 we show that if

$$x^{(k)}(s) = \begin{cases} 0 & \text{when } s \notin [\frac{1}{2}(1-\frac{1}{k}), \frac{1}{2}(1+\frac{1}{k})], \\ k & \text{otherwise,} \end{cases}$$

then $Ax^{(k)} = y^{(k)} \to g \in K$ as $k \to \infty$, where

$$g(t) = \begin{cases} -t & \text{when } t < \frac{1}{2}, \\ 1-t & \text{when } t > \frac{1}{2}. \end{cases}$$

However, $g \notin A(H)$, and hence $A(H)$ is not closed. In general, if $y = Ax \in A(H)$, then y is differentiable almost everywhere and $y' = [x - \mu(x)] \in H$. Thus we can define a new energy inner product on the range space given by

$$\langle y, v\rangle_E = \int_\Omega \left[y(t)\overline{v}(t) + y'(t)\overline{v}'(t)\right]dt$$

for each $y \in A(H)$ *and* $v \in A(H)$. *Indeed, it can be shown that the space*

$$K_E = \{y \mid y \in L^2(\Omega),\ y' \in L^2(\Omega)\} = W^1(\Omega)$$

with inner product $\langle \cdot,\cdot \rangle_E$ *is a Hilbert space. If we define* $A_E \in \mathscr{L}(H, K_E)$ *by setting* $A_E x = Ax$ *for each* $x \in H$*, then*

$$A_E(H) = \{y \mid y \in \mathscr{C}_0^1(\Omega) \text{ and } y' \in L^2(\Omega)\}$$

is a closed subspace of K_E. *Suppose* $y^{(k)} = A_E x^{(k)}$, *and suppose that* $\|y^{(k)} - g\|_E \to 0$ *as* $k \to \infty$. *Since*

$$\|y^{(k)} - g\|_E^2 = \|y^{(k)} - g\|^2 + \|y^{(k)\prime} - g'\|^2,$$

it follows that $y^{(k)} \to g$ *in* $L^2(\Omega)$ *and also that* $y^{(k)\,\prime} \to g'$ *in* $L^2(\Omega)$. *Note that*

$$\begin{aligned}
|y^{(k)}(t) - [g(t) - g(0)]| &\le \int_{(0,t)} |y^{(k)\prime}(s) - g'(s)| ds \\
&\le \left[\int_{(0,t)} |y^{(k)\prime}(s) - g'(s)|^2 ds\right]^{\frac{1}{2}} \left[\int_{(0,t)} 1^2 ds\right]^{\frac{1}{2}} \\
&\le \|y^{(k)\prime} - g'\|
\end{aligned}$$

for almost all $t \in [0,1]$, *and since we also know that* $\|y^{(k)\prime} - g'\| \to 0$ *as* $k \to \infty$ *it follows that* $y^{(k)}(t)$ *converges uniformly to* $g(t) - g(0)$. *Note also that* $y^{(k)}(1) = 0$ *for all* k, *and hence* $g(1) = g(0)$. *Because*

$$\begin{aligned}
\|y^{(k)} - [g - g(0)]\|^2 &= \int_{[0,1]} |y^{(k)}(t) - [g(t) - g(0)]|^2 dt \\
&\le \|y^{(k)\prime} - g'\|^2 \\
&\to 0
\end{aligned}$$

and $\|y^{(k)} - g\| \to 0$ *as* $k \to \infty$ *we know that* $g(0) = 0$. *If we set* $f = g'$, *then* $\mu(f) = 0$ *and hence* $A_E f = g$. *Therefore,* $g \in A_E(H)$, *and hence* $A_E(H)$ *is closed.*

We can use Fourier series to show that the ideas of Example 8.6 can also be expressed via an infinite matrix representation.

Example 8.7. *Let* $\Omega = [0,1]$ *and* $H = K = L^2(\Omega)$, *and let* $A \in \mathscr{L}(H,K)$ *be the mapping defined in Example* 8.6. *For each* $m = 0, \pm 1, \pm 2, \ldots$ *let* $e_m : [0,1] \to \mathbb{C}$ *be defined by the formula*

$$e_m(s) = e^{2m\pi i s}.$$

The functions $\{e_m\}_{m=-\infty}^{+\infty}$ *form an orthonormal basis for* $L^2(\Omega)$. *In Problem 8.17 we show that for each* $k = 1, 2, \ldots$ *the functions* $x^{(k)}$ *and* $y^{(k)}$ *given by*

$$x^{(k)}(s) = \begin{cases} 0 & \text{when } s \notin [\frac{1}{2}(1 - \frac{1}{k}), \frac{1}{2}(1 + \frac{1}{k})], \\ k & \text{otherwise} \end{cases}$$

and $y^{(k)} = Ax^{(k)}$ *can be represented by the Fourier series*

$$x^{(k)} = \sum_{m=-\infty}^{\infty} \xi_m^{(k)} e_m \quad \text{and} \quad y^{(k)} = \sum_{m=-\infty}^{\infty} \eta_m^{(k)} e_m,$$

where the coefficients for $x^{(k)}$ *are* $\xi_0^{(k)} = 1$ *and* $\xi_m^{(k)} = (-1)^m \frac{k}{m\pi} \sin \frac{m\pi}{k}$ *for* $m \neq 0$ *and those for* $y^{(k)}$ *are* $\eta_0^{(k)} = 0$ *and* $\eta_m^{(k)} = (-1)^m \frac{k}{2m^2\pi^2 i} \sin \frac{m\pi}{k}$ *for* $m \neq 0$*. Since*

$$Ae_0(t) = 0 \quad \text{and} \quad Ae_m(t) = \frac{1}{2m\pi i}[e_m(t) - e_0(t)]$$

for each $m = \pm 1, \pm 2, \ldots$ *it follows that the operator equation* $y^{(k)} = Ax^{(k)}$ *can be rewritten in matrix form as*

$$\begin{bmatrix} \eta_0^{(k)} \\ \eta_{-1}^{(k)} \\ \eta_1^{(k)} \\ \eta_{-2}^{(k)} \\ \eta_2^{(k)} \\ \vdots \end{bmatrix} = \begin{bmatrix} 0 & \frac{1}{2\pi i} & \frac{-1}{2\pi i} & \frac{-1}{4\pi i} & \frac{1}{4\pi i} & \cdots \\ 0 & \frac{-1}{2\pi i} & 0 & 0 & 0 & \cdots \\ 0 & 0 & \frac{1}{2\pi i} & 0 & 0 & \cdots \\ 0 & 0 & 0 & \frac{-1}{4\pi i} & 0 & \cdots \\ 0 & 0 & 0 & 0 & \frac{1}{4\pi i} & \cdots \\ \vdots & \vdots & \vdots & \vdots & \vdots & \ddots \end{bmatrix} \begin{bmatrix} \xi_0^{(k)} \\ \xi_{-1}^{(k)} \\ \xi_1^{(k)} \\ \xi_{-2}^{(k)} \\ \xi_2^{(k)} \\ \vdots \end{bmatrix}. \tag{8.12}$$

By taking the limit as $k \to \infty$ *we note that*

$$\xi_m^{(k)} \to \xi_m = (-1)^m \quad \text{and} \quad \eta_m^{(k)} \to \eta_m = (-1)^m \frac{1}{2m\pi i},$$

and hence the components of $x^{(k)}$ *and* $y^{(k)}$ *converge to well-defined limiting components. We also note that the limiting form of the corresponding matrix equation*

$$\begin{bmatrix} 0 \\ \frac{1}{2\pi i} \\ \frac{-1}{2\pi i} \\ \frac{-1}{4\pi i} \\ \frac{1}{4\pi i} \\ \vdots \end{bmatrix} = \begin{bmatrix} 0 & \frac{1}{2\pi i} & \frac{-1}{2\pi i} & \frac{-1}{4\pi i} & \frac{1}{4\pi i} & \cdots \\ 0 & \frac{-1}{2\pi i} & 0 & 0 & 0 & \cdots \\ 0 & 0 & \frac{1}{2\pi i} & 0 & 0 & \cdots \\ 0 & 0 & 0 & \frac{-1}{4\pi i} & 0 & \cdots \\ 0 & 0 & 0 & 0 & \frac{1}{4\pi i} & \cdots \\ \vdots & \vdots & \vdots & \vdots & \vdots & \ddots \end{bmatrix} \begin{bmatrix} 1 \\ -1 \\ -1 \\ 2 \\ 2 \\ \vdots \end{bmatrix} \tag{8.13}$$

seems meaningful at an intuitive level. Nevertheless, there is a serious problem. Although the vector g *on the left-hand side correctly represents the function*

$$g(t) = \begin{cases} -t & \text{if } t \in (0, \frac{1}{2}), \\ 1 - t & \text{if } t \in (\frac{1}{2}, 1) \end{cases}$$

in $K = L^2([0,1])$*, it is clear that the vector* f *on the right-hand side does not represent a function in* $H = L^2([0,1])$*. We may wonder how it is possible to obtain an equation such as* (8.13) *in which the left-hand side is well defined and the right-hand side is not. The answer lies in our failure to select appropriate measurement scales in the respective domain and range spaces to describe the operator* A*. From the infinite vector representation it is not difficult to see that*

$$\|y\|^2 = \sum_{m \neq 0} |\eta_m^{(k)}|^2 = \sum_{m \neq 0} \frac{|\xi_m|^2}{4m^2\pi^2} \leq \frac{1}{2\pi^2} \cdot \sum_{m=1}^{\infty} \frac{1}{m^2} \cdot \|x\|^2$$

and hence that

$$||Ax|| \le \frac{1}{2\sqrt{3}}||x||.$$

Thus the operator A is bounded above. However, we can also see that

$$Ae_m = \frac{1}{2m\pi i}e_m$$

for each natural number m. Therefore,

$$||Ae_m|| = \frac{1}{2m\pi},$$

and hence, for any $\epsilon > 0$, we can always find an element $x \in H$ for which $0 < ||Ax|| < \epsilon||x||$. Therefore, the operator A is not bounded below on $[A^{-1}(\{0\})]^{\perp}$. This means we can construct corresponding sequences $\{x^{(k)}\}$ and $\{y^{(k)}\} = \{Ax^{(k)}\}$ such that $||x^{(k)}|| \to \infty$ while $||y^{(k)}||$ remains bounded and for which there is some $g \in K$ such that $||y^{(k)} - g|| \to 0$ as $k \to \infty$. Thus $A(H)$ is not closed. This is precisely the difficulty arising in equation (8.13). The problem is resolved if we use a more restrictive definition of the image space. If we replace $K = L^2(\Omega)$ with the space $K_E = W^1(\Omega)$ where we only allow functions $y = \sum_{m=-\infty}^{\infty} \eta_m e_m$ for which

$$\sum_{m=-\infty}^{\infty} (1 + 4m^2\pi^2)|\eta_m|^2 < \infty$$

and where the inner product of $y = \sum_{m=-\infty}^{\infty} \eta_m e_m$ and $z = \sum_{m=-\infty}^{\infty} \zeta_m e_m$ is defined by

$$\langle y, z\rangle = \sum_{m=-\infty}^{\infty} (1 + 4m^2\pi^2)\eta_m\overline{\zeta_m},$$

then with the new measurement scale we can show that for each $y \in K_E$ the matrix equation $y = Ax$ given by

$$\begin{bmatrix} \eta_0 \\ \eta_{-1} \\ \eta_1 \\ \eta_{-2} \\ \eta_2 \\ \vdots \end{bmatrix} = \begin{bmatrix} 0 & \frac{1}{2\pi i} & \frac{-1}{2\pi i} & \frac{-1}{4\pi i} & \frac{1}{4\pi i} & \cdots \\ 0 & \frac{-1}{2\pi i} & 0 & 0 & 0 & \cdots \\ 0 & 0 & \frac{1}{2\pi i} & 0 & 0 & \cdots \\ 0 & 0 & 0 & \frac{-1}{4\pi i} & 0 & \cdots \\ 0 & 0 & 0 & 0 & \frac{1}{4\pi i} & \cdots \\ \vdots & \vdots & \vdots & \vdots & \vdots & \ddots \end{bmatrix} \begin{bmatrix} \xi_0 \\ \xi_{-1} \\ \xi_1 \\ \xi_{-2} \\ \xi_2 \\ \vdots \end{bmatrix}$$

has a unique solution $x \in H$.

8.4 ▪ Motivating Applications

There are two applications that attracted our attention. In the first instance we consider a general problem of input retrieval in infinite dimensional linear systems, and in the second instance we discuss mean transition times for a singularly perturbed Markov process. Note once again that the important general properties of infinite dimensional spaces are listed in Section 8.5.

8.4.1 ▪ Input retrieval in linear control systems

Let H be a Banach space and let $A \in \mathscr{L}(H)$ be a bounded linear map on H. Suppose there exists some $\omega > 0$ and further suppose that for each ϵ with $0 < \epsilon < \omega$ we can find $M_\epsilon > 0$ such that

$$||(sI - A)^{-1}|| \leq \frac{M_\epsilon}{|s|}$$

for all $s \in \mathbb{C}$ with $|\arg s| < \frac{\pi}{2} + \omega - \epsilon$. Then A generates a bounded holomorphic semigroup e^{At} in the region $|\arg t| < \omega$ and the resolvent of A is given by the formula

$$(sI - A)^{-1} = \int_0^\infty e^{-st} e^{At} dt$$

for $s \in \mathbb{C}$ with $\Re(s) > 0$. Thus the resolvent of A can be interpreted as the Laplace transform of the semigroup generated by A. The integral in the above expression is a Bochner integral (see the bibliographic notes). If $r_\sigma > 0$ is the spectral radius of A, then

$$(sI - A)^{-1} = \frac{1}{s}\left[I + \frac{A}{s} + \left(\frac{A}{s}\right)^2 + \cdots \right] \tag{8.14}$$

for all $s \in \mathbb{C}$ with $|s| > r_\sigma$ (see Problem 8.20). Now suppose that G and K are Banach spaces and that $B \in \mathscr{L}(G, H)$ and $C \in \mathscr{L}(H, K)$ are bounded linear transformations. Let $u : [0, \infty) \to G$ be an analytic function defined by

$$u(t) = u_0 + u_1 t + \frac{u_2 t^2}{2!} + \cdots$$

for all $t \in [0, \infty)$, where $\{u_j\} \subset G$ and $||u_j|| \leq a^{j+1}$ for some $a \in \mathbb{R}$ with $a > 0$. The Laplace transform of u will be

$$U(s) = \frac{1}{s}\left[u_0 + \frac{u_1}{s} + \frac{u_2}{s^2} + \cdots \right]$$

for $|s| > a$. We consider an infinite dimensional linear control system

$$x' = Ax + Bu,$$
$$y = Cx,$$

where $u = u(t)$ is the input, $x = x(t)$ is the state, and $y = y(t)$ is the output and where we assume that the system is initially at rest. Thus we assume $x(0) = 0$. If the input to the system is assumed to be analytic (as described above), it follows (see Problem 8.21) that the output from the system is determined by the formula

$$y(t) = \int_0^t Ce^{A(t-\tau)} Bu(\tau) d\tau \tag{8.15}$$

or equivalently by the corresponding Laplace transform formula

$$Y(s) = C(sI - A)^{-1} BU(s).$$

The latter formula will be well defined in the region $|s| > \max[r_\sigma, a]$ by the series expansion

$$Y(s) = \frac{1}{s}\left[CB + \frac{CAB}{s} + \frac{CA^2B}{s^2} + \cdots \right] U(s).$$

Thus the problem of input retrieval can be formulated as a power series inversion problem with

$$U(s) = s\left[CB + \frac{CAB}{s} + \frac{CA^2B}{s^2} + \dots\right]^{-1} Y(s).$$

If we write $z = 1/s$ and define $A_0 = CB$ and $A_1 = CAB$, then we can certainly find the desired inverse operator if we can find an expression for $(A_0 + A_1 z)^{-1}$ in some region $0 < |z| < r$. We are particularly interested in the case where $A_0 = CB$ is singular.

8.4.2 ▪ Singularly perturbed Markov processes

The intrinsic structure of a Markov process[8] can be substantially changed by a small perturbation. For instance, the perturbation may introduce state transitions that are not possible in the original unperturbed process. For a Markov process, defined by a transition kernel, T, it is known that the mean first passage times between states can be calculated by finding the linear operator

$$[I - T + T^\infty]^{-1},$$

where $T^\infty = \lim_{n\to\infty} T^n$ is suitably defined. We introduce the topic by considering a finite dimensional problem on a discrete state space and then move on to an analogous infinite dimensional problem on a continuous state space.

Let $T_\epsilon : \mathbb{R}^{1\times(r+1)} \to \mathbb{R}^{1\times(r+1)}$ be a perturbed transition kernel defined by the linear combination of transition kernels

$$T_\epsilon(\pi) = \pi[(1-\epsilon)I + \epsilon P],$$

where $I \in \mathbb{R}^{(r+1)\times(r+1)}$ is the identity matrix, $\epsilon \in (0,1]$ is the perturbation parameter, and

$$P = \begin{bmatrix} 1 & 0 & 0 & \cdots & 0 & 0 \\ \frac{1}{2} & \frac{1}{2} & 0 & \cdots & 0 & 0 \\ \frac{1}{3} & \frac{1}{3} & \frac{1}{3} & \cdots & 0 & 0 \\ \vdots & \vdots & \vdots & \ddots & \vdots & \vdots \\ \frac{1}{r} & \frac{1}{r} & \frac{1}{r} & \cdots & \frac{1}{r} & 0 \\ \frac{1}{r+1} & \frac{1}{r+1} & \frac{1}{r+1} & \cdots & \frac{1}{r+1} & \frac{1}{r+1} \end{bmatrix} = \begin{bmatrix} L_r & 0^T \\ \frac{1}{r+1}\mathbf{1} & \frac{1}{r+1} \end{bmatrix} \in \mathbb{R}^{(r+1)\times(r+1)},$$

where $0, \mathbf{1} \in \mathbb{R}^{1\times r}$, and we use the notation $0 = [0,\dots,0] \in \mathbb{R}^{1\times n}$ and $\mathbf{1} = [1,\dots,1] \in \mathbb{R}^{1\times n}$ for each $n \in \mathbb{N}$. The chain T_ϵ is a perturbation of the identity. It is a singular perturbation because the chain changes radically for $\epsilon \neq 0$. When $\epsilon = 0$ the transition kernel is simply an identity transformation, and the initial state does not change. If we regard the state space as the set of numbers

$$S = \left\{0, \frac{1}{r}, \frac{2}{r}, \dots, \frac{r-1}{r}, 1\right\},$$

then the perturbed transformation T_ϵ allows leakage back to the zero state. Indeed,

$$T_\epsilon^n(\pi) \to T_\epsilon^\infty(\pi) = e_1 \in \mathbb{R}^{1\times(r+1)}$$

as $n \to \infty$ for all probability vectors $\pi \in \mathbb{R}^{1\times(r+1)}$, where we use the notation $e_1 = [1,0,\dots,0] \in \mathbb{R}^{1\times n}$ for each $n \in \mathbb{N}$. Thus the invariant measure for the perturbed chain

[8]Because of a preference for operator notation consistent with functional analysis literature, the notation for Markov processes introduced here is independent of that used in earlier chapters. However, it is self-contained in this section.

T_ϵ lies entirely at zero. To find the fundamental matrix we must essentially solve the equation

$$[I - T_\epsilon + T_\epsilon^\infty](\xi) = \eta$$

for each $\eta \in \mathbb{R}^{1\times(r+1)}$. Define $T_0, T_1 : \mathbb{R}^{1\times(r+1)} \to \mathbb{R}^{1\times(r+1)}$ by setting $A_0(\xi) = \xi R_0$ and $A_1(\xi) = \xi R_1$, where

$$R_0 = \begin{bmatrix} 1 & 0 & \cdots & 0 \\ 1 & 0 & \cdots & 0 \\ \vdots & \vdots & \ddots & \vdots \\ 1 & 0 & \cdots & 0 \end{bmatrix} = [\mathbf{1}^T 0^T \cdots 0^T] \in \mathbb{R}^{(r+1)\times(r+1)} \quad \text{and} \quad R_1 = I - P,$$

where $\mathbf{1} \in \mathbb{R}^{1\times(r+1)}$ and $0 \in \mathbb{R}^{1\times(r+1)}$. The equation can now be rewritten as

$$(A_0 + \epsilon A_1)(\xi) = \eta, \tag{8.16}$$

where A_0 is a singular transformation. To solve the equation we decompose both ξ and η into two parts. If $M = A_0^{-1}(\{0\})$ is the null space of A_0 and $N = A_1(M)$ is the image of M under A_1, then we can define $\mu = \xi - \langle \xi, \mathbf{1}\rangle e_1 \in M$ and $\nu = \eta - \langle \nu, \mathbf{1}\rangle e_1 \in N$, where $\langle \cdot, \cdot \rangle$ denotes the usual Euclidean inner product. Hence we can write

$$\xi = \mu + \langle \xi, \mathbf{1}\rangle e_1 \quad \text{and} \quad \eta = \nu + \langle \nu, \mathbf{1}\rangle e_1,$$

where $\langle \xi, \mathbf{1}\rangle e_1 \in M^c$ and $\langle \nu, \mathbf{1}\rangle e_1 \in N^c$. Our single equation (8.16) now generates two separate equations

$$\langle \xi, \mathbf{1}\rangle e_1 R_0 = \langle \nu, \mathbf{1}\rangle e_1 \quad \text{and} \quad \epsilon \mu R_1 = \nu.$$

If we define

$$Q = \begin{bmatrix} 1 & 1 & \frac{1}{2} & \cdots & \frac{1}{r-1} & \frac{1}{r} \\ 0 & 0 & \frac{1}{2} & \cdots & \frac{1}{r-1} & \frac{1}{r} \\ 0 & 0 & 0 & \cdots & \frac{1}{r-1} & \frac{1}{r} \\ \vdots & \vdots & \vdots & \ddots & \vdots & \vdots \\ 0 & 0 & 0 & \cdots & 0 & \frac{1}{r} \\ 0 & 0 & 0 & \cdots & 0 & 0 \end{bmatrix} = \begin{bmatrix} e_1^T & L_r^T \\ 0 & 0 \end{bmatrix} \in \mathbb{R}^{(r+1)\times(r+1)},$$

then the full solution can be written as

$$\xi = \mu + \langle \nu, \mathbf{1}\rangle e_1, \quad \text{where} \quad \mu = \frac{1}{\epsilon}[-\langle \nu(I-Q), \mathbf{1}\rangle e_1 + \nu(I-Q)],$$

which clearly has a pole of order 1 at $\epsilon = 0$. If the operator $T : \mathbb{R}^{1\times(r+1)} \to \mathbb{R}^{1\times(r+1)}$ is defined by $T(\pi) = \pi P$, then we have the transition formula

$$[T\pi]_j = \sum_{k=j}^{r} \frac{\pi_k}{k+1}$$

for each $j = 0, 1, \ldots, r$. We want to write the formula in a different way. If we define the cumulative probability by setting $\xi_0 = 0$ and $\xi_j = \pi_0 + \pi_1 + \cdots + \pi_{j-1}$ for $1 \le j \le r+1$, then summing the above equations gives

$$[T\xi]_j = \xi_j + (j+1) \sum_{k=j+1}^{r} \frac{\Delta \xi_k}{k+1}.$$

We are now able to consider an analogous infinite dimensional problem. Let $X = \mathscr{C}([0,1])$ be the Banach space of continuous complex-valued functions on $[0,1]$ and $X^* = \text{rca}([0,1])$ be the corresponding adjoint space of regular countably additive complex-valued measures on $[0,1]$. Define a continuous state Markov process $T : X^* \to X^*$ by the formula

$$T\xi([0,t]) = \xi([0,t]) + t\int_{(t,1]} \frac{d\xi([0,s])}{s}$$

for $t \in [0,1)$ with $T\xi([0,1]) = \xi([0,1])$. Consider the transformation $T_\epsilon : X^* \to X^*$ defined by

$$T_\epsilon = (1-\epsilon)I + \epsilon T,$$

where $I : X^* \to X^*$ is the identity transformation. Once again the transformation T_ϵ is a perturbation of the identity that allows a small probability of transition between states. Mean transition times are determined by the operator

$$[I - T_\epsilon + T_\epsilon^\infty]^{-1},$$

where $T_\epsilon^\infty = \lim_{n\to\infty} T_\epsilon^n$, and intuitively we expect these times to increase as ϵ decreases to zero. We can see that

$$dT\xi([0,t]) = \left(\int_{(t,1]} \frac{d\xi([0,s])}{s}\right) dt,$$

and if we define $E : X \to X$ by setting

$$E\varphi(s) = \frac{1}{s}\int_{[0,s]} \varphi(t)dt$$

for each $\varphi \in X$, then it follows that

$$\begin{aligned}
\langle T\xi, \varphi\rangle &= \int_{[0,1]} \overline{\varphi(t)} dT\xi([0,t]) \\
&= \int_{[0,1]} \overline{\varphi(t)} \left(\int_{(t,1]} \frac{d\xi([0,s])}{s}\right) dt \\
&= \int_{[0,1]} \left(\frac{1}{s}\int_{[0,s]} \overline{\varphi(t)} dt\right) d\xi([0,s]) \\
&= \int_{[0,1]} \overline{E\varphi(s)} d\xi([0,s]) \\
&= \langle \xi, E\varphi\rangle.
\end{aligned}$$

Thus $T = E^*$. For each $n = 0, 1, \ldots$ it is not difficult to show that

$$E^{n+1}\varphi(s) = \int_{[0,s]} w_n(s,t)\varphi(t)dt,$$

where

$$w_n(s,t) = \frac{1}{n!}\frac{[\ln(s/t)]^n}{s}.$$

Note that $w_n(s,t) \geq 0$ for $t \in (0,s]$ with

$$\int_{[0,s]} w_n(s,t)dt = 1$$

and that $w_n(s,t) \downarrow 0$ uniformly in t for $t \in [\sigma, s]$ for each $\sigma > 0$ as $n \to \infty$. It follows that $E^{n+1}\varphi(s) \to \varphi(0)\chi_{[0,1]}(s)$ for each $s \in [0,1]$, where we have written $\chi_{[0,1]}$ for the characteristic function of the interval $[0,1]$. Hence we deduce that

$$\langle T^{n+1}\xi, \varphi \rangle = \langle \xi, E^{n+1}\varphi \rangle \to \xi([0,1])\overline{\varphi(0)}$$

for each $\varphi \in X$. If we define the Dirac measure $\delta \in X^*$ by the formula $\langle \delta, \varphi \rangle = \overline{\varphi(0)}$, then we can say that

$$T^{n+1}\xi \to T^{\infty}\xi = \xi([0,1])\delta$$

in the weak* sense. Let $\varphi \in X$ be any fixed test function, and let τ be a positive real number. We can find $N \in \mathbf{N}$ such that

$$|\langle T^k \xi, \varphi \rangle - \xi([0,1])\overline{\varphi(0)}| < \tau$$

for all $k \geq N+1$. It follows that

$$
\begin{aligned}
&|\langle T_\epsilon^{n+1}\xi, \varphi \rangle - \xi([0,1])\overline{\varphi(0)}| \\
&\leq \sum_{k=0}^{n+1} \binom{n+1}{k} (1-\epsilon)^{n+1-k}\epsilon^k |\langle T^k\xi, \varphi \rangle - \xi([0,1])\overline{\varphi(0)}| \\
&\leq \sum_{k=0}^{N} \binom{n+1}{k} (1-\epsilon)^{n+1-k}\epsilon^k |\langle T^k\xi, \varphi \rangle - \xi([0,1])\overline{\varphi(0)}| \\
&\qquad + \sum_{k=N+1}^{n} \binom{n+1}{k} (1-\epsilon)^{n+1-k}\epsilon^k \tau \\
&\leq \sum_{k=0}^{N} \binom{n+1}{k} (1-\epsilon)^{n+1-k}\epsilon^k |\langle T^k\xi, \varphi \rangle - \xi([0,1])\overline{\varphi(0)}| + \tau
\end{aligned}
$$

and hence

$$\limsup_{n \to \infty} |\langle T_\epsilon^{n+1}\xi, \varphi \rangle - \xi([0,1])\overline{\varphi(0)}| \leq \tau.$$

Since τ is arbitrary, it follows that $\langle T_\epsilon^{n+1}\xi, \varphi \rangle \to \xi([0,1])\overline{\varphi(0)}$ for each $\varphi \in X$. Thus we also have

$$T_\epsilon^{n+1}\xi \to T_\epsilon^{\infty}\xi = \xi([0,1])\delta$$

in the weak* sense. Hence we have $T_\epsilon^{\infty} = T^{\infty}$. The equation

$$[I - T_\epsilon + T_\epsilon^{\infty}]\xi = \eta$$

can be rewritten as

$$[T^{\infty} + \epsilon(I - T)]\xi = \eta,$$

and if we set $A_0 = T^{\infty}$ and $A_1 = I - T$, then it takes the form

$$(A_0 + A_1\epsilon)\xi = \eta, \tag{8.17}$$

where A_0 is singular. The null space of A_0 is given by

$$M = A_0^{-1}(\{0\}) = \{\mu \mid \mu([0,1]) = 0\},$$

and the projection $P_M : X^* \to X^*$ is defined by

$$\mu = P_M \xi = \xi - \xi([0,1])\delta$$

for each $\xi \in X^*$. We wish to find a simple description for the space $N = A_1(M)$. On the one hand, if $\nu = (I-T)\mu$, then

$$\langle \nu, \varphi \rangle = \langle \mu, \varphi - E\varphi \rangle$$

for $\varphi \in X$. It follows that

$$\nu([0,1]) = \langle \nu, \chi_{[0,1]} \rangle = \langle \mu, \chi_{[0,1]} - E\chi_{[0,1]} \rangle = 0$$

since $E\chi_{[0,1]} = \chi_{[0,1]}$. On the other hand, suppose $\nu([0,1]) = 0$. If we set $\psi = \varphi - E\varphi \in X$, then $\psi \in X$ and $\psi(0) = 0$. By solving an elementary differential equation it can be seen that $\varphi - E\varphi(1)\chi_{[0,1]} = \psi - F\psi$, where

$$F\psi(s) = \int_{(s,1]} \frac{\psi(t)}{t} dt.$$

Note that $F\psi(0) = E\varphi(1) - \varphi(0)$ is well defined. Define $\langle \mu, \psi \rangle = \langle \nu, \psi - F\psi \rangle$ for each $\psi \in X$ with $\psi(0) = 0$. Since $\langle \nu, \chi_{[0,1]} \rangle = 0$, we deduce that

$$\langle \nu, \varphi \rangle = \langle \nu, \varphi - E\varphi(1)\chi_{[0,1]} \rangle = \langle \nu, \psi - F\psi \rangle = \langle \mu, \psi \rangle = \langle \mu, \varphi - E\varphi \rangle$$

for each $\varphi \in X$. Therefore, $\nu = (I-T)\mu$, and hence

$$N = A_1(M) = \{\nu \mid \nu([0,1]) = 0\},$$

and the projection $Q_N : X^* \to X^*$ is defined by

$$\nu = Q_N \eta = \eta - \eta([0,1])\delta$$

for each $\eta \in X^*$. By applying an appropriate decomposition to equation (8.17) with $\mu = P_M \xi \in M$ and $\nu = Q_N \eta \in N$ and by noting that $T^\infty \delta = \delta$ and $(I-T)\mu = \mu(I-E)$, we obtain

$$[T^\infty + \epsilon(I-T)](\mu + \xi([0,1])\delta) = \epsilon\mu(I-E) + \xi([0,1])\delta$$

on the left-hand side and

$$\eta = \nu + \eta([0,1])\delta$$

on the right-hand side. By equating corresponding terms we have

$$\epsilon\mu(I-E) = \nu$$

and

$$\xi([0,1])\delta = \eta([0,1])\delta.$$

The former equation means that $\epsilon\langle \mu, \varphi - E\varphi \rangle = \langle \nu, \varphi \rangle$ for each $\varphi \in X$ and could be rewritten in the form $\epsilon\langle \mu, \psi \rangle = \langle \nu, \psi - F\psi \rangle$ for each $\psi \in X$ with $\psi(0) = 0$. Thus

$$\epsilon\mu = \nu(I-F).$$

Since $\xi = \mu + \xi([0,1])\delta$, the solution is given by

$$\begin{aligned}\xi &= \frac{1}{\epsilon}\nu(I-F) + \eta([0,1])\delta \\ &= \frac{1}{\epsilon}Q_N\eta(I-F) + (I-Q_N)\eta.\end{aligned}$$

As expected there is a pole of order one at $\epsilon = 0$.

8.5 ▪ Review of Banach and Hilbert Spaces

Although the two previous sections have made some reference to infinite dimensional spaces, the focus has been on particular examples and applications that could be used to motivate a more systematic discussion of operator perturbation. Before we begin such a discussion it is convenient to review some of the more important general properties of infinite dimensional spaces. More details can be found in Chapter 9 and in a number of excellent texts mentioned in the bibliographic notes at the end of the chapter.

8.5.1 ▪ Banach spaces

A normed linear space is a vector space X with a real-valued function $||\cdot|| : X \to [0,\infty)$ called the norm such that

1. $||x|| \geq 0$ for all $x \in X$ with $||x|| = 0$ if and only if $x = 0$;
2. $||x+y|| \leq ||x|| + ||y||$ for all $x, y \in X$; and
3. $||\alpha x|| = |\alpha| \cdot ||x||$ for all $x \in X$ and $\alpha \in \mathbb{C}$.

We will not present an in-depth treatment of normed linear space theory, but we will define the relevant terminology and summarize the important theorems that underlie our methods. The importance of normed linear spaces in mathematical analysis is closely linked to the existence of bounded linear functionals. In this regard the Hahn–Banach theorem is fundamental.

Theorem 8.5 (Hahn–Banach). *Let X be a normed linear space, and let f be a bounded linear functional defined on a subspace M of X satisfying $f(m) \leq k \cdot ||m||$ for some $k \in (0,\infty)$ and all $m \in M$. Then there is an extension F of f from M to X such that $F(x) \leq k \cdot ||x||$ for all $x \in X$.*

Definition 8.1. *An infinite sequence $\{x_n\}_{n\in\mathbb{N}}$ in a normed linear space X is said to converge to a vector $x \in X$ if $||x_n - x|| \to 0$ as $n \to \infty$.*

Definition 8.2. *A sequence $\{x_n\}_{n\in\mathbb{N}}$ in a normed space is said to be a Cauchy sequence if $||x_n - x_m|| \to 0$ as $m, n \to \infty$. That is, given $\delta > 0$, there is a number $N = N(\delta) \in \mathbb{N}$ such that $||x_n - x_m|| < \delta$ for all $m, n > N$.*

Definition 8.3. *A normed linear space X is said to be complete if every Cauchy sequence $\{x_n\}_{n\in\mathbb{N}}$ in X converges to a limit $x \in X$. A normed linear space X that is complete is called a Banach space. If X is a Banach space, it can be shown that the space X^* of all bounded linear functionals $f : X \to \mathbb{R}$ is also a Banach space. We will say that X^* is the dual space to X, and we will normally use the notation $x^* \in X^*$ to denote the elements of X^*.*

Definition 8.4. *Let X and Y be Banach spaces, and let $A \in \mathscr{L}(X,Y)$. Let X^* and Y^* denote the dual spaces. The adjoint operator $A^* : Y^* \to X^*$ is defined by the equation*

$$\langle x, A^* y^* \rangle = \langle Ax, y \rangle,$$

where we have used the notation $\langle x, x^* \rangle$ to denote the value at the point $x \in X$ of the linear functional $x^* \in X^*$. An alternative equivalent notation is $\langle x, x^* \rangle = x^*(x)$.

It can be shown that $A^* \in \mathscr{L}(Y^*, X^*)$ is a bounded linear operator. The notion of orthogonality in Banach space can be introduced as a relationship between the original Banach space X and the dual space X^*.

Definition 8.5. *The vectors $x \in X$ and $x^* \in X^*$ are said to be orthogonal if $\langle x, x^* \rangle = 0$. If $S \subset X$, then the orthogonal complement $S^\perp \subset X^*$ of S is the set of all $x^* \in X^*$ such that $\langle s, x^* \rangle = 0$ for all $s \in S$.*

If X and Y are Banach spaces and $A \in \mathscr{L}(X,Y)$, then for each subset $S \subset Y$ we use the notation

$$A^{-1}(S) = \{x \mid Ax \in S\}$$

for the inverse image of S under A.

Theorem 8.6. *Let X and Y be Banach spaces, and let $A \in \mathscr{L}(X,Y)$. Then*

$$[A(X)]^\perp = [A^*]^{-1}(\{0\}).$$

Proof: Let $y^* \in [A^*]^{-1}(\{0\})$, and let $y \in A(X)$. Then $A^* y^* = 0$ and $y = Ax$ for some $x \in X$, and the equality $0 = \langle x, A^* y^* \rangle = \langle Ax, y^* \rangle = \langle y, y^* \rangle$ shows that $y^* \in A(X)^\perp$. Thus $[A^*]^{-1}(\{0\}) \subset [A(X)]^\perp$. On the other hand, if $y^* \in [A(X)]^\perp$, then for every $x \in X$ we have $\langle Ax, y^* \rangle = 0$. Hence $\langle x, A^* y^* \rangle = 0$ and thus $A(X)]^\perp \subset [A^*]^{-1}(\{0\})$. □

The open mapping theorem is a cornerstone of modern analysis. It provides a topological characterization for a bounded linear mapping of X onto Y.

Theorem 8.7 (Banach). *Let X and Y be Banach spaces, and let $A \in \mathscr{L}(X,Y)$. If $A(X) = Y$, then A maps every open set $U \subseteq X$ onto an open set $V = A(U) \subseteq Y$.*

Although it is essentially a corollary to the open mapping theorem, the Banach inverse theorem is an equally important and celebrated result. It tells us that if a bounded linear mapping is invertible, then the inverse mapping is also a bounded linear map.

Theorem 8.8 (Banach). *Let X and Y be Banach spaces, and let $A \in \mathscr{L}(X,Y)$. If A is a one-to-one mapping of X onto Y, then the inverse mapping $A^{-1} \in \mathscr{L}(Y,X)$.*

Corollary 8.1. *Let X and Y be Banach spaces, and let $A \in \mathscr{L}(X,Y)$. Assume that $A(X)$ is closed. Then there is a constant $\epsilon > 0$ such that for each $y \in A(X)$ we can find $x \in A^{-1}(\{y\})$ satisfying $||y|| \geq \epsilon ||x||$.*

The next result is the dual of Theorem 8.6, but it is much deeper. The proof depends on both the Banach inverse theorem and the Hahn–Banach theorem.

Theorem 8.9. *Let X and Y be Banach spaces, and let $A \in \mathscr{L}(X,Y)$. If $A(X)$ is closed, then $A^*(Y) = [A^{-1}(\{0\})]^\perp$.*

8.5.2 ▪ Hilbert spaces

A linear space H with a complex-valued inner product $\langle \cdot,\cdot \rangle : H \times H \to \mathbb{C}$ such that

1. $\langle x,y\rangle = \overline{\langle y,x\rangle}$ for all $x,y \in H$,
2. $\langle x+y,z\rangle = \langle x,z\rangle + \langle y,z\rangle$ for all $x,y,z \in H$,
3. $\langle \alpha x,y\rangle = \alpha\langle x,y\rangle$ for all $x,y \in H$ and $\alpha \in \mathbb{C}$, and
4. $\langle x,x\rangle \geq 0$ and $\langle x,x\rangle = 0$ if and only if $x = 0$

is called an inner product space. The associated norm $||\cdot|| : H \to [0,\infty)$ is defined by $||x|| = \langle x,x\rangle^{1/2}$ for each $x \in H$. An inner product space that is complete is called a Hilbert space. The fundamental structural theorem in Hilbert space is the projection theorem.

Theorem 8.10. *Let H be a Hilbert space, and let $M \subseteq H$ be a closed subspace of H. For each $x \in H$ there is a unique element $x_M \in M$ such that $||x - x_M|| \leq ||x - m||$ for all $m \in M$. Furthermore, $\langle x - x_M, m\rangle = 0$ for all $m \in M$.*

The projection theorem allows us to decompose each element into complementary orthogonal components.

Definition 8.6. *Let H be a Hilbert space, and let $M \subseteq H$ be a closed subspace of H. For each $x \in H$ let $x_M \in M$ be the unique projection of x into M. Let $P_M : H \mapsto M \subseteq H$ be defined by $P_M x = x_M$ for all $x \in H$. The operator $P_M \in \mathscr{L}(H)$ is called the projection operator onto the closed subspace M, and each $x \in H$ can be written in the form $x = P_M x + (I - P_M)x = x_M + x_M^{\perp}$. The operator $P_{M^\perp} = I - P_M \in \mathscr{L}(H)$ is the projection operator onto the closed subspace $M^\perp$.*

We have the following useful corollary to the projection theorem.

Corollary 8.2. *Let H be a Hilbert space, and let M be a closed subspace. Each vector $x \in H$ can be written uniquely in the form $x = x_M + x_M^{\perp}$. Furthermore,*

$$\langle x,u\rangle_H = \langle x_M, u_M\rangle_H + \langle x_M^{\perp}, u_M^{\perp}\rangle_H$$

for each $x,u \in H$. We will say that H is the direct sum of M and $M^\perp$, and we will write $H = M \oplus M^\perp$.

A final important result for Hilbert spaces is the Riesz–Fréchet representation theorem. For each $y \in H$ the functional $f_y : H \to \mathbb{C}$ defined by $f_y(x) = \langle x,y\rangle$ is a bounded linear functional on H. It can be shown that $||f_y|| = ||y||$ and that all linear functionals on H take this form.

Theorem 8.11. *If $f : H \to \mathbb{C}$ is a bounded linear functional, there exists a unique vector $y \in H$ such that for all $x \in H$ we have $f(x) = \langle x,y\rangle$. Furthermore, $||f|| = ||y||$, and every $y \in H$ determines a unique bounded linear functional this way.*

This theorem allows us to argue that $H^* \cong H$. In particular it shows us that if H, K are Hilbert spaces over the field of complex numbers and if $A \in \mathscr{L}(H,K)$ is a bounded linear operator, then the Hilbert space adjoint operator $A^* \in \mathscr{L}(K,H)$ is defined by the relationship $\langle Ax,y\rangle_K = \langle x,A^*y\rangle_H$ for each $x \in H$ and $y \in K$. Now if M is a closed subspace of H, we can write $H = M \oplus M^\perp$, and if $P_M, P_{M^\perp} \in \mathscr{L}(H)$ are the corresponding projection operators, then

$$\langle P_M u, v\rangle = \langle P_M u, P_M v + P_{M^\perp} v\rangle = \langle P_M u, P_M v\rangle = \langle P_M u + P_{M^\perp} u, P_M v\rangle = \langle u, P_M v\rangle$$

for each $u, v \in H$, and hence $P_M^* = P_M$. That is, the projection operator $P_M \in \mathscr{L}(H)$ is self-adjoint.

8.6 ▪ Inversion of Linearly Perturbed Operators on Hilbert Spaces

Let H and K be Hilbert spaces, and consider bounded but not necessarily compact linear operators $A_0 \in \mathscr{L}(H,K)$ and $A_1 \in \mathscr{L}(H,K)$. Let $A(z) = A_0 + A_1 z$ be a linear perturbation of A_0 that depends on a single complex parameter $z \in \mathbb{C}$. When A_0 is nonsingular the Neumann expansion (see the bibliographic notes) can be used to calculate $(A_0 + A_1 z)^{-1}$.

Lemma 8.12 (Neumann). *Let $A_0 \in \mathscr{L}(H,K)$ and $A_1 \in \mathscr{L}(H,K)$, and suppose that A_0 is* one-to-one *with $A_0(H) = K$. Thus we suppose A_0^{-1} is well defined. Let $A(z) = A_0 + A_1 z$, where $z \in \mathbb{C}$. Then for some $b > 0$ we have that $A(z)^{-1}$ is well defined for $|z| < b$ with*

$$A(z)^{-1} = \sum_{j=0}^{\infty} (-1)^j (A_0^{-1} A_1)^j A_0^{-1} z^j.$$

When A_0 is singular we consider three different situations:

- A_0 is not one-to-one.
- $A_0(H)$ is closed but $A_0(H) \neq K$.
- $A_0(H)$ is not closed.

We outline the procedure when A_0 is not one-to-one. Let $M = A_0^{-1}(\{0\})$ and $N = A_1(M)$. If there is some $z_0 \neq 0$ for which $A(z_0)^{-1}$ is well defined, then calculation of $(A_0 + A_1 z)^{-1} \in \mathscr{L}(K,H)$ can be reduced to a term in z^{-1} plus a similar projected calculation of $(A_{0,22} + A_{1,22} z)^{-1} \in \mathscr{L}(N^\perp, M^\perp)$ where $A_{0,22}, A_{1,22} \in \mathscr{L}(M^\perp, N^\perp)$. If $A_{0,22}$ is nonsingular, the Neumann expansion can be applied to the projected problem and the original inverse can be represented on a region $0 < |z| < b$ by a convergent Laurent series with a pole of order 1 at the origin. If $A_{0,22}$ is not one-to-one, then the reduction procedure can be applied again. Thus the procedure is essentially recursive. If the procedure terminates after a finite number of steps, then the inverse operator $A(z)^{-1}$ is defined on some region $0 < |z| < b$ by a convergent Laurent series with a finite order pole. It is possible that the procedure will not terminate and that a general Laurent series representation may not be found. The other cases described above are manipulated so that a similar reduction procedure can be used. The method is not restricted to Fredholm operators.

We also consider unbounded operators. When $A_0 : \mathscr{D}(A_0) \subset H \to K$ is a densely defined and closed unbounded linear operator we show that by changing to a standard Sobolev topology on H we can replace A_0 by a bounded operator and apply the previous results. Several pertinent examples will be presented.

8.6.1 ▪ The unperturbed mapping is not one-to-one

We assume A_0 is not one-to-one. Thus A_0 is singular. The following lemma establishes the basis for the inversion procedure.

Lemma 8.13. *Let H and K be Hilbert spaces, and let $A_0, A_1 \in \mathcal{L}(H,K)$ be bounded linear maps. For each $z \in \mathbb{C}$ define $A(z) \in \mathcal{L}(H,K)$ by $A(z) = A_0 + A_1 z$. Suppose $M = A_0^{-1}(\{0\}) \neq \{0\}$, and let $N = A_1(M) \subset K$. If $A(z_0)^{-1}$ is well defined for some $z_0 \neq 0$, then A_1 is bounded below on M and N is a closed subspace of K.*

Proof: By the Banach inverse theorem the map $(A_0 + A_1 z_0)$ is bounded below on H. Therefore, we can find $\epsilon > 0$ such that

$$||(A_0 + A_1 z_0)x|| \geq \epsilon ||x||$$

for all $x \in H$. Since $A_0 m = 0$, it follows that

$$||A_1 m|| \geq \frac{\epsilon}{|z_0|} ||m||$$

for all $m \in M$. If $\{n_r\}$ is a Cauchy sequence in $N = A_1(M)$, then $n_r = A_1 m_r$, where $\{m_r\}$ is a corresponding sequence in M. Because A_1 is bounded below on M, the sequence $\{m_r\}$ must also be a Cauchy sequence. If $m_r \to m$ and $n_r \to n$, then $A_1 m = n$. Thus $n \in A_1(M) = N$. □

The orthogonal decomposition

Since $M = A_0^{-1}(\{0\})$ is closed and since the orthogonal complement $M^\perp$ is also closed, it follows that $H_1 = M$ and $H_2 = M^\perp$ are each Hilbert spaces. Let $P \in \mathcal{L}(H,H)$ denote the natural projection onto the subspace $M \subset H$, and define associated self-adjoint mappings $P_i \in \mathcal{L}(H,H_i)$ for $i = 1,2$ by setting $P_1 = P$ and $P_2 = I - P$. Define $R \in \mathcal{L}(H, H_1 \times H_2)$ by the formula

$$Rx = \begin{pmatrix} P_1 x \\ P_2 x \end{pmatrix}$$

for each $x \in H$. Note that $R^* = (P_1, P_2) \in \mathcal{L}(H_1 \times H_2, H)$. Since $\langle Rx_1, Rx_2 \rangle = \langle x_1, x_2 \rangle$ for each $x_1, x_2 \in H$ the mapping R defines a unitary equivalence between H and $H_1 \times H_2$. In the same way note that $N = A_1(M)$ is closed, and since $N^\perp$ is also closed, it follows that $K_1 = N$ and $K_2 = N^\perp$ are each Hilbert spaces. Let $Q \in \mathcal{L}(K,K)$ denote the natural projection onto the subspace $N \subset K$, and define associated self-adjoint mappings $Q_j \in \mathcal{L}(K,K_j)$ for $j = 1,2$ by setting $Q_1 = Q$ and $Q_2 = I - Q$. Define $S \in \mathcal{L}(K, K_1 \times K_2)$ by the formula

$$Sy = \begin{pmatrix} Q_1 y \\ Q_2 y \end{pmatrix}$$

for each $y \in K$. Note also that $S^* = (S_1, S_2)$. The mapping S defines a unitary equivalence between K and $K_1 \times K_2$. Now partition the operators A_0 and A_1 in the form

$$SA_0R^* = \begin{pmatrix} 0 & A_{0,12} \\ 0 & A_{0,22} \end{pmatrix} \quad \text{and} \quad SA_1R^* = \begin{pmatrix} A_{1,11} & A_{1,12} \\ 0 & A_{1,22} \end{pmatrix},$$

where $A_{0,ij}, A_{1,ij} \in \mathcal{L}(H_i, K_j)$ and where we note that $A_{0,11} = Q_1 A_0 P_1 = 0$, $A_{0,12} = Q_1 A_0 P_2$, $A_{0,21} = Q_2 A_0 P_1 = 0$, $A_{0,22} = Q_2 A_0 P_2$, $A_{1,11} = Q_1 A_1 P_1$, $A_{1,12} = Q_1 A_1 P_2$, $A_{1,21} = Q_2 A_1 P_1 = 0$, and $A_{1,22} = Q_2 A_1 P_2$.

Remark 8.1. *Recall that if A_0 is not one-to-one and $(A_0 + A_1 z_0)^{-1}$ exists for some $z_0 \in \mathbb{C}$ with $z_0 \neq 0$, then A_1 is bounded below on H_1. Equivalently we can say that $A_{1,11} \in \mathcal{L}(H_1, K_1)$ is bounded below. It follows that $A_{1,11}$ is a one-to-one mapping of H_1 onto K_1.*

The basic inversion formula

We use the notation introduced above.

Theorem 8.14. *Let $A_0 \in \mathcal{L}(H,K)$ with $H_1 = A_0^{-1}(\{0\}) \neq \{0\}$. Suppose $A_{1,11} \in \mathcal{L}(H_1,K_1)$ is a one-to-one mapping of H_1 onto $K_1 = A_1(H_1)$. The mapping $A(z) \in \mathcal{L}(H,K)$ is a one-to-one mapping of H onto K if and only if $z \neq 0$ and $(A_{0,22}+A_{1,22}z) \in \mathcal{L}(H_2,K_2)$ is a one-to-one mapping of $H_2 = H_1^{\perp}$ onto $K_2 = K_1^{\perp}$. In this case*

$$\begin{aligned} A(z)^{-1} &= P_1 S A_{1,11}^{-1} Q_1/z \\ &\quad + \left[P_2 - P_1 A_{1,11}^{-1}(A_{0,12}+A_{1,12}z)/z\right](A_{0,22}+A_{1,22}z)^{-1}Q_2. \end{aligned} \tag{8.18}$$

Proof: Since

$$A(z) = S^* \begin{pmatrix} A_{1,11}z & A_{0,12}+A_{1,12}z \\ 0 & A_{0,22}+A_{1,22}z \end{pmatrix} R,$$

where R and S are unitary operators, it follows that $A(z)^{-1}$ exists if and only if

$$\begin{pmatrix} A_{1,11}z & A_{0,12}+A_{1,12}z \\ 0 & A_{0,22}+A_{1,22}z \end{pmatrix}^{-1}$$

exists. Let $x = R\xi$ and $y = S\eta$. The system of equations $A(z)x = y$ has a unique solution $x \in H$ for each $y \in K$ if and only if the system of equations

$$\begin{aligned} (A_{1,11}z)\xi_1 + (A_{0,12}+A_{1,12}z)\xi_2 &= \eta_1, \\ (A_{0,22}+A_{1,22}z)\xi_2 &= \eta_2 \end{aligned}$$

has a unique solution $\xi \in H_1 \times H_2$ for each $\eta \in K_1 \times K_2$. The latter system can be rewritten as

$$\begin{aligned} (A_{0,22}+A_{1,22}z)\xi_2 &= \eta_2, \\ (A_{1,11}z)\xi_1 &= \eta_1 - (A_{0,12}+A_{1,12}z)\xi_2, \end{aligned}$$

and so there is a unique solution if and only if $z \neq 0$ and $A_{1,11}$ is a one-to-one mapping of H_1 onto K_1 and $(A_{0,22}+A_{1,22}z)$ is a one-to-one mapping of H_2 onto K_2. Therefore,

$$\begin{aligned} \xi_2 &= (A_{0,22}+A_{1,22}z)^{-1}\eta_2, \\ \xi_1 &= A_{1,11}^{-1}\left[\eta_1 - (A_{0,12}+A_{1,12}z)\xi_2\right]/z, \end{aligned}$$

and hence, by back substitution, $x = P_1\xi_1 + P_2\xi_2$ gives

$$x = \left\{P_1 A_{1,11}^{-1} Q_1/z + \left[P_2 - P_1 A_{1,11}^{-1}(A_{0,12}+A_{1,12}z)/z\right](A_{0,22}+A_{1,22}z)^{-1}Q_2\right\}y.$$

Thus we obtain the desired formula for $A(z)^{-1}$. □

Remark 8.2. *If $A_{0,22} \in \mathcal{L}(H_2,K_2)$ is a one-to-one mapping of H_2 onto K_2, then $A_{0,22}{}^{-1}$ is well defined and for some real number $b > 0$ the operator $(A_{0,22}+A_{1,22}z) \in \mathcal{L}(H_2,K_2)$ is defined by a convergent Neumann series in the region $|z| < b$. Thus the operator $A(z)^{-1}$ is defined in the region $0 < |z| < b$ by a convergent Laurent series with a pole of order 1 at $z = 0$.*

We illustrate our results with some examples.

Example 8.8 (discrete spectrum). *Each element in the space $L^2([0,1])$ can be represented by a Fourier series and defined by a countably infinite discrete spectrum. A bounded linear operator on any subspace of $L^2([0,1])$ can be regarded as a linear transformation on a discrete spectrum. Let $H = H^2([0,1]) \cap H_0^1([0,1])$ be the Hilbert space of measurable functions $x : [0,1] \to \mathbb{C}$ with*

$$\int_{[0,1]} \left[|x(t)|^2 + |x'(t)|^2 + |x''(t)|^2\right] dt < \infty,$$

and $x(0) = x(1) = 0$ and with inner product given by

$$\langle x_1, x_2 \rangle_H = \int_{[0,1]} \left[x_1(t)\overline{x}_2(t) + x_1{}'(t)\overline{x}_2{}'(t) + x_1{}''(t)\overline{x}_2{}''(t)\right] dt.$$

Let $K = L^2([0,1])$ be the Hilbert space of measurable functions $y : [0,1] \to \mathbb{C}$. Define $A_0, A_1 \in \mathcal{L}(H,K)$ by setting

$$A_0 x = x'' + \pi^2 x \quad \text{and} \quad A_1 x = x$$

for all $x \in H$. Note that $\|x''\|_K^2 \le \|x\|_H^2$. For each $y \in K$ and $z \in \mathbb{C}$ we wish to find $x \in H$ to solve the differential equation

$$[x''(t) + \pi^2 x(t)] + z x(t) = y(t).$$

This equation can be written in the form $(A_0 + A_1 z)x = y$, and hence the solution is given by $x = (A_0 + A_1 z)^{-1} y$, provided the inverse exists. If $\boldsymbol{e}_k : [0,1] \to \mathbb{C}$ is defined by

$$\boldsymbol{e}_k(t) = \sqrt{2} \sin k\pi t$$

for each $k = 1, 2, \ldots$ and all $t \in [0,1]$, then each $x \in H$ can be written as $x = \sum_{k=1}^{\infty} x_k \boldsymbol{e}_k$ where $x_k \in \mathbb{C}$ and $\sum_{k=1}^{\infty}(1 + \pi^2 k^2 + \pi^4 k^4)|x_k|^2 < \infty$ and each $y \in K$ can be written as $y = \sum_{k=1}^{\infty} y_k \boldsymbol{e}_k$ where $y_k \in \mathbb{C}$ and $\sum_{k=1}^{\infty} |y_k|^2 < \infty$. The operator A_0 is singular because $A_0 \boldsymbol{e}_1 = 0$. Nevertheless, $(A_0 + A_1 z)$ is nonsingular for $0 < |z| < 3\pi^2$, and equating coefficients in the respective Fourier series gives the solution

$$x_1 = y_1/z \ \text{ and } \ x_k = (-1)y_k/[\pi^2(k^2-1) - z] \quad \text{for } k \ge 2.$$

By writing the solution in the form

$$\begin{aligned} x &= \frac{y_1 \boldsymbol{e}_1}{z} - \sum_{k=2}^{\infty} \frac{y_k \boldsymbol{e}_k}{\pi^2(k^2-1)} \left[1 + \frac{z}{\pi^2(k^2-1)} + \cdots\right] \\ &= y_1 \boldsymbol{e}_1 \frac{1}{z} - \sum_{k=2}^{\infty} \frac{y_k \boldsymbol{e}_k}{\pi^2(k^2-1)} 1 - \sum_{k=2}^{\infty} \frac{y_k \boldsymbol{e}_k}{[\pi^2(k^2-1)]^2} z - \cdots \end{aligned}$$

for $0 < |z| < 3\pi^2$ we can see that the expansion is a Laurent series with a pole of order 1 at $z = 0$.

Example 8.9 (continuous spectrum). *Each element in the space $L^2(\mathbb{R})$ can be represented by a Fourier integral and defined by a continuously distributed spectral density. A bounded linear operator on $L^2(\mathbb{R})$ can be regarded as a linear transformation on a continuous spectrum. Let*

$$w(t) = \frac{2\sin(u_0 t)}{t},$$

where $u_0 \in \mathbb{R}$ *and* $u_0 > 0$. *Define* $A_0 : L^2(\mathbb{R}) \to L^2(\mathbb{R})$ *by the formula*

$$A_0 x(t) = x(t) - [x * w](t) = x(t) - \frac{1}{\pi}\int_{\mathbb{R}} x(\tau) w(t-\tau)\,d\tau$$

for all $t \in \mathbb{R}$. *The Fourier cosine and sine transforms are defined by*

$$\mathscr{F}_c[p](u) = \frac{1}{\pi}\int_{\mathbb{R}} p(t)\cos(ut)\,dt \quad \text{and} \quad \mathscr{F}_s[p](u) = \frac{1}{\pi}\int_{\mathbb{R}} p(t)\sin(ut)\,dt$$

for each $p \in L^2(\mathbb{R})$. *It is well known that* p *can be reconstructed by the formula*

$$p(t) = \int_{\mathbb{R}} [\mathscr{F}_c[p](u)\cos(ut) + \mathscr{F}_s[p](u)\sin(ut)]\,dt$$

and that the correspondence $p \in L^2(\mathbb{R}) \Leftrightarrow (\mathscr{F}_c[p], \mathscr{F}_s[p]) \in L^2(\mathbb{R}) \times L^2(\mathbb{R})$ *is unique. If* $p, q \in L^2(\mathbb{R})$, *then*

$$\begin{aligned}\mathscr{F}_c[p * q](u) &= \mathscr{F}_c[p](u)\mathscr{F}_c[q](u) - \mathscr{F}_s[p](u)\mathscr{F}_s[q](u) \quad \text{and} \\ \mathscr{F}_s[p * q](u) &= \mathscr{F}_c[p](u)\mathscr{F}_s[q](u) + \mathscr{F}_s[p](u)\mathscr{F}_c[q](u).\end{aligned}$$

Since $\mathscr{F}_c[w](u) = \chi_{(-u_0,u_0)}(u)$ *and* $\mathscr{F}_s[w](u) = 0$, *it follows that*

$$\begin{aligned}\mathscr{F}_c[A_0 x](u) &= \mathscr{F}_c[x](u) - \mathscr{F}_c[x * w](u) = \mathscr{F}_c[x](u)\left[1 - \chi_{(-u_0,u_0)}(u)\right] \quad \text{and} \\ \mathscr{F}_s[A_0 x](u) &= \mathscr{F}_s[x](u) - \mathscr{F}_s[x * w](u) = \mathscr{F}_s[x](u)\left[1 - \chi_{(-u_0,u_0)}(u)\right]\end{aligned}$$

for each $x \in L^2(\mathbb{R})$. *Define* $A_1 : L^2(\mathbb{R}) \to L^2(\mathbb{R})$ *by* $A_1 x = x$ *for all* $x \in L^2(\mathbb{R})$ *and consider the equation* $(A_0 + A_1 z)x = y$. *The solution is given by* $x = (A_0 + A_1 z)^{-1} y$ *provided the inverse exists. Taking a Fourier cosine transform of the original equation gives*

$$\mathscr{F}_c[x](u)\left[(1+z) - \chi_{(-u_0,u_0)}(u)\right] = \mathscr{F}_c[y](u),$$

and hence

$$\begin{aligned}\mathscr{F}_c[x](u) &= \mathscr{F}_c[y](u)\chi_{(-u_0,u_0)}(u)\cdot\frac{1}{z} + \mathscr{F}_c[y](u)\left[1 - \chi_{(-u_0,u_0)}(u)\right]\cdot\frac{1}{1+z} \\ &= \mathscr{F}_c[y * w](u)\cdot\frac{1}{z} + [\mathscr{F}_c[y](u) - \mathscr{F}_c[y * w](u)]\cdot[1 - z + z^2 - \cdots]\end{aligned}$$

for $|z| < 1$. *In similar fashion a Fourier sine transform of the original equation gives*

$$\mathscr{F}_s[x](u)\left[(1+z) - \chi_{(-u_0,u_0)}(u)\right] = \mathscr{F}_s[y](u)$$

from which it follows that

$$\mathscr{F}_s[x](u) = \mathscr{F}_s[y * w](u)\cdot\frac{1}{z} + [\mathscr{F}_s[y](u) - \mathscr{F}_s[y * w](u)]\cdot[1 - z + z^2 - \cdots]$$

for $|z| < 1$. *Therefore the solution is*

$$x(t) = (y * w)(t)\cdot\frac{1}{z} + [y(t) - (y * w)(t)]\cdot[1 - z + z^2 - \cdots]$$

for $|z| < 1$. *Note that the Laurent series has a pole of order* 1 *provided* $(y * w) \neq 0$. *By considering the Fourier transforms it can be seen that* $(y * w) = 0$ *if and only if* $\mathscr{F}_c[y](u) = 0$ *and* $\mathscr{F}_s[y](u) = 0$ *for almost all* $u \in (-u_0, u_0)$.

Remark 8.3. *If* $A(z_0) \in \mathscr{L}(H,K)$ *is nonsingular, then* $(A_{0,22} + A_{1,22}z_0) \in \mathscr{L}(H_2,K_2)$ *is also nonsingular. If* $A_{0,22} \in \mathscr{L}(H_2,K_2)$ *is onto but not one-to-one, then Theorem 8.14 can be applied to the operator* $(A_{0,22} + A_{1,22}z)$. *Thus the procedure is essentially recursive.*

Example 8.10. *Let* $u : [-\pi,\pi] \to \mathbb{R}$ *be defined by*

$$u(t) = \begin{cases} 1 & \text{for } t \in (-\pi, 0), \\ 0 & \text{for } t = -\pi, 0, \pi, \\ -1 & \text{for } t \in (0, \pi) \end{cases}$$

for all $t \in [-\pi,\pi]$. *Let* $H = K = L^2([-\pi,\pi])$. *Define* $A_0 : H \to K$ *by setting*

$$\begin{aligned} A_0x(t) &= \frac{1}{16}\Big[(x * u)(t) + (x * u)(-t)\Big] \\ &= \frac{1}{16}\int_{-\pi}^{\pi} x(s)\Big[u(t-s) + u(-t-s)\Big]ds \\ &= -\frac{1}{8}\Big[X(t) + X(-t)\Big] + \frac{1}{8}\Big[X(t-\pi) + X(-t+\pi)\Big], \end{aligned}$$

where $X(t) = \int_{[0,t]} x(s)ds$ *and where we have used the periodic extensions of* $x(t)$, $u(t)$ *as required in the convolution integral. The functions*

$$e_0 = 1,\ e_1(t) = \cos t,\ f_1(t) = \sin t,\ e_2(t) = \cos 2t,\ f_2(t) = \sin 2t, \ldots$$

form an orthogonal basis for L, and hence we can represent each element $f \in L$ *as an infinite sequence*

$$f = (a_0, a_1, b_1, a_2, b_2, \ldots) \Leftrightarrow a_0 + \sum_{n=1}^{\infty}\left[a_n e_n + b_n f_n\right]$$

of Fourier coefficients. Note that

$$A_0 e_n = 0, \quad A_0 f_{2m} = 0, \quad \text{and} \quad A_0 f_{2m-1} = e_{2m-1}/(2m-1)$$

for all $m,n \in \mathbb{N}$. *The null space* $M = A_0^{-1}(\{0\})$ *is defined by*

$$M = \{x \mid x \in H \text{ and } x = (a_0, a_1, 0, a_2, b_2, a_3, 0, a_4, b_4, \ldots)\},$$

and the orthogonal complement $M^\perp = A_0^{-1}(\{0\})^\perp$ *is defined by*

$$M^\perp = \{x \mid x \in H \text{ and } x = (0, 0, b_1, 0, 0, 0, b_3, 0, 0, \ldots)\}.$$

Both M *and* $M^\perp$ *are infinite dimensional spaces. In terms of Fourier coefficients the mapping* $A_0 \in \mathscr{L}(H,K)$ *can be described by the relationship*

$$A_0(a_0, a_1, b_1, a_2, b_2, a_3, b_3, a_4, b_4, \ldots) = (0, b_1, 0, 0, 0, b_3/3, 0, 0, 0, \ldots).$$

Let $A_1 = I$. *The perturbed operator* $(A_0 + A_1z) : H \to K$ *can be defined by an equivalent transformation* $(A_0 + A_1z) : \ell^2 \to \ell^2$ *using the formula*

$$\begin{aligned} &(A_0 + A_1z)(a_0, a_1, b_1, a_2, b_2, a_3, b_3, a_4, b_4, a_5, b_5, \ldots) \\ &\quad = (a_0z, b_1 + a_1z, b_1z, a_2z, b_2z, b_3/3 + a_3z, b_3z, a_4z, b_4z, \ldots), \end{aligned}$$

where a_0, a_n, and b_n are the usual Fourier coefficients. Solving a simple set of equations shows that the equivalent inverse transformation $(A_0+A_1z)^{-1}:\ell^2\to\ell^2$ is defined by

$$(A_0+A_1z)^{-1}(c_0,c_1,d_1,c_2,d_2,c_3,d_3,c_4,d_4,c_5,d_5,\dots)$$
$$=\left(\frac{c_0}{z},\frac{c_1}{z}-\frac{d_1}{z^2},\frac{d_1}{z},\frac{c_2}{z},\frac{d_2}{z},\frac{c_3}{z}-\frac{d_3}{3z^2},\frac{d_3}{z},\frac{c_4}{z},\frac{d_4}{z},\dots\right),$$

where c_0, c_n, and b_n are the usual Fourier coefficients. Thus, the inverse operator has a pole of order 2 at the origin. Write $H=M\times M^\perp$ and $K=N\times N^\perp$, where $N=A_1(M)=M$ and $N^\perp=M^\perp$. Now, using an infinite dimensional matrix notation,

$$(A_0+A_1z)=\left[\begin{array}{cccccc|ccc} z&0&0&0&0&\cdots&0&0&\cdots\\ 0&z&0&0&0&\cdots&1&0&\cdots\\ 0&0&z&0&0&\cdots&0&0&\cdots\\ 0&0&0&z&0&\cdots&0&0&\cdots\\ 0&0&0&0&z&\cdots&0&\frac{1}{3}&\cdots\\ \vdots&\vdots&\vdots&\vdots&\vdots&\ddots&\vdots&\vdots&\vdots\\ \hline 0&0&0&0&0&\cdots&z&0&\cdots\\ 0&0&0&0&0&\cdots&0&z&\cdots\\ \vdots&\vdots&\vdots&\vdots&\vdots&\vdots&\vdots&\vdots&\ddots \end{array}\right]=\left[\begin{array}{c|c} Iz & A_{0,12}\\ \hline 0 & Iz\end{array}\right],$$

and hence

$$(A_0+A_1z)^{-1}=\left[\begin{array}{c|c} I\cdot\frac{1}{z} & -A_{0,12}\cdot\frac{1}{z^2}\\ \hline 0 & I\cdot\frac{1}{z}\end{array}\right].$$

In the previous example the image space K for the mapping A_0 could be chosen differently. Since $A_0 f_{2m-1}=e_{2m-1}/(2m-1)$, it follows that A_0 is not bounded below. Thus the image set $A_0(H)$ is not closed in K. We could change this by choosing a more restrictive image space. Thus, if we choose the image space $K_E=H^1([-\pi,\pi])\subset K$, then

$$K_E=\left\{y\mid y=(c_0,c_1,d_1,c_2,d_2,\dots)\Leftrightarrow y=c_0+\sum_{n=1}^{\infty}[c_ne_n+d_nf_n]\right\},$$

where

$$c_0^2+\sum_{n=0}^{\infty}(1+n^2)\left[c_n^2+d_n^2\right]<\infty,$$

and if $y=(c_0,c_1,d_1,c_2,d_2\dots),z=(p_0,p_1,q_1,p_2,q_2,\dots)\in K_E$, then the inner product is given by

$$\langle y,z\rangle_E=\sum_{m=1}^{\infty}(1+n^2)[c_np_n+d_nq_n].$$

Now $A_0\in\mathcal{L}(H,K_E)$ is bounded below, and hence $A_0(H)$ is closed in K_E. Although it is not necessary in this particular example that $A_0(H)$ be closed, there are situations where such a closure may be desirable.

Remark 8.4. *If the procedure described in Theorem 8.14 is applied recursively to generate a sequence $M_1^\perp\supset M_2^\perp\supset\cdots$ of complementary spaces and if $M_n^\perp$ is finite dimensional for some*

$n \in \mathbb{N}$, then the recursive procedure terminates after a finite number of steps and the Laurent series has a finite order pole and converges on some region $0 < |z| < b$.

Remark 8.5. *If the action of the operators is restricted to a finite dimensional subspace for the purpose of numerical calculation, then the Laurent series for the inverse of the perturbed restricted operator has at most a finite order pole.*

The recursive procedure may continue indefinitely as the following example shows.

Example 8.11. *Consider the mappings on ℓ^2 defined by the infinite matrices*

$$A_0 = \left[\begin{array}{cc} 0 & A_{0,12} \\ 0 & A_{0,22} \end{array}\right] = \left[\begin{array}{c|ccccc} 0 & 1 & 0 & 0 & \cdots \\ \hline 0 & 0 & 1 & 0 & \cdots \\ 0 & 0 & 0 & 1 & \cdots \\ 0 & 0 & 0 & 0 & \cdots \\ \vdots & \vdots & \vdots & \vdots & \ddots \end{array}\right]$$

and

$$A_1 = \left[\begin{array}{cc} A_{1,11} & A_{1,12} \\ 0 & A_{1,22} \end{array}\right] = \left[\begin{array}{c|ccccc} 1 & 0 & 0 & 0 & \cdots \\ \hline 0 & 1 & 0 & 0 & \cdots \\ 0 & 0 & 1 & 0 & \cdots \\ 0 & 0 & 0 & 1 & \cdots \\ \vdots & \vdots & \vdots & \vdots & \ddots \end{array}\right] = I$$

and the linearly perturbed infinite matrix

$$A(z) = \left[\begin{array}{cc} A_{1,11}z & A_{0,12} + A_{1,12}z \\ 0 & A_{0,22} + A_{1,22}z \end{array}\right] = \left[\begin{array}{c|ccccc} z & 1 & 0 & 0 & \cdots \\ \hline 0 & z & 1 & 0 & \cdots \\ 0 & 0 & z & 1 & \cdots \\ 0 & 0 & 0 & z & \cdots \\ \vdots & \vdots & \vdots & \vdots & \ddots \end{array}\right] = (A_0 + Iz).$$

The reduced problem to calculate $(A_{0,22} + A_{1,22}z)^{-1}$ is the same as the original problem to calculate $A(z)^{-1}$. By an elementary calculation

$$(A_0 + Iz)^{-1} = \left[\begin{array}{ccccc} z^{-1} & -z^{-2} & z^{-3} & -z^{-4} & \cdots \\ 0 & z^{-1} & -z^{-2} & z^{-3} & \cdots \\ 0 & 0 & z^{-1} & -z^{-2} & \cdots \\ 0 & 0 & 0 & z^{-1} & \cdots \\ \vdots & \vdots & \vdots & \vdots & \ddots \end{array}\right]$$

$$= I \cdot \frac{1}{z} + (-1)A_0 \cdot \frac{1}{z^2} + (-1)^2 A_0^2 \cdot \frac{1}{z^3} + \cdots.$$

In general, this series does not converge near $z = 0$, but if we wish to compute $(A_0 + Iz)^{-1}y$ where $y = \sum_{j=1}^{n} y_j e_j$ for some natural number $n \in \mathbb{N}$, then only the first n terms of the expansion are nonzero and the series converges for all $z \neq 0$ with a pole of order at most n at the origin.

If $A_0 \in \mathcal{L}(H,K)$ and $A_0(H) = K$ but A_0 is not one-to-one, then some further remarks are in order. If we write $H = H_1 \times H_2$ where $H_1 = A_0^{-1}(\{0\})$ and $H_2 = H_1^{\perp}$ and $K = K_1 \times K_2$

where $K_1 = A_1(H_1)$ and $K_2 = K_1^{\perp}$, then the restricted mapping $A_0 |_{H_2,K} \in \mathscr{L}(H_2,K)$ is one-to-one and onto. It follows that the mapping $A_{0,22} = A_0 |_{H_2,K_2} \in \mathscr{L}(H_2,K_2)$ must be onto, but it will be one-to-one only if $K_2 = \{0\}$, in which case $A_{0,22}^{-1}$ is well defined and the process terminates. If $K_2 \neq \{0\}$, then the reduced problem will be to calculate $(A_{0,22} + A_{1,22}z)^{-1}$ where $A_{0,22}(H_2) = K_2$ but $A_{0,22}$ is not one-to-one. Thus the original problem has been reduced to an equivalent problem on smaller spaces.

8.6.2 ▪ The unperturbed mapping is one-to-one, and has closed range, but is not onto

Let $A_0 \in \mathscr{L}(H,K)$. Assume A_0 is one-to-one and $A_0(H)$ is closed but $A_0(H) \neq K$. Thus A_0 is singular. The Hilbert space adjoint $A_0{}^* \in \mathscr{L}(K,H)$ is defined by the relationship $\langle x, A_0{}^*y\rangle = \langle A_0x, y\rangle$ for all $x \in H$ and $y \in K$. The following standard result is used.

Lemma 8.15. *Let $A_0 \in \mathscr{L}(H,K)$ and let $A_0{}^* \in \mathscr{L}(K,H)$ denote the Hilbert space adjoint. If $A_0(H)$ is closed but $A_0(H) \neq K$, then $[A_0{}^*]^{-1}(\{0\}) = A_0(H)^{\perp} \neq \{0\}$, and hence $A_0{}^*$ is not one-to-one.*

Proof: Let $y \in K$ and define $\varphi_y : H \to \mathbb{C}$ by the formula

$$\varphi_y(x) = \langle A_0x, y\rangle$$

for each $x \in H$. The functional φ_y is a bounded linear functional on H, and hence there is a unique element $z_y \in H$ such that $\varphi_y(x) = \langle x, z_y\rangle$. We define $A_0^* : K \to H$ by the formula

$$A_0^*y = z_y$$

for all $y \in K$. If $y \in A_0(H)^{\perp}$, then $\langle x, z_y\rangle = \langle A_0x, y\rangle = 0$ for all $x \in H$. Hence $A_0^*y = z_y = 0$. Thus $A_0(H)^{\perp} \subseteq (A_0^*)^{-1}(\{0\})$. On the other hand, if $y \in (A_0^*)^{-1}(\{0\})$, then $\langle A_0x, y\rangle = 0$ for all $x \in H$ and it follows that $y \in A_0(H)^{\perp}$. Thus $(A_0^*)^{-1}(\{0\}) \subseteq A_0(H)^{\perp}$. Hence $(A_0^*)^{-1}(\{0\}) = A_0(H)^{\perp} \neq \{0\}$. □

Remark 8.6. *If $A^{-1} \in \mathscr{L}(K,H)$ is well defined, then $[A^*]^{-1} = [A^{-1}]^* \in \mathscr{L}(H,K)$ is also well defined.*

Lemma 8.15 and Remark 8.6 provide a basis for the inversion procedure when $A_0(H)$ is closed but $A_0(H) \neq K$.

Proposition 8.1. *Let $A_0 \in \mathscr{L}(H,K)$ with $A_0{}^{-1}(\{0\}) = \{0\}$ and with $A_0(H)$ closed but $A_0(H) \neq K$. If the inverse operator $A(z_0)^{-1} = (A_0 + A_1z_0)^{-1}$ is well defined for some $z_0 \neq 0$, then*

$$[A(z_0)^*]^{-1} = (A_0{}^* + A_1{}^*\overline{z_0})^{-1} = [A(z_0)^{-1}]^*$$

is also well defined. If Theorem 8.14 can be applied to show that for some $b > 0$ the inverse operator $[A(z)^*]^{-1}$ is well defined for $0 < |z| < b$, then $A(z)^{-1} = [\{A(z)^*\}^{-1}]^*$ is also well defined for $0 < |z| < b$.

Proof: Apply the original inversion formula to the adjoint operator $A(z)^*$ and recover the desired series from the formula $A(z)^{-1} = [\{A(z)^*\}^{-1}]^*$. □

8.6.3 ▪ The unperturbed mapping is one-to-one but has nonclosed range

In this section we show that by modifying the topology of the range space we can ensure that the range space is closed. The previous inversion procedures can then be used. We begin with an important observation.

Lemma 8.16. *If $A_0 \in \mathscr{L}(H,K)$ and $A_0(H)$ is not closed, then A_0 is not bounded below.*

Proof: If $y \in \overline{A_0(H)} \setminus A_0(H)$, then we can find $\{x_n\} \in H$ such that $y_n = A_0 x_n$ and $\|y_n - y\|_K \to 0$ as $n \to \infty$. Let us suppose that $\|x_n\|_H \leq k$ for some $k > 0$ and all $n \in \mathbb{N}$. Since $\|x_n\|_H$ is bounded it follows from the Eberlein–Shmulyan theorem (see the bibliographic notes) that we can find a weakly convergent subsequence $\{x_{n(m)}\}$ and some $x \in H$ such that $x_{n(m)}$ converges weakly to x. That is, we can find $x \in H$ such that

$$\langle q, A_0 x_{n(m)}\rangle_K = \langle A_0^* q, x_{n(m)}\rangle_H \to \langle A_0^* q, x\rangle_H = \langle q, A_0 x\rangle_K$$

for all $q \in K$ as $m \to \infty$. Since $\langle q, y_{n(m)}\rangle_K \to \langle q, y\rangle_K$ it follows that $\langle q, A_0 x - y\rangle_K = 0$ for all $q \in K$ and hence that $A_0 x = y$. This is a contradiction and so the assumption must be wrong. Hence we can find a subsequence $\{x_{r(m)}\}$ with $\|x_{r(m)}\|_H \geq m$ for all m. Choose an arbitrary real number $\delta > 0$. Since $A_0 x_{r(m)} = y_{r(m)}$ and $\|y_{r(m)} - y\|_K \to 0$ as $m \to \infty$ it follows that

$$\|A_0 x_{r(m)}\|_K \leq \frac{(\|y\|_K + \delta)}{m} \|x_{r(m)}\|_H$$

when m is sufficiently large. Hence A_0 is not bounded below. □

When $A_0(H)$ is not closed in K the essence of the difficulty is that K is an inappropriate image space because the topology allows images of divergent sequences to converge. We restrict the image space with a new topology that excludes the unwanted limit points.

Definition 8.7. *Let $M = A_0(\{0\})^{-1}$ be the null space of A_0. Let $\langle \cdot, \cdot \rangle_E : A_0(H) \times A_0(H) \to \mathbb{C}$ be defined by the formula*

$$\langle y, v\rangle_E = \langle y, v\rangle_K + \langle x_M^{\perp}, u_M^{\perp}\rangle_H$$

for each $y, v \in A_0(H)$ where $x_M^{\perp}, u_M^{\perp} \in M^{\perp}$ are the uniquely defined elements with $A_0 x_M^{\perp} = y$ and $A_0 u_M^{\perp} = v$.

With the new inner product and the associated new topology on $A_0(H)$ we have the following result.

Lemma 8.17. *The space $K_E = \{A_0(H), \langle \cdot, \cdot \rangle_E\}$ is a Hilbert space.*

Remark 8.7. *The new inner product is simply a more appropriate measurement tool on the space $A_0(H)$ in relation to the operator A_0. One could argue that the elements of the space $K_E = A_0(H)$ remain unchanged.*

The mapping $A_{0,E} \in \mathscr{L}(H, K_E)$ defined by $A_{0,E} x = A_0 x$ for all $x \in H$ is onto but not necessarily one-to-one. Of course it may well be true that K_E can be regarded as a closed subspace of some larger Hilbert space K' in which case the mapping $A_{0,E} \in \mathscr{L}(H, K')$ is no longer onto. In any case the original inversion formulae can now be applied to the operator $A_{0,E} \in \mathscr{L}(H, K')$.

Example 8.12 (a modified integral operator). *Let $H = K = L^2([0,1])$. Note that the space $L^2([0,1])$ can be generated by the limits of all Cauchy sequences of continuous functions $\{x_n\} \in \mathscr{C}_0([0,1])$ in $L^2([0,1])$ satisfying $x_n(0) = x_n(1) = 0$. Define $A_0 \in \mathscr{L}(H,K)$ by setting $A_0 x(t) = \mathscr{X}(1) - X(t)$, where*

$$X(t) = \int_0^t x(s)ds \quad \text{and} \quad \mathscr{X}(u) = \int_0^u X(t)dt.$$

If we define $x_n \in H$ by

$$x_n(s) = \sin n\pi s,$$

then $\|x_n\| = 1/\sqrt{2}$ for all $n \in \mathbb{N}$, but we have

$$A_0 x_n(t) = \frac{\cos n\pi t}{n\pi},$$

and hence $\|A_0 x_n\| \to 0$ as $n \to \infty$. Therefore A_0 is not bounded below and $A_0(H)$ is not closed in K. For instance, if we define $y_\infty \in K$ by the formula

$$\begin{aligned} y_\infty(t) &= \begin{cases} \frac{1}{2} & \text{for } 0 < t < \frac{1}{2}, \\ -\frac{1}{2} & \text{for } \frac{1}{2} < t < 1 \end{cases} \\ &= \frac{2}{\pi}\left[\cos \pi t - \frac{\cos 3\pi t}{3} + \frac{\cos 5\pi t}{5} - \cdots \right], \end{aligned}$$

then $y_\infty \notin A_0(H)$. However, the functions

$$y_n(t) \begin{cases} \frac{1}{2} & \text{for } t \in [0, \frac{n-1}{2n}), \\ -nt + \frac{n}{2} & \text{for } t \in [\frac{n-1}{2n}, \frac{n+1}{2n}], \\ -\frac{1}{2} & \text{for } (\frac{n+1}{2n}, 1] \end{cases}$$

are given by $y_n = Ax_n$, where

$$x_n(t) = \begin{cases} 0 & \text{for } t \in [0, \frac{n-1}{2n}), \\ -n & \text{for } t \in [\frac{n-1}{2n}, \frac{n+1}{2n}], \\ 0 & \text{for } (\frac{n+1}{2n}, 1], \end{cases}$$

and hence $y_n \in A_0(H)$ with $\|y_\infty - y_n\|_K \to 0$ as $n \to \infty$. Hence $y_\infty \in \overline{A_0(H)} \setminus A_0(H)$. In general, there are many nondifferentiable functions on the boundary of the set $A_0(H)$. If we define a new inner product in $A_0(H)$ according to the formula

$$\begin{aligned} \langle y, v \rangle_E &= \langle y, v \rangle_K + \langle x, u \rangle_H \\ &= \int_{[0,1]} \left[y(t)\overline{v(t)} + y'(t)\overline{v'(t)} \right] dt, \end{aligned}$$

where x, u are the unique solutions to $y = A_0 x$ and $v = A_0 u$, then nondifferentiable functions such as y_∞ are removed from the boundary of the image space and $A_0(H)$ is now closed. Indeed, since

$$\|y_m - y_n\|_E^2 \geq \|x_m - x_n\|_H^2 = m^2\left[\frac{1}{m} - \frac{1}{n}\right] + (n-m)^2 \frac{1}{n} = n - m$$

when $m < n$, *it follows that* $\{y_n\}_{n\in\mathbb{N}}$ *is no longer a Cauchy sequence. The image space* $K_E = A_0(H)$ *now consists of those functions* $y \in L^2([0,1])$ *with generalized derivative* $y' \in L^2([0,1])$ *such that* $\int_0^1 y(t)dt = 0$, *and with* $\|y\|_E^2 = \|y\|_2^2 + \|y'\|_2^2$.

Without loss of generality we may therefore suppose that $A \in \mathscr{L}(H,K)$, where $\mathscr{R}(A) \subseteq K$ is a closed subspace.

8.6.4 ▪ The case where the unperturbed mapping is unbounded

In this section we will show that by changing the topology in the domain space we can ensure that the operator is bounded. Previous inversion techniques can then be used. We begin with some basic definitions and standard results. Let H and K be Hilbert spaces, and let $A_0 : \mathscr{D}(A_0) \subseteq H \to K$ be a linear operator defined on a linear subspace $\mathscr{D}(A_0) \subseteq H$.

Definition 8.8. *The operator* A_0 *is densely defined if* $\mathscr{D}(A_0)$ *is a dense subset of* H. *That is, for each* $x \in H$ *and each* $\epsilon > 0$ *there exists* $u = u(x,\epsilon) \in \mathscr{D}(A_0)$ *with* $\|u - x\| < \epsilon$.

Lemma 8.18. *Let* $y \in K$ *and let* A_0 *be densely defined. If* $\exists\ z \in H$ *such that* $\langle y, A_0 x\rangle_K = \langle z, x\rangle_H$ *for all* $x \in \mathscr{D}(A_0)$, *then* z *is uniquely defined.*

Definition 8.9. *Let* $y \in K$ *and let* A_0 *be densely defined. Let*

$$\mathscr{D}(A_0^*) = \{y \mid \exists\ z \in H \text{ with } \langle y, A_0 x\rangle_K = \langle z, x\rangle_H \text{ for all } x \in \mathscr{D}(A_0)\}$$

and define $A_0^* : \mathscr{D}(A_0^*) \subseteq K \mapsto H$ *by setting* $A_0^* y = z$.

Definition 8.10. *The set* $G(A_0) = \{(x, A_0 x) \mid x \in \mathscr{D}(A_0)\} \subseteq H \times K$ *is called the graph of the operator* A_0. *If* $G(A_0)$ *is closed, then we say that* A_0 *is a closed linear operator.*

Lemma 8.19. *If* A_0 *is a closed operator, then, for each sequence* $\{x_n\}_{n\in\mathbb{N}} \in \mathscr{D}(A_0)$ *with* $x_n \to x$ *and* $A_0 x_n \to y$ *as* $n \to \infty$, *it follows that* $x \in \mathscr{D}(A_0)$ *and* $A_0 x = y$.

Lemma 8.20. *If* A_0 *is densely defined, then* A_0^* *is a closed linear operator. If* A_0 *is closed, then* A_0^* *is densely defined.*

Proof: Let $G(A_0^*) = \{(y, A_0^* y),\ y \in \mathscr{D}(A_0^*)\}$ be the graph of A_0^*, and suppose $\{y_n\}_{n\in\mathbb{N}} \in \mathscr{D}(A_0^*)$ with $y_n \to y$ and $A_0^* y_n \to x$ as $n \to \infty$. If $u \in \mathscr{D}(A_0)$, then $\langle y_n, A_0 u\rangle_K = \langle A_0^* y_n, u\rangle_H$ and by taking limits as $n \to \infty$ it follows that $\langle y, A_0 u\rangle_K = \langle x, u\rangle_H$. Therefore, $A_0^* x = y$. Hence $G(A_0^*)$ is closed.

Let $V \in \mathscr{L}(H \times K, K \times H)$ be defined by $V(x,y) = (-y, x)$. Since $G(A_0)$ is closed, it follows that $G(A_0^*)^\perp = VG(A_0)$. If $k \in \mathscr{D}(A_0^*)^\perp$ and $y \in \mathscr{D}(A_0^*)$, then

$$\langle (k,0), (y, A_0^* y)\rangle_{K\times H} = \langle k, y\rangle_K + \langle 0, A_0^* y\rangle_H = 0.$$

Therefore, $(k,0) \in G(A_0^*)^\perp = VG(A_0)$, and hence $k = -T0 = 0$. □

Theorem 8.21 (J. von Neumann). *If* $A_0 : \mathscr{D}(A_0) \subseteq H \to K$ *is densely defined and closed, then the operators* $A_0^* A_0$ *and* $A_0 A_0^*$ *are self-adjoint with* $(I + A_0^* A_0)^{-1} \in \mathscr{L}(H)$ *and* $(I + A_0 A_0^*)^{-1} \in \mathscr{L}(K)$.

Proof: Let $h \in H$. Since $H \times K = G(A_0) \oplus VG(A_0^*)$, it follows that there is a uniquely determined decomposition

$$(h,0) = (x, A_0 x) + (-A_0^* y, y)$$

where $x \in \mathscr{D}(A_0)$ and $y \in \mathscr{D}(A_0^*)$. Thus $h = x - A_0^* y$ and $0 = A_0 x + y$. Therefore, $x \in \mathscr{D}(A_0^* A_0)$ and $(I + A_0^* A_0)x = h$. Because the decomposition is unique, the element x is uniquely determined by h, and so the inverse operator $(I + A_0^* A_0)^{-1} \in \mathscr{L}(H)$. For $u, v \in H$ let

$$p = (I + A_0^* A_0)^{-1} u \quad \text{and} \quad q = (I + A_0^* A_0)^{-1} v.$$

Therefore, $p, q \in \mathscr{D}(A_0^* A_0)$. Since

$$\begin{aligned}\langle u, (I + A_0^* A_0)^{-1} v\rangle_H &= \langle (I + A_0^* A_0) p, q\rangle_H \\ &= \langle p, q\rangle_H + \langle A_0^* A_0 p, q\rangle_H = \langle p, q\rangle_H + \langle A_0 p, A_0 q\rangle_K\end{aligned}$$

and

$$\begin{aligned}\langle (I + A_0^* A_0)^{-1} u, v\rangle_H &= \langle p, (I + A_0^* A_0) q\rangle_H \\ &= \langle p, q\rangle_H + \langle p, A_0^* A_0 q\rangle_H = \langle p, q\rangle_H + \langle A_0 p, A_0 q\rangle_K,\end{aligned}$$

it follows that

$$\langle u, (I + A_0^* A_0)^{-1} v\rangle_H = \langle (I + A_0^* A_0)^{-1} u, v\rangle_H,$$

and hence $(I + A_0^* A_0)^{-1}$ is self adjoint. Since $(I + A_0^* A_0)^{-1}$ is everywhere defined and self-adjoint, Lemma 8.20 shows that it is closed. The closed graph theorem shows that $(I + A_0^* A_0)^{-1}$ is bounded. It follows that the inverse operator $(I + A_0^* A_0)$ and the associated operator $A_0^* A_0$ are self-adjoint, too.

Because A_0 is closed we have $(A_0^*)^* = A_0$. By applying similar arguments it follows that $A_0 A_0^* = (A_0^*)^* A_0^*$ is self-adjoint and that $(I + A_0 A_0^*)^{-1} \in \mathscr{L}(K)$. □

The energy space

Let $A_0 : \mathscr{D}(A_0) \subseteq H \to K$ be a densely defined and closed linear operator. For each $\varphi, \psi \in \mathscr{D}(A_0)$ define a new inner product

$$\langle \varphi, \psi\rangle_E = \langle \varphi, \psi\rangle_H + \langle A_0 \varphi, A_0 \psi\rangle_K$$

with a norm $||\varphi||_E = [\langle \varphi, \varphi\rangle_E]^{1/2}$. The energy space $H_E = (\mathscr{D}(A_0), \langle\cdot,\cdot\rangle_E)$ is a Hilbert space. We denote the new mapping by $A_{0,E} : H_E \to K$. In practice the operator A_0 may be defined on a dense subset $\mathscr{C} \subset H$ but may not be closed. In such cases the set $H_E \subset H$ is defined as the completion of $\mathscr{C}$ in the new norm. The point $x \in H$ will belong to H_E if there exists a sequence $\{\varphi_n\} \in \mathscr{C}$ with $||\varphi_n - x||_E \to 0$ as $n \to \infty$. Thus we must also have $y \in K$ with $||A_0 \varphi_n - y||_K \to 0$. The completion is guaranteed if we allow the limit process to define an appropriate equivalence class.

Lemma 8.22. *The mapping $A_{0,E} : H_E \to K$ is a bounded linear mapping. That is, $A_{0,E} \in \mathscr{L}(H_E, K)$.*

Remark 8.8. *$A_0(H)$ is closed if and only if A_0 is bounded below on $\mathscr{D}(A_0)$.*

Lemma 8.23. *The new adjoint mapping $A_{0,E}{}^* \in \mathscr{L}(K, H_E)$ is defined in terms of the original adjoint mapping $A_0{}^* : \mathscr{D}(A_0{}^*) \subset K \to H$ by the formulae*

$$A_{0,E}{}^* = A_0{}^*(I + A_0 A_0{}^*)^{-1} = (I + A_0{}^* A_0)^{-1} A_0{}^*.$$

Proof: For each $y \in A_{0,E}(H)$ and each $x \in H_E$ we have

$$\langle y, A_0 x\rangle_K = \langle A_0^* y, x\rangle_H = \langle A_0^* y, x\rangle_E - \langle A_0 A_0^* y, A_0 x\rangle_K.$$

Rearranging this equation gives

$$\langle (I + A_0 A_0^*) y, A_0 x\rangle_K = \langle A_0^* y, x\rangle_E,$$

and if we write $z = (I + A_0 A_0^*) y$, then we have

$$\langle z, A_0 x\rangle_K = \langle A_0^*(I + A_0 A_0^*)^{-1} z, x\rangle_E,$$

and hence $A_{0,E}^* = A_0^*(I + A_0 A_0^*)^{-1}$. From this formula it follows that

$$(I + A_0^* A_0) A_{0,E}^* = (I + A_0^* A_0) A_0^* (I + A_0 A_0^*)^{-1} = A_0^*(I + A_0 A_0^*)(I + A_0 A_0^*)^{-1} = A_0^*$$

and hence that $A_{0,E}^* = (I + A_0^* A_0)^{-1} A_0^*$. □

Since the operator $A_{0,E} : H_E \to K$ is a bounded linear mapping, the original inversion formula can now be applied.

Example 8.13 (the differentiation operator). *Let $H = L^2([0,1])$, and define $A_0 \varphi(t) = \varphi'(t)$ for all $\varphi \in \mathscr{C}_0^1([0,1])$ and all $t \in [0,1]$. For each $\{\varphi_n\} \in \mathscr{C}_0^1([0,1])$ with*

$$\int_{[0,1]} \left\{ |\varphi_m(t) - \varphi_n(t)|^2 + |\varphi_m'(t) - \varphi_n'(t)|^2 \right\} dt \to 0$$

as $m, n \to \infty$ there exist functions x and y such that

$$\int_{[0,1]} |\varphi_n(t) - x(t)|^2 dt \to 0 \quad \text{and} \quad \int_{[0,1]} |\varphi_n'(t) - y(t)|^2 dt \to 0$$

as $n \to \infty$. We say $y = x'$ is the generalized derivative of x. Note that

$$||x||^2 = \int_0^1 \left| \int_0^t x'(s) ds \right|^2 dt \le \frac{1}{2} ||x'||^2.$$

The Hilbert space H_E is the completion of the space $\mathscr{C}_0^1([0,1])$ with the inner product

$$\langle x, u\rangle_E = \int_0^1 [x(t)\overline{u}(t) + x'(t)\overline{u}'(t)] dt$$

and the norm

$$||x||_E = \left[\int_0^1 \{ |x(t)|^2 + |x'(t)|^2 \} dt \right]^{1/2}.$$

It can be shown that

$$H_E = \{x \mid x \in \mathscr{C}_0^0([0,1]) \text{ and } x' \in L^2([0,1])\}.$$

The space $H_E = H_0^1([0,1])$ is an elementary example of a Sobolev space. Define the generalized differentiation operator $A_{0,E} : H_E \to K$ by the formula $A_{0,E} x = \lim_{n\to\infty} A_0 \varphi_n$, where

$\varphi_n \in \mathscr{C}_0^1([0,1])$ and $\varphi_n \to x$ in H_E as $n \to \infty$. Thus $A_{0,E}x = x'$ is simply the generalized derivative. It follows from the inequality above that $A_{0,E}$ is bounded below and hence $A_{0,E}(H_E)$ is closed. It is also obvious that $\|A_{0,E}x\| \le \|x\|_E$ and so $A_{0,E} \in \mathscr{L}(H_E, K)$. For the original mapping $A_0 : \mathscr{C}_0^1([0,1]) \subset L^2([0,1]) \to L^2([0,1])$ consider the adjoint mapping ${A_0}^*$. If ${A_0}^*\eta = \xi$, then

$$\int_0^1 \varphi'(t)\overline{\eta}(t)dt = \int_0^1 \varphi(t)\overline{\xi}(t)dt \quad \Rightarrow \quad \int_0^1 \varphi'(t)\left[\overline{\eta}(t) + \int_0^t \overline{\xi}(s)ds\right]dt = 0$$

for all $\varphi \in \mathscr{C}_0^1([0,1])$. Hence η is differentiable and $\xi = -\eta' = {A_0}^*\eta$. Now consider the adjoint of the generalized mapping. If ${A_{0,E}}^*\eta = \zeta$, then

$$\int_0^1 \varphi'(t)\overline{\eta}(t)dt = \int_0^1 [\varphi(t)\overline{\zeta}(t) + \varphi'(t)\overline{\zeta}'(t)]dt,$$

and therefore

$$\int_0^1 \varphi'(t)\left[\overline{\eta}(t) - \overline{\zeta}'(t) + \int_0^t \overline{\zeta}(s)ds\right]dt = 0$$

for all $\varphi \in \mathscr{C}_0^1([0,1])$. Hence ζ' is differentiable and $\zeta - \zeta'' = -\eta'$. It follows that

$$(I + {A_0}^*A_0){A_{0,E}}^* = {A_0}^* \quad \Leftrightarrow \quad {A_{0,E}}^* = (I + {A_0}^*A_0)^{-1}{A_0}^*.$$

Example 8.14. *We now reconsider Example 8.13. Each element $x \in H_E = H_0^1([0,1]) \subseteq L^2([0,1])$ can be represented by a Fourier sine series*

$$x = \sum_{k=1}^{\infty} x_k e_k,$$

where $e_k(t) = \sqrt{2}\sin k\pi t$ and $\sum_{k=1}^{\infty}(1 + \pi^2 k^2)x_k^2 < \infty$. In Fourier series terminology the extended mapping $A_{0,E} : H_0^1([0,1]) \to L^2([0,1])$ is defined by the formula

$$A_{0,E}x = \sum_{k=1}^{\infty} k\pi x_k f_k,$$

where $f_k(t) = \sqrt{2}\cos k\pi t$. Each element $y \in \mathscr{D}(A_0^*) \subseteq L^2([0,1])$ can be represented by a Fourier cosine series

$$y = \sum_{k=0}^{\infty} y_k f_k,$$

where $f_0(t) = 1$ and $\sum_{k=0}^{\infty}(1 + \pi^2 k^2)y_k^2 < \infty$. The original adjoint mapping is represented as a Fourier series by the formula

$$A_0^* y = \sum_{k=1}^{\infty} k\pi y_k e_k.$$

The self-adjoint mappings $(I + A_0^*A_0)^{-1}$ and $(I + A_0A_0^*)^{-1}$ are given by

$$(I + A_0^*A_0)^{-1}x = \sum_{k=1}^{\infty} \frac{x_k}{1 + k^2\pi^2} e_k \quad \text{and} \quad (I + A_0A_0^*)^{-1}y = \sum_{k=0}^{\infty} \frac{y_k}{1 + k^2\pi^2} f_k$$

for each $x, y \in L^2([0,1])$. *The new adjoint mapping* $A_{0,E}^* : L^2([0,1]) \to H_0^1([0,1])$ *is given by the formula*

$$A_{0,E}^* y = \sum_{k=1}^{\infty} \frac{k\pi y_k}{1+k^2\pi^2} e_k$$

for each $y \in L^2([0,1])$.

We therefore argue that the inversion of a linearly perturbed unbounded linear operator can be reduced to the inversion of a linearly perturbed bounded linear operator. In so doing we assume that the perturbation is a perturbation to the modified operator in the new topology. If the perturbation is given as an unbounded perturbation, then we must modify the topology in such a way that both the unperturbed operator and the perturbation are reduced to bounded operators.

8.7 ▪ Inversion of Linearly Perturbed Operators on Banach Spaces

Let H and K be Banach spaces, and let $A_0, A_1 \in \mathscr{L}(H,K)$ be bounded linear maps. Let $z \in \mathbb{C}$ be a complex variable. We wish to consider the linearly perturbed operator $A(z) = A_0 + A_1 z \in \mathscr{L}(H,K)$. It is clear that the operator-valued function $A : \mathbb{C} \to \mathscr{L}(H,K)$ is analytic everywhere. Under what circumstances can we find an analytic expression in a region $0 < |z| < r$ for the inverse operator $A(z)^{-1}$?

8.7.1 ▪ Regular perturbations

If $A_0(H) = K$ and $A_0^{-1}(\{0\}) = \{0\}$, then $A_0^{-1} \in \mathscr{L}(K,H)$ is well defined. The sequence $\{X_j\} \subset \mathscr{L}(K,H)$ defined by

$$X_j = (-1)^j (A_0^{-1} A_1)^j A_0^{-1}$$

for each $j \in \mathbb{Z}_+$ is a solution to each of the linear systems

$$\begin{array}{rcl} A_0X_0 &=& I, \\ A_1X_0 + A_0X_1 &=& 0, \\ A_1X_1 + A_0X_2 &=& 0, \\ A_1X_2 + A_0X_3 &=& 0, \\ \vdots && \vdots \end{array} \quad \text{and} \quad \begin{array}{rcl} X_0A_0 &=& I, \\ X_0A_1 + X_1A_0 &=& 0, \\ X_1A_1 + X_2A_0 &=& 0, \\ X_2A_1 + X_3A_0 &=& 0, \\ \vdots && \vdots \end{array} \tag{8.19}$$

and furthermore, from the definition, it follows that $||X_j|| \leq ||A_0^{-1}||^{j+1} ||A_1||^j$ for all $j \in \mathbb{Z}_+$. If we define $r = 1/R$ where $R = ||A_0^{-1}|| \cdot ||A_1||$, then

$$(A_0 + A_1 z)^{-1} = X_0 + X_1 z + X_2 z^2 + \cdots$$

for all $z \in \mathbb{C}$ with $|z| < r$. The Maclaurin series expansion for the inverse operator is known as the Neumann expansion. The equations (8.19) are usually referred to as the *fundamental equations* for inversion of a regular perturbation. Unlike the finite dimensional case, we must use two sided equations to define an inverse operator in an infinite dimensional Banach space as the following example shows.

An operator that is not invertible

Let $A \in \mathcal{L}(\ell_1)$ be a bounded linear operator on the set of vectors $v = [v_i]$ such that $\|v\|_1 = \sum_i |v_i| < \infty$. If ℓ_∞ denotes the set of vectors $u = [u_i]$ with bounded elements, then we can define a bounded linear operator $u^T \in \mathcal{L}(\ell_1, \mathbb{R})$ by the formula $u^T e_i = u_i$ where $e_1, e_2, \ldots$ are the standard unit vectors in ℓ_1. We can now represent each $A \in \mathcal{L}(\ell_1)$ as an infinite dimensional matrix $A = [a_{ij}]$ where $a_{ij} = e_i^T A e_j$ and $|a_{ij}|$ is bounded. Define

$$A = \begin{bmatrix} 0 & 1 & 0 & 0 & \cdots \\ 0 & 0 & 1 & 0 & \cdots \\ 0 & 0 & 0 & 1 & \cdots \\ \vdots & \vdots & \vdots & \vdots & \ddots \end{bmatrix} \quad \text{and} \quad X = \begin{bmatrix} 0 & 0 & 0 & 0 & \cdots \\ 1 & 0 & 0 & 0 & \cdots \\ 0 & 1 & 0 & 0 & \cdots \\ \vdots & \vdots & \vdots & \vdots & \ddots \end{bmatrix}.$$

Both A and X are elements of $\mathcal{L}(\ell_1)$, and we have $AX = I$ but $XA \neq I$. The mapping A is *surjective* but not *injective*. That is, A maps ℓ_1 onto ℓ_1, but the mapping is not one-to-one.

8.7.2 ▪ Necessary conditions for a singular perturbation with a first order pole

Let us now consider the possibility that the inverse operator $A(z)^{-1}$ can be represented in a neighborhood of $z = 0$ by a Laurent series with a pole of order 1. If we assume a relationship of the form

$$(A_0 + A_1 z)^{-1} = \frac{1}{z}\left(X_0 + X_1 z + X_2 z^2 + \cdots\right)$$

that is valid for some deleted neighborhood $0 < |z| < r$, then the sequence $\{X_j\} \subset \mathcal{L}(K,H)$ must satisfy the equations

$$\begin{array}{rcl} A_0X_0 &=& 0, \\ A_1X_0 + A_0X_1 &=& I, \\ A_1X_1 + A_0X_2 &=& 0, \\ A_1X_2 + A_0X_3 &=& 0, \\ \vdots && \vdots \end{array} \quad \text{and} \quad \begin{array}{rcl} X_0A_0 &=& 0, \\ X_0A_1 + X_1A_0 &=& I, \\ X_1A_1 + X_2A_0 &=& 0, \\ X_2A_1 + X_3A_0 &=& 0, \\ \vdots && \vdots \end{array} \tag{8.20}$$

and since $\|X_j\| \cdot |z|^j \to 0$ as $j \to \infty$ for all z with $0 < |z| < r$, we must have $\|X_j\| \leq R^{j+1}$ for some $R > 0$. We will refer to (8.20) as the *fundamental equations* for inversion of a singular perturbation with a pole of order 1. Our investigation will focus on the role of these fundamental equations.

Remark 8.9. *Consider the Hilbert space formula* (8.18) *in the case where A_{022}^{-1} is well defined. By applying the Neumann expansion to the term $(A_{022} + A_{122}z)^{-1}$ we have*

$$X_0 = P_1 A_{111}^{-1} Q_1 - P_1 A_{111}^{-1} A_{012} A_{022}^{-1} Q_2$$

and

$$X_j = (P_2 - P_1 A_{111}^{-1} A_{112}) A_{022}^{-1} [-A_{122} A_{022}^{-1}]^{j-1} Q_2 \\ + P_1 A_{111}^{-1} A_{012} A_{022}^{-1} [-A_{122} A_{022}^{-1}]^j Q_2$$

for $j \geq 1$. Since

$$A_0 = (Q_1 A_{012} + Q_2 A_{022}) P_2 \quad \text{and} \quad A_1 = Q_1 A_{111} P_1 + (Q_1 A_{112} + Q_2 A_{122}) P_2,$$

we can use an explicit calculation to check that the equations (8.20) *are satisfied.*

8.7.3 ▪ The main results

The main aim of this subsection is to provide two important results, Theorems 8.24 and 8.25, which we shall state now and briefly discuss but prove later by a rather indirect and extended argument. Indeed, we shall use Subsection 8.7.4 to prove a sequence of intermediate results (Lemma 8.26, Theorem 8.27, and Corollary 8.3) before we return to the task of proving the two important results. The intermediate results allow us to partition the domain and range into corresponding subspaces and thereby rewrite the relevant operators and the determining equations in partitioned form. This allows us to establish Theorems 8.24 and 8.25 in Remark 8.11 by simply extracting key points from our intervening discussion.

Theorem 8.24. *Let H,K be Banach spaces, let $A_0,A_1 \in \mathscr{L}(H,K)$ be bounded linear maps, and suppose that $M = A_0^{-1}(\{0\}) \neq \{0\}$. Let $z \in \mathbb{C}$ and let $N = A_1(M)$. The operator $A(z)^{-1} \in \mathscr{L}(K,H)$ is well defined and analytic on a region $0 < |z| < r$ with a pole of order 1 if and only if there exist bounded linear operators $X_0,X_1 \in \mathscr{L}(K,H)$ that satisfy the right-hand determining equations*

$$A_0X_0 = 0 \quad \text{and} \quad A_1X_0 + A_0X_1 = I \tag{8.21}$$

and bounded linear operators $Y_0,Y_1 \in \mathscr{L}(K,H)$ that satisfy the left-hand determining equations

$$Y_0A_0 = 0 \quad \text{and} \quad Y_0A_1 + Y_1A_0 = I. \tag{8.22}$$

If (8.21) and (8.22) are satisfied, then $X_0 = Y_0$.

Theorem 8.25. *There exist bounded linear operators $X_0,X_1 \in \mathscr{L}(K,H)$ that satisfy the equations (8.21) and bounded linear operators $Y_0,Y_1 \in \mathscr{L}(K,H)$ that satisfy the equations (8.22) if and only if there exist linear projections $P \in \mathscr{L}(H,M)$ mapping H onto the null space $M = A_0^{-1}(\{0\}) \neq \{0\}$ with $A_0P = 0$ and $Q \in \mathscr{L}(K,N)$ mapping K onto the image $N = A_1(M)$ with $QA_0 = 0$ and such that A_0 is bounded below on $M^c = (I-P)(H)$ and A_1 is bounded below on $M = P(H)$.*

The conditions of the above theorem imply that $H = P(H) \oplus (I-P)(H) = M \oplus M^c$, that $K = Q(K) \oplus (I-Q)(K) = N \oplus N^c$, and that the bounded linear operators $A_{022} = A_0|_{M^c,N^c} \in \mathscr{L}(M^c,N^c)$ and $A_{111} = A_1|_{M.N} \in \mathscr{L}(M,N)$ are bounded below and hence invertible. The subspaces M and M^c are closed in H, and the subspaces N and N^c are closed in K. In general, there is no guarantee in a Banach space that a given subspace can be complemented. The significance of this result is that if there exists an inverse operator $A(z)^{-1}$ in the form of a Laurent series with a pole of order 1 at the origin, if $M = A_0^{-1}(\{0\})$ denotes the null space of A_0, and if $N = A_1(M)$ is the image under A_1 of M, then M and N are closed subspaces and there must be complementary closed subspaces M^c such that $H = M \oplus M^c$ and N^c such that $K = N \oplus N^c$.

Remark 8.10. *The main results describe necessary and sufficient conditions for representation of the inverse operator $A(z)^{-1}$ by a Laurent series with a pole of order 1 at $z = 0$. The same conditions applied to special augmented operators are necessary and sufficient for representation of the inverse operator $A(z)^{-1}$ by a Laurent series with a higher order pole at $z = 0$. These results will be described later.*

8.7.4 ▪ Singular perturbation with a first order pole

We argued in an earlier section that existence of a Laurent series representation with a pole of order 1 at $z=0$ for the inverse operator $A(z)^{-1}$ implies a solution to the fundamental equations (8.20). We now consider the converse implication. We have the following simple but powerful result.

Lemma 8.26. *If $X_0, X_1 \in \mathcal{L}(K,H)$ satisfy the right-hand determining equations (8.21) and $Y_0, Y_1 \in \mathcal{L}(K,H)$ satisfy the left-hand determining equations (8.22), then $Y_0 = X_0$ and the operators $P = X_0A_1 \in \mathcal{L}(H,H)$ and $Q = A_1X_0 \in \mathcal{L}(K,K)$ are projection operators with $P^2 = P$ and $Q^2 = Q$. Furthermore, we have $P(H) = A_0^{-1}(\{0\}) = M$ with $A_0P = 0$ and $Q(K) = A_1A_0^{-1}(\{0\}) = A_1(M) = N$ with $QA_0 = 0$.*

Proof: It is convenient to introduce the notation

$$\mathcal{A}_1 = \begin{bmatrix} A_0 & 0 \\ A_1 & A_0 \end{bmatrix}, \quad \mathcal{X}_1 = \begin{bmatrix} X_0 & 0 \\ X_1 & X_0 \end{bmatrix}, \quad \mathcal{Y}_1 = \begin{bmatrix} Y_0 & 0 \\ Y_1 & Y_0 \end{bmatrix},$$

and

$$\mathcal{I}_1 = \begin{bmatrix} 0 & 0 \\ I & 0 \end{bmatrix},$$

from which it follows that the given equations can be written more compactly in the form

$$\mathcal{A}_1\mathcal{X}_1 = \mathcal{I}_1 \quad \text{and} \quad \mathcal{Y}_1\mathcal{A}_1 = \mathcal{I}_1.$$

It is now quite straightforward to verify that

$$\begin{bmatrix} 0 & 0 \\ X_0 & 0 \end{bmatrix} = \mathcal{I}_1\mathcal{X}_1 = (\mathcal{Y}_1\mathcal{A}_1)\mathcal{X}_1 = \mathcal{Y}_1(\mathcal{A}_1\mathcal{X}_1) = \mathcal{Y}_1\mathcal{I}_1 = \begin{bmatrix} 0 & 0 \\ Y_0 & 0 \end{bmatrix}.$$

Thus $Y_0 = X_0$. Define $P = X_0A_1 \in \mathcal{L}(H,H)$ and $Q = A_1X_0 \in \mathcal{L}(K,K)$. Since $X_0A_0 = 0$, we have

$$P^2 = X_0A_1 \cdot X_0A_1 = X_0(I - A_0X_1)A_1 = X_0A_1 - X_0A_0 \cdot X_1A_1 = X_0A_1 = P$$

and

$$Q^2 = A_1X_0 \cdot A_1X_0 = A_1X_0(I - A_0X_1) = A_1X_0 - A_1 \cdot X_0A_0 \cdot X_1 = A_1X_0 = Q.$$

If $\xi \in P(H)$, then $\xi = Px = X_0A_1x$ and hence $A_0\xi = A_0X_0A_1x = 0$. Thus $\xi \in A_0^{-1}(\{0\})$. On the other hand, if $\xi \in A_0^{-1}(\{0\})$, then $\xi = P\xi + (I-P)\xi = X_0A_1\xi + Y_1A_0\xi = X_0A_1\xi = P\xi \in P(H)$ because $A_0\xi = 0$. Thus $P(H) = A_0^{-1}(\{0\}) = M$. If $\zeta \in Q(K)$, then $\zeta = A_1X_0y = A_1x$, where $A_0x = A_0X_0y = 0$. Hence $\zeta \in A_1A_0^{-1}(\{0\})$. Conversely if $\zeta \in A_1A_0^{-1}(\{0\})$, then $\zeta = A_1\xi$, where $\xi \in P(H)$. Thus $\zeta = A_1X_0A_1x = A_1X_0y \in Q(K)$. Thus $Q(K) = A_1A_0^{-1}(\{0\}) = N$. Finally, we note that $A_0P = A_0X_0A_1 = 0$ and $QA_0 = A_1X_0A_0 = 0$. □

In view of the previous result we define Banach spaces $H_1 = (P_1(H), \|\cdot\|_H)$, $H_2 = (P_2(H), \|\cdot\|_H)$, $K_1 = (Q_1(K), \|\cdot\|_K)$, and $K_2 = (Q_2(K), \|\cdot\|_K)$, where we have written $P_1 = P$, $P_2 = I - P$, $Q_1 = Q$, and $Q_2 = I - Q$ for convenience. We also define auxiliary

Banach spaces $H_1 \times H_2 = (H_1 \times H_2, ||\cdot||_{H_1} + ||\cdot||_{H_2})$ and $K_1 \times K_2 = (K_1 \times K_2, ||\cdot||_{K_1} + ||\cdot||_{K_2})$. We note that if $||x_n||_H \to 0$ as $n \to \infty$, then

$$||(P_1 x_n, P_2 x_n)||_{H_1 \times H_2} = ||P_1 x_n||_H + ||P_2 x_n||_H \le 2||x_n||_H \to 0.$$

On the other hand, if $||(P_1 x_n, P_2 x_n)||_{H_1 \times H_2} \to 0$ as $n \to \infty$, then

$$||x_n||_H = ||P_1 x_n + P_2 x_n||_H \le ||P_1 x_n||_H + ||P_2 x_n||_H = ||(P_1 x_n, P_2 x_n)||_{H_1 \times H_2} \to 0.$$

Thus the topologies on H and $H_1 \times H_2$ are equivalent. A similar argument shows that the topologies on K and $K_1 \times K_2$ are also equivalent. We can reformulate the original problem in terms of equivalent operators $A_j \in \mathscr{L}(H_1 \times H_2, K_1 \times K_2)$, $X_j \in \mathscr{L}(K_1 \times K_2, H_1 \times H_2)$, and $Y_j \in \mathscr{L}(K_1 \times K_2, H_1 \times H_2)$, defined, respectively, by $A_j(P_1 x, P_2 x) = (Q_1 A_j x, Q_2 A_j x)$, $X_j(Q_1 y, Q_2 y) = (P_1 X_j y, P_2 X_j y)$, and $Y_j(Q_1 y, Q_2 y) = (P_1 X_j y, P_2 X_j y)$ for each $j = 0, 1$. For convenience we use the same symbols to denote the new operators. These operators can be represented in augmented matrix form as

$$A_j = \begin{bmatrix} A_{j11} & A_{j12} \\ A_{j21} & A_{j22} \end{bmatrix}, \quad X_j = \begin{bmatrix} X_{j11} & X_{j12} \\ X_{j21} & X_{j22} \end{bmatrix}, \quad \text{and} \quad Y_j = \begin{bmatrix} Y_{j11} & Y_{j12} \\ Y_{j21} & Y_{j22} \end{bmatrix},$$

where $A_{jrs} \in \mathscr{L}(H_r, K_s)$, $X_{jsr} \in \mathscr{L}(K_s, H_r)$, and $Y_{jsr} \in \mathscr{L}(K_s, H_r)$ are defined by the relevant projection operators. We write $x \in H \Leftrightarrow (\xi_1, \xi_2) \in H_1 \times H_2$ and $y \in K \Leftrightarrow (\zeta_1, \zeta_2) \in K_1 \times K_2$ and consider the various component operators. For the operator A_0 we have

1. $A_{011}(\xi_1) = Q_1 A_0 P_1(x) = A_1 X_0 A_0 P_1(x) = 0$ and hence $A_{011} = 0$;
2. $A_{012}(\xi_2) = Q_1 A_0 P_2(x) = A_1 X_0 A_0 P_2(x) = 0$ and hence $A_{012} = 0$;
3. $A_{021}(\xi_1) = Q_2 A_0 P_1(x) = Q_2 A_0 X_0 A_1(x) = 0$ and hence $A_{021} = 0$; and
4. $A_{022}(\xi_2) = Q_2 A_0 P_2(x) = (I - A_1 X_0) A_0 P_2(x) = A_0 P_2(x) = A_0(\xi_2)$ and hence $A_{022} = A_0|_{(H_2,K_2)}$ is the restriction of A_0 to $\mathscr{L}(H_2, K_2)$.

Therefore we write

$$A_0 = \begin{bmatrix} 0 & 0 \\ 0 & A_{022} \end{bmatrix}.$$

For the operator A_1 we calculate

1. $A_{111}(\xi_1) = Q_1 A_1 P_1(x) = A_1 X_0 A_1 X_0 A_1(x)$ and since $Q_1 = A_1 X_0$ is a projection it follows that $A_{111}(\xi) = A_1 X_0 A_1(x) = A_1 P_1(x) = A_1(\xi_1)$, and hence $A_{111} = A_1|_{(H_1,K_1)}$ is the restriction of A_1 to $\mathscr{L}(H_1, K_1)$;
2. $A_{112}(\xi_2) = Q_1 A_1 P_2(x) = A_1 X_0 A_1 (I - X_0 A_1)(x) = Q_1 A_1(x) - Q_1^2 A_1(x) = 0$ and hence $A_{112} = 0$;
3. $A_{121}(\xi_1) = Q_2 A_1 P_1(x) = A_0 X_1 A_1 X_0 A_1(x) = A_0 X_1 (I - A_0 X_1) A_1(x) = Q_2 A_1(x) - Q_2^2 A_1(x) = 0$ and hence $A_{121} = 0$; and
4. $A_{122}(\xi_2) = Q_2 A_1 P_2(x) = (I - A_1 X_0) A_1 (I - X_0 A_1)(x)$, from which it follows that $A_{122}(\xi_2) = A_1(x) - 2A_1 X_0 A_1(x) + A_1 (X_0 A_1)^2(x) = A_1(x) - A_1 X_0 A_1(x) = A_1 P_2(x) = A_1(\xi_2)$, and hence $A_{122} = A_1|_{(H_2,K_2)}$ is the restriction of A_1 to $\mathscr{L}(H_2, K_2)$.

Therefore, we write

$$A_1 = \begin{bmatrix} A_{111} & 0 \\ 0 & A_{122} \end{bmatrix}.$$

For the operator X_0 we find

1. $X_{011}(\zeta_1) = P_1X_0Q_1(y) = X_0A_1X_0A_1X_0(y) = X_0A_1X_0(y) = X_0(\zeta_1)$ and hence $X_{011} = X_0|_{K_1,H_1}$ is the restriction of X_0 to $\mathscr{L}(K_1, H_1)$;
2. $X_{012}(\zeta_2) = P_1X_0Q_2(y) = P_1X_0A_0X_1(y) = 0$ and hence $X_{012} = 0$;
3. $X_{021}(\zeta_1) = P_2X_0Q_1(y) = Y_1A_0X_0Q_1(y) = 0$ and hence $X_{021} = 0$; and
4. $X_{022}(\zeta_2) = P_2X_0Q_2(y) = Y_1A_0X_0Q_2(y) = 0$ and hence $X_{022} = 0$.

Therefore, we write

$$Y_0 = X_0 = \begin{bmatrix} X_{011} & 0 \\ 0 & 0 \end{bmatrix}.$$

For the operators X_1 and Y_1 there are no obvious simplifications at this stage, and hence we write

$$X_1 = \begin{bmatrix} X_{111} & X_{112} \\ X_{121} & X_{122} \end{bmatrix} \quad \text{and} \quad Y_1 = \begin{bmatrix} Y_{111} & Y_{112} \\ Y_{121} & Y_{122} \end{bmatrix}.$$

In the augmented matrix notation the two equations for system (8.21) become

$$A_0X_0 = 0 \Leftrightarrow \begin{bmatrix} 0 & 0 \\ 0 & A_{022} \end{bmatrix} \begin{bmatrix} X_{011} & 0 \\ 0 & 0 \end{bmatrix} = \begin{bmatrix} 0 & 0 \\ 0 & 0 \end{bmatrix}$$

and

$$A_1X_0 + A_0X_1 = I \Leftrightarrow$$

$$\begin{bmatrix} A_{111} & 0 \\ 0 & A_{122} \end{bmatrix} \begin{bmatrix} X_{011} & 0 \\ 0 & 0 \end{bmatrix} + \begin{bmatrix} 0 & 0 \\ 0 & A_{022} \end{bmatrix} \begin{bmatrix} X_{111} & X_{112} \\ X_{121} & X_{122} \end{bmatrix} = \begin{bmatrix} I & 0 \\ 0 & I \end{bmatrix}.$$

By considering the equations for the various components we can see that our transformations have reduced the system to three equations

$$A_{111}X_{011} = I, \quad A_{022}X_{121} = 0, \quad \text{and} \quad A_{022}X_{122} = I. \tag{8.23}$$

In the augmented matrix notation the two equations for the system (8.22) become

$$X_0A_0 = 0 \Leftrightarrow \begin{bmatrix} X_{011} & 0 \\ 0 & 0 \end{bmatrix} \begin{bmatrix} 0 & 0 \\ 0 & A_{022} \end{bmatrix} = \begin{bmatrix} 0 & 0 \\ 0 & 0 \end{bmatrix}$$

and

$$X_0A_1 + Y_1A_0 = I \Leftrightarrow$$

$$\begin{bmatrix} X_{011} & 0 \\ 0 & 0 \end{bmatrix} \begin{bmatrix} A_{111} & 0 \\ 0 & A_{122} \end{bmatrix} + \begin{bmatrix} Y_{111} & Y_{112} \\ Y_{121} & Y_{122} \end{bmatrix} \begin{bmatrix} 0 & 0 \\ 0 & A_{022} \end{bmatrix} = \begin{bmatrix} I & 0 \\ 0 & I \end{bmatrix}.$$

By considering the various components it follows, once again, that our transformations have reduced the system to three equations

$$X_{011}A_{111} = I, \quad Y_{112}A_{022} = 0, \quad \text{and} \quad Y_{122}A_{022} = I. \tag{8.24}$$

From equations (8.23) and (8.24) we have $A_{111}X_{011} = I$ and $X_{011}A_{111} = I$. Thus it is necessary and sufficient that $A_{111} \in \mathscr{L}(H_1,K_1)$ is one-to-one and onto and in this case $X_{011} = A_{111}^{-1}$. Equations (8.23) and (8.24) also show us that

$$Y_{122} = Y_{122}(A_{022}X_{122}) = (Y_{122}A_{022})X_{122} = X_{122},$$

and hence $A_{022}X_{122} = I$ and $X_{122}A_{022} = I$. Therefore, it is necessary and sufficient that $A_{022} \in \mathscr{L}(H_2,K_2)$ is one-to-one and onto and in this case $X_{122} = A_{022}^{-1}$. Finally, it follows that $X_{121} = 0$ and $Y_{112} = 0$. We can summarize these results in the following theorem.

Theorem 8.27. *If $X_0, X_1 \in \mathscr{L}(K,H)$ satisfy the right-hand determining equations (8.21) and $Y_0, Y_1 \in \mathscr{L}(K,H)$ satisfy the left-hand determining equations (8.22), then we can define projections $P \in \mathscr{L}(H,H)$ and $Q \in \mathscr{L}(K,K)$ by the formulae $P = X_0A_1$ and $Q = A_1X_0$. If we define $P_1 = P$, $P_2 = I - P$, $Q_1 = Q$, and $Q_2 = I - Q$, then we can also define Banach spaces $H_1 = P_1(H) = A_0^{-1}(\{0\}) = M$, $H_2 = P_2(H) = M^c$, $K_1 = Q_1(K) = A_1A_0^{-1}(\{0\}) = N$, $K_2 = Q_2(K) = N^c$ and represent the given mappings in the form $A_j \in \mathscr{L}(H_1 \times H_2, K_1 \times K_2)$, where*

$$A_0 = \begin{bmatrix} 0 & 0 \\ 0 & A_{022} \end{bmatrix} \quad \text{and} \quad A_1 = \begin{bmatrix} A_{111} & 0 \\ 0 & A_{122} \end{bmatrix}$$

and where $A_{022} \in \mathscr{L}(H_2,K_2)$ and $A_{111} \in \mathscr{L}(H_1,K_1)$ are each one-to-one and onto. Furthermore, if we represent the solutions as mappings in the form $X_j \in \mathscr{L}(K_1 \times K_2, H_1 \times H_2)$ and $Y_j \in \mathscr{L}(K_1 \times K_2, H_1 \times H_2)$, then

$$Y_0 = X_0 = \begin{bmatrix} A_{111}^{-1} & 0 \\ 0 & 0 \end{bmatrix}, \quad X_1 = \begin{bmatrix} X_{111} & X_{112} \\ 0 & A_{022}^{-1} \end{bmatrix}, \quad \text{and} \quad Y_1 = \begin{bmatrix} Y_{111} & 0 \\ Y_{121} & A_{022}^{-1} \end{bmatrix},$$

where X_{111}, X_{112}, Y_{111}, and Y_{121} are undetermined.

Corollary 8.3. *Suppose $X_0, X_1 \in \mathscr{L}(K,H)$ satisfy the right-hand determining equations (8.21) and $Y_0, Y_1 \in \mathscr{L}(K,H)$ satisfy the left-hand determining equations (8.22). If we define $Z_0, Z_1 \in \mathscr{L}(K,H)$ by setting $Z_0 = X_0 = Y_0$ and $Z_1 = Y_1A_0X_1$ and if we represented these operators as mappings in the form $Z_j \in \mathscr{L}(K_1 \times K_2, H_1 \times H_2)$, we then have*

$$Z_0 = \begin{bmatrix} A_{111}^{-1} & 0 \\ 0 & 0 \end{bmatrix} \quad \text{and} \quad Z_1 = \begin{bmatrix} 0 & 0 \\ 0 & A_{022}^{-1} \end{bmatrix}$$

and Z_0, Z_1 satisfy both sets of determining equations (8.21) and (8.22).

Proof: We have

$$Z_1 = \begin{bmatrix} Y_{111} & 0 \\ Y_{121} & A_{022}^{-1} \end{bmatrix} \begin{bmatrix} 0 & 0 \\ 0 & A_{022} \end{bmatrix} \begin{bmatrix} X_{111} & X_{112} \\ 0 & A_{022}^{-1} \end{bmatrix} = \begin{bmatrix} 0 & 0 \\ 0 & A_{022}^{-1} \end{bmatrix}.$$

The determining equations can be verified by substituting the expressions for the partitioned operators. □

Now that we have obtained a clear view of the underlying structure, we can formulate the sufficient conditions in a more basic form. In Lemma 8.26 and Theorem 8.27 the existence of solutions of the equations (8.21) and (8.22) was shown to be a sufficient condition to construct the two related projections that define the desired complementation process. Suppose we assume instead the existence of linear projections $P \in \mathscr{L}(H,H_1)$, where

$H_1 = M = A_0^{-1}(\{0\})$ with $PA_0 = 0$, and $Q \in \mathscr{L}(K, K_1)$, where $K_1 = N = A_1 A_0^{-1}(\{0\})$ with $A_0 Q = 0$ such that A_0 is bounded below on $H_2 = (I-P)(H) = M^c$ and A_1 is bounded below on $H_1 = M$. We use the same notation as before and similar reasoning to show that $A_j \in \mathscr{L}(H_1 \times H_2, K_1 \times K_2)$ for each $j = 0, 1$ can be represented in the form

$$A_0 = \begin{bmatrix} 0 & 0 \\ 0 & A_{022} \end{bmatrix} \quad \text{and} \quad A_1 = \begin{bmatrix} A_{111} & 0 \\ 0 & A_{122} \end{bmatrix},$$

where $A_{022}^{-1}, A_{111}^{-1}$ are well defined. In particular, we note that $PA_0 = 0$ and $A_0 Q = 0$ implies $A_{011} = 0$, $A_{012} = 0$ and $A_{021} = 0$. We also note that $A_1(I-P)\xi_2 = 0$ implies $A_{112} = 0$ and $(I-Q)A_1\xi_1 = (I-Q)\zeta_1 = 0$ implies $A_{121} = 0$. If we define operators $X_j, Y_j \in \mathscr{L}(K_1 \times K_2, H_1 \times H_2)$ for each $j = 0, 1$ by the formulae

$$Y_0 = X_0 = \begin{bmatrix} A_{111}^{-1} & 0 \\ 0 & 0 \end{bmatrix}, \quad X_1 = \begin{bmatrix} X_{111} & X_{112} \\ 0 & A_{022}^{-1} \end{bmatrix}, \quad \text{and} \quad Y_1 = \begin{bmatrix} Y_{111} & 0 \\ Y_{121} & A_{022}^{-1} \end{bmatrix},$$

where X_{111}, X_{112}, Y_{111}, and Y_{121} are unspecified, then the operators X_0, X_1 solve the equations (8.21) and the operators Y_0, Y_1 solve the equations (8.22). If we set $X_{111} = 0$, $X_{112} = 0$, $Y_{111} = 0$, and $Y_{121} = 0$, then $X_0 = Y_0 = Z_0$ and $X_1 = Y_1 = Z_1$ are solutions to both (8.21) and (8.22).

We return to the original question which we now state in terms of the reformulated operators. Let $A_j \in \mathscr{L}(H_1 \times H_2, K_1 \times K_2)$ be given by

$$A_0 = \begin{bmatrix} 0 & 0 \\ 0 & A_{022} \end{bmatrix} \quad \text{and} \quad A_1 = \begin{bmatrix} A_{111} & 0 \\ 0 & A_{122} \end{bmatrix},$$

where A_{022}^{-1} and A_{111}^{-1} are well defined. Can we find $\{X_j\} \subset \mathscr{L}(K_1 \times K_2, H_1 \times H_2)$ such that

$$(A_0 + A_1 z)^{-1} = \frac{1}{z}\left(X_0 + X_1 z + X_2 z^2 + \cdots\right)$$

for some deleted neighborhood $0 < |z| < r$? It is now straightforward to answer this question in the affirmative. Indeed, we can see from the Neumann expansion that

$$\begin{aligned}
(A_0 + A_1 z)^{-1} &= \begin{bmatrix} A_{111} z & 0 \\ 0 & A_{022} + A_{122} z \end{bmatrix}^{-1} \\
&= \begin{bmatrix} A_{111}^{-1} \cdot \dfrac{1}{z} & 0 \\ 0 & (A_{022} + A_{122} z)^{-1} \end{bmatrix} \\
&= \begin{bmatrix} A_{111}^{-1} \cdot \dfrac{1}{z} & 0 \\ 0 & A_{022}^{-1} + (-1) A_{022}^{-1} A_{122} A_{022}^{-1} \cdot z + \cdots \end{bmatrix} \\
&= \begin{bmatrix} A_{111}^{-1} & 0 \\ 0 & 0 \end{bmatrix} \cdot \frac{1}{z} + \begin{bmatrix} 0 & 0 \\ 0 & A_{022}^{-1} \end{bmatrix} \\
&\qquad + \begin{bmatrix} 0 & 0 \\ 0 & (-1) A_{022}^{-1} A_{122} A_{022}^{-1} \end{bmatrix} \cdot z + \cdots \\
&= \frac{1}{z}\left(X_0 + X_1 z + X_2 z^2 + \cdots\right)
\end{aligned}$$

as required, where

$$X_0 = \begin{bmatrix} A_{111}^{-1} & 0 \\ 0 & 0 \end{bmatrix}, \quad X_1 = \begin{bmatrix} 0 & 0 \\ 0 & A_{022}^{-1} \end{bmatrix},$$

and

$$X_j = \begin{bmatrix} 0 & 0 \\ 0 & (-1)^{j-1}(A_{022}^{-1}A_{122})^{j-1}A_{022}^{-1} \end{bmatrix}$$

for each $j \geq 2$. If we define $R = ||A_{022}^{-1}|| \cdot ||A_{122}||$ and set $r = 1/R$, then the series converges for $0 < |z| < r$.

There is an alternative, but equivalent, form for the question. Can we find operators $\{X_j\} \subset \mathscr{L}(K_1 \times K_2, H_1 \times H_2)$ to solve the equations

$$\begin{array}{rcl} A_0X_0 & = & 0, \\ A_1X_0 + A_0X_1 & = & I, \\ A_1X_1 + A_0X_2 & = & 0, \\ A_1X_2 + A_0X_3 & = & 0, \\ \vdots & & \vdots \end{array} \quad \text{and} \quad \begin{array}{rcl} X_0A_0 & = & 0, \\ X_0A_1 + X_1A_0 & = & I, \\ X_1A_1 + X_2A_0 & = & 0, \\ X_2A_1 + X_3A_0 & = & 0, \\ \vdots & & \vdots \end{array}$$

with $||X_j|| < Cr^{j+1}$ for all $j = 0, 1, \ldots$, for some $C, r > 0$? Once again the answer is clear. We can represent the system of right-hand inverse equations in augmented matrix form as

$$\left[\begin{array}{ccccccccc|cc} 0 & 0 & 0 & 0 & 0 & 0 & 0 & 0 & \cdots & 0 & 0 \\ 0 & A_{022} & 0 & 0 & 0 & 0 & 0 & 0 & \cdots & 0 & 0 \\ A_{111} & 0 & 0 & 0 & 0 & 0 & 0 & 0 & \cdots & I & 0 \\ 0 & A_{122} & 0 & A_{022} & 0 & 0 & 0 & 0 & \cdots & 0 & I \\ 0 & 0 & A_{111} & 0 & 0 & 0 & 0 & 0 & \cdots & 0 & 0 \\ 0 & 0 & 0 & A_{122} & 0 & A_{022} & 0 & 0 & \cdots & 0 & 0 \\ 0 & 0 & 0 & 0 & A_{111} & 0 & 0 & 0 & \cdots & 0 & 0 \\ 0 & 0 & 0 & 0 & 0 & A_{122} & 0 & A_{022} & \cdots & 0 & 0 \\ \vdots & \vdots & \vdots & \vdots & \vdots & \vdots & \vdots & \vdots & & \vdots & \vdots \end{array}\right],$$

which we reduce using elementary row operations to give

$$\left[\begin{array}{ccccccccc|cc} I & 0 & 0 & 0 & 0 & 0 & 0 & 0 & \cdots & A_{111}^{-1} & 0 \\ 0 & I & 0 & 0 & 0 & 0 & 0 & 0 & \cdots & 0 & 0 \\ 0 & 0 & I & 0 & 0 & 0 & 0 & 0 & \cdots & 0 & 0 \\ 0 & 0 & 0 & I & 0 & 0 & 0 & 0 & \cdots & 0 & A_{022}^{-1} \\ 0 & 0 & 0 & 0 & I & 0 & 0 & 0 & \cdots & 0 & 0 \\ 0 & 0 & 0 & 0 & 0 & I & 0 & 0 & \cdots & 0 & (-1)A_{022}^{-1}A_{122}A_{022}^{-1} \\ 0 & 0 & 0 & 0 & 0 & 0 & I & 0 & \cdots & 0 & 0 \\ 0 & 0 & 0 & 0 & 0 & 0 & 0 & I & \cdots & 0 & (-1)^2(A_{022}^{-1}A_{122})^2A_{022}^{-1} \\ \vdots & \vdots & \vdots & \vdots & \vdots & \vdots & \vdots & \vdots & & \vdots & \vdots \end{array}\right].$$

By transposing the system of left-hand inverse equations and applying analogous row operations and subsequently transposing again we obtain a similar reduction for the left-hand inverse equations. The reduced equations define a unique solution and allow us to construct the reformulated inverse operator. While our transformations have resulted in an elegant separation, it is clear that we can convert the solution of the separated problem into a solution for the original problem by applying the inverse transformations. Thus we have the original mappings represented in the form

$$A_0 = (I-Q)A_{022}(I-P) \quad \text{and} \quad A_1 = QA_{111}P + (I-Q)A_{122}(I-P)$$

with the original solutions given by

$$X_0 = PA_{111}^{-1}Q \quad \text{and} \quad X_j = (I-P)(A_{022}^{-1}A_{122})^{j-1}A_{022}^{-1}(I-Q)$$

for each $j \geq 1$. Since P and Q are projections, it follows that $||X_0|| \leq ||A_{111}^{-1}||$ and $||X_j|| \leq ||A_{022}^{-1}||^j ||A_{122}||^{j-1}$ for $j \geq 1$, and hence if we let $R = ||A_{022}^{-1}|| \cdot ||A_{122}||$ and set $r = 1/R$, then

$$(A_0 + A_1 z)^{-1} = \frac{1}{z}\left(X_0 + X_1 z + X_2 z^2 + \cdots\right)$$

for $0 < |z| < r$.

Remark 8.11. *It is important to summarize what we have done. Theorem 8.27 shows us that a solution to the determining equations implies existence of two related projections. The subsequent discussion shows us that these projections enable us to construct the inverse operator* $A(z)^{-1}$. *Since we already know from Subsection 8.7.2 that existence of the inverse operator implies a solution to the fundamental equations, we have now established Theorem 8.24. We have observed in Theorem 8.27 that the determining equations imply the existence of two related projections. The discussion following Theorem 8.27 also shows us that existence of the two projections allows us to construct the inverse operator, and this, in turn, allows us to solve the fundamental equations. Thus we have also established Theorem 8.25.*

8.7.5 ▪ Singular perturbations with higher order poles

Similar results can be established for singular perturbations where the inverse operator has a higher order pole. Although these results can be obtained directly, we will use certain special augmented operators to show that only the first order theory is required. We will consider the particular case of a second order pole and simply assert that similar methods can be applied to higher order poles. If we assume that

$$(A_0 + A_1 z)^{-1} = \frac{1}{z^2}\left(X_0 + X_1 z + X_2 z^2 + \cdots\right)$$

on some deleted neighborhood $0 < |z| < r$, then the sequence $\{X_j\} \subset \mathscr{L}(K,H)$ must satisfy the equations

$$\begin{array}{rcl} A_0X_0 &=& 0, \\ A_1X_0 + A_0X_1 &=& 0, \\ A_1X_1 + A_0X_2 &=& I, \\ A_1X_2 + A_0X_3 &=& 0, \\ \vdots & & \vdots \end{array} \quad \text{and} \quad \begin{array}{rcl} X_0A_0 &=& 0, \\ X_0A_1 + X_1A_0 &=& 0, \\ X_1A_1 + X_2A_0 &=& I, \\ X_2A_1 + X_3A_0 &=& 0, \\ \vdots & & \vdots \end{array} \tag{8.25}$$

and we must have $||X_j|| \cdot |z|^j \to 0$ as $j \to \infty$ for all $|z| < r$. If we use the augmented matrix notation $\mathscr{A}_j \in \mathscr{L}(H \times H, K \times K)$ for each $j = 0, 1$, where

$$\mathscr{A}_0 = \begin{bmatrix} A_0 & 0 \\ A_1 & A_0 \end{bmatrix} \quad \text{and} \quad \mathscr{A}_1 = \begin{bmatrix} 0 & A_1 \\ 0 & 0 \end{bmatrix},$$

and $\mathscr{X}_j \in \mathscr{L}(K \times K, H \times H)$ for each $j = 0, 1, 2, \ldots$, where

$$\mathscr{X}_0 = \begin{bmatrix} X_0 & 0 \\ X_1 & X_0 \end{bmatrix}, \quad \mathscr{X}_1 = \begin{bmatrix} X_2 & X_1 \\ X_3 & X_2 \end{bmatrix}, \quad \mathscr{X}_2 = \begin{bmatrix} X_4 & X_3 \\ X_5 & X_4 \end{bmatrix}, \ldots,$$

and if we write

$$\mathscr{I} = \begin{bmatrix} I & 0 \\ 0 & I \end{bmatrix},$$

then the above equations can be rewritten in the equivalent form

$$\begin{array}{rcl} \mathscr{A}_0\mathscr{X}_0 & = & 0, \\ \mathscr{A}_1\mathscr{X}_0 + \mathscr{A}_0\mathscr{X}_1 & = & \mathscr{I}, \\ \mathscr{A}_1\mathscr{X}_1 + \mathscr{A}_0\mathscr{X}_2 & = & 0, \\ \mathscr{A}_1\mathscr{X}_2 + \mathscr{A}_0\mathscr{X}_3 & = & 0, \\ \vdots & & \vdots \end{array} \quad \text{and} \quad \begin{array}{rcl} \mathscr{X}_0\mathscr{A}_0 & = & 0, \\ \mathscr{X}_0\mathscr{A}_1 + \mathscr{X}_1\mathscr{A}_0 & = & \mathscr{I}, \\ \mathscr{X}_1\mathscr{A}_1 + \mathscr{X}_2\mathscr{A}_0 & = & 0, \\ \mathscr{X}_2\mathscr{A}_1 + \mathscr{X}_3\mathscr{A}_0 & = & 0, \\ \vdots & & \vdots \end{array} \tag{8.26}$$

where we must have $||\mathscr{X}_j|| \cdot |z|^j \to 0$ as $j \to \infty$ for all $|z| < r$. In the first instance we have the following result.

Theorem 8.28. *The representation*

$$(A_0 + A_1 z)^{-1} = \frac{1}{z^2}\left(X_0 + X_1 z + X_2 z^2 + \cdots\right)$$

is valid on the deleted neighborhood $0 < |z| < r$ *if and only if the representation*

$$(\mathscr{A}_0 + \mathscr{A}_1 z)^{-1} = \frac{1}{z}\left(\mathscr{X}_0 + \mathscr{X}_1 z + \mathscr{X}_2 z^2 + \cdots\right)$$

is valid on the deleted neighborhood $0 < |z| < r$.

We can use this result to write the following analogues of Theorem 8.27 and Corollary 8.3.

Theorem 8.29. *If* $X_0, X_1, X_2, X_3 \in \mathscr{L}(K, H)$ *satisfy the right-hand determining equations*

$$A_0X_0 = 0, \quad A_1X_0 + A_0X_1 = 0, \quad A_1X_1 + A_0X_2 = I, \quad A_1X_2 + A_0X_3 = 0, \tag{8.27}$$

and $Y_0, Y_1, Y_2, Y_3 \in \mathscr{L}(K, H)$ *satisfy the left-hand determining equations*

$$Y_0A_0 = 0, \quad Y_0A_1 + Y_1A_0 = 0, \quad Y_1A_1 + Y_2A_0 = I, \quad Y_2A_1 + Y_3A_0 = 0, \tag{8.28}$$

then we can define projections $\mathscr{P} \in \mathscr{L}(\mathscr{H}, \mathscr{H})$ *and* $\mathscr{Q} \in \mathscr{L}(\mathscr{K}, \mathscr{K})$, *where* $\mathscr{H} = H \times H$ *and* $\mathscr{K} = K \times K$, *by the formulae* $\mathscr{P} = \mathscr{X}_0\mathscr{A}_1$ *and* $\mathscr{Q} = \mathscr{A}_1\mathscr{X}_0$. *If we define* $\mathscr{P}_1 = \mathscr{P}$, $\mathscr{P}_2 = \mathscr{I} - \mathscr{P}$, $\mathscr{Q}_1 = \mathscr{Q}$, *and* $\mathscr{Q}_2 = \mathscr{I} - \mathscr{Q}$, *then we can also define Banach spaces* $\mathscr{H}_1 = \mathscr{P}_1(\mathscr{H}) = \mathscr{A}_0^{-1}(\{0\}) = \mathscr{M}$, $\mathscr{H}_2 = \mathscr{P}_2(\mathscr{H}) = \mathscr{M}^c$, $\mathscr{K}_1 = \mathscr{Q}_1(\mathscr{K}) = \mathscr{A}_1\mathscr{A}_0^{-1}(\{0\}) = \mathscr{N}$, $\mathscr{K}_2 = \mathscr{Q}_2(\mathscr{K}) = \mathscr{N}^c$ *and represent the given mappings in the form* $\mathscr{A}_j \in \mathscr{L}(\mathscr{H}_1 \times \mathscr{H}_2, \mathscr{K}_1 \times \mathscr{K}_2)$, *where*

$$\mathscr{A}_0 = \begin{bmatrix} 0 & 0 \\ 0 & \mathscr{A}_{022} \end{bmatrix} \quad \text{and} \quad \mathscr{A}_1 = \begin{bmatrix} \mathscr{A}_{111} & 0 \\ 0 & \mathscr{A}_{122} \end{bmatrix}$$

and where $\mathscr{A}_{022} \in \mathscr{L}(\mathscr{H}_2,\mathscr{K}_2)$ and $\mathscr{A}_{111} \in \mathscr{L}(\mathscr{H}_1,\mathscr{K}_1)$ are each one-to-one and onto. Furthermore, if we represent the solutions as mappings in the form $\mathscr{X}_j \in \mathscr{L}(\mathscr{K}_1\times\mathscr{K}_2,\mathscr{H}_1\times\mathscr{H}_2)$ and $\mathscr{Y}_j \in \mathscr{L}(\mathscr{K}_1\times\mathscr{K}_2,\mathscr{H}_1\times\mathscr{H}_2)$, then

$$\mathscr{Y}_0 = \mathscr{X}_0 = \begin{bmatrix} \mathscr{A}_{111}^{-1} & 0 \\ 0 & 0 \end{bmatrix}, \quad \mathscr{X}_1 = \begin{bmatrix} \mathscr{X}_{111} & \mathscr{X}_{112} \\ 0 & \mathscr{A}_{022}^{-1} \end{bmatrix}, \text{ and } \mathscr{Y}_1 = \begin{bmatrix} \mathscr{Y}_{111} & 0 \\ \mathscr{Y}_{121} & \mathscr{A}_{022}^{-1} \end{bmatrix},$$

where $\mathscr{X}_{111}$, $\mathscr{X}_{112}$, $\mathscr{Y}_{111}$, and $\mathscr{Y}_{121}$ are undetermined.

Corollary 8.4. *Suppose $X_0,X_1,X_2,X_3 \in \mathscr{L}(K,H)$ satisfy the right-hand determining equations (8.27) and $Y_0,Y_1,Y_2,Y_3 \in \mathscr{L}(K,H)$ satisfy the left-hand determining equations (8.28). If we define $\mathscr{Z}_0,\mathscr{Z}_1 \in \mathscr{L}(\mathscr{K},\mathscr{H})$ by setting $\mathscr{Z}_0 = \mathscr{X}_0 = \mathscr{Y}_0$ and $\mathscr{Z}_1 = \mathscr{Y}_1\mathscr{A}_0\mathscr{X}_1$ and if we represent these operators as mappings in the form $\mathscr{Z}_j \in \mathscr{L}(\mathscr{K}_1\times\mathscr{K}_2,\mathscr{H}_1\times\mathscr{H}_2)$, then*

$$\mathscr{Z}_0 = \begin{bmatrix} Z_0 & 0 \\ Z_1 & Z_0 \end{bmatrix} \quad \text{and} \quad \mathscr{Z}_1 = \begin{bmatrix} Z_2 & Z_1 \\ Z_3 & Z_2 \end{bmatrix},$$

where $Z_0 = X_0 = Y_0$, $Z_1 = X_1 = Y_1$, $Z_2 = Y_2A_0X_2$, and $Z_3 = (-1)Y_2A_1X_2$ satisfy both sets of determining equations (8.27) and (8.28).

Proof: To begin we note that

$$\mathscr{X}_0 = \begin{bmatrix} X_0 & 0 \\ X_1 & X_0 \end{bmatrix} \quad \text{and} \quad \mathscr{Y}_0 = \begin{bmatrix} Y_0 & 0 \\ Y_1 & Y_0 \end{bmatrix},$$

and since $\mathscr{Z}_0 = \mathscr{X}_0 = \mathscr{Y}_0$, it follows that

$$\mathscr{Z}_0 = \begin{bmatrix} Z_0 & 0 \\ Z_1 & Z_0 \end{bmatrix},$$

where $Z_0 = X_0 = Y_0$ and $Z_1 = X_1 = Y_1$. Next we have

$$\mathscr{X}_1 = \begin{bmatrix} X_2 & X_1 \\ X_3 & X_2 \end{bmatrix} \quad \text{and} \quad \mathscr{Y}_1 = \begin{bmatrix} Y_2 & Y_1 \\ Y_3 & Y_2 \end{bmatrix},$$

and since $\mathscr{Z}_1 = \mathscr{Y}_1\mathscr{A}_0\mathscr{X}_1$, it follows that

$$\begin{aligned}
\mathscr{Z}_1 &= \begin{bmatrix} Y_2 & Y_1 \\ Y_3 & Y_2 \end{bmatrix}\begin{bmatrix} A_0 & 0 \\ A_1 & A_0 \end{bmatrix}\begin{bmatrix} X_2 & X_1 \\ X_3 & X_2 \end{bmatrix} \\
&= \begin{bmatrix} Y_2A_0 + Y_1A_3 & Y_1A_0 \\ Y_3A_0 + Y_2A_1 & Y_2A_0 \end{bmatrix}\begin{bmatrix} X_2 & X_1 \\ X_3 & X_2 \end{bmatrix} \\
&= \begin{bmatrix} I & Y_1A_0 \\ 0 & Y_2A_0 \end{bmatrix}\begin{bmatrix} X_2 & X_1 \\ X_3 & X_2 \end{bmatrix} \\
&= \begin{bmatrix} X_2 + Y_1A_0X_3 & X_1 + Y_1A_0X_2 \\ Y_2A_0X_3 & Y_2A_0X_2 \end{bmatrix}.
\end{aligned}$$

By applying the determining equations (8.27) we note that

$$Y_1A_0X_2 = -Y_0A_1X_2 = (-1)^2Y_0A_0X_3 = 0.$$

We also use (8.28) to observe that

$$X_2 + Y_1A_0X_3 = X_2 - Y_1A_1X_2 = (I - Y_1A_1)X_2 = Y_2A_0X_2,$$

and finally from (8.27) we note that

$$Y_2A_0X_3 = (-1)Y_2A_1X_2.$$

Hence we have

$$\mathscr{X}_1 = \begin{bmatrix} Y_2A_0X_2 & X_1 \\ (-1)Y_2A_1X_2 & Y_2A_0X_2 \end{bmatrix},$$

and so we obtain

$$\mathscr{X}_1 = \begin{bmatrix} Z_2 & Z_1 \\ Z_3 & Z_2 \end{bmatrix},$$

where $Z_2 = Y_2A_0X_2$ and $Z_3 = (-1)Y_2A_1X_2$. By substituting these expressions into the determining equations and using some elementary matrix algebra and the fact that the X_j and Y_j satisfy (8.27) and (8.28) we can now show that the Z_j satisfy both (8.27) and (8.28). □

Although the augmentation is a very convenient way to formulate the necessary and sufficient conditions, the solution is best computed directly from the original equations.

Example 8.15. *Let*

$$A_0 = \begin{bmatrix} 1 & 0 \\ 1 & 0 \end{bmatrix} \quad \textit{and} \quad A_1 = \begin{bmatrix} 1 & 1 \\ 0 & 1 \end{bmatrix}.$$

The equation

$$A_0X_0 = I \quad \Leftrightarrow \quad [A_0 \mid I] = \left[\begin{array}{cc|cc} 1 & 0 & 1 & 0 \\ 1 & 0 & 0 & 1 \end{array}\right]$$

can be reduced using elementary row operations to

$$\left[\begin{array}{cc|cc} 1 & 0 & 1 & 0 \\ 0 & 0 & 1 & -1 \end{array}\right],$$

which has no solution. The equations

$$A_0X_0 = 0 \quad \textit{and} \quad A_1X_0 + A_0X_1 = I$$

can be written in the form

$$\left[\begin{array}{cc|c} A_0 & 0 & 0 \\ A_1 & A_0 & I \end{array}\right] = \left[\begin{array}{cccc|cc} 1 & 0 & 0 & 0 & 0 & 0 \\ 1 & 0 & 0 & 0 & 0 & 0 \\ 1 & 1 & 1 & 0 & 1 & 0 \\ 0 & 1 & 1 & 0 & 0 & 1 \end{array}\right]$$

and then reduced to the equivalent form

$$\left[\begin{array}{cccc|cc} 1 & 0 & 0 & 0 & 0 & 0 \\ 0 & 1 & 1 & 0 & 0 & 1 \\ 0 & 0 & 0 & 0 & 1 & -1 \\ 0 & 0 & 0 & 0 & 0 & 0 \end{array}\right],$$

which also has no solution. However, the augmented equations

$$\mathscr{A}_0\mathscr{X}_0 = 0 \quad \textit{and} \quad \mathscr{A}_1\mathscr{X}_0 + \mathscr{A}_0\mathscr{X}_1 = \mathscr{I}$$

can be written as

$$\left[\begin{array}{cc|c} \mathcal{A}_0 & 0 & 0 \\ \mathcal{A}_1 & \mathcal{A}_0 & \mathcal{I} \end{array}\right] = \left[\begin{array}{cccccccc|cccc} 1 & 0 & 0 & 0 & 0 & 0 & 0 & 0 & 0 & 0 & 0 & 0 \\ 1 & 0 & 0 & 0 & 0 & 0 & 0 & 0 & 0 & 0 & 0 & 0 \\ 1 & 1 & 1 & 0 & 0 & 0 & 0 & 0 & 0 & 0 & 0 & 0 \\ 0 & 1 & 1 & 0 & 0 & 0 & 0 & 0 & 0 & 0 & 0 & 0 \\ 0 & 0 & 1 & 1 & 1 & 0 & 0 & 0 & 1 & 0 & 0 & 0 \\ 0 & 0 & 0 & 1 & 1 & 0 & 0 & 0 & 0 & 1 & 0 & 0 \\ 0 & 0 & 0 & 0 & 1 & 1 & 1 & 0 & 0 & 0 & 1 & 0 \\ 0 & 0 & 0 & 0 & 0 & 1 & 1 & 0 & 0 & 0 & 0 & 1 \end{array}\right]$$

and reduced to

$$\left[\begin{array}{cccccccc|cccc} 1 & 0 & 0 & 0 & 0 & 0 & 0 & 0 & 0 & 0 & 0 & 0 \\ 0 & 1 & 0 & 0 & 0 & 0 & 0 & 0 & -1 & 1 & 0 & 0 \\ 0 & 0 & 1 & 0 & 0 & 0 & 0 & 0 & 1 & -1 & 0 & 0 \\ 0 & 0 & 0 & 1 & 0 & 0 & 0 & 0 & 0 & 1 & -1 & 1 \\ 0 & 0 & 0 & 0 & 1 & 0 & 0 & 0 & 0 & 0 & 1 & -1 \\ 0 & 0 & 0 & 0 & 0 & 1 & 1 & 0 & 0 & 0 & 0 & 1 \\ 0 & 0 & 0 & 0 & 0 & 0 & 0 & 0 & 0 & 0 & 0 & 0 \\ 0 & 0 & 0 & 0 & 0 & 0 & 0 & 0 & 0 & 0 & 0 & 0 \end{array}\right],$$

from which it follows that

$$X_0 = \begin{bmatrix} 0 & 0 \\ -1 & 1 \end{bmatrix}, \quad X_1 = \begin{bmatrix} 1 & -1 \\ 0 & 1 \end{bmatrix}, \quad \textit{and} \quad X_2 = \begin{bmatrix} 0 & 0 \\ s & t \end{bmatrix},$$

where s and t are arbitrary parameters. Clearly

$$A_0X_0 = \begin{bmatrix} 1 & 0 \\ 1 & 0 \end{bmatrix}\begin{bmatrix} 0 & 0 \\ -1 & 1 \end{bmatrix} = \begin{bmatrix} 0 & 0 \\ 0 & 0 \end{bmatrix},$$

$$A_0X_1 + A_1X_0 = \begin{bmatrix} 1 & 0 \\ 1 & 0 \end{bmatrix}\begin{bmatrix} 1 & -1 \\ 0 & 1 \end{bmatrix} + \begin{bmatrix} 1 & 1 \\ 0 & 1 \end{bmatrix}\begin{bmatrix} 0 & 0 \\ -1 & 1 \end{bmatrix} = \begin{bmatrix} 0 & 0 \\ 0 & 0 \end{bmatrix},$$

and finally

$$A_0X_2 + A_1X_1 = \begin{bmatrix} 1 & 0 \\ 1 & 0 \end{bmatrix}\begin{bmatrix} 0 & 0 \\ s & t \end{bmatrix} + \begin{bmatrix} 1 & 1 \\ 0 & 1 \end{bmatrix}\begin{bmatrix} 1 & -1 \\ 0 & 1 \end{bmatrix} = \begin{bmatrix} 1 & 0 \\ 0 & 1 \end{bmatrix}.$$

This confirms a second order pole. By extending the elimination to include more equations it is easy to see that

$$X_0 = \begin{bmatrix} 0 & 0 \\ -1 & 1 \end{bmatrix}, \quad X_1 = \begin{bmatrix} 1 & -1 \\ 0 & 1 \end{bmatrix}, \quad \textit{and} \quad X_j = \begin{bmatrix} 0 & 0 \\ 0 & 0 \end{bmatrix}$$

for all $j \geq 2$. Thus

$$(A_0 + A_1 z)^{-1} = \frac{1}{z^2}\left(\begin{bmatrix} 0 & 0 \\ -1 & 1 \end{bmatrix} + \begin{bmatrix} 1 & -1 \\ 0 & 1 \end{bmatrix} z\right).$$

Of course in this case our answer can be verified by elementary matrix algebra.

8.8 ▪ Polynomial and Analytic Perturbations

In this section we show that inversion of a polynomial perturbation is equivalent to inversion of a corresponding linear perturbation. The main ideas are encapsulated in a sequence of preliminary results and an important theorem. The proofs are essentially an exercise in elementary algebraic manipulation and have been left as a collection of exercises for the reader. Subsequently we extend these arguments to include analytic perturbations.

8.8.1 ▪ Polynomial perturbations

Let H and K be Hilbert spaces, and let $\{A_i\}_{i=0}^{\infty} \subseteq \mathscr{L}(H,K)$. Define $\mathscr{A}_0^{(k)} \in \mathscr{L}(H^k,K^k)$ by setting

$$\mathscr{A}_0^{(k)}(X) = \begin{bmatrix} A_0 & 0 & \cdots & 0 & 0 \\ A_1 & A_0 & \cdots & 0 & 0 \\ \vdots & \vdots & & \vdots & \vdots \\ A_{k-2} & A_{k-3} & \cdots & A_0 & 0 \\ A_{k-1} & A_{k-2} & \cdots & A_1 & A_0 \end{bmatrix} \begin{bmatrix} x_0 \\ x_1 \\ \vdots \\ x_{k-2} \\ x_{k-1} \end{bmatrix}$$
$$= \begin{bmatrix} A_0 x_0 \\ A_1 x_0 + A_0 x_1 \\ \vdots \\ A_{k-2} x_0 + \cdots + A_0 x_{k-2} \\ A_{k-1} x_0 + \cdots + A_0 x_{k-1} \end{bmatrix}$$

and $\mathscr{A}_r^{(k)} \in \mathscr{L}(H^k,K^k)$ in general for $r \geq 1$ by setting

$$\mathscr{A}_r^{(k)}(X) = \begin{bmatrix} A_{rk} & A_{rk-1} & \cdots & A_{(r-1)k+2} & A_{(r-1)k+1} \\ A_{rk+1} & A_{rk} & \cdots & A_{(r-1)k+3} & A_{(r-1)k+2} \\ \vdots & \vdots & & \vdots & \vdots \\ A_{(r+1)k-2} & A_{(r+1)k-3} & \cdots & A_{rk} & A_{rk-1} \\ A_{(r+1)k-1} & A_{(r+1)k-2} & \cdots & A_{rk+1} & A_{rk} \end{bmatrix} \begin{bmatrix} x_0 \\ x_1 \\ \vdots \\ x_{k-2} \\ x_{k-1} \end{bmatrix}$$
$$= \begin{bmatrix} A_{rk} x_0 + \cdots + A_{(r-1)k+1} x_{k-1} \\ A_{rk+1} x_0 + \cdots + A_{(r-1)k+2} x_{k-1} \\ \vdots \\ A_{(r+1)k-2} x_0 + \cdots + A_{rk-1} x_{k-1} \\ A_{(r+1)k-1} x_0 + \cdots + A_{rk} x_{k-1} \end{bmatrix}$$

for each $X \in H^k$. For convenience, we use a standard block matrix representation. We also define $\mathscr{D} : \mathscr{L}(H,K) \to \mathscr{L}(H^k,K^k)$ by setting

$$\mathscr{D}(A) = \begin{bmatrix} A & 0 & \cdots & 0 & 0 \\ 0 & A & \cdots & 0 & 0 \\ \vdots & \vdots & & \vdots & \vdots \\ 0 & 0 & \cdots & A & 0 \\ 0 & 0 & \cdots & 0 & A \end{bmatrix}$$

for each $A \in \mathscr{L}(H,K)$. For any Hilbert space H and each $z \in \mathbb{C}$ define $\mathscr{Z}(z) \in \mathscr{L}(H^k,H^k)$ by the formula

$$\mathscr{Z}(z) = \begin{bmatrix} 0 & 0 & \cdots & 0 & zI \\ I & 0 & \cdots & 0 & 0 \\ \vdots & \vdots & & \vdots & \vdots \\ 0 & 0 & \cdots & 0 & 0 \\ 0 & 0 & \cdots & I & 0 \end{bmatrix} = [E_2, E_3, \dots, E_k, zE_1].$$

We will normally write $\mathscr{Z}(z) = \mathscr{Z}$. We note that

$$\begin{aligned} \mathscr{Z}^2 &= [E_3, E_4, \dots, zE_1, zE_2], \\ \mathscr{Z}^3 &= [E_4, E_5, \dots, zE_2, zE_3], \\ &\vdots \\ \mathscr{Z}^{k-1} &= [E_k, zE_1, \dots, zE_{k-2}, zE_{k-1}], \end{aligned}$$

and finally

$$\mathscr{Z}^k = z[E_1, E_2, \dots, E_{k-1}, E_k] = z\mathscr{I}.$$

In general, we can see that

$$\mathscr{Z}^{rk+s} = z^r[E_{s+1}, E_{s+2}, \dots, E_k, zE_1, \dots, zE_s]$$

for $r = 0, 1, \dots$ and $s = 0, 1, \dots, k-1$. Note that the complex number $w \in \mathbb{C}$ is an eigenvalue of $\mathscr{Z}$ if and only if w satisfies the equation $w^k = z$. If $v \in H$ and we define

$$V = \begin{bmatrix} w^{k-1} e^{2(k-1)\pi/k} v \\ w^{k-2} e^{2(k-2)\pi/k} v \\ \vdots \\ w e^{2\pi/k} v \\ v \end{bmatrix},$$

then it is easily seen that $\mathscr{Z}V = wV$. It follows that $||Z|| = |z|^{1/k}$. Let $\{A_i\}_{i=0}^{\infty} \subseteq \mathscr{L}(H,K)$. We now have the following results. Proofs of Lemmas 8.30 and 8.32 are left as exercises for the reader (see Problems 8.22 and 8.23, respectively).

Lemma 8.30. *The identity*

$$\sum_{i=0}^{\infty} \mathscr{D}(A_i)\mathscr{Z}^i = \sum_{r=0}^{\infty} \mathscr{A}_r^{(k)} z^r$$

is valid for each $z \in \mathbb{C}$.

Lemma 8.31. *The series*

$$\sum_{i=0}^{\infty} \mathscr{D}(A_i)\mathscr{Z}^i$$

converges for $|z| < \epsilon^{1/k}$ *if and only if the series* $\sum_{i=0}^{\infty} A_i z^i$ *converges for* $|z| < \epsilon$.

Proof: If the series $\sum_{i=0}^{\infty} \mathscr{D}(A_i)\mathscr{Z}^i$ converges for $||\mathscr{Z}|| < \epsilon$, then it converges absolutely for $||\mathscr{Z}|| < \epsilon$. Since $||\mathscr{D}(A_i)|| = ||A_i||$ and $||\mathscr{Z}|| = |z|^{1/k}$, it follows that the series $\sum_{i=0}^{\infty} A_i z^{i/k}$ converges absolutely for $|z|^{1/k} < \epsilon$ and hence that the series $\sum_{i=0}^{\infty} A_i z^i$ converges for $|z| < \epsilon$. It is easily seen that the reverse implication is also true. □

Lemma 8.32. *The identity*

$$\left(\sum_{i=0}^{\infty} \mathscr{D}(A_i)\mathscr{Z}^i\right)\left(\sum_{i=0}^{\infty} \mathscr{D}(X_i)\mathscr{Z}^i\right) = \mathscr{Z}^m$$

is valid for some nonnegative integer m if and only if the identity

$$\left(\sum_{i=0}^{\infty} A_i z^i\right)\left(\sum_{i=0}^{\infty} X_i z^i\right) = z^m I$$

is also valid.

In stating the next result it is useful to extend our previous notation. For each $s \in \{0,1,\ldots,k-1\}$ define $\mathscr{X}_i^{(k,s)} \in \mathscr{L}(H^k, K^k)$ by setting

$$\mathscr{X}_0^{(k,s)} = \begin{bmatrix} 0 & 0 & \cdots & 0 & 0 & 0 & \cdots & 0 \\ \vdots & \vdots & & \vdots & \vdots & \vdots & & \vdots \\ 0 & 0 & \cdots & 0 & 0 & 0 & \cdots & 0 \\ X_0 & 0 & \cdots & 0 & 0 & 0 & \cdots & 0 \\ X_1 & X_0 & \cdots & 0 & 0 & 0 & \cdots & 0 \\ \vdots & \vdots & & \vdots & \vdots & \vdots & & \vdots \\ X_{s-2} & X_{s-3} & \cdots & X_0 & 0 & 0 & \cdots & 0 \\ X_{s-1} & X_{s-2} & \cdots & X_1 & X_0 & 0 & \cdots & 0 \end{bmatrix},$$

$$\mathscr{X}_1^{(k,s)} = \begin{bmatrix} X_s & X_{s-1} & \cdots & X_0 & 0 & \cdots & 0 & 0 \\ X_{s+1} & X_s & \cdots & X_1 & X_0 & \cdots & 0 & 0 \\ \vdots & \vdots & & \vdots & \vdots & & \vdots & \vdots \\ X_{k-1} & X_{k-2} & \cdots & X_{k-s-1} & X_{k-s-2} & \cdots & X_0 & 0 \\ X_k & X_{k-1} & \cdots & X_{k-s} & X_{k-s-1} & \cdots & X_1 & X_0 \\ \vdots & \vdots & & \vdots & \vdots & & \vdots & \vdots \\ X_{s+k-2} & X_{s+k-3} & \cdots & X_{k-2} & X_{k-3} & \cdots & X_{s+1} & X_s \\ X_{s+k-1} & X_{s+k-2} & \cdots & X_{k-1} & X_{k-2} & \cdots & X_{s+2} & X_{s+1} \end{bmatrix},$$

and

$$\mathscr{X}_r^{(k,s)} = \begin{bmatrix} X_{rk+s} & X_{rk+s-1} & \cdots & X_{(r-1)k+s+2} & X_{(r-1)k+s+1} \\ X_{rk+s+1} & X_{rk+s} & \cdots & X_{(r-1)k+s+3} & X_{(r-1)k+s+2} \\ \vdots & \vdots & & \vdots & \vdots \\ X_{(r+1)k+s-2} & X_{(r+1)k+s-3} & \cdots & X_{rk+s} & X_{rk+s-1} \\ X_{(r+1)k+s-1} & X_{(r+1)k+s-2} & \cdots & X_{rk+s+1} & X_{rk+s} \end{bmatrix}$$

for $r > 1$. Note that with the new definition we have $\mathscr{X}_r^{(k,0)} = \mathscr{X}_r^{(k)}$.

Theorem 8.33. *The inverse operator*

$$\left(A_0 + \cdots + A_k z^k\right)^{-1} \in \mathcal{L}(K, H)$$

is given by the formula

$$\left(A_0 + \cdots + A_k z^k\right)^{-1} = \frac{1}{z^{rk+s}}(X_0 + X_1 z + \cdots),$$

where $r \in \{0, 1, \ldots\}$ *and* $s \in \{0, 1, \ldots, k-1\}$ *if and only if the inverse operator*

$$\left(\mathscr{A}_0^{(k)} + \mathscr{A}_1^{(k)} z\right)^{-1} \in \mathcal{L}(K^k, H^k)$$

is given by the formula

$$\left(\mathscr{A}_0^{(k)} + \mathscr{A}_1^{(k)} z\right)^{-1} = \frac{1}{z^r}\left(\mathscr{X}_0^{(k)} + \mathscr{X}_1^{(k)} z + \cdots\right)$$

when $s = 0$ *and by the formula*

$$\left(\mathscr{A}_0^{(k)} + \mathscr{A}_1^{(k)} z\right)^{-1} = \frac{1}{z^{r+1}}\left(\mathscr{X}_0^{(k,s)} + \mathscr{X}_1^{(k,s)} z + \cdots\right)$$

when $s \in \{1, 2, \ldots, k-1\}$.

The reader is invited to supply the proof of this theorem in Problem 8.26.

8.8.2 ▪ Analytic perturbations

If the coefficients $\{A_i\}_{i=0}^{\infty}$ of the analytic perturbation

$$A(z) = \sum_{i=0}^{\infty} A_i z^i$$

satisfy a finite order linear recursion, then multiplication by a polynomial will reduce the series $\sum_{i=1}^{\infty} A_i z^i$ to a polynomial which can then be inverted.

Let

$$A(z) = \sum_{i=0}^{\infty} A_i z^i$$

be an analytic perturbation of A_0 which converges in the region $|z| < r$. If the inverse $[A(z_0)]^{-1}$ of the analytic perturbation is well defined for some $z_0 \neq 0$ with $|z_0| < r$, then by the Banach inverse theorem we can find $\epsilon > 0$ such that

$$||A(z_0)x|| \geq \epsilon ||x||$$

for all $x \in H$. Because the power series converges at z_0 we can find m such that

$$\left\| A(z_0) - \sum_{i=0}^{m} A_i z_0^i \right\| < \frac{\epsilon}{2},$$

and hence

$$\left\| \left[\sum_{i=0}^{m} A_i z_0^i \right] x \right\| \geq \frac{\epsilon}{2} ||x||$$

for all $x \in H$. It follows that $[A_m(z_0)]^{-1} = [\sum_{i=0}^{m} A_i z_0^i]^{-1}$ is well defined. Since $A_m(z)$ is a polynomial perturbation and since $[A_m(z_0)]^{-1}$ is well defined for some $z_0 \neq 0$, we can use our previous methods to calculate $[A_m(z)]^{-1}$, and we have

$$[A(z)]^{-1} = [A_m(z) + R_m(z)]^{-1} = [I + (A_m(z))^{-1} R_m(z)]^{-1} [A_m(z)]^{-1}.$$

In particular, if we restrict the action of the perturbed operator to a finite dimensional domain, we can see that the reduction process described in Theorem 8.14 and applied to appropriate augmented operators must terminate after a finite number of steps.

8.9 ▪ Problems

Problem 8.1. Let $\{A_j\}_{j=0,1,\ldots}$ be a sequence of square matrices $A_j : \mathbb{C}^m \to \mathbb{C}^m$ such that A_0 is nonsingular. Show that the linear systems

$$A_0 X_0 = I_m \quad \text{and} \quad \sum_{j=0}^{k} A_{k-j} X_j = 0 \quad \text{for } k = 1,2,\ldots \tag{8.29}$$

and

$$Y_0 A_0 = I_m \quad \text{and} \quad \sum_{j=0}^{k} Y_j A_{k-j} = 0 \quad \text{for } k = 1,2,\ldots \tag{8.30}$$

each have uniquely defined solutions $\{X_j\}$ and $\{Y_j\}$, respectively, and furthermore that $X_j = Y_j$ for all $j = 0,1,\ldots$. *Hint: Show that for each $k = 1,2,\ldots$ the systems* (8.29) *and* (8.30) *can be rewritten in the form*

$$\mathscr{A}_0^{(k)} \mathscr{X}_0^{(k)} = I_{km} \quad \textit{and} \quad \mathscr{Y}_0^{(k)} \mathscr{A}_0^{(k)} = I_{km}. \tag{8.31}$$

Problem 8.2. Prove Theorem 8.1. *Hint: Let $s = \max\{\|A_0^{-1}\|, \|A_0^{-1}\| r^2 + r\} + 1$ and show that $\|X_j\| < s^{j+1}$.*

Problem 8.3. Let $\{A_j\}_{j=0}^{\infty}$ be a sequence of square matrices $A_j : \mathbb{C}^m \to \mathbb{C}^m$ such that A_0 is nonsingular. Suppose that for some integer $n > 0$ we have

$$A_{j+n+1} = \sum_{k=1}^{n+1} \alpha_k A_{j+n+1-k} \quad \text{for each } j = 1,1,\ldots. \tag{8.32}$$

Define a finite sequence $\{B_j\}$ by setting

$$B_0 = A_0 \quad \text{and} \quad B_j = A_j - \sum_{k=1}^{j} \alpha_k A_{j-k} \quad \text{for each} \quad j = 1,2,\ldots,n, \tag{8.33}$$

and use the notation $B(z) = \sum_{j=0}^{n} B_j z^j$ to denote the associated power series. Show that

$$[A(z)]^{-1} = \left(1 - \sum_{k=1}^{n} \alpha_k z^k\right) [B(z)]^{-1}.$$

Problem 8.4. Let $\mathscr{A}_0^{(j)}$ be defined as in (8.10). Define $\Delta : \mathbb{Z}^+ \to \mathbb{Z}^+$ by the formula

$$\Delta(k) = \begin{cases} \operatorname{rank} \mathscr{A}_0^{(k+1)} & \text{if } k = 0, \\ \operatorname{rank} \mathscr{A}_0^{(k+1)} - \operatorname{rank} \mathscr{A}_0^{(k)} & \text{if } k = 1, 2, \ldots. \end{cases}$$

Show that $\Delta(k+1) \geq \Delta(k)$ for all $k \in \mathbb{Z}^+$.

Problem 8.5. Let $\{A_j\}_{j=0}^{\infty}$ be a finite sequence of square matrices $A_j : \mathbb{C}^m \to \mathbb{C}^m$ such that A_0 is nonsingular, and let $A(z) = \sum_{j=0}^{n} A_j z^j$. Show that we can write $[A(z)]^{-1} = \sum_{j=0}^{\infty} X_j z^j$, where the sequence $\{X_j\}$ satisfies a finite recursion of the form

$$X_{j+nh} = \sum_{k=1}^{h} \xi_k X_{j+nh-nk}$$

for each $j = 0, 1, 2, \ldots$, where $h \leq nm$. *Hint: Let* $\mathscr{L} = \mathscr{A}_0^{(n)}$ *and* $\mathscr{U} = \mathscr{A}_1^{(n)}$ *and show that for each* $i = 1, 2, \ldots$ *the equations* (8.29) *can be written in the form*

$$\begin{bmatrix} \mathscr{L} & 0 & 0 & \cdots & 0 \\ \mathscr{U} & \mathscr{L} & 0 & \cdots & 0 \\ 0 & \mathscr{U} & \mathscr{L} & \cdots & 0 \\ \vdots & \vdots & \vdots & \ddots & \vdots \\ 0 & 0 & 0 & \cdots & \mathscr{L} \end{bmatrix} \begin{bmatrix} \mathscr{X}_0^{(n)} \\ \mathscr{X}_1^{(n)} \\ \mathscr{X}_2^{(n)} \\ \vdots \\ \mathscr{X}_i^{(n)} \end{bmatrix} = \begin{bmatrix} I_{nm} \\ 0 \\ 0 \\ \vdots \\ 0 \end{bmatrix}. \tag{8.34}$$

Hence show that for each $j = 0, 1, \ldots, i$ *the solution can be written in the form*

$$\mathscr{X}_j^{(n)} = (-1)^j \left(\mathscr{L}^{-1}\mathscr{U}\right)^j \mathscr{L}^{-1}.$$

Use the Cayley–Hamilton theorem to deduce that the $\mathscr{X}_j^{(n)}$ *satisfy a finite recursion. By applying this recursion to each component of* $\mathscr{X}_j^{(n)}$, *establish the desired result. Note that this recursion may not be the simplest such recursion.*

Problem 8.6. Let $\{A_j\}_{j=0}^{\infty}$ be a sequence of square matrices $A_j : \mathbb{C}^m \to \mathbb{C}^m$ such that A_0 is singular. If we can find nonsingular matrices $F : \mathbb{C}^m \to \mathbb{C}^m$ and $G : \mathbb{C}^m \to \mathbb{C}^m$ such that

$$F^{-1}A_0G = \begin{bmatrix} I_{m_1} & 0 \\ 0 & 0 \end{bmatrix} \quad \text{and} \quad F^{-1}A_1G = \begin{bmatrix} A'_{111} & A'_{112} \\ A'_{121} & I_{m_2} \end{bmatrix},$$

where $m_1 > 0$, $m_2 > 0$, and $m_1 + m_2 = m$, show that the linear systems

$$A_0X_0 = 0, \quad A_1X_0 + A_0X_1 = I_m, \quad \text{and} \quad \sum_{j=0}^{k} A_{k-j}X_j = 0 \quad \text{for each } k = 2, 3, \ldots \tag{8.35}$$

and

$$Y_0A_0 = 0, \quad Y_0A_1 + Y_1A_0 = I_m, \quad \text{and} \quad \sum_{j=0}^{k} Y_jA_{k-j} = 0 \quad \text{for each } k = 2, 3, \ldots \tag{8.36}$$

each have uniquely defined solutions $\{X_j\}$ and $\{Y_j\}$, respectively, and furthermore $X_j = Y_j$ for all $j = 0, 1, \ldots$. *Hint: Define* $A_j' = F^{-1} A_j G$, $X_j' = G^{-1} X_j F$, *and* $Y_j' = G^{-1} Y_j F$, *and show that for all* $k = 0, 1, \ldots$ *and any sequence* $\{\alpha_k\}$ *of complex numbers we have*

$$\sum_{j=0}^{k} A_{k-j} X_j = \alpha_k I_m \quad \Leftrightarrow \quad \sum_{j=0}^{k} A_{k-j}' X_j' = \alpha_k I_m$$

and

$$\sum_{j=0}^{k} Y_j A_{k-j} = \alpha_k I_m \quad \Leftrightarrow \quad \sum_{j=0}^{k} Y_j' A_{k-j}' = \alpha_k I_m .$$

Hence deduce that the original system has a unique solution if and only if the modified system has a unique solution. Without loss of generality assume that

$$A_0 = \begin{bmatrix} I_{m_1} & 0 \\ 0 & 0 \end{bmatrix} \quad \text{and} \quad A_1 = \begin{bmatrix} A_{111} & A_{112} \\ A_{121} & I_{m_2} \end{bmatrix},$$

and hence use the first two equations in (8.35) *to deduce that*

$$X_0 = \begin{bmatrix} 0 & 0 \\ 0 & I_{m_2} \end{bmatrix}.$$

Now assume that X_j *is uniquely defined for* $j = 0, 1, \ldots, k-1$, *and hence show that* X_k *is also uniquely defined. For an arbitrarily chosen value of* k *use the equations*

$$\mathscr{A}_0^{(k+1)} \mathscr{X}_0^{(k+1)} = \mathscr{Y}_0^{(k+1)} \mathscr{A}_0^{(k+1)} = \mathscr{J},$$

where $\mathscr{A}_0^{(k+1)}$ *is defined by* (8.9) *and* $\mathscr{J} : \mathbb{C}^{(k+1)m} \to \mathbb{C}^{(k+1)m}$ *is given by*

$$\mathscr{J} = \begin{bmatrix} 0 & 0 & 0 & \cdots & 0 & 0 \\ I_m & 0 & 0 & \cdots & 0 & 0 \\ 0 & I_m & 0 & \cdots & 0 & 0 \\ \vdots & \vdots & \vdots & \ddots & \vdots & \vdots \\ 0 & 0 & 0 & \cdots & 0 & 0 \\ 0 & 0 & 0 & \cdots & I_m & 0 \end{bmatrix}$$

to deduce that $\mathscr{X}_0^{(k)} = \mathscr{Y}_0^{(k)}$.

Problem 8.7. For the partitioned matrix

$$A = \begin{bmatrix} A_{11} & A_{12} \\ A_{21} & A_{22} \end{bmatrix}$$

show that $||A_{pq}|| \le ||A|| \le \sum_{p,q} ||A_{pq}||$.

Problem 8.8. Prove Theorem 8.2. *Hint: Define* A_j' *and* X_j' *as in Problem 8.6, and show first that* $||X_j'|| < s^{j+1}$, *where* $s = \max\{6r^4 + r, 6r^2 + r, 1\} + 1$. *Use the method of Problem 8.2 and the result of Problem 8.7.*

Problem 8.9. Let $\{A_j\}_{j=0}^{\infty}$ be a finite sequence of square matrices $A_j : \mathbb{C}^m \to \mathbb{C}^m$ such that A_0 is singular but such that we can find nonsingular matrices $F : \mathbb{C}^m \to \mathbb{C}^m$ and $G : \mathbb{C}^m \to \mathbb{C}^m$ with

$$F^{-1}A_0G = \begin{bmatrix} I_{m_1} & 0 \\ 0 & 0 \end{bmatrix} \quad \text{and} \quad F^{-1}A_1G = \begin{bmatrix} A'_{111} & A'_{112} \\ A'_{121} & I_{m_2} \end{bmatrix},$$

where $m_1 > 0$, $m_2 > 0$, and $m_1 + m_2 = m$. From Problems 8.6 and 8.8 it follows that there is a uniquely defined power series $X(z) = \sum_{j=0}^{\infty} X_j z^j$ with a positive radius of convergence such that $[A(z)]^{-1} = X(z)/z$ inside the circle of convergence provided $z \neq 0$. Show that the coefficients X_j satisfy a finite recursion in the form

$$X_{j+nh} = \sum_{k=1}^{h} \xi_k X_{j+nh-nk}$$

for each $j = 0, 1, 2, \ldots$, where $h \leq nm$. *Hint: Without loss of generality assume that*

$$A_0 = \begin{bmatrix} I_{m_1} & 0 \\ 0 & 0 \end{bmatrix} \quad \textit{and} \quad A_1 = \begin{bmatrix} A_{111} & A_{112} \\ A_{121} & I_{m_2} \end{bmatrix},$$

and write

$$A_j = \begin{bmatrix} A_{j11} & A_{j12} \\ A_{j21} & A_{j22} \end{bmatrix}.$$

Define new sequences

$$B_j = \begin{bmatrix} A_{j+1,22} & A_{j21} \\ A_{j+1,12} & A_{j11} \end{bmatrix} \quad \textit{and} \quad W_j = \begin{bmatrix} X_{j21} & X_{j22} \\ X_{j+1,11} & X_{j+1,12} \end{bmatrix}$$

for each $j = 0, 1, \ldots,$ and let

$$\mathcal{L} = \mathcal{B}_0^{(n)}, \quad \mathcal{U} = \mathcal{B}_1^{(n)}, \quad \textit{and} \quad \mathcal{K} = \begin{bmatrix} 0 & I_{m_2} & 0 & 0 & \cdots & 0 & 0 \\ I_{m_1} & 0 & 0 & 0 & \cdots & 0 & 0 \\ 0 & 0 & 0 & I_{m_2} & \cdots & 0 & 0 \\ 0 & 0 & I_{m_1} & 0 & \cdots & 0 & 0 \\ \vdots & \vdots & \vdots & \vdots & \ddots & 0 & 0 \\ 0 & 0 & 0 & 0 & \cdots & 0 & I_{m_2} \\ 0 & 0 & 0 & 0 & \cdots & I_{m_1} & 0 \end{bmatrix}.$$

Show that for each $i = 1, 2, \ldots$ the equations (8.35) *can be rewritten in the form*

$$\begin{bmatrix} \mathcal{L} & 0 & 0 & \cdots & 0 \\ \mathcal{U} & \mathcal{L} & 0 & \cdots & 0 \\ 0 & \mathcal{U} & \mathcal{L} & \cdots & 0 \\ \vdots & \vdots & \vdots & \ddots & \vdots \\ 0 & 0 & 0 & \cdots & \mathcal{L} \end{bmatrix} \begin{bmatrix} \mathcal{W}_0^{(n)} \\ \mathcal{W}_1^{(n)} \\ \mathcal{W}_2^{(n)} \\ \vdots \\ \mathcal{W}_i^{(n)} \end{bmatrix} = \begin{bmatrix} \mathcal{K} \\ 0 \\ 0 \\ \vdots \\ 0 \end{bmatrix}, \tag{8.37}$$

and hence deduce that the solution is given by $\mathcal{W}_j^{(n)} = (-1)^j \left(\mathcal{L}^{-1}\mathcal{U}\right)^j \mathcal{L}^{-1}\mathcal{K}$.

Problem 8.10. Let $\{A_j\}_{j=0}^{\infty}$ be a sequence of square matrices $A_j : \mathbb{C}^m \to \mathbb{C}^m$ such that A_0 is singular but such that condition (8.8) is not satisfied. Define $\mathscr{A}_j^{(k)}$ using (8.9) and (8.10). Let p be the smallest positive integer for which we can find nonsingular matrices $\mathscr{F} : \mathbb{C}^{pm} \to \mathbb{C}^{pm}$ and $\mathscr{G} : \mathbb{C}^{pm} \to \mathbb{C}^{pm}$ such that

$$\mathscr{F}^{-1}\mathscr{A}_0^{(p)}\mathscr{G} = \begin{bmatrix} I_{m_1} & 0 \\ 0 & 0 \end{bmatrix} \quad \text{and} \quad \mathscr{F}^{-1}\mathscr{A}_1^{(p)}\mathscr{G} = \begin{bmatrix} \mathscr{A}'_{111} & \mathscr{A}'_{112} \\ \mathscr{A}'_{121} & I_{m_2} \end{bmatrix}$$

where $m_1 > 0$, $m_2 > 0$ and $m_1 + m_2 = pm$. Show that the linear systems

$$\sum_{j=0}^{p} A_{p-j}X_j = I_m \quad \text{and} \quad \sum_{j=0}^{k} A_{k-j}X_j = 0 \quad \text{for } k \neq p \tag{8.38}$$

and

$$\sum_{j=0}^{p} Y_j A_{p-j} = I_m \quad \text{and} \quad \sum_{j=0}^{k} Y_j A_{k-j} = 0 \quad \text{for } k \neq p \tag{8.39}$$

have uniquely defined solutions $\{X_j\}$ and $\{Y_j\}$, respectively, and furthermore that $X_j = Y_j$. *Hint: Show that equations* (8.38) *and* (8.39) *can be rewritten in the form*

$$\mathscr{A}_0^{(p)}\mathscr{X}_0^{(p)} = 0, \quad \mathscr{A}_1^{(p)}\mathscr{X}_0^{(p)} + \mathscr{A}_0^{(p)}\mathscr{X}_1^{(p)} = I_m, \quad \text{and} \quad \sum_{j=0}^{k} \mathscr{A}_{k-j}^{(p)}\mathscr{X}_j^{(p)} = 0 \tag{8.40}$$

for each $k = 2, 3, \ldots$ *and*

$$\mathscr{Y}_0^{(p)}\mathscr{A}_0^{(p)} = 0, \quad \mathscr{Y}_0^{(p)}\mathscr{A}_1^{(p)} + \mathscr{Y}_1^{(p)}\mathscr{A}_0^{(p)} = I_m, \quad \text{and} \quad \sum_{j=0}^{k} \mathscr{Y}_j^{(p)}\mathscr{A}_{k-j}^{(p)} = 0 \tag{8.41}$$

for each $k = 2, 3, \ldots$.

Problem 8.11. Prove Theorem 8.3. *Hint: Use equations* (8.40) *and* (8.41).

Problem 8.12. Let $\{A_j\}_{j=0}^{\infty}$ be a finite sequence of square matrices $A_j : \mathbb{C}^m \to \mathbb{C}^m$ such that the conditions of Problem 8.11 are satisfied. From Problems 8.10 and 8.11 it follows that there is a uniquely defined power series $X(z) = \sum_{j=0}^{\infty} X_j z^j$ with a positive radius of convergence such that $[A(z)]^{-1} = X(z)/z^p$ inside the circle of convergence provided $z \neq 0$. Show that the coefficients X_j satisfy a finite recursion in the form

$$X_{j+qph} = \sum_{k=1}^{h} \xi_k X_{j+qph-qpk}$$

for each $j = 0, 1, 2, \ldots$, where $h \leq qpm$ and q is the unique integer such that $qp \geq n > (q-1)p$.

Problem 8.13. Let $\{A_j\}_{j=0}^{\infty}$ be a sequence of square matrices $A_j : \mathbb{C}^m \to \mathbb{C}^m$ with $\Delta(k)$ defined as in Problem 8.4. Define $\mathscr{A}_j^{(k)}$ using (8.9) and (8.10). Observe that A_0 is nonsingular if and only if $\Delta(0) = m$. If $\Delta(0) < m$, show that we can find nonsingular matrices $\mathscr{F} : \mathbb{C}^{pm} \to \mathbb{C}^{pm}$ and $\mathscr{G} : \mathbb{C}^{pm} \to \mathbb{C}^{pm}$ such that

$$\mathscr{F}^{-1}\mathscr{A}_0^{(p)}\mathscr{G} = \begin{bmatrix} I_{m_1} & 0 \\ 0 & 0 \end{bmatrix} \quad \text{and} \quad \mathscr{F}^{-1}\mathscr{A}_1^{(p)}\mathscr{G} = \begin{bmatrix} \mathscr{A}'_{111} & \mathscr{A}'_{112} \\ \mathscr{A}'_{121} & I_{m_2} \end{bmatrix},$$

where $m_1 > 0$, $m_2 > 0$, and $m_1 + m_2 = pm$ if and only if $\Delta(p) = m$.

Problem 8.14. Let

$$A_0 = \begin{bmatrix} 1 & 0 \\ 0 & 0 \end{bmatrix}, \quad A_1 = \begin{bmatrix} 1 & 1 \\ 0 & 0 \end{bmatrix}, \quad \text{and} \quad A_2 = \begin{bmatrix} 1 & 0 \\ 0 & 1 \end{bmatrix},$$

and define $A(z) = A_0 + A_1 z + A_2 z^2$. Calculate $[A(z)]^{-1}$ near $z = 0$. Determine the order of the pole at $z = 0$ for the inverse matrix $[A(z)]^{-1}$ and find a recursive relationship for the coefficients of the corresponding power series. *Hint: Consider* $\operatorname{rank}\mathscr{A}_0^{(2p)} - \operatorname{rank}\mathscr{A}_0^{(p)}$.

Problem 8.15. Consider $(A_0 + A_1 z)^{-1}$, where

$$A_0 = \begin{bmatrix} 1 & 2 \\ 1 & 2 \end{bmatrix} \quad \text{and} \quad A_1 = \begin{bmatrix} 1 & 3 \\ 0 & 1 \end{bmatrix}.$$

Show that $A_0^{-1}(\{0\}) = \{x \mid x_1 + 2x_2 = 0\}$ and that $A_1 A_0^{-1}(\{0\}) = \{y \mid y_1 - y_2 = 0\}$. Hence find unitary matrices P and Q such that

$$A_0' = Q^* A_0 P = \begin{bmatrix} 0 & a'_{012} \\ 0 & a'_{022} \end{bmatrix} \quad \text{and} \quad A_1' = Q^* A_1 P = \begin{bmatrix} a'_{111} & a'_{112} \\ 0 & a'_{122} \end{bmatrix},$$

and use these transformations to show that

$$(A_0 + A_1 z)^{-1} = P(A_0' + A_1' z)^{-1} Q^* = \frac{1}{z^2}\begin{bmatrix} 2+z & -2-3z \\ -1 & 1+z \end{bmatrix}.$$

Problem 8.16. Let $\Omega = [0,1]$, and let $H = K = L^2(\Omega)$. For each $x \in H$ define $\mu(x) = \int_\Omega x(s)ds$. Let $A \in \mathscr{L}(H,K)$ be defined by

$$Ax(t) = \int_{(0,t)} [x(s) - \mu(x)]ds \quad \forall\, x \in H,\ t \in [0,1].$$

If

$$x^{(k)}(s) = \begin{cases} 0 & \text{when } s \notin [\frac{1}{2}(1-\frac{1}{k}), \frac{1}{2}(1+\frac{1}{k})], \\ k & \text{otherwise} \end{cases}$$

and $y^{(k)} = Ax^{(k)}$, find an expression for $y^{(k)}(t)$, for all $t \in [0,1]$. If we define

$$g(t) = \begin{cases} -t & \text{when } t < \frac{1}{2}, \\ 1-t & \text{otherwise,} \end{cases}$$

show that $g \in K$ and that

$$\|y^{(k)} - g\|^2 = \frac{1}{12k} \to 0$$

as $k \to \infty$. Show that $g \notin A(H)$, and hence deduce that $A(H)$ is not closed. *Hint: To show that $g \notin A(H)$ it is necessary to show that there is no function $x \in H$ with*

$$\int_{(\frac{1}{2}-\delta,\frac{1}{2}+\delta)} x(s)ds = 1$$

for all $\delta > 0$.

Problem 8.17. Let $\Omega = [0,1]$. For each $m = 0, \pm 1, \pm 2, \ldots$ let $e_m : [0,1] \to \mathbb{C}$ be defined by setting

$$e_m(s) = e^{2\pi i m s}.$$

The functions $\{e_m\}_{m=-\infty}^{+\infty}$ form an orthonormal basis for $L^2(\Omega)$, and each $f \in L^2(\Omega)$ can be written as a Fourier series

$$f = \sum_{m=-\infty}^{\infty} \varphi_m e_m,$$

where the Fourier coefficients are given by

$$\varphi_m = \langle f, e_m \rangle = \int_{[0,1]} f(s) e^{2\pi i m s} ds$$

for each $m \in \mathbb{Z}$. Define $x^{(k)} : [0,1] \to \mathbb{C}$ for each $k = 1, 2, \ldots$ and $y^{(k)} = Ax^{(k)}$ as in Problem 8.16. Show that

$$x^{(k)} = 1 + \sum_{m=-\infty}^{\infty} (-1)^m \frac{k}{m\pi} \sin \frac{m\pi}{k} e_m,$$

and hence deduce that

$$\|x^{(k)}\|^2 = 1 + \frac{2k^2}{\pi^2} \sum_{m=1}^{\infty} \frac{\sin^2 \frac{m\pi}{k}}{m^2} = k.$$

Show also that

$$y^{(k)} = \sum_{m=-\infty}^{\infty} (-1)^m \frac{k}{2m^2\pi^2 i} \sin \frac{m\pi}{k} e_m.$$

Problem 8.18. Let $x^{(k)}$, $y^{(k)}$, and g be the functions defined in Problem 8.16. Use the Fourier series representation to show that if we choose $k = k_R$ sufficiently large such that

$$\frac{\sin \frac{m\pi}{k}}{\frac{m\pi}{k}} \geq \frac{1}{\sqrt{2}}$$

for all $m \leq R - 1$, then it follows that

$$\|x^{(k)}\|^2 \geq R$$

whenever $k \geq k_R$. Hence, deduce that $\|x^{(k)}\| \to \infty$ as $k \to \infty$. On the other hand, if we choose $\delta > 0$ and $R = R(\delta)$ so that

$$\sum_{m=R}^{\infty} \frac{1}{m^2} \leq \pi^2 \delta$$

and choose $k = k_R$ so large that for each $m = 1, 2, \ldots, R-1$ we have

$$\left| 1 - \frac{\sin \frac{m\pi}{k}}{\frac{m\pi}{k}} \right|^2 \leq 6\delta$$

whenever $k \geq k_R$, show that $\|g - y^{(k)}\|^2 \leq \delta$ whenever $k \geq k_R$. Hence, deduce that $y^{(k)} \to g$ in K as $k \to \infty$.

Problem 8.19. Let

$$x = \begin{bmatrix} \xi_0 \\ \xi_{-1} \\ \xi_1 \\ \xi_{-2} \\ \xi_2 \\ \vdots \end{bmatrix} \quad \text{and} \quad y = \begin{bmatrix} \eta_0 \\ \eta_{-1} \\ \eta_1 \\ \eta_{-2} \\ \eta_2 \\ \vdots \end{bmatrix}$$

such that

$$\|x\|^2 = \sum_{m=-\infty}^{\infty} |\xi_m|^2 < \infty \quad \text{and} \quad \|y\|_E^2 = \sum_{m=-\infty}^{\infty} (1 + 4m^2\pi^2)|\eta_m|^2 < \infty.$$

Show that for each $y \in K_E = \{y \mid \|y\|_E^2 < \infty\}$ the matrix equation $y = Ax$ given by

$$\begin{bmatrix} \eta_0 \\ \eta_{-1} \\ \eta_1 \\ \eta_{-2} \\ \eta_2 \\ \vdots \end{bmatrix} = \begin{bmatrix} 0 & \frac{1}{2\pi i} & \frac{-1}{2\pi i} & \frac{-1}{4\pi i} & \frac{1}{4\pi i} & \cdots \\ 0 & \frac{-1}{2\pi i} & 0 & 0 & 0 & \cdots \\ 0 & 0 & \frac{1}{2\pi i} & 0 & 0 & \cdots \\ 0 & 0 & 0 & \frac{-1}{4\pi i} & 0 & \cdots \\ 0 & 0 & 0 & 0 & \frac{1}{4\pi i} & \cdots \\ \vdots & \vdots & \vdots & \vdots & \vdots & \ddots \end{bmatrix} \begin{bmatrix} \xi_0 \\ \xi_{-1} \\ \xi_1 \\ \xi_{-2} \\ \xi_2 \\ \vdots \end{bmatrix}$$

has a unique solution $x \in H = \{x \mid \|x\|^2 < \infty\}$.

Problem 8.20. Prove that the expansion (8.14) holds. *Hint: See Yosida [163, pp. 132–135].*

Problem 8.21. Prove that the formula (8.15) holds. *Hint: See Kato [99, pp. 493–494].*

Problem 8.22. Verify the identity described in Lemma 8.30.

Problem 8.23. Prove Lemma 8.32. *Hint: Try it first with $m = 1$, $k = 2$, and $A(z) = A_0 + A_1 z$.*

Problem 8.24. Let

$$A_0 = \begin{bmatrix} 0 & 1 & 0 \\ 0 & 0 & 0 \\ 0 & 0 & 0 \end{bmatrix}, \quad A_1 = \begin{bmatrix} 1 & 1 & 0 \\ 0 & 0 & 0 \\ 0 & 0 & 1 \end{bmatrix}, \quad \text{and} \quad A_2 = \begin{bmatrix} 0 & 0 & 0 \\ 0 & 1 & -1 \\ 0 & 0 & 0 \end{bmatrix},$$

and suppose that

$$A_{k+3} = \frac{5}{6}A_{k+2} - \frac{1}{6}A_k$$

for each $k \in \mathbb{N}$. Find the circle of convergence for the series

$$A(z) = A_0 + A_1 z + A_2 z^2 + A_3 z^3 + \cdots,$$

and find a Laurent series for $A(z)^{-1}$ in some region $0 < |z| < s$. *Hint: First calculate* $B(z) = A(z)[1 - 5z/6 + z^3/6]$.

Problem 8.25. The recursive analytic perturbation in Problem 8.24 is generated by an underlying quadratic perturbation $P(z) = A_0 + A_1 z + A_2 z^2$. Use augmented matrices and the method of Theorem 8.33 to find a linear perturbation that is equivalent to $P(z)$ and hence verify the result in this case.

Problem 8.26. Prove Theorem 8.33.

8.10 ▪ Bibliographic Notes

The classic spectral theory for a bounded linear operator is presented elegantly in Kato [99, pp. 178–179]. Much of the work on perturbed linear operators has been restricted to matrix operators [13, 66, 69, 71, 141, 157], classes of differential operators [100, 157], or Fredholm operators [66] and has often concentrated on analysis of the eigenspaces [99, 109]. The local theory of regular analytic matrix functions developed in Gohberg et al. [69] uses a canonical system of root functions to compute a representation of the Laurent principal part of the inverse operator near an isolated singular point. In this finite dimensional analysis the determinant of the matrix function plays a key diagnostic role. Although an earlier, elegant, exposition in Vishik and Lyusternik [157] is more general in scope, the inversion formulae are developed for singularities on finite dimensional subspaces. To extend the finite dimensional theory to more general classes of operators, some of the familiar algebraic techniques must be discarded or revised.

The spectral theory for general linear operator pencils on Banach space was developed in Stummel [149] and is well described in the book by Gohberg et al. [66, pp. 49–54]. These authors assume that the resolvent is analytic in some annular region and use contour integrals to construct key projection operators that are then used to establish well-known spectral separation properties. These developments apply to bounded but not necessarily compact linear operators and include the case where the null space is nontrivial for the unperturbed operator but becomes trivial under perturbation. We consider essentially the same situation but from the viewpoint of the fundamental equations. These equations were proposed by Sain and Massey [135] and later used by Howlett [84] to solve the problem of input retrieval in finite dimensional linear control systems. The fundamental equations were also central to the PhD thesis by Avrachenkov [8] on analytic perturbations and their application and to subsequent works by Howlett and Avrachenkov [86] and Howlett et al. [5, 88, 87, 85] on operator perturbation.

Our approach to the inversion of linear pencils on Hilbert space was inspired by the work of Schweitzer and Stewart [141] on a corresponding matrix inversion problem, but our technique depends on a geometric separation of the underlying spaces. The separation mimics the algebraic separation employed by Howlett [84] for matrix operators but does not depend directly on other established perturbation techniques. For this reason we defer to [8, 13, 66, 99] for a more comprehensive review of the literature. Our work relies heavily on standard functional analysis, for which we cite the classic texts by Courant and Hilbert [44], Dunford and Schwartz [51], Hewitt and Stromberg [79], Luenberger [117], Singer [145], and Yosida [163]. For a general discussion about semigroups we refer to the classic texts by Kato [99] and Yosida [163]. In particular, the theory of one parameter semigroups is described clearly and concisely in Kato [99, pp. 479–495]. For more information about the Bochner integral consult Yosida [163, pp. 132–135]. We refer the reader to Courant and Hilbert [44, pp. 18, 140–142] for further discussion of the Neumann expansion. The reader is referred to Yosida [163, pp. 141–145] for more information about the Eberlein–Shmulyan theorem. In fact, to make the book as self-contained as possible, we have included an additional chapter, Chapter 9, where we present a systematic introduction to the background material from functional analysis.

The return to an algebraic spectral separation technique for the inversion of linear pencils on Banach space resulted from a chance observation that the fundamental equations could be used to define the required projection operators. The separation was described by Howlett et al. [85] for first order poles and later extended by Albrecht et al. [5] to higher order poles. Recent investigations indicate that the fundamental equations can also be used to achieve the required spectral separation near an isolated essential singularity.

The reader can find detailed information about input retrieval in finite dimensional linear control systems in [84, 135].

Chapter 9

Background on Hilbert Spaces and Fourier Analysis

To help make this book more self-contained and assist students with, perhaps, insufficient knowledge of functional analysis to easily follow Chapter 8, we include this appendix. In the overall context of this book our real aim is to provide a solid basis for discussion of the inversion of perturbed linear operators on infinite dimensional vector spaces.

In particular, we introduce the general properties and key structural theorems of Hilbert space by considering two special spaces of square integrable functions. The integrals used here are Lebesgue integrals, but our presentation does not rely on any a priori knowledge of the Lebesgue theory. We assume that the reader is familiar with the Riemann theory of integration on the space of continuous functions with compact support. We will show how a Euclidean space of continuous functions can be extended to define a complete space of square integrable functions.

From a philosophical point of view one could argue that the development of the Lebesgue integral was a consequence of the search for a deeper understanding of the Fourier representation theory. In particular it could be said that the unsatisfactory nature of the pointwise convergence theory for Fourier series was a primary motivation for the generalized notions of function convergence that led, on the one hand, to an elegant theorem of Fejér, that the Fourier series for a continuous function converges everywhere in the sense of Cesàro and, on the other hand, to the deeply satisfying result of Lebesgue that the Fourier series for a square integrable function converges in the mean square sense. We acknowledge this rich history and explore the fundamental structures of Hilbert space via the Fourier series and Fourier integral representations. In the overall context of this book our real aim is to provide a solid basis for discussion of the inversion of perturbed linear operators on infinite dimensional vector spaces.

9.1 ▪ The Hilbert Space $L^2([-\pi, \pi])$

The space $L^2 = L^2([-\pi, \pi])$ is the Hilbert space of real-valued square integrable functions on the finite length closed interval $[-\pi, \pi]$. We will show that L^2 is a linear space over the field $\mathbb{R}$ of real numbers. Consider the space $\mathscr{C} = \mathscr{C}([-\pi, \pi])$ of all continuous real-valued functions on the closed interval $[-\pi, \pi]$. If we define a norm $||\cdot|| : \mathscr{C} \to \mathbb{R}$ by the formula

$$||f||^2 = \int_{[-\pi,\pi]} [f(t)]^2 \, dt \; < \infty$$

for all $f \in \mathscr{C}$, then

1. $||f|| \geq 0$ and $||f|| = 0$ if and only if $f = 0$,
2. $||f+g|| \leq ||f|| + ||g||$ (the triangle inequality), and
3. $||cf|| = |c|\ ||f||$ for all $c \in \mathbb{R}$,

and hence the important properties of a norm are all satisfied. Because the norm also satisfies the property

$$\begin{aligned}
||f+g||^2 + ||f-g||^2 &= \int_{[-\pi,\pi]} \left[(f(t)+g(t))^2 + (f(t)-g(t))^2\right] dt \\
&= 2\int_{[-\pi,\pi]} \left[f(t)^2 + g(t)^2\right] dt \\
&= 2\left[||f||^2 + ||g||^2\right],
\end{aligned}$$

there is a well-defined inner product $\langle \cdot,\cdot \rangle : \mathscr{C} \times \mathscr{C} \to \mathbb{R}$ given by

$$\begin{aligned}
\langle f,g \rangle &= \frac{1}{4}\left[||f+g||^2 - ||f-g||^2\right] \\
&= \frac{1}{4}\int_{[-\pi,\pi]} \left[[f(t)+g(t)]^2 - [f(t)-g(t)]^2\right] dt \\
&= \int_{[-\pi,\pi]} f(t)g(t)\, dt.
\end{aligned}$$

The important properties of an inner product are

1. $\langle f+g,h \rangle = \langle f,h \rangle + \langle g,h \rangle$;
2. $\langle cf,g \rangle = c\,\langle f,g \rangle$ for all $c \in \mathbb{R}$;
3. $\langle f,g \rangle = \langle g,f \rangle$; and
4. $\langle f,f \rangle = ||f||^2$.

With these definitions of norm and inner product the space $\mathscr{C}$ becomes a Euclidean space $\mathscr{C}_E = \mathscr{C}_E([-\pi,\pi])$. We will show that the Euclidean space $\mathscr{C}_E$ can be extended to form a Hilbert space. We begin by introducing the concept of a null set.

Definition 9.1. *A subset $E \subseteq [-\pi,\pi]$ is said to be a null set if there exists a sequence $\{f_n\} = \{f_n\}_{n\in\mathbb{N}} \subseteq \mathscr{C}_E$ of nonnegative continuous functions such that $f_n(t) \to \infty$ as $n \to \infty$ for each $t \in E$ and such that*

$$||f_n|| \leq L < \infty$$

for some $L \in \mathbb{R}$ and all $n \in \mathbb{N}$.

Lemma 9.1. *If $\{E_m\} \subseteq [-\pi,\pi]$ is a sequence of null sets, then the set*

$$E = \bigcup_{m\in\mathbb{N}} E_m \subseteq [-\pi,\pi]$$

is also a null set.

Proof: For each $m \in \mathbb{N}$ let $\{f_{m,n}\} \subseteq \mathscr{C}_E$ be a sequence of nonnegative functions with $f_{m,n}(t) \to \infty$ as $n \to \infty$ for each $t \in E_m$ and with

$$||f_{m,n}|| \leq L_m < \infty.$$

Define $f_\ell : [-\pi,\pi] \to \mathbb{R}$ for each $\ell \in \mathbb{N}$ by setting

$$f_\ell(t) = \sum_{m=1}^{\ell} \frac{1}{2^m L_m} f_{m,\ell-m+1}(t)$$

for each $t \in [-\pi,\pi]$. Hence we have a sequence of nonnegative functions $f_\ell \in \mathscr{C}_E$ with $f_\ell(t) \to \infty$ for each $t \in E$ and

$$||f_\ell|| \leq \sum_{m=1}^{\ell} \frac{1}{2^m L_m} ||f_{m,\ell-m+1}|| \leq 1$$

for each $\ell \in \mathbb{N}$. □

Example 9.1. *The set E of rational numbers on the interval $[-\pi,\pi]$ is a null set. Let $\{r_m\}$ be an ordered list of all rational numbers in the interval $[-\pi,\pi]$. Define $p : \mathbb{R} \setminus \{0\} \to \mathbb{R}$ by the formula*

$$p(t) = \frac{1}{2^{1/2}|t|^{1/4}(1+|t|)^{1/2}},$$

and for each $n \in \mathbb{N}$ let $p_n : \mathbb{R} \to \mathbb{R}$ be defined by

$$p_n(t) = \begin{cases} n & \text{when } p(t) > n, \\ p(t) & \text{otherwise.} \end{cases}$$

Define $f_{m,n} : [-\pi,\pi] \to \mathbb{R}$ for each $m,n \in \mathbb{N}$ by the formula

$$f_{m,n}(t) = p_n(t - r_m)$$

when $t \in [-\pi,\pi]$. It follows that $f_{m,n} \in \mathscr{C}_E$ and

$$||f_{m,n}||^2 < \int_{\mathbb{R}} [p_n(t)]^2 dt < \int_{\mathbb{R}} [p(t)]^2\, dt = 2\int_{(0,\infty)} \frac{du}{1+u^2} = \pi.$$

For each $\ell \in \mathbb{N}$ let $f_\ell \in \mathscr{C}_E$ be defined by the formula

$$f_\ell(t) = \sum_{m=1}^{\ell} \frac{1}{2^m \pi^{1/2}} f_{m,\ell-m+1}(t).$$

It follows from the definitions of the various functions that $f_\ell(r_m) \to \infty$ for each $m \in \mathbb{N}$ as $\ell \to \infty$. On the other hand,

$$||f_\ell|| \leq \sum_{m=1}^{\ell} \frac{1}{2^m \pi^{1/2}} ||f_{m,\ell-m+1}|| \leq 1.$$

Thus E is a null set.

Definition 9.2. *If $f, g : [-\pi, \pi] \to \mathbb{R}$ and if $f(t) = g(t)$ for all $t \in [-\pi, \pi] \setminus E$ where E is a null set, then we say that $f(t) = g(t)$ almost everywhere on $[-\pi, \pi]$.*

Definition 9.3. *We say that $\{f_n\} \subseteq \mathscr{C}_E$ is a Cauchy sequence in $\mathscr{C}_E$ if, for each $\epsilon > 0$, we can find $N = N(\epsilon)$ such that*

$$||f_n - f_m|| < \epsilon$$

whenever $m, n \geq N$.

The fundamental mathematical problem with the space $\mathscr{C}_E$ is that it is not complete. There are Cauchy sequences $\{f_n\} \subseteq \mathscr{C}_E$ of continuous functions that do not converge in the mean square sense to a continuous limit function $f \in \mathscr{C}_E$. That is, there may be no $f \in \mathscr{C}_E$ such that $||f_n - f|| \to 0$ as $n \to \infty$. We wish to extend $\mathscr{C}_E$ to a larger space that is complete. The abstract idea behind our extension is that every Cauchy sequence defines a unique element in a larger space. The concrete manifestation of this idea is that we can use an elementary argument to construct a representative limit function. The limit function may remain undefined on some null set but is otherwise unique. Of course it is important to note that the limit function need not be continuous. The extension procedure is quite general and in principle is the same procedure used to extend the set of rational numbers to a complete set of real numbers. We begin by showing that a Cauchy sequence in $\mathscr{C}_E$ has a subsequence that converges in pointwise fashion to a well-defined limit at all points other than those contained in some unspecified null set.

Lemma 9.2. *If $\{f_n\} \subseteq \mathscr{C}_E$ is a Cauchy sequence, then there exists a subsequence $\{f_{n(k)}\} = \{f_{n(k)}\}_{k \in \mathbb{N}} \subseteq \mathscr{C}_E$ and a function $f : [-\pi, \pi] \to \mathbb{R}$ such that*

$$f(t) = \lim_{k \to \infty} f_{n(k)}(t)$$

for almost all $t \in [-\pi, \pi]$.

Proof: For each $k \in \mathbb{N}$ choose $n(k)$ such that $||f_m - f_n|| < 2^{-k}$ when $m, n \geq n(k)$. Let $g_k, h_k \in \mathscr{C}_E$ be defined by

$$g_k = f_{n(k)} \quad \text{and} \quad h_k = f_{n(1)} + \sum_{j=1}^{n(k)-1} |f_{n(j+1)} - f_{n(j)}|.$$

We note that $\{h_k(t)\} \subseteq \mathscr{C}_E$ is an increasing sequence for each $t \in [-\pi, \pi]$ and that

$$||h_k|| \leq ||f_{n(1)}|| + \sum_{j=1}^{n(k)-1} ||f_{n(j+1)} - f_{n(j)}|| \leq ||f_{n(1)}|| + 1$$

for all $k \in \mathbb{N}$. Thus there is a null set $E \subseteq [-\pi, \pi]$ and a function $h : [-\pi, \pi] \to \mathbb{R}$ such that $h_k(t) \to h(t)$ when $t \in [-\pi, \pi] \setminus E$. It follows from the definitions that the sequence $\{[g_k(t) + h_k(t)]\} \subseteq \mathscr{C}_E$ is also an increasing sequence for each $t \in [-\pi, \pi]$ and that

$$||[g_k + h_k]|| \leq 2||h_k|| \leq 2(||f_{n(1)}|| + 1).$$

Hence there is a function $s : [-\pi, \pi] \to \mathbb{R}$ such that $[g_k(t) + h_k(t)] \to s(t)$ when $t \in [-\pi, \pi] \setminus E$. By subtraction it follows that $g_k(t) \to s(t) - h(t)$ when $t \in [-\pi, \pi] \setminus E$. We define $f : [-\pi,\pi] \to \mathbb{R}$ by setting

$$f(t) = \begin{cases} s(t) - h(t) & \text{when } t \in [-\pi,\pi] \setminus E, \\ 0 & \text{when } t \in E. \end{cases}$$

This completes the proof. □

For each Cauchy sequence we will show that the limit function is uniquely defined up to some unspecified null set. We have the following results.

Lemma 9.3. *Let $\{g_n\} \subseteq \mathscr{C}_E$ be a sequence of nonnegative functions with $\|g_n\| \to 0$ as $n \to \infty$. Suppose $g_n(t) \to g(t)$ for all $t \in [-\pi,\pi] \setminus E$ for some null set E. If we define $G = \{t \mid g(t) > 0\}$, then G is a null set.*

Proof: If there is a subsequence $\{g_{n(k)}\}$ with $\|g_{n(k)}\| = 0$ for all $k \in \mathbb{N}$, then $g_{n(k)}(t) = 0$, and hence $g(t) = 0$ for all $t \in [-\pi,\pi]$. Hence we suppose, without loss of generality, that $\|g_n\| > 0$ for all $n \in \mathbb{N}$. Let $h_n = g_n/\|g_n\|$. Then $\{h_n\} \in \mathscr{C}_E$ with $\|h_n\| = 1$ for all $n \in \mathbb{N}$. Since $g_n(t) \to g(t) > 0$ when $t \in G \setminus E$, it follows that $h_n(t) \to \infty$ when $t \in G \setminus E$. Hence $G \setminus E$ is null, and since E is also null, it follows that $G = (G \setminus E) \cup E$ is null. □

Corollary 9.1. *Let $\{f_n\} \in \mathscr{C}_E$ and $\{f_n^\star\} \in \mathscr{C}_E$ be Cauchy sequences with $\|f_n - f_n^\star\| \to 0$ as $n \to \infty$. If $f_n(t) \to f(t)$ for all $t \in [-\pi,\pi] \setminus E$ where E is a null set and $f_n^\star(t) \to f^\star(t)$ for all $t \in [-\pi,\pi] \setminus E^\star$ where $E^\star$ is a null set, then $f^\star(t) = f(t)$ almost everywhere.*

Proof: If we define $\{g_n\} \subseteq \mathscr{C}_E$ by setting $g_n = |f_n - f_n^\star|$, then it follows that $\|g_n\| \to 0$ and $g_n(t) \to |f^\star(t) - f(t)|$ as $n \to \infty$ for all $t \in [-\pi,\pi] \setminus (E \cup E^\star)$. By Lemma 9.3 it follows that $|f^\star(t) - f(t)| = 0$ almost everywhere. □

9.1.1 ▪ Defining the elements of $L^2([-\pi,\pi])$

The set of all Cauchy sequences $\{f_n\} \in \mathscr{C}_E$ can be classified according to the equivalence relation

$$\{f_n\} \equiv \{g_n\} \quad \Leftrightarrow \quad \lim_{n\to\infty} \|f_n - g_n\| = 0.$$

We write $\mathscr{E}(\{f_n\})$ to denote the equivalence class containing $\{f_n\}$. Corollary 9.1 shows that this class defines a unique collection

$$\mathscr{E}(f) = \{f^\star \mid f^\star(t) = f(t) \text{ almost everywhere}\}$$

of limit functions represented by a nominal function f from the class. For this reason we will refer to $\mathscr{E}(f)$ as the limit class represented by the function f. The set of all limit classes $\mathscr{E}(f)$ is a linear space with the definitions

$$\mathscr{E}(f) + \mathscr{E}(g) = \mathscr{E}(f+g) \quad \text{and} \quad c\,\mathscr{E}(f) = \mathscr{E}(cf)$$

for each $c \in \mathbb{R}$. Since

$$\big|\, \|f_m\| - \|f_n\| \,\big| \le \|f_m - f_n\|,$$

it follows that $\{\|f_n\|\} \subseteq \mathbb{R}$ is a Cauchy sequence of real numbers with a unique limit. We define

$$\|\mathscr{E}(f)\| = \lim_{n\to\infty} \|f_n\| \tag{9.1}$$

and note that the definition does not depend on the choice of representative sequence $\{f_n\}$ from the class $\mathscr{E}(f)$. For example, if $\{f_n^\star\} \in \mathscr{E}(f)$, then

$$\lim_{n\to\infty} \|f_n\| \le \lim_{n\to\infty} \|f_n - f_n^\star\| + \lim_{n\to\infty} \|f_n^\star\| = \lim_{n\to\infty} \|f_n^\star\|.$$

A similar argument shows that

$$\lim_{n\to\infty} \|f_n^\star\| \le \lim_{n\to\infty} \|f_n\|,$$

and hence the limits are equal. It follows from the definition that

$$\|\mathscr{E}(f)\| = \lim_{n\to\infty} \|f_n\| \ge 0$$

and

$$\|\mathscr{E}(f)\| = 0 \Leftrightarrow \lim_{n\to\infty} \|f_n\| = 0 \Leftrightarrow f(t) = 0 \text{ almost everywhere } \Leftrightarrow \mathscr{E}(f) = \mathscr{E}(0).$$

It is also true that

$$\|\mathscr{E}(f) + \mathscr{E}(g)\| = \lim_{n\to\infty} \|f_n + g_n\| \le \lim_{n\to\infty} [\|f_n\| + \|g_n\|] = \|\mathscr{E}(f)\| + \|\mathscr{E}(g)\|$$

and

$$\|\mathscr{E}(cf)\| = \lim_{n\to\infty} c\ \|f_n\| = c \lim_{n\to\infty} \|f_n\| = c\ \|\mathscr{E}(f)\|$$

for each $c \in \mathbb{R}$. Since

$$\begin{aligned}
\|\mathscr{E}(f) + \mathscr{E}(g)\|^2 + \|\mathscr{E}(f) - \mathscr{E}(g)\|^2 &= \lim_{n\to\infty} \left[\|f_n + g_n\|^2 + \|f_n - g_n\|^2\right] \\
&= \lim_{n\to\infty} 2\left[\|f_n\|^2 + \|g_n\|^2\right] \\
&= 2\left[\|\mathscr{E}(f)\|^2 + \|\mathscr{E}(g)\|^2\right],
\end{aligned}$$

it follows that the scalar product

$$\langle\, \mathscr{E}(f), \mathscr{E}(g)\,\rangle = \frac{1}{4}\left[\|\mathscr{E}(f+g)\|^2 - \|\mathscr{E}(f-g)\|^2\right] \tag{9.2}$$

is also well defined.

The space of all limit classes $\mathscr{E}(f)$ with the norm and scalar product defined above will be denoted by $L^2 = L^2([-\pi,\pi])$. The following result shows that each element $\mathscr{E}(f) = \mathscr{E}(\{f_n\}) \in L^2$ can be interpreted as the limit of the Cauchy sequence $\{\mathscr{E}(f_n)\}$ in the extended space.

Lemma 9.4. *Let $\{f_n\} \subseteq \mathscr{C}_E$ be a Cauchy sequence, and let $f \in \mathscr{E}(f) = \mathscr{E}(\{f_n\})$ be a nominal representative from the corresponding limit class. For each $m \in \mathbb{N}$ let $\{f_{m,n}\}_{n\in\mathbb{N}} \subseteq \mathscr{C}_E$ be the Cauchy sequence defined by $f_{m,n} = f_m$ for all $n \in \mathbb{N}$ with limit class $\mathscr{E}(f_m) = \mathscr{E}(\{f_m, f_m, \ldots\})$. Then $\|\mathscr{E}(f_m) - \mathscr{E}(f)\| \to 0$ as $m \to \infty$.*

Proof: For each fixed $m \in \mathbb{N}$ the Cauchy sequence $\{f_{m,n}\}_{n\in\mathbb{N}} = \{f_m, f_m, \ldots\} \subseteq \mathscr{C}_E$ satisfies $\|f_{m,n} - f_m\| = 0$ for all $n \in \mathbb{N}$ and hence converges (trivially) in L^2 to the limit function f_m. Indeed, we have $f_{m,n}(t) \to f_m(t)$ as $n \to \infty$ for all $t \in [-\pi,\pi]$. Thus $f_m \in \mathscr{E}(f_m) = \mathscr{E}(\{f_{m,n}\})$, and our definition of the norm on L^2 gives

$$\|\mathscr{E}(f_m)\| = \lim_{n\to\infty} \|f_{m,n}\| = \|f_m\|.$$

Since $f_n \in \mathscr{C}_E$ for each $n \in \mathbb{N}$, we can also define a Cauchy sequence $\{g_{m,n}\}_{n\in\mathbb{N}} \in \mathscr{C}_E$ for each $m \in \mathbb{N}$ using the formula $g_{m,n} = f_{m,n} - f_n$ for each $n \in \mathbb{N}$. Clearly $g_{m,n}(t) \to f_m(t) - f(t)$ for almost all $t \in [-\pi,\pi]$. Thus $f_m - f \in \mathscr{E}(\{g_{m,n}\})$ and our definition of the norm on L^2 gives

$$||\mathscr{E}(f_m - f)|| = \lim_{n\to\infty} ||g_{m,n}|| = \lim_{n\to\infty} ||f_m - f_n||,$$

and hence

$$\lim_{m\to\infty} ||\mathscr{E}(f_m) - \mathscr{E}(f)|| = \lim_{m,n\to\infty} ||f_m - f_n|| = 0.$$

Thus the equivalence class $\mathscr{E}(f)$ can be regarded as the limit as $n \to \infty$ of the sequence of equivalence classes $\{\mathscr{E}(f_n)\}$ in L^2. □

Before we show that L^2 is complete, it is convenient to simplify our notation and to discuss the integral interpretation of our definitions.

9.1.2 ▪ Understanding the elements of $L^2([-\pi,\pi])$

For each element $f \in \mathscr{C}_E$ the norm is defined as an integral by the formula

$$||f||^2 = \int_{[-\pi,\pi]} [f(t)]^2 \, dt.$$

We will extend this integral interpretation of the norm to all elements in L^2. If $\{f_n\} \subseteq \mathscr{C}_E$ is a Cauchy sequence, there is a uniquely defined limit class $\mathscr{E}(\{f_n\}) = \mathscr{E}(f)$ that we can identify by a nominal representative function $f : [-\pi,\pi] \to \mathbb{R}$. Thus we have a one-to-one correspondence

$$\mathscr{E}(\{f_n\}) \Leftrightarrow \mathscr{E}(f) \Leftrightarrow f$$

that identifies each element $\mathscr{E}(f)$ with a real-valued function f. Henceforth we will interpret each limit class $\mathscr{E}(f) \in L^2$ as a function and simply write $f \in L^2$. Thus, if $\{f_n\} \subseteq \mathscr{C}_E$ is a Cauchy sequence with limit class $\mathscr{E}(f)$ and nominal representative function f, we write

$$||f||^2 = \lim_{n\to\infty} ||f_n||^2 = \lim_{n\to\infty} \int_{[-\pi,\pi]} [f_n(t)]^2 \, dt.$$

In the new notation Lemma 9.4 can be rewritten as $||f_m - f|| \to 0$ as $m \to \infty$. Hence it is natural to write

$$||f||^2 = \int_{[-\pi,\pi]} [f(t)]^2 \, dt.$$

Since

$$\langle f, g \rangle = \frac{1}{4}\left[||f+g||^2 - ||f-g||^2\right]$$

when $f, g \in \mathscr{C}_E$, we can use the same idea to extend the integral definition of the scalar product. Hence, if $\{f_n\} \subseteq \mathscr{C}_E$ and $\{g_n\} \subseteq \mathscr{C}_E$ are Cauchy sequences with limit classes $\mathscr{E}(f)$ and $\mathscr{E}(g)$, respectively, then we represent the classes with the nominal representative functions f and g, respectively, and write

$$\begin{aligned}\langle f,g\rangle &= \frac{1}{4}\left[\|f+g\|^2-\|f-g\|^2\right]\\ &= \frac{1}{4}\int_{[-\pi,\pi]}\left[[f(t)+g(t)]^2-[f(t)-g(t)]^2\right]dt\\ &= \lim_{n\to\infty}\frac{1}{4}\int_{[-\pi,\pi]}\left[[f_n(t)+g_n(t)]^2-[f_n(t)-g_n(t)]^2\right]dt\\ &= \lim_{n\to\infty}\frac{1}{4}\left[\|f_n+g_n\|^2-\|f_n-g_n\|^2\right]\\ &= \lim_{n\to\infty}\langle f_n,g_n\rangle.\end{aligned}$$

If $\chi:[-\pi,\pi]\to\mathbb{R}$ is defined by $\chi(t)=1$ for all $t\in[-\pi,\pi]$, then for each $f\in\mathscr{C}_E$ we have

$$\langle f,\chi\rangle=\int_{[-\pi,\pi]}f(t)\chi(t)\,dt=\int_{[-\pi,\pi]}f(t)\,dt.$$

We also extend this interpretation to all $f\in L^2$. In general, if S is some subset of $[-\pi,\pi]$, if $\chi_S:[-\pi,\pi]\to\mathbb{R}$ is defined by

$$\chi_S(t)=\begin{cases}1 & \text{when } t\in S,\\ 0 & \text{when } t\notin S,\end{cases}$$

and if $\chi_S\in L^2$, then we say that S is measurable, and we define

$$\mu(S)=\langle\chi_S,\chi\rangle=\int_{[-\pi,\pi]}\chi_S(t)\chi(t)\,dt=\int_S dt$$

to be the measure of the set S. For any given function $f\in L^2$ and each $\alpha\in\mathbb{R}$ we can define the subset $S_f(\alpha)\subseteq[-\pi,\pi]$ by setting

$$S_f(\alpha)=\{t\mid f(t)>\alpha\}.$$

If $S_f(\alpha)$ is a measurable set for each $\alpha\in\mathbb{R}$, then we say that f is a measurable function.

9.1.3 ▪ The completeness of $L^2([-\pi,\pi])$

Finally, we must show that L^2 is complete. Suppose $\{f^{(k)}\}$ is a Cauchy sequence in L^2. That is, we suppose

$$\|f^{(k)}-f^{(\ell)}\|\to 0$$

as $k,\ell\to\infty$. On the other hand, we must remember that for each $k\in\mathbb{N}$ the element $f^{(k)}\in L^2$ is the limit of a Cauchy sequence $\{f_n^{(k)}\}_{n\in\mathbb{N}}\subseteq\mathscr{C}_E$. Choose $n(k)$ so that

$$\|f_n^{(k)}-f_{n(k)}^{(k)}\|<\frac{1}{k}$$

when $n\geq n(k)$. If we write $g^{(k)}=f_{n(k)}^{(k)}\in\mathscr{C}_E$, then the above inequality can be rewritten as

$$\|f_n^{(k)}-g^{(k)}\|<\frac{1}{k} \tag{9.3}$$

when $n > n(k)$. From (9.3) we deduce that

$$\|f^{(k)} - g^{(k)}\| = \lim_{n\to\infty} \|f_n^{(k)} - g^{(k)}\| \le \frac{1}{k}, \tag{9.4}$$

and hence for fixed $k, \ell \in \mathbb{N}$ it follows that

$$\begin{aligned}\|g^{(k)} - g^{(\ell)}\| &\le \|g^{(k)} - f^{(k)}\| + \|f^{(k)} - f^{(\ell)}\| + \|f^{(\ell)} - g^{(\ell)}\| \\ &\le \|f^{(k)} - f^{(\ell)}\| + \frac{1}{k} + \frac{1}{\ell}.\end{aligned} \tag{9.5}$$

Because the sequence $\{f^{(k)}\}_{k\in\mathbb{N}}$ is a Cauchy sequence in L^2, it follows from (9.5) that $\{g^{(k)}\}$ is a Cauchy sequence in $\mathscr{C}_E$. Hence there is an element $g \in L^2$ such that $\|g^{(k)} - g\| \to 0$ as $k \to \infty$. We will show that our given Cauchy sequence $\{f^{(k)}\}$ converges to g. Thus we must show that

$$\|f^{(k)} - g\| \to 0$$

as $k \to \infty$. From (9.4), we have

$$\begin{aligned}\|f^{(k)} - g\| &\le \|f^{(k)} - g^{(k)}\| + \|g^{(k)} - g\| \\ &\le \|g^{(k)} - g\| + \frac{1}{k},\end{aligned} \tag{9.6}$$

and since (9.5) implies

$$\|g^{(k)} - g\| = \lim_{\ell\to\infty} \|g^{(k)} - g^{(\ell)}\| \le \lim_{\ell\to\infty} \left(\|f^{(k)} - f^{(\ell)}\| + \frac{1}{\ell}\right) + \frac{1}{k} = \lim_{\ell\to\infty} \|f^{(k)} - f^{(\ell)}\| + \frac{1}{k},$$

it follows from (9.6) that

$$\|f^{(k)} - g\| \le \lim_{\ell\to\infty} \|f^{(k)} - f^{(\ell)}\| + \frac{2}{k}.$$

By taking the limit as $k \to \infty$ we obtain

$$\lim_{k\to\infty} \|f^{(k)} - g\| \le \lim_{k,\ell\to\infty} \left[\|f^{(k)} - f^{(\ell)}\| + \frac{2}{\ell}\right] = 0.$$

Thus we have shown that the space L^2 with the given norm and scalar product is complete, and hence it is a Hilbert space.

9.2 ▪ The Fourier Series Representation on $\mathscr{C}([-\pi,\pi])$

In this section we will show that if we relax our intuitive ideas of pointwise convergence, then every continuous function can be represented by a Fourier series. Note that in this section we consider pointwise convergence on the space of $\mathscr{C}([-\pi,\pi])$ of continuous functions on the interval $[-\pi,\pi]$ and will not use the topology of the Euclidean space $\mathscr{C}_E$. We begin by considering the simple trigonometric series

$$\mathscr{D}_n(t) = \frac{1}{2} + \cos t + \cos 2t + \cdots + \cos nt$$

for all $t \in [-\pi, \pi]$ and each $n \in \mathbb{N}$. If we multiply both sides by $2\sin(t/2)$, then we obtain

$$2\sin\frac{t}{2}\cdot\mathscr{D}_n(t) = \sin\frac{t}{2} + \left[\sin\frac{3t}{2} - \sin\frac{t}{2}\right] + \left[\sin\frac{5t}{2} - \sin\frac{3t}{2}\right] + \cdots$$
$$\cdots + \left[\sin\frac{(2n+1)t}{2} - \sin\frac{(2n-1)t}{2}\right]$$

and hence deduce that

$$\mathscr{D}_n(t) = \begin{cases} \frac{\sin\frac{(2n+1)t}{2}}{2\sin\frac{t}{2}} & \text{when } t \neq 0, \\ \frac{2n+1}{2} & \text{when } t = 0. \end{cases}$$

The functions $\mathscr{D}_n(t)$ are known as the Dirichlet kernels, and they have some interesting properties. In the first place we can integrate term by term to see that

$$\int_{[-\pi,\pi]} \mathscr{D}_n(t)\,dt = \pi \tag{9.7}$$

for all $n \in \mathbb{N}$. In the second place we can use an elementary Maclaurin series argument to see that

$$\left|\frac{1}{2\sin\frac{t}{2}} - \frac{1}{t}\right| = \left|\frac{t - 2\sin\frac{t}{2}}{t\sin\frac{t}{2}}\right| < \frac{t}{12} \to 0$$

as $t \to 0$. It follows that for any $\epsilon > 0$ we can choose $\delta = \delta(\epsilon) > 0$ such that

$$\left|\mathscr{D}_n(t) - \frac{\sin\frac{(2n+1)t}{2}}{t}\right| < \epsilon$$

for all $|t| < \delta$. If we define

$$I_n(\delta) = \int_{|t|<\delta} \mathscr{D}_n(t)\,dt,$$

it follows from the uniform convergence on $[-\delta, \delta]$ that

$$\lim_{n\to\infty} I_n(\delta) = \lim_{n\to\infty}\int_{|t|<\delta} \frac{\sin\frac{(2n+1)t}{2}}{t}\,dt$$

$$= \lim_{n\to\infty}\int_{|s|<\frac{(2n+1)\delta}{2}} \frac{\sin s}{s}\,ds = \int_{(-\infty,\infty)} \frac{\sin s}{s}\,ds.$$

Now a separate calculation shows us that

$$\int_{(0,\infty)} e^{-ps}\sin s\,ds = \frac{p}{s^2+p^2},$$

and hence by integrating both sides with respect to p and interchanging the order of integration on the left-hand side we obtain

$$\begin{aligned}\int_{(0,\infty)} \frac{\sin s}{s}\, ds &= \int_{(0,\infty)} \left[\int_{(0,\infty)} e^{-sp}\, dp\right] \sin s\, ds \\ &= \int_{(0,\infty)} \left[\int_{(0,\infty)} e^{-ps} \sin s\, ds\right] \\ &= \int_{(0,\infty)} \frac{p}{s^2+p^2}\, dp \\ &= \frac{\pi}{2}. \end{aligned} \tag{9.8}$$

We deduce that

$$\lim_{n\to\infty} I_n(\delta) = \pi.$$

Hence we can use (9.7) and (9.8) to conclude that

$$\int_{\delta<|t|<\pi} \mathscr{D}_n(t)\, dt \to 0$$

as $n \to \infty$ for each fixed $\delta \in (0,\pi)$. Thus the entire effective mass of the Dirichlet kernel $\mathscr{D}_n(t)$ appears to move toward $t = 0$ as $n \to \infty$. Some typical Dirichlet kernels are shown in Figure 9.1.

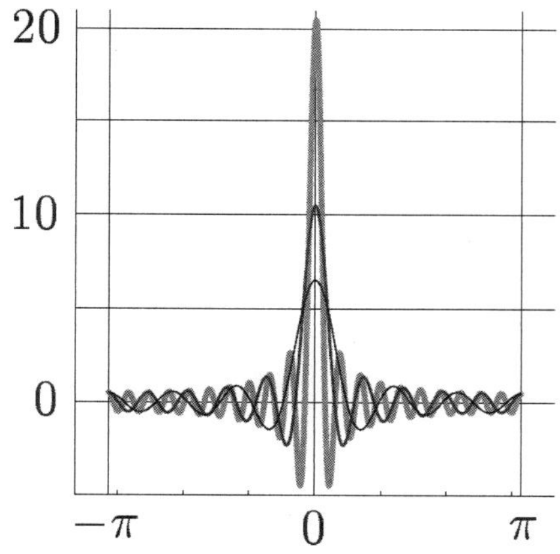

Figure 9.1. *The Dirichlet kernels $\mathscr{D}_n(t)$ for $n = 7,\ 10,\ 20$*

However, we also note that for each fixed value of $n \in \mathbb{N}$ the value $\mathscr{D}_n(t)$ oscillates between

$$\pm \frac{1}{2\sin\frac{t}{2}},$$

and for each fixed value of $t \in \mathbb{R}$ with $t \neq 0$ it is certainly not true that $\mathscr{D}_n(t) \to 0$ as $n \to \infty$. It follows that the sum of the Fourier series

$$\mathscr{D}(t) = \frac{1}{2} + \sum_{n=1}^{\infty} \cos nt$$

cannot be defined by taking the pointwise limit of the sequence $\{\mathscr{D}_n(t)\}$ of Dirichlet kernels as $n \to \infty$. Despite the factual observation that the sequence diverges, there is a strong suggestion that some sort of convergence is taking place. The inspirational step forward is to discard the sequence $\{\mathscr{D}_n(t)\}$ of Dirichlet kernels in favor of the sequence $\{\mathscr{F}_m(t)\}$ of the averages of the Dirichlet kernels. Thus we define

$$\mathscr{F}_m(t) = \frac{1}{m+1}\left[\frac{1}{2} + \mathscr{D}_1(t) + \mathscr{D}_2(t) + \cdots + \mathscr{D}_m(t)\right]$$

for each $m \in \mathbb{N}$. The new functions are called the Fejér kernels. From our earlier formulae we have

$$\mathscr{F}_m(t) = \frac{1}{2(m+1)\sin\frac{t}{2}}\left[\sin\frac{t}{2} + \sin\frac{3t}{2} + \cdots + \sin\frac{(2m+1)t}{2}\right],$$

and hence

$$\begin{aligned} 4\sin^2\frac{t}{2}\cdot\mathscr{F}_m(t) &= \frac{1}{m+1}\frac{1}{m+1}\sum_{k=0}^{m}(\cos kt - \cos(k+1)t) \\ &= \frac{1}{m+1}[1-\cos(m+1)t] \\ &= \frac{2\sin^2\frac{(m+1)t}{2}}{m+1}, \end{aligned}$$

from which it follows that

$$\mathscr{F}_m(t) = \begin{cases} \frac{1}{2(m+1)}\left(\frac{\sin\frac{(m+1)t}{2}}{\sin\frac{t}{2}}\right)^2 & \text{when } t \neq 0, \\ \frac{m+1}{2} & \text{when } t = 0. \end{cases}$$

Hence $\mathscr{F}_m(t) \geq 0$ for all t. In terms of the basic trigonometric elements we can see that

$$\begin{aligned} \mathscr{F}_m(t) &= \frac{1}{m+1}\left[\frac{1}{2} + \left(\frac{1}{2} + \cos t\right) + \left(\frac{1}{2} + \cos t + \cos 2t\right) + \cdots \right. \\ &\qquad\qquad \left. \cdots + \left(\frac{1}{2} + \cos t + \cos 2t + \cdots + \cos mt\right)\right] \\ &= \frac{1}{m+1}\left[\frac{m+1}{2} + m\cos t + (m-1)\cos 2t + \cdots + \cos mt\right] \\ &= \frac{1}{2} + \sum_{\ell=1}^{m}\left(1 - \frac{\ell}{m+1}\right)\cos \ell t, \end{aligned}$$

from which it follows directly that

$$\int_{[-\pi,\pi]} \mathscr{F}_m(t)\, dt = \pi.$$

Suppose $0 < \delta < \pi$. For $|t| > \delta$ we have

$$0 \leq \mathscr{F}_m(t) \leq \frac{1}{2(m+1)\sin^2\frac{\delta}{2}},$$

and hence $\mathscr{F}_m(t) \to 0$ uniformly on the set $[-\pi,\pi]\setminus[-\delta,\delta]$ as $m \to \infty$. Therefore,

$$0 \leq \int_{\delta<|t|<\pi} \mathscr{F}_m(t)\,dt < \frac{(\pi-\delta)}{(m+1)\sin^2\frac{\delta}{2}} \to 0$$

as $m \to \infty$, and, of course, this means

$$J_m(\delta) = \int_{|t|<\delta} \mathscr{F}_m(t)\,dt \to \pi$$

as $m \to \infty$. Some typical Fejér kernels are shown in Figure 9.2.

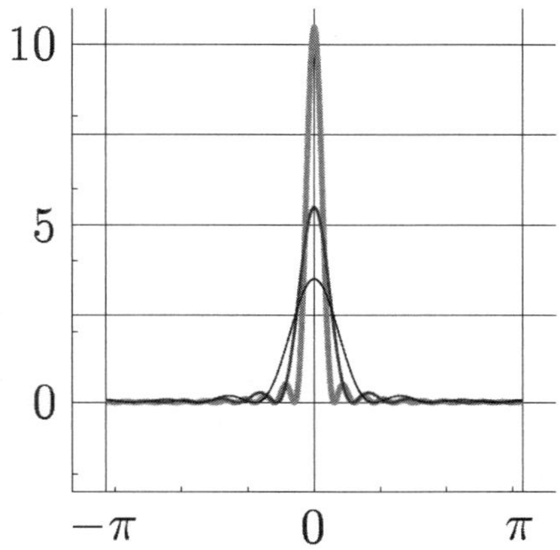

Figure 9.2. *The Fejér kernels $\mathscr{F}_m(t)$ for $n = 7$, 10, 20*

In summary, the key properties of the Fejér kernels are

$$\int_{[-\pi,\pi]} \mathscr{F}_m(t)\,dt = \pi, \tag{9.9}$$

$$0 \leq \mathscr{F}_m(t) \leq \frac{1}{2(m+1)\sin^2\frac{\delta}{2}} \quad \text{when } |t| \geq \delta, \tag{9.10}$$

and

$$\left|\int_{\delta<|t|<\pi} \mathscr{F}_m(t)\,dt\right| < \frac{(\pi-\delta)}{(m+1)\sin^2\frac{\delta}{2}} \to 0 \quad \text{as } m \to \infty. \tag{9.11}$$

We conclude that when m is large the Fejér kernel is very nearly an impulse of strength π located at the origin. That is, the entire area under the graph becomes concentrated around $t = 0$. Let $f \in \mathscr{C}$, and consider the Fejér integral

$$\sigma_m[f](t) = \frac{1}{\pi}\int_{[-\pi,\pi]} f(\tau)\mathscr{F}_m(t-\tau)\,d\tau$$

at the point $\tau = t$. Because the area under the graph of the Fejér kernel $\mathscr{F}_m(t-\tau)$ becomes concentrated at $\tau = t$, we could expect $\sigma_m[f](t)$ to converge to $f(t)$ as m increases. This is indeed the case. Suppose $f \in \mathscr{C}$. Since f is uniformly continuous on $[-\pi,\pi]$, there is some finite constant K such that $|f(\tau)| \leq K$ for all $\tau \in [-\pi, \pi]$ and for each fixed $\epsilon > 0$ we can find $\delta = \delta(\epsilon)$ so that

$$|f(t) - f(\tau)| < \epsilon$$

whenever $t, \tau \in [-\pi, \pi]$ and $|\tau - t| < \delta$. Therefore,

$$\begin{aligned}
|f(t) - \sigma_m[f](t)| &= \left| f(t) - \frac{1}{\pi} \int_{[-\pi,\pi]} f(\tau) \mathscr{F}_m(t-\tau)\, d\tau \right| \\
&= \left| \frac{1}{\pi} \int_{[-\pi,\pi]} (f(t) - f(\tau)) \mathscr{F}_m(t-\tau)\, d\tau \right| \\
&\leq \frac{1}{\pi} \int_{[-\pi,\pi]} |f(t) - f(\tau)| \mathscr{F}_m(t-\tau)\, d\tau \\
&\leq \frac{\epsilon}{\pi} \int_{|\tau - t| < \delta} \mathscr{F}_m(t-\tau)\, d\tau + \frac{2K}{\pi} \int_{\delta < |\tau - t| < \pi} \mathscr{F}_m(t-\tau)\, d\tau \\
&\leq \frac{\epsilon}{\pi} \cdot \pi + \frac{2K}{\pi} \cdot \frac{(\pi - \delta)}{(m+1)\sin^2 \frac{\delta}{2}} \\
&< 2\epsilon
\end{aligned}$$

for m sufficiently large. Since $\epsilon > 0$ is arbitrary, it follows that $\sigma_m[f](t)$ converges uniformly to $f(t)$ on $[-\pi, \pi]$ as $m \to \infty$. Thus

$$f(t) = \lim_{m \to \infty} \frac{1}{\pi} \int_{[-\pi,\pi]} f(\tau) \mathscr{F}_m(t-\tau)\, d\tau$$

for all $t \in [-\pi, \pi]$. Since

$$\begin{aligned}
\mathscr{F}_m(t-\tau) &= \frac{1}{2} + \left(1 - \frac{1}{m+1}\right)\cos(t-\tau) + \left(1 - \frac{2}{m+1}\right)\cos 2(t-\tau) + \cdots \\
&\qquad \cdots + \left(1 - \frac{m}{m+1}\right)\cos m(t-\tau) \\
&= \frac{1}{2} + \left(1 - \frac{1}{m+1}\right)[\cos t \cos \tau + \sin t \sin \tau] \\
&\qquad + \left(1 - \frac{2}{m+1}\right)[\cos 2t \cos 2\tau + \sin t \sin 2\tau] + \cdots \\
&\qquad \cdots + \left(1 - \frac{m}{m+1}\right)[\cos mt \cos m\tau + \sin mt \sin m\tau] \\
&= \frac{1}{2} + \sum_{\ell=1}^{m} \left(1 - \frac{\ell}{m+1}\right)[\cos \ell t \cos \ell \tau + \sin \ell t \sin \ell \tau],
\end{aligned}$$

the Fejér integral can be written in the form

$$\sigma_m[f](t) = a_0 + \sum_{\ell=1}^{m} \left(1 - \frac{\ell}{m+1}\right)[a_\ell \cos \ell t + b_\ell \sin \ell t],$$

where

$$a_0 = \frac{1}{2\pi}\int_{[-\pi,\pi]} f(\tau)\,d\tau$$

and where

$$a_\ell = \frac{1}{\pi}\int_{[-\pi,\pi]} f(\tau)\cos\ell\tau\,d\tau \quad\text{and}\quad b_\ell = \frac{1}{\pi}\int_{[-\pi,\pi]} f(\tau)\sin\ell\tau\,d\tau$$

for each $\ell = 1,2,\ldots,m$. It follows that

$$f(t) = \lim_{m\to\infty}\left[a_0 + \sum_{\ell=1}^{m}\left(1 - \frac{\ell}{m+1}\right)[a_\ell\cos\ell t + b_\ell\sin\ell t]\right]. \tag{9.12}$$

If we define the partial sums of the Fourier series

$$S[f](t) = a_0 + \sum_{\ell=1}^{\infty}[a_\ell\cos\ell t + b_\ell\sin\ell t]$$

by the formulae

$$S_0[f](t) = a_0 \quad\text{and}\quad S_n[f](t) = a_0 + \sum_{\ell=1}^{n}[a_\ell\cos\ell t + b_\ell\sin\ell t]$$

for each $n\in\mathbb{N}$, then

$$f(t) = \lim_{m\to\infty}\frac{1}{m+1}[S_0[f](t) + S_1[f](t) + \cdots + S_m[f](t)].$$

Thus, for each $f\in\mathscr{C}$ and each point $t\in[-\pi,\pi]$, the average of the sequence of partial sums of the Fourier series for f converges to the value $f(t)$. Thus we have established the famous theorem of Fejér.

Theorem 9.5. *Let $f\in\mathscr{C}([-\pi,\pi])$. Then*

$$\begin{aligned} f(t) &= \lim_{m\to\infty}\frac{1}{m+1}[S_0[f](t) + S_1[f](t) + \cdots + S_m[f](t)] \\ &= \lim_{m\to\infty}\left[a_0 + \sum_{\ell=1}^{m}\left(1 - \frac{\ell}{m+1}\right)[a_\ell\cos\ell t + b_\ell\sin\ell t]\right] \end{aligned}$$

for each $t\in[-\pi,\pi]$. We say that the Fourier series

$$S[f](t) = a_0 + \sum_{\ell=1}^{\infty}[a_\ell\cos\ell t + b_\ell\sin\ell t]$$

converges everywhere to $f(t)$ in the sense of Cesàro.

Of course we should point out the Fourier series for a piecewise smooth function converges at each point to the average of the left- and right-hand limits. The first proof of pointwise convergence, due to Dirichlet, for a continuous function with at most a finite number of local extrema, was established in 1829, well before the Fejér theorem in 1904. The Dirichlet conditions are sufficient for pointwise convergence but are not necessary. For this reason the study of Fourier series continued unabated until the convergence questions were finally settled by Lebesgue in 1905.

9.3 ▪ Fourier Series Representation on $L^2([-\pi,\pi])$

In the previous section we discussed the pointwise convergence of Fourier series and proved that the Fourier series for a continuous function converges in the sense of Cesàro to the value of the function at every point. Despite this very satisfying result it is nevertheless true that there are continuous functions for which the sequence of partial sums of the Fourier series does not converge. Indeed, the theory of Fourier series and the representation of functions by Fourier series was never properly understood until the notion of pointwise convergence was discarded. In this section we will show that the natural setting for Fourier series is the Hilbert space $L^2([-\pi,\pi])$.

Recall that if $f\in\mathscr{C}$, then

$$|f(t)-\sigma_m[f](t)|\le\frac{\epsilon}{\pi}\cdot\pi+\frac{2K}{\pi}\cdot\frac{(\pi-\delta)}{(m+1)\sin^2\frac{\delta}{2}}\le 2\epsilon$$

for all $t\in[-\pi,\pi]$ provided $m\in\mathbb{N}$ is sufficiently large. Thus, if $f\in\mathscr{C}_E$, then

$$\|f-\sigma_m[f]\|^2=\int_{[-\pi,\pi]}|f(t)-\sigma_m[f](t)|^2\,dt\le 8\epsilon^2\pi$$

when m is sufficiently large. Thus $\|f-\sigma_m[f]\|\to 0$ as $m\to\infty$. Therefore, provided $f\in\mathscr{C}_E$, the sequence $\{\sigma_m[f]\}_{m\in\mathbb{N}}$ converges to f in L^2. Despite the convergence of the sequence of arithmetic means of the partial sums, we can show that in L^2 the original sequence of partial sums provides a better approximation. Indeed, for every $f\in L^2$, we will show that the best possible approximation to f in L^2 by a trigonometric polynomial of degree m is given by the polynomial $S_m[f]\in\mathscr{C}_E$ defined by

$$S_m[f](t)=a_0c_0(t)+\sum_{\ell=1}^{m}[a_\ell c_\ell(t)+b_\ell s_\ell(t)],$$

where

$$a_0=\frac{1}{2\pi}\langle f,c_0\rangle,\text{ and where } a_\ell=\frac{1}{\pi}\langle f,c_\ell\rangle \text{ and } b_\ell=\frac{1}{\pi}\langle f,s_\ell\rangle \text{ for each } \ell\in\mathbb{N}$$

are the Fourier coefficients for f, and where, for convenience, we have defined the function $c_0:[-\pi,\pi]\to\mathbb{R}$, and the functions $c_\ell:[-\pi,\pi]\to\mathbb{R}$ and $s_\ell:[-\pi,\pi]\to\mathbb{R}$ for each $\ell\in\mathbb{N}$, by the formulae $c_0(t)=1$, $c_\ell(t)=\cos\ell t$, and $s_\ell(t)=\sin\ell t$ for all $t\in[-\pi,\pi]$.

Lemma 9.6. *Let $f\in L^2$, and let*

$$P_m[\alpha,\beta](t)=\alpha_0c_0(t)+\sum_{\ell=1}^{m}[\alpha_\ell c_\ell(t)+\beta_\ell s_\ell(t)]$$

be a trigonometric polynomial of degree m. Let $E_m[f,\alpha,\beta]=\|f-P_m[\alpha,\beta]\|^2$ denote the mean square error for the approximation of f by $P_m[\alpha,\beta]$ in L^2. The minimum value of $E_m[f,\alpha,\beta]$ is attained when $\alpha_0=a_0$, and when $\alpha_\ell=a_\ell$ and $\beta_\ell=b_\ell$ for all $\ell=1,2,\dots,m$, where a_ℓ and b_ℓ are the Fourier coefficients of f. In this case $P_m[\alpha,\beta]=S_m[f]$ and the mean square error is

$$E_m[a,b]=\|f\|^2-\pi\left(2a_0^2+\sum_{\ell=1}^{m}[a_\ell^2+b_\ell^2]\right).$$

Proof: We note first that $\langle c_0, c_\ell \rangle = \langle c_0, s_\ell \rangle = 0$ for each $\ell = 1,2,\ldots,m$, $\langle c_k, s_\ell \rangle = 0$ for each $k,\ell = 1,2,\ldots,m$, and $\langle c_k, c_\ell \rangle = \langle s_k, s_\ell \rangle = 0$ for each $k,\ell = 1,2,\ldots,m$ provided $k \neq \ell$. We have

$$
\begin{aligned}
E_m[\alpha,\beta] &= \|f - P_m\|^2 \\
&= \left\| f - \left(\alpha_0 c_0 + \sum_{\ell=1}^{m} [\alpha_\ell c_\ell + \beta_\ell s_\ell] \right) \right\|^2 \\
&= \langle f, f \rangle - 2\left(\alpha_0 \langle f, c_0 \rangle + \sum_{\ell=1}^{m} [\alpha_\ell \langle f, c_\ell \rangle + \beta_\ell \langle f, s_\ell \rangle] \right) \\
&\qquad + \alpha_0^2 \|c_0\|^2 + \sum_{\ell=1}^{m} [\alpha_\ell^2 \|c_\ell\|^2 + \beta_\ell^2 \|s_\ell\|^2] \\
&= \|f\|^2 - 2\pi\left(2a_0\alpha_0 + \sum_{\ell=1}^{m} [a_\ell \alpha_\ell + b_\ell \beta_\ell] \right) + \pi\left(2\alpha_0^2 + \sum_{\ell=1}^{m} [\alpha_\ell^2 + \beta_\ell^2] \right).
\end{aligned}
$$

We can minimize this quadratic expression in α and β by setting

$$
\frac{\partial E_m}{\partial \alpha_0} = 0, \quad \frac{\partial E_m}{\partial \alpha_\ell} = 0, \text{ and } \frac{\partial E_m}{\partial \beta_\ell} = 0
$$

for each $\ell = 1,2,\ldots,m$. This gives $\alpha_0 = a_0$, $\alpha_\ell = a_\ell$, and $\beta_\ell = b_\ell$ for each $\ell = 1,2,\ldots,m$. Therefore, the best approximation by a trigonometric polynomial of degree m is obtained when $P_m[\alpha,\beta] = S_m[f]$. The minimum mean square error is given by

$$
\|f - S_m[f]\|^2 = \|f\|^2 - \pi\left(2a_0^2 + \sum_{\ell=1}^{m} [a_\ell^2 + b_\ell^2] \right) \geq 0. \tag{9.13}
$$

This completes the proof. □

Corollary 9.2. *For each $f \in L^2$ we have*

$$
\pi\left(2a_0^2 + \sum_{\ell=1}^{\infty} [a_\ell^2 + b_\ell^2] \right) \leq \|f\|^2. \tag{9.14}
$$

This is Bessel's inequality.

When $f \in \mathscr{C}_E$ we have $\|f - S_m[f]\| \leq \|f - \sigma_m[f]\|$, and since $\|f - \sigma_m[f)]\| \to 0$ as $m \to \infty$ it follows that $\|f - S_m[f]\| \to 0$ as $m \to \infty$. Thus, for all $f \in \mathscr{C}_E$, the sequence $\{S_m[f]\}$ of partial sums of the Fourier series converges in L^2 to f. Therefore, we can write

$$
f = S[f] = a_0 c_0 + \sum_{\ell=1}^{\infty} [a_\ell c_\ell + b_\ell s_\ell]
$$

in L^2 when $f \in \mathscr{C}_E$, but we must remember that this representation does not imply pointwise convergence. Since $\|f - S_m[f]\| \to 0$ as $m \to \infty$, equation (9.13) now shows us that

$$
\|f\|^2 = \pi\left(2a_0^2 + \sum_{\ell=1}^{\infty} [a_\ell^2 + b_\ell^2] \right)
$$

for all $f \in \mathscr{C}_E$. In general, when $f \in L^2$, we know that there is a Cauchy sequence $\{f_n\} \subseteq \mathscr{C}_E$ with $||f_n - f|| \to 0$ as $n \to \infty$. Choose $\epsilon > 0$, and choose $k = k(\epsilon)$ so that

$$||f_k - f|| < \frac{\epsilon}{2}.$$

Since $f_k \in \mathscr{C}_E$, we can find $m = m(k)$ so that

$$||S_m[f_k] - f_k|| < \frac{\epsilon}{2}.$$

Therefore, $||S_m[f_k] - f|| < \epsilon$. However, Lemma 9.6 tells us that the best trigonometric polynomial approximation of degree m to f in L^2 is given by $S_m[f]$. Hence

$$||S_m[f] - f|| < \epsilon.$$

Since $\epsilon > 0$ is arbitrary, it follows that $||S_m[f] - f|| \to 0$ as $m \to \infty$. Therefore, we write

$$f = S[f] = a_0 c_0 + \sum_{\ell=1}^{\infty} [a_\ell c_\ell + b_\ell s_\ell] \tag{9.15}$$

in L^2 when $f \in L^2$ provided we remember, once again, that this representation does not imply pointwise convergence. From (9.13) it follows that

$$||f||^2 = \pi \left(2a_0^2 + \sum_{\ell=1}^{\infty} [a_\ell^2 + b_\ell^2] \right) \tag{9.16}$$

for all $f \in L^2$. This equation is Parseval's identity, and it tells us that the square of the magnitude of the function f is, except for a scale factor, the sum of the squares of the Fourier coefficients. If

$$S[f] = a_0 c_0 + \sum_{\ell=1}^{\infty} [a_\ell c_\ell + b_\ell s_\ell] \quad \text{and} \quad S[g] = \alpha_0 c_0 + \sum_{\ell=1}^{\infty} [\alpha_\ell c_\ell + \beta_\ell s_\ell],$$

we can use Parseval's identity to show that

$$\begin{aligned}
\langle f, g \rangle &= \frac{1}{4}[||f+g||^2 - ||f-g||^2] \\
&= \frac{\pi}{4} \left[2(a_0 + \alpha_0)^2 + \sum_{\ell=1}^{\infty} [(a_\ell + \alpha_\ell)^2 + (b_\ell + \beta_\ell)^2] \right. \\
&\qquad \left. - 2(a_0 - \alpha_0)^2 - \sum_{\ell=1}^{\infty} [(a_\ell - \alpha_\ell)^2 + (b_\ell - \beta_\ell)^2] \right] \\
&= \pi \left[2a_0\alpha_0 + \sum_{\ell=1}^{\infty} [a_\ell \alpha_\ell + b_\ell \beta_\ell] \right].
\end{aligned}$$

Thus, except for a scale factor, the inner product of f and g is the sum of the products of the corresponding Fourier coefficients. We can describe L^2 as the linear space of all Fourier series $S[f] = a_0 c_0 + \sum_{\ell=1}^{\infty} [a_\ell c_\ell + b_\ell s_\ell]$ for which $\sum_{\ell=1}^{\infty} [a_\ell^2 + b_\ell^2] < \infty$. A typical Fourier approximation is shown in Figure 9.3. It is intuitively clear from the graphs that the partial sum $S_m[f]$ provides a better approximation than the average of the partial sums $\sigma_m[f]$, but the average is seemingly much smoother.

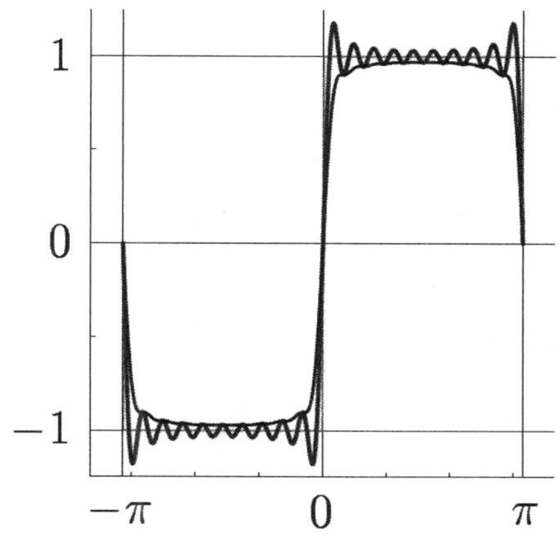

Figure 9.3. *The Fourier approximations $S_{20}[f]$ and $\sigma_{20}[f]$ when $f(t) = sign(t)$*

9.4 ▪ The Space ℓ^2

Every element f in the Hilbert space $L^2([-\pi, \pi])$ can be represented as a linear combination of the basic element c_0 and the basic elements c_ℓ and s_ℓ for each $\ell \in \mathbb{N}$. Since

$$\langle c_0, c_\ell \rangle = 0 \quad \text{and} \quad \langle c_0, s_\ell \rangle = 0$$

for $\ell \in \mathbb{N}$, since

$$\langle c_k, s_\ell \rangle = 0$$

for all $k, \ell \in \mathbb{N}$, and since

$$\langle c_k, c_\ell \rangle = 0 \quad \text{and} \quad \langle s_k, s_\ell \rangle = 0$$

for all $k, \ell \in \mathbb{N}$ provided $k \neq \ell$, we say that the countable set $\{c_0, c_1, s_1, c_2, s_2, \ldots\}$ is a set of orthogonal functions in L^2, and because every element of L^2 can be represented as a linear combination of these basic elements we say that the set is a complete orthogonal set. It is often helpful to normalize the basis elements. If we define functions $e_\ell \in L^2$ for each $\ell \in \mathbb{N}$ by setting

$$e_1 = \frac{c_0}{\sqrt{2\pi}}, \quad e_{2\ell} = \frac{c_\ell}{\sqrt{\pi}}, \quad \text{and} \quad e_{2\ell+1} = \frac{s_\ell}{\sqrt{\pi}},$$

then $||e_\ell|| = 1$ for each $\ell \in \mathbb{N}$ and $\langle e_k, e_\ell \rangle = 0$ for all $k, \ell \in \mathbb{N}$ with $k \neq \ell$. The Fourier series can now be rewritten in the form

$$\begin{aligned} f &= a_0 c_0 + \sum_{\ell=1}^{\infty} [a_\ell c_\ell + b_\ell s_\ell] \\ &= \sqrt{2\pi}\, a_0\, e_1 + \sum_{\ell=1}^{\infty} [\sqrt{\pi}\, a_\ell e_{2\ell} + \sqrt{\pi}\, b_\ell e_{2\ell+1}] \\ &= \varphi_1 e_1 + \sum_{\ell=1}^{\infty} [\varphi_{2\ell} e_{2\ell} + \varphi_{2\ell+1} e_{2\ell+1}], \end{aligned}$$

where

$$\varphi_1 = \sqrt{2\pi}\, a_0 = \frac{1}{\sqrt{2\pi}}\langle f, c_0 \rangle = \langle f, e_1 \rangle$$

and

$$\varphi_{2\ell} = \sqrt{\pi}\, a_\ell = \frac{1}{\sqrt{\pi}}\langle f, c_\ell \rangle = \langle f, e_{2\ell} \rangle,$$

$$\varphi_{2\ell+1} = \sqrt{\pi}\, b_\ell = \frac{1}{\sqrt{\pi}}\langle f, s_\ell \rangle = \langle f, e_{2\ell+1} \rangle$$

for each $\ell \in \mathbb{N}$. Of course, we can now write this expression more compactly in the form

$$f = \sum_{k=1}^{\infty} \varphi_k e_k, \tag{9.17}$$

where

$$\varphi_k = \langle f, e_k \rangle = \int_{[-\pi,\pi]} f(t) e_k(t) dt \tag{9.18}$$

for all $k \in \mathbb{N}$. In the new notation Parseval's identity becomes

$$\|f\|^2 = \pi\left(2a_0^2 + \sum_{\ell=1}^{\infty}[a_\ell^2 + b_\ell^2]\right) = \sum_{k=1}^{\infty} \varphi_k^2. \tag{9.19}$$

If

$$f = \sum_{k=1}^{\infty} \varphi_k e_k \quad \text{and} \quad g = \sum_{k=1}^{\infty} \psi_k e_k,$$

then the scalar product is given by

$$\begin{aligned}
\langle f, g \rangle &= \frac{1}{4}\left[\|f+g\|^2 - \|f-g\|^2\right] \\
&= \frac{1}{4}\sum_{k=1}^{\infty}[(\varphi_k + \psi_k)^2 - (\varphi_k - \psi_k)^2] \\
&= \sum_{k=1}^{\infty} \varphi_k \psi_k.
\end{aligned} \tag{9.20}$$

The Fourier series representation uniquely defines each element $f = (\varphi_1, \varphi_2, \ldots) \in L^2$ via an infinite sequence of real number components. Equations (9.19) and (9.20) show that the square of the norm of a function is the sum of the squares of the components and that the scalar product of two functions is the sum of the products of corresponding components. The real space ℓ^2 is the set of all infinite sequences $x = \{\xi_k\}_{k \in \mathbb{N} \subseteq \mathbb{R}}$ of real numbers such that $\|x\|^2 = \sum_{k=1}^{\infty} \xi_k^2 < \infty$ with the scalar product of $x = \{\xi_k\}$ and $y = \{\eta_k\}$ defined by $\langle x, y \rangle = \sum_{k=1}^{\infty} \xi_k \eta_k$. The one-to-one correspondence $f \longleftrightarrow (\varphi_1, \varphi_2, \ldots)$ defines an isometric isomorphism between $L^2([-\pi,\pi])$ and ℓ^2.

9.5 ▪ The Hilbert Space $H_0^1([-\pi,\pi])$

The space $H_0^1 = H_0^1([-\pi,\pi])$ is the Hilbert space of real-valued continuous functions f with $f(-\pi) = f(\pi) = 0$ and with generalized derivative $f' \in L^2([-\pi,\pi])$. The space

$\mathscr{C}_0^1 = \mathscr{C}_0^1([-\pi,\pi])$ is the space of all real-valued functions which are continuous and have a continuous first derivative on $[-\pi,\pi]$ and for which $f(-\pi)=f(\pi)=0$ and $f'(-\pi)=f'(\pi)=0$. If we define a norm $\|\cdot\|_1 : \mathscr{C}_0^1 \to \mathbb{R}$ by the formula

$$\|f\|_1^2 = \int_{[-\pi,\pi]} \left[[f(t)]^2 + [f'(t)]^2\right] dt = \|f\|^2 + \|f'\|^2 < \infty$$

for all $f \in \mathscr{C}_0^1$, then the important properties of a norm are all satisfied. Because the norm also satisfies the property

$$\|f+g\|_1^2 + \|f-g\|_1^2 = 2\left[\|f\|_1^2 + \|g\|_1^2\right],$$

there is a well-defined inner product $\langle \cdot,\cdot \rangle_1 : \mathscr{C}_0^1 \times \mathscr{C}_0^1 \to \mathbb{R}$ given by the formula

$$\begin{aligned}
\langle f,g \rangle_1 &= \frac{1}{4}\left[\|f+g\|_1^2 - \|f-g\|_1^2\right] \\
&= \frac{1}{4}\int_{[-\pi,\pi]} \left[[f(t)+g(t)]^2 + [f'(t)+g'(t)]^2\right] dt \\
&\qquad - \frac{1}{4}\int_{[-\pi,\pi]} [f(t)-g(t)]^2 + [f'(t)-g'(t)]^2 dt \\
&= \int_{[-\pi,\pi]} \left[f(t)g(t) + f'(t)g'(t)\right] dt \\
&= \langle f,g \rangle + \langle f',g' \rangle.
\end{aligned}$$

The important properties of an inner product are all satisfied. With these definitions of norm and inner product the space $\mathscr{C}_0^1$ becomes a Euclidean space $(\mathscr{C}_0^1)_E = (\mathscr{C}_0^1)_E([-\pi,\pi])$. We will show that the Euclidean space $(\mathscr{C}_0^1)_E$ can be extended to form the Hilbert space H_0^1.

9.5.1 ▪ The generalized derivative

We begin by explaining the concept of a generalized derivative. If $f, g \in L^2([-\pi,\pi])$ and

$$\langle f,\varphi' \rangle = \int_{[-\pi,\pi]} f(t)\varphi'(t)\, dt = (-1)\int_{[-\pi,\pi]} g(t)\varphi(t)\, dt = (-1)\langle g,\varphi \rangle$$

for all test functions $\varphi \in (\mathscr{C}_0^1)_E$, then we say that g is the generalized derivative of f and we write $g = f'$. Note that if we also have $f \in (\mathscr{C}_0^1)_E$, then integration by parts shows us that

$$\begin{aligned}
\langle f,\varphi' \rangle = \int_{[-\pi,\pi]} f(t)\varphi'(t)\, dt &= [f(t)\varphi(t)]\,\big|_{-\pi}^{\pi} - \int_{[-\pi,\pi]} f'(t)\varphi(t)\, dt \\
&= (-1)\int_{[-\pi,\pi]} f'(t)\varphi(t)\, dt = (-1)\langle f',\varphi \rangle,
\end{aligned}$$

and hence the generalized derivative extends our original concept of differentiation. Since $g \in L^2$, it follows that

$$\left|\int_{(s,t)} g(\tau)\, d\tau\right|^2 \le \left(\int_{(s,t)} [g(\tau)]^2\, d\tau\right)\left(\int_{(s,t)} 1^2\, d\tau\right) \le \|g\|^2\,(t-s) < \infty$$

for all s,t with $-\pi \le s \le t \le \pi$. Hence the function $G:[-\pi,\pi]\to\mathbb{R}$ given by

$$G(t)=\begin{cases} (-1)\displaystyle\int_{(t,0)} g(\tau)\,d\tau & \text{for } -\pi\le t<0,\\ \displaystyle\int_{(0,t)} g(\tau)\,d\tau & \text{for } 0\le t\le\pi \end{cases}$$

is well defined. For each s,t with $-\pi\le s\le t\le\pi$ the previous inequality tells us that

$$|G(t)-G(s)|^2 \le \|g\|^2\,(t-s)\to 0$$

as $|t-s|\to 0$. Thus $G\in\mathscr{C}([-\pi,\pi])$. Indeed, the inequality establishes that G is uniformly continuous on $[-\pi,\pi]$. Since G is a primitive of g, the fundamental theorem of calculus tells us the G is differentiable almost everywhere with $G'(t)=g(t)$ for almost all $t\in[-\pi,\pi]$. Now

$$\begin{aligned}\langle G,\varphi'\rangle &= \int_{[-\pi,\pi]} G(t)\varphi'(t)\,dt = G(t)\varphi(t)\Big|_{-\pi}^{\pi} - \int_{[-\pi,\pi]} g(t)\varphi(t)\,dt\\ &= (-1)\langle g,\varphi\rangle = \langle f,\varphi'\rangle,\end{aligned}$$

and hence $\langle f-G,\varphi'\rangle=0$ for all $\varphi\in(\mathscr{C}_0^1)_E$. Thus there is some $c\in\mathbb{R}$ such that $f(t)-G(t)=c$ for almost all $t\in[-\pi,\pi]$. Hence we can see that if $f\in L^2$ has a generalized derivative $g\in L^2$, then $f(t)=G(t)+c$ for almost all $t\in[-\pi,\pi]$ where G is a primitive of g.

9.5.2 ▪ Defining the elements of $H_0^1([-\pi,\pi])$

Let $\{f_n\}\subseteq(\mathscr{C}_0^1)_E$ be a Cauchy sequence with $\|f_m-f_n\|_1\to 0$ as $m,n\to\infty$. Since $\|f_m-f_n\|_1^2=\|f_m-f_n\|^2+\|f_m'-f_n'\|^2$, it follows that $\{f_n\}$ and $\{f_n'\}$ are Cauchy sequences in L^2, and hence there exist functions $f,g\in L^2$ such that $f_n\to f$ and $f_n'\to g$. Since

$$\langle f,\varphi'\rangle = \lim_{n\to\infty}\langle f_n,\varphi'\rangle = (-1)\lim_{n\to\infty}\langle f_n',\varphi\rangle = (-1)\langle g,\varphi\rangle$$

for all $\varphi\in(\mathscr{C}_0^1)_E$, it follows that $g=f'$. From the previous subsection we can see that $f(t)=G(t)+c$ for almost all $t\in[-\pi,\pi]$. Thus we may suppose, without loss of generality, that $f\in\mathscr{C}$. Indeed, we note that f is uniformly continuous on $[-\pi,\pi]$. Suppose we choose a subsequence $\{f_{n(k)}\}_{k\in\mathbb{N}}$ which converges in pointwise fashion almost everywhere in $[-\pi,\pi]$. Since $\|f_{n(k)}\|\le\|g\|+1$ for k sufficiently large, we know that

$$\left|f_{n(k)}(t)-f_{n(k)}(s)\right|^2 = \left|\int_{(s,t)} f_{n(k)}(\tau)\,d\tau\right|^2 \le \|f_{n(k)}\|^2(t-s) \le (\|g\|+1)(t-s)$$

for all $s<t$, and hence the sequence $\{f_{n(k)}\}$ is uniformly continuous on $[-\pi,\pi]$. Choose $\epsilon>0$ and $\delta=\delta(\epsilon)>0$ such that

$$|f_{n(k)}(t)-f_{n(k)}(s)|<\epsilon \quad\text{and}\quad |f(t)-f(s)|<\epsilon$$

whenever $|t-s|<\delta$. If we choose s such that $s>\pi-\delta$ and $f_{n(k)}(s)\to f(s)$ as $k\to\infty$, then it follows that

$$\begin{aligned}|f(\pi)-f_{n(k)}(\pi)| &= |f(\pi)-f(s)|+|f(s)-f_{n(k)}(s)|+|f_{n(k)}(s)-f_{n(k)}(\pi)|\\ &\le |f(s)-f_{n(k)}(s)|+2\epsilon,\end{aligned}$$

and since $f_{n(k)}(\pi)=0$ and $|f(s)-f_{n(k)}(s)|\to 0$ as $k\to\infty$, it follows that $|f(\pi)|\le 2\epsilon$. Since $\epsilon>0$ is arbitrary, we see that $f(\pi)=0$. A similar argument shows that $f(-\pi)=0$. Therefore, we can describe $H_0^1([-\pi,\pi])$ as the linear space of all functions $f\in\mathscr{C}$ with $f(-\pi)=f(\pi)=0$ and such that f is differentiable almost everywhere with derivative $f'\in L^2$.

9.5.3 ▪ The completeness of $H_0^1([-\pi,\pi])$

Let $\{f_n\}\subseteq H_0^1([-\pi,\pi])$ be a Cauchy sequence with $\|f_m-f_n\|_1\to 0$ as $m,n\to\infty$. Since $\|f_m-f_n\|_1^2=\|f_m-f_n\|^2+\|f_m'-f_n'\|^2$, it follows that $\{f_n\}$ and $\{f_n'\}$ are Cauchy sequences in L^2, and hence there exist functions $f,g\in L^2$ such that $f_n\to f$ and $f_n'\to g$. Since

$$\langle f,\varphi'\rangle=\lim_{n\to\infty}\langle f_n,\varphi'\rangle=(-1)\lim_{n\to\infty}\langle f_n',\varphi\rangle=(-1)\lim_{n\to\infty}\langle g,\varphi\rangle$$

for all $\varphi\in(\mathscr{C}_0^1)_E$, it follows that $g=f'$. Our earlier arguments can be used to show that $f\in H_0^1([-\pi,\pi])$ and that $\|f_n-f\|_1\to 0$ as $n\to\infty$. It follows that $H_0^1([-\pi,\pi])$ is complete and hence is a Hilbert space. Note that $H_0^1([-\pi,\pi])$ is an elementary example of a Sobolev space.

9.6 ▪ Fourier Series in $H_0^1([-\pi,\pi])$

Let $f\in H_0^1([-\pi,\pi])$. Since $f,f'\in L^2$, we can represent both functions by their respective Fourier series:

$$f=a_0c_0+\sum_{\ell=1}^{\infty}[a_\ell c_\ell+b_\ell s_\ell]$$

and

$$f'=a_0'c_0+\sum_{\ell=1}^{\infty}[a_\ell' c_\ell+b_\ell' s_\ell].$$

Clearly

$$a_0'=\frac{1}{2\pi}\langle f',c_0\rangle=(-1)\frac{1}{2\pi}\langle f,c_0'\rangle=(-1)\frac{1}{2\pi}\langle f,0\rangle=0$$

since $c_0(t)=1$ for all $t\in[-\pi,\pi]$. We also have

$$a_\ell'=\frac{1}{\pi}\langle f',c_\ell\rangle=(-1)\frac{1}{\pi}\langle f,c_\ell'\rangle=\frac{1}{\pi}\langle f,\ell s_\ell'\rangle=\ell b_\ell$$

and

$$b_\ell'=\frac{1}{\pi}\langle f',s_\ell\rangle=(-1)\frac{1}{\pi}\langle f,s_\ell'\rangle=(-1)\frac{1}{\pi}\langle f,\ell c_\ell'\rangle=(-1)\ell a_\ell$$

for each $\ell\in\mathbb{N}$. Hence Parseval's identity becomes

$$\|f\|_1^2=\|f\|^2+\|f'\|^2=\pi\left(2a_0^2+\sum_{\ell=1}^{\infty}(1+\ell^2)[a_\ell^2+b_\ell^2]\right)<\infty.$$

In the case where $f,g\in H_0^1([-\pi,\pi])$ a similar argument shows us that if

$$f=a_0c_0+\sum_{\ell=1}^{\infty}[a_\ell c_\ell+b_\ell s_\ell]\quad\text{and}\quad g=\alpha_0c_0+\sum_{\ell=1}^{\infty}[\alpha_\ell c_\ell+\beta_\ell s_\ell],$$

then

$$\langle f, g \rangle_1 = \langle f, g \rangle + \langle f', g' \rangle = \pi \left(2a_0\alpha_0 + \sum_{\ell=1}^{\infty} (1+\ell^2)[a_\ell \alpha_\ell + b_\ell \beta_\ell] \right) < \infty.$$

In Fourier series terminology $H_0^1([-\pi, \pi])$ is the collection of all Fourier series

$$S[f] = a_0 c_0 + \sum_{\ell=1}^{\infty} [a_\ell c_\ell + b_\ell s_\ell]$$

with real coefficients such that

$$\sum_{\ell=1}^{\infty} (1+\ell^2)[a_\ell^2 + b_\ell^2] < \infty.$$

9.7 ▪ The Complex Hilbert Space $L^2([-\pi, \pi])$

It is possible to extend $L^2 = L^2([-\pi, \pi])$ to a space of complex-valued square integrable functions on the finite length closed interval $[-\pi, \pi]$. The extended space is a linear space over the field $\mathbb{C}$ of complex numbers. Because the extended complex space includes the original real space as a subspace, we will not use any distinguishing notation. Consider the space $\mathscr{C} = \mathscr{C}([-\pi, \pi])$ of all complex-valued continuous functions on the closed interval $[-\pi, \pi]$. If we define a norm $||\cdot|| : \mathscr{C} \to \mathbb{R}$ by the formula

$$\begin{aligned} ||f||^2 &= \int_{[-\pi,\pi]} |f(t)|^2 \, dt = \int_{[-\pi,\pi]} |p(t) + iq(t)|^2 \, dt \\ &= \int_{[-\pi,\pi]} \left[[p(t)]^2 + [q(t)]^2 \right] dt < \infty \end{aligned}$$

for all $f = p + iq \in \mathscr{C}$, where $p, q \in \mathscr{C}$ are real-valued functions, then

1. $||f|| \geq 0$ and $||f|| = 0$ if and only if $f = 0$,
2. $||f + g|| \leq ||f|| + ||g||$ (the triangle inequality), and
3. $||cf|| = |c| \, ||f||$ for all $c \in \mathbb{C}$,

and the important properties of a norm are all satisfied. The additional property

$$\begin{aligned} ||f+g||^2 + ||f-g||^2 &= \int_{[-\pi,\pi]} \left[|f(t)+g(t)|^2 + |f(t)-g(t)|^2 \right] dt \\ &= \int_{[-\pi,\pi]} \Big[[p(t)+r(t)]^2 + [q(t)+s(t)]^2 \\ &\qquad + [p(t)-r(t)]^2 + [q(t)-s(t)]^2 \Big] dt \\ &= 2\int_{[-\pi,\pi]} \left[[p(t)]^2 + [q(t)]^2 + [r(t)]^2 + [s(t)]^2 \right] dt \\ &= 2\int_{[-\pi,\pi]} \left[|f(t)|^2 + |g(t)|^2 \right] dt \\ &= 2\left[||f||^2 + ||g||^2 \right] \end{aligned}$$

is also satisfied, and hence there is a well-defined inner product $\langle \cdot , \cdot \rangle : \mathscr{C} \times \mathscr{C} \to \mathbb{C}$ given by

$$\begin{aligned}
\langle f,g \rangle &= \frac{1}{4}\left[\|f+g\|^2 - \|f-g\|^2\right] + \frac{i}{4}\left[\|f+ig\|^2 - \|f-ig\|^2\right] \\
&= \frac{1}{4}\int_{[-\pi,\pi]} \Big[[p(t)+r(t)]^2 + [q(t)+s(t)]^2 - [p(t)-r(t)]^2 - [q(t)-s(t)]^2 \\
&\qquad + i[p(t)-s(t)]^2 + i[q(t)+r(t)]^2 - i[p(t)+s(t)]^2 - i[q(t)-r(t)]^2\Big]\, dt \\
&= \int_{[-\pi,\pi]} [\, p(t)r(t) + q(t)s(t) + i[q(t)r(t) - p(t)s(t)]\,]\, dt \\
&= \int_{[-\pi,\pi]} [p(t)+iq(t)][r(t)-is(t)]\, dt \\
&= \int_{[-\pi,\pi]} f(t)\overline{g(t)}\, dt,
\end{aligned}$$

where $f = p+iq, g = r+is \in \mathscr{C}$, and $p,q,r,s \in \mathscr{C}$ are real-valued functions. The important properties of an inner product in a complex space are

1. $\langle f+g,h \rangle = \langle f,h \rangle + \langle g,h \rangle$;
2. $\langle cf,g \rangle = c\,\langle f,g \rangle$ for all $c \in \mathbb{C}$;
3. $\langle f,g \rangle = \overline{\langle g,f \rangle}$; and
4. $\langle f,f \rangle = \|f\|^2$.

With these definitions of norm and inner product the space $\mathscr{C}$ becomes a complex Euclidean space $\mathscr{C}_E = \mathscr{C}_E([-\pi,\pi])$. The complex Euclidean space $\mathscr{C}_E$ can be extended to form a complex Hilbert space $L^2[-\pi,\pi]$ using the same methodology used for the corresponding real spaces.

9.8 ▪ Fourier Series in the Complex Space $L^2([-\pi,\pi])$

By applying the formulae from the previous section and using the earlier Fourier series results it is easy to see that if $f = p+iq$, then the Fourier coefficients of f are given by $a_0 = a_0[p] + ia_0[q]$ and by $a_\ell = a_\ell[p] + ia_\ell[q]$ and $b_\ell = b_\ell[p] + ib_\ell[q]$ for each $\ell \in \mathbb{N}$. Thus the norm of f is given by

$$\begin{aligned}
\|f\|^2 &= \|p\|^2 + \|q\|^2 \\
&= \pi\left(2(a_0[p])^2 + \sum_{\ell=1}^{\infty}[(a_\ell[p])^2 + (b_\ell[p])^2]\right) \\
&\quad + \pi\left(2(a_0[q])^2 + \sum_{\ell=1}^{\infty}\left[(a_\ell[q])^2 + (b_\ell[q])^2\right]\right) \\
&= \pi\left(2|a_0|^2 + \sum_{\ell=1}^{\infty}\left[|a_\ell|^2 + |b_\ell|^2\right]\right).
\end{aligned}$$

If $g = r+is$ and the Fourier coefficients are denoted by $\alpha_0 = a_0[r] + ia_0[s]$ and by $\alpha_\ell = a_\ell[r] + ia_\ell[s]$ and $\beta_\ell = b_\ell[r] + ib_\ell[s]$ for each $\ell \in \mathbb{N}$, then the inner product of f

and g can be calculated from

$$\begin{aligned}
\langle f,g\rangle &= \langle p,r\rangle + \langle q,s\rangle + i\langle q,r\rangle - i\langle p,s\rangle \\
&= \pi\left(2a_0[p]a_0[r] + \sum_{\ell=1}^{\infty}[a_\ell[p]a_\ell[r] + b_\ell[p]b_\ell[r]]\right) \\
&\quad + \pi\left(2a_0[q]a_0[s] + \sum_{\ell=1}^{\infty}[a_\ell[q]a_\ell[s] + b_\ell[q]b_\ell[s]]\right) \\
&\quad + i\pi\left(2a_0[q]a_0[r] + \sum_{\ell=1}^{\infty}[a_\ell[q]a_\ell[r] + b_\ell[q]b_\ell[r]]\right) \\
&\quad - i\pi\left(2a_0[p]a_0[r] + \sum_{\ell=1}^{\infty}[a_\ell[p]a_\ell[r] + b_\ell[p]b_\ell[r]]\right) \\
&= \pi\left(2a_0\overline{\alpha_0} + \sum_{\ell=1}^{\infty}\left[a_\ell\overline{\alpha_\ell} + b_\ell\overline{\beta_\ell}\right]\right).
\end{aligned}$$

9.9 ▪ The Hilbert Space $L^2(\mathbb{R})$

The space $L^2 = L^2(\mathbb{R})$ is the Hilbert space of real-valued square integrable functions on the real line $\mathbb{R}$. We note that L^2 is a linear space over the field $\mathbb{R}$ of real numbers. Consider the space $\mathscr{C} = \mathscr{C}_0(\mathbb{R})$ of all continuous real-valued functions with compact support. That is, for each $f \in \mathscr{C}$, we can find a finite closed interval $[a,b]$, which depends on the particular choice of f, such that $f(t) = 0$ provided $t \notin [a,b]$. If we define a norm $||\cdot|| : \mathscr{C} \to \mathbb{R}$ by the formula

$$||f||^2 = \int_{\mathbb{R}} [f(t)]^2\, dt < \infty$$

for all $f \in \mathscr{C}$, then the important properties of a norm are all satisfied. The norm also satisfies the property

$$\begin{aligned}
||f+g||^2 + ||f-g||^2 &= \int_{\mathbb{R}} \left[(f(t)+g(t))^2 + (f(t)-g(t))^2\right] dt \\
&= 2\int_{\mathbb{R}} \left[f(t)^2 + g(t)^2\right] dt \\
&= 2\left[||f||^2 + ||g||^2\right],
\end{aligned}$$

and hence there is a well-defined inner product $\langle \cdot,\cdot \rangle : \mathscr{C} \times \mathscr{C} \to \mathbb{R}$ given by

$$\begin{aligned}
\langle f,g\rangle &= \frac{1}{4}\left[||f+g||^2 - ||f-g||^2\right] \\
&= \frac{1}{4}\int_{\mathbb{R}} \left[[f(t)+g(t)]^2 - [f(t)-g(t)]^2\right] dt \\
&= \int_{\mathbb{R}} f(t)g(t)\, dt.
\end{aligned}$$

With these definitions of norm and inner product the space $\mathscr{C}$ becomes a Euclidean space $\mathscr{C}_E = \mathscr{C}_E(\mathbb{R})$. We will show that the Euclidean space $\mathscr{C}_E$ can be extended to form a

Hilbert space. The methods used here are formally the same as those used earlier in Section 9.1, and so we mostly restrict our attention to the points where some interpretation is required. Once again the fundamental mathematical problem with the space $\mathscr{C}_E$ is that it is not complete. We will extend $\mathscr{C}_E$ to a larger space that is complete. Once again it can be shown that a Cauchy sequence in $\mathscr{C}_E$ has a subsequence that converges in pointwise fashion to a well-defined limit at all points other than those contained in some unspecified null set.

Lemma 9.7. *If $\{f_n\} \subseteq \mathscr{C}_E$ is a Cauchy sequence, then there exist a subsequence $\{f_{n(k)}\} = \{f_{n(k)}\}_{k\in\mathbb{N}} \subseteq \mathscr{C}_E$ and a function $f : \mathbb{R} \to \mathbb{R}$ such that*

$$f(t) = \lim_{k\to\infty} f_{n(k)}(t)$$

for almost all $t \in \mathbb{R}$.

Proof: For each $k \in \mathbb{N}$ choose $n(k)$ such that $\|f_m - f_n\| < 2^{-k}$ when $m, n \geq n(k)$. Let $g_k, h_k \in \mathscr{C}_E$ be defined by

$$g_k = f_{n(k)} \quad \text{and} \quad h_k = f_{n(1)} + \sum_{j=1}^{n(k)-1} |f_{n(j+1)} - f_{n(j)}|.$$

We note that $\{h_k(t)\} \subseteq \mathscr{C}_E$ is an increasing sequence for each $t \in \mathbb{R}$ and that

$$\|h_k\| \leq \|f_{n(1)}\| + \sum_{j=1}^{n(k)-1} \|f_{n(j+1)} - f_{n(j)}\| \leq \|f_{n(1)}\| + 1$$

for all $k \in \mathbb{N}$. Thus there is a null set $E \subseteq \mathbb{R}$ and a function $h : \mathbb{R} \to \mathbb{R}$ such that $h_k(t) \to h(t)$ when $t \in \mathbb{R} \setminus E$. It follows from the definitions that the sequence $\{[g_k(t) + h_k(t)]\} \subseteq \mathscr{C}_E$ is also an increasing sequence for each $t \in \mathbb{R}$ and that

$$\|[g_k + h_k]\| \leq 2\|h_k\| \leq 2(\|f_{n(1)}\| + 1).$$

Hence there is a function $s : \mathbb{R} \to \mathbb{R}$ such that $[g_k(t) + h_k(t)] \to s(t)$ when $t \in \mathbb{R} \setminus E$. By subtraction it follows that $g_k(t) \to s(t) - h(t)$ when $t \in \mathbb{R} \setminus E$. We define $f : \mathbb{R} \to \mathbb{R}$ by setting

$$f(t) = \begin{cases} s(t) - h(t) & \text{when } t \in \mathbb{R} \setminus E, \\ 0 & \text{when } t \in E. \end{cases}$$

This completes the proof. □

For each Cauchy sequence the limit function is uniquely defined up to some unspecified null set. The arguments are similar to those used previously. We simply state the key results without proof.

Lemma 9.8. *Let $\{g_n\} \subseteq \mathscr{C}_E$ be a sequence of nonnegative functions with $\|g_n\| \to 0$ as $n \to \infty$. Suppose $g_n(t) \to g(t)$ for all $t \in \mathbb{R} \setminus E$ for some null set E. If we define $G = \{t \mid g(t) > 0\}$, then G is a null set.*

Corollary 9.3. *Let $\{f_n\} \in \mathscr{C}_E$ and $\{f_n^\star\} \in \mathscr{C}_E$ be Cauchy sequences with $||f_n - f_n^\star|| \to 0$ as $n \to \infty$. If $f_n(t) \to f(t)$ for all $t \in \mathbb{R} \setminus E$ where E is a null set and $f_n^\star(t) \to f^\star(t)$ for all $t \in \mathbb{R} \setminus E^\star$ where $E^\star$ is a null set, then $f^\star(t) = f(t)$ almost everywhere.*

9.9.1 ▪ Defining the elements of $L^2(\mathbb{R})$

The set of all Cauchy sequences $\{f_n\} \in \mathscr{C}_E$ can be classified according to the equivalence relation

$$\{f_n\} \equiv \{g_n\} \quad \Leftrightarrow \quad \lim_{n\to\infty} ||f_n - g_n|| = 0.$$

We write $\mathscr{E}(\{f_n\})$ to denote the equivalence class containing $\{f_n\}$. Corollary 9.3 shows that this class defines a unique collection

$$\mathscr{E}(f) = \{f^\star \mid f^\star(t) = f(t) \text{ almost everywhere}\}$$

of limit functions represented by a nominal function f from the class. For this reason we will refer to $\mathscr{E}(f)$ as the limit class represented by the function f. The set of all limit classes $\mathscr{E}(f)$ is a linear space with the obvious definitions, and, as before, we define

$$||\mathscr{E}(f)|| = \lim_{n\to\infty} ||f_n|| \tag{9.21}$$

and note that the definition does not depend on the choice of representative sequence $\{f_n\}$ from the class $\mathscr{E}(f)$. For example, if $\{f_n^\star\} \in \mathscr{E}(f)$, then

$$\lim_{n\to\infty} ||f_n|| \le \lim_{n\to\infty} ||f_n - f_n^\star|| + \lim_{n\to\infty} ||f_n^\star|| = \lim_{n\to\infty} ||f_n^\star||.$$

A similar argument shows that

$$\lim_{n\to\infty} ||f_n^\star|| \le \lim_{n\to\infty} ||f_n||,$$

and hence the limits are equal. It follows from the definition that

$$||\mathscr{E}(f)|| = \lim_{n\to\infty} ||f_n|| \ge 0$$

and

$$||\mathscr{E}(f)|| = 0 \Leftrightarrow \lim_{n\to\infty} ||f_n|| = 0 \Leftrightarrow f(t) = 0 \text{ almost everywhere} \Leftrightarrow \mathscr{E}(f) = \mathscr{E}(0).$$

It is also true that

$$||\mathscr{E}(f) + \mathscr{E}(g)|| = \lim_{n\to\infty} ||f_n + g_n|| \le \lim_{n\to\infty} [||f_n|| + ||g_n||] = ||\mathscr{E}(f)|| + ||\mathscr{E}(g)||$$

and

$$||\mathscr{E}(cf)|| = \lim_{n\to\infty} c\ ||f_n|| = c \lim_{n\to\infty} ||f_n|| = c\ ||\mathscr{E}(f)||$$

for each $c \in \mathbb{R}$. Since

$$\begin{aligned}
||\mathscr{E}(f) + \mathscr{E}(g)||^2 + ||\mathscr{E}(f) - \mathscr{E}(g)||^2 &= \lim_{n\to\infty} \left[||f_n + g_n||^2 + ||f_n - g_n||^2\right] \\
&= \lim_{n\to\infty} 2\left[||f_n||^2 + ||g_n||^2\right] \\
&= 2\left[||\mathscr{E}(f)||^2 + ||\mathscr{E}(g)||^2\right],
\end{aligned}$$

it follows that the scalar product

$$\langle \mathscr{E}(f), \mathscr{E}(g) \rangle = \frac{1}{4}\left[\|\mathscr{E}(f+g)\|^2 - \|\mathscr{E}(f-g)\|^2\right] \tag{9.22}$$

is also well defined.

The space of all limit classes $\mathscr{E}(f)$ with the norm and scalar product defined above will be denoted by $L^2 = L^2(\mathbb{R})$. The following result shows that each element $\mathscr{E}(f) = \mathscr{E}(\{f_n\}) \in L^2$ can be interpreted as the limit of the Cauchy sequence $\{\mathscr{E}(f_n)\}$ in the extended space. Once again the proof is similar to an earlier proof and is omitted.

Lemma 9.9. *Let $\{f_n\} \subseteq \mathscr{C}_E$ be a Cauchy sequence, and let $f \in \mathscr{E}(f) = \mathscr{E}(\{f_n\})$ be a nominal representative from the corresponding limit class. For each $m \in \mathbb{N}$ let $\{f_{m,n}\}_{n\in\mathbb{N}} \subseteq \mathscr{C}_E$ be the Cauchy sequence defined by $f_{m,n} = f_m$ for all $n \in \mathbb{N}$ with limit class $\mathscr{E}(f_m) = \mathscr{E}(\{f_m, f_m, \ldots\})$. Then $\|\mathscr{E}(f_m) - \mathscr{E}(f)\| \to 0$ as $m \to \infty$.*

Before we show that L^2 is complete it is convenient to simplify our notation and to discuss the integral interpretation of our definitions.

9.9.2 ▪ Understanding the elements of $L^2(\mathbb{R})$

For each element $f \in \mathscr{C}_E$ the norm is defined as an integral by the formula

$$\|f\|^2 = \int_{\mathbb{R}} [f(t)]^2\, dt.$$

We will extend this integral interpretation of the norm to all elements in L^2. If $\{f_n\} \subseteq \mathscr{C}_E$ is a Cauchy sequence, there is a uniquely defined limit class $\mathscr{E}(\{f_n\}) = \mathscr{E}(f)$ that we can identify by a nominal representative function $f : \mathbb{R} \to \mathbb{R}$. Thus we have a one-to-one correspondence

$$\mathscr{E}(\{f_n\}) \Leftrightarrow \mathscr{E}(f) \Leftrightarrow f$$

that identifies each element $\mathscr{E}(f)$ with a real-valued function f. Henceforth we will interpret each limit class $\mathscr{E}(f) \in L^2$ as a function and simply write $f \in L^2$. Thus, if $\{f_n\} \subseteq \mathscr{C}_E$ is a Cauchy sequence with limit class $\mathscr{E}(f)$ and nominal representative function f, we write

$$\|f\|^2 = \lim_{n\to\infty} \|f_n\|^2 = \lim_{n\to\infty} \int_{\mathbb{R}} [f_n(t)]^2\, dt.$$

In the new notation Lemma 9.9 can be rewritten as $\|f_m - f\| \to 0$ as $m \to \infty$. Hence it is natural to write

$$\|f\|^2 = \int_{\mathbb{R}} [f(t)]^2\, dt.$$

Since

$$\langle f, g \rangle = \frac{1}{4}\left[\|f+g\|^2 - \|f-g\|^2\right]$$

when $f, g \in \mathscr{C}_E$, we can use the same idea to extend the integral definition of the scalar product. Hence, if $\{f_n\} \subseteq \mathscr{C}_E$ and $\{g_n\} \subseteq \mathscr{C}_E$ are Cauchy sequences with limit classes

$\mathscr{E}(f)$ and $\mathscr{E}(g)$ and nominal representative functions f and g, respectively, we write

$$\begin{aligned}\langle f, g \rangle &= \frac{1}{4}\left[\|f+g\|^2 - \|f-g\|^2\right]\\ &= \frac{1}{4}\int_{[-\pi,\pi]} \left[[f(t)+g(t)]^2 - [f(t)-g(t)]^2\right] dt\\ &= \lim_{n\to\infty} \frac{1}{4}\int_{[-\pi,\pi]} \left[[f_n(t)+g_n(t)]^2 - [f_n(t)-g_n(t)]^2\right] dt\\ &= \lim_{n\to\infty} \frac{1}{4}\left[\|f_n+g_n\|^2 - \|f_n-g_n\|^2\right]\\ &= \lim_{n\to\infty} \langle f_n, g_n \rangle.\end{aligned}$$

In general, if S is a subset of $\mathbb{R}$, then the characteristic function $\chi_S : \mathbb{R} \to \mathbb{R}$ is defined by

$$\chi_S(t) = \begin{cases} 1 & \text{when } t \in S,\\ 0 & \text{when } t \notin S.\end{cases}$$

If $S(n) = S \cap [-n, n]$ and if $\chi_{S(n)} \in L^2$ for all $n \in \mathbb{N}$, then S is measurable and

$$\mu(S) = \lim_{n\to\infty} \langle \chi_{S(n)}, \chi_{[-n,n]} \rangle = \lim_{n\to\infty} \int_{S(n)} dt = \int_S dt$$

is said to be the measure of the set S. Note that it is possible to have $\mu(S) = +\infty$. For any given function $f \in L^2$ and each $\alpha \in \mathbb{R}$ we can define the subset $S_f(\alpha) \subseteq \mathbb{R}$ by setting

$$S_f(\alpha) = \{t \mid f(t) > \alpha\}.$$

If $S_f(\alpha)$ is a measurable set for each $\alpha \in \mathbb{R}$, then f is said to be a measurable function.

9.9.3 ▪ The completeness of $L^2(\mathbb{R})$

Finally, we note that L^2 is complete. That is, every Cauchy sequence $\{f^{(k)}\} \subseteq L^2$ has a uniquely defined limit $f \in L^2$. The argument is formally the same as an earlier argument in Subsection 9.1.3 and is omitted.

9.10 ▪ The Fourier Integral Representation on $\mathscr{C}_0(\mathbb{R})$

If we relax our intuitive ideas about pointwise convergence, then every continuous function with compact support can be represented by a Fourier integral. Nevertheless in this section we will consider pointwise convergence on the space $\mathscr{C}_0 = \mathscr{C}_0(\mathbb{R})$ of continuous functions on the real line with compact support and will not use the topology of the Euclidean space $\mathscr{C}_E$. We begin by considering the simple trigonometric integral

$$\mathscr{D}_U(t) = \int_{(0,U)} \cos ut \, du = \begin{cases} \dfrac{\sin Ut}{t} & \text{for } t \neq 0,\\ U & \text{for } t = 0.\end{cases}$$

The functions $\mathscr{D}_U(t)$ are modified Dirichlet kernels with characteristics similar to those of the Dirichlet kernels considered earlier. We note that

$$\int_{\mathbb{R}} \mathscr{D}_U(t)dt = \int_{\mathbb{R}} \frac{\sin Ut}{t} dt = \int_{\mathbb{R}} \frac{\sin s}{s} ds = \pi$$

for all $U > 0$. For each $\delta > 0$ we have

$$J_U(\delta) = \int_{(-\delta,\delta)} \mathscr{D}_U(t)dt = \int_{(-\delta,\delta)} \frac{\sin Ut}{t} dt = \int_{(-U\delta,U\delta)} \frac{\sin s}{s} ds \to \int_{\mathbb{R}} \frac{\sin s}{s} ds = \pi$$

as $U \to \infty$. Hence we conclude that

$$\int_{\delta<|t|<\infty} \mathscr{D}_U(t)dt \to 0$$

as $U \to \infty$ for each fixed $\delta > 0$. Thus the entire effective mass of the Dirichlet kernel appears to move toward $t = 0$ as $U \to \infty$. However, we also observe that for each fixed $t \neq 0$ the value $\mathscr{D}_U(t)$ oscillates between

$$\pm \frac{1}{t}$$

as $U \to \infty$, and hence there is no well-defined pointwise limit. Thus we cannot define the integral

$$\int_{\mathbb{R}} \cos ut \, du$$

by taking the pointwise limit of the sequence $\{\mathscr{D}_U(t)\}_{U \in \mathbb{R}^+}$ of Dirichlet kernels as $U \to \infty$. Although it is certainly not true that $\mathscr{D}_U(t) \to 0$ as $U \to \infty$ for $t \neq 0$, there is an intuitive idea that some form of convergence is occurring. The way forward is to discard the sequence $\{\mathscr{D}_U(t)\}_{U \in \mathbb{R}^+}$ of Dirichlet kernels in favor of the sequence $\{\mathscr{F}_V(t)\}_{V \in \mathbb{R}^+}$ of the averages of the Dirichlet kernels. Thus we define

$$\begin{aligned}
\mathscr{F}_V(t) &= \frac{1}{V} \int_{(0,V)} \mathscr{D}_U(t) \, dU \\
&= \frac{1}{V} \int_{(0,V)} \frac{\sin Ut}{t} \, dU \\
&= \frac{1}{Vt^2} [1 - \cos Vt] \\
&= \frac{2 \sin^2 \frac{Vt}{2}}{Vt^2}.
\end{aligned}$$

In deference to the earlier work on Fourier series we shall refer to these new functions as modified Fejér kernels. We observe that

$$\begin{aligned}\int_{\mathbb{R}} \mathscr{F}_V(t)\,dt &= \int_{\mathbb{R}} \left(\frac{1}{V}\int_{(0,V)} \frac{\sin Ut}{t}\,dU\right)dt \\ &= \frac{1}{V}\int_{(0,V)} \left(\int_{\mathbb{R}} \frac{\sin Ut}{t}\,dt\right)dU \\ &= \frac{1}{V}\int_{(0,V)} \left(\int_{\mathbb{R}} \frac{\sin s}{s}\,ds\right)dU \\ &= \frac{1}{V}\int_{(0,V)} \pi\,dU \\ &= \pi.\end{aligned}$$

We note also, for each fixed value $\delta > 0$, that

$$0 \le \mathscr{F}_V(t) \le \frac{2}{V\delta^2}$$

when $|t| > \delta$ and hence conclude that the sequence $\{\mathscr{F}_V(t)\}_{V\in\mathbb{R}^+}$ of Fejér kernels converges uniformly to zero in the region $|t| > \delta$. Finally,

$$0 \le \int_{|t|>\delta} \mathscr{F}_V(t)\,dt \le \int_{|t|>\delta} \frac{2}{Vt^2}\,dt = \frac{4}{V\delta} \to 0$$

as $V \to \infty$. These properties confirm that when $V > 0$ is very large, the Fejér kernel $\mathscr{F}_V(t)$ is very nearly an impulse of strength π located at the origin. Let $f \in \mathscr{C}_0$, and consider the associated Fejér integral

$$\iota_V[f](t) = \frac{1}{\pi}\int_{\mathbb{R}} f(\tau)\mathscr{F}_V(t-\tau)\,d\tau$$

at the point $\tau = t$. Because the area under the graph $y = \mathscr{F}_V(t-\tau)$ becomes concentrated at $\tau = t$ we could expect $\iota_V(t)$ to converge to $f(t)$ as V increases. This is the case. Since $f \in \mathscr{C}_0$, we can find a finite constant K such that $|f(\tau)| < K$ for all $\tau \in \mathbb{R}$, and for each $\epsilon > 0$ we can find $\delta = \delta(\epsilon) > 0$ such that

$$|f(t) - f(\tau)| < \epsilon$$

whenever $|t - \tau| < \delta$. Therefore,

$$\begin{aligned}|f(t) - \iota_V[f](t)| &= \left| f(t) - \frac{1}{\pi}\int_{\mathbb{R}} f(\tau)\mathscr{F}_V(t-\tau)\,d\tau \right| \\ &= \left| \frac{1}{\pi}\int_{\mathbb{R}} (f(t) - f(\tau))\mathscr{F}_V(t-\tau)\,d\tau \right|\end{aligned}$$

$$\begin{aligned}
&\leq \frac{1}{\pi}\int_{\mathbb{R}} |f(t)-f(\tau)|\mathscr{F}_V(t-\tau)\,d\tau \\
&\leq \frac{\epsilon}{\pi}\int_{|t-\tau|<\delta} \mathscr{F}_V(t-\tau)\,d\tau + \frac{2K}{\pi}\int_{\delta<|t-\tau|} \mathscr{F}_V(t-\tau)\,d\tau \\
&\leq \frac{\epsilon}{\pi}\cdot\pi + \frac{2K}{\pi}\cdot\frac{4}{V\delta} \\
&< 2\epsilon
\end{aligned}$$

when V is sufficiently large. Since $\epsilon > 0$ is arbitrary, it follows that $\iota_V[f](t)$ converges uniformly to $f(t)$ on $\mathbb{R}$ as $V \to \infty$. Thus

$$f(t) = \lim_{V\to\infty} \frac{1}{\pi}\int_{\mathbb{R}} f(\tau)\mathscr{F}_V(t-\tau)\,d\tau$$

for all $t \in \mathbb{R}$. Since

$$\begin{aligned}
\mathscr{F}_V(t-\tau) &= \frac{1}{V}\int_{(0,V)}\left[\int_{(0,U)} \cos u(t-\tau)\,du\right]dU \\
&= \frac{1}{V}\left\{\left[U\int_{(0,U)} \cos u(t-\tau)\,du\right]\Bigg|_0^V - \int_{(0,V)} U\cos U(t-\tau)\,dU\right\} \\
&= \frac{1}{V}\left\{V\int_{(0,V)} \cos u(t-\tau) - \int_{(0,V)} u\cos u(t-\tau)\,du\right\} \\
&= \int_{(0,V)}\left(1-\frac{u}{V}\right)\cos u(t-\tau)\,du,
\end{aligned}$$

it follows that

$$\begin{aligned}
\iota_V[f](t) &= \frac{1}{\pi}\int_{\mathbb{R}} f(\tau)\left[\int_{(0,V)}\left(1-\frac{u}{V}\right)\cos u(t-\tau)\,du\right]d\tau \\
&= \frac{1}{\pi}\int_{(0,V)}\left(1-\frac{u}{V}\right)\left[\int_{\mathbb{R}} f(\tau)[\cos ut\cos u\tau + \sin ut\sin u\tau]\,d\tau\right]du \\
&= \int_{(0,V)}\left(1-\frac{u}{V}\right)[A(u)\cos ut + B(u)\sin ut]\,du,
\end{aligned}$$

where

$$A(u) = \frac{1}{\pi}\int_{\mathbb{R}} f(\tau)\cos u\tau\,d\tau \quad \text{and} \quad B(u) = \frac{1}{\pi}\int_{\mathbb{R}} f(\tau)\sin u\tau\,d\tau$$

are the Fourier cosine and sine transforms. Since $f \in \mathscr{C}_0$ is continuous with compact support, it is clear that these two transforms are well-defined continuous functions. It follows that

$$\begin{aligned}
f(t) &= \lim_{V\to\infty}\int_{(0,V)}\left(1-\frac{u}{V}\right)[A(u)\cos ut + B(u)\sin ut]\,du \\
&= \int_{(0,\infty)} [A(u)\cos ut + B(u)\sin ut]\,du,
\end{aligned}$$

where the integral on the right-hand side is the Cesàro integral. If we interpret the desired Fourier integral representation

$$I[f](t)=\int_{(0,\infty)}[A(u)\cos ut+B(u)\sin ut]du$$

in terms of the partial Fourier integrals

$$I_U[f](t)=\int_{(0,U)}[A(u)\cos ut+B(u)\sin ut]\,du,$$

then we have shown that

$$f(t)=\lim_{V\to\infty}\frac{1}{V}\int_{(0,V)}I_U[f](t)\,dU$$

for each $f\in\mathscr{C}_0$. Thus, for each $f\in\mathscr{C}_0$ and each $t\in\mathbb{R}$, the average of the partial Fourier integrals converges to $f(t)$. Thus we have established a Fourier integral representation theorem that is analogous to the famous Fejér theorem.

Theorem 9.10. *Let $f\in\mathscr{C}_0(\mathbb{R})$. Then*

$$\begin{aligned}f(t)&=\lim_{V\to\infty}\frac{1}{V}\int_{(0,V)}I_U[f](t)\,dU\\&=\pi\lim_{V\to\infty}\int_{(0,V)}\left(1-\frac{u}{V}\right)[A(u)\cos ut+B(u)\sin ut]\,du\end{aligned}$$

for each $t\in\mathbb{R}$. We say that the Fourier integral

$$I[f](t)=\int_{(0,U)}[A(u)\cos ut+B(u)\sin ut]\,du$$

converges everywhere to $f(t)$ in the sense of Cesàro.

It is certainly true, for piecewise smooth functions with compact support, that the Fourier integral converges to the value of the function everywhere. These conditions are due to Dirichlet. However, there are no succinct necessary and sufficient conditions for pointwise convergence. Indeed, with Fourier integrals, as with Fourier series, we find that true understanding of the Fourier integral representation is achieved only when we relinquish our desire for pointwise convergence.

9.11 ▪ The Fourier Integral Representation on $L^2(\mathbb{R})$

Despite the Cesàro convergence of the Fourier integrals on $\mathscr{C}_0$, there are continuous functions for which the Fourier integral does not converge in the usual pointwise sense. In this section we will show that the natural setting for Fourier integrals is the Hilbert space $L^2(\mathbb{R})$.

Recall that if $f \in \mathscr{C}_0$, then for arbitrarily chosen $\epsilon > 0$ we showed that

$$|f(t) - \iota_V[f](t)| < 2\epsilon$$

for all $t \in \mathbb{R}$ provided V is sufficiently large. Let us choose $T > 0$ so that $f(t) = 0$ for all $t \notin [-T, T]$. Choose $S > T$. For $t > S$ we have

$$\begin{aligned}
[\iota_V[f](t)]^2 &= \left[\frac{1}{\pi}\int_{(-T,T)} f(\tau)\mathscr{F}_V(t-\tau)\,d\tau\right]^2 \\
&\le \frac{1}{\pi^2}\cdot\int_{(-T,T)} [f(\tau)]^2\,d\tau \cdot \int_{(-T,T)} [\mathscr{F}_V(t-\tau)]^2\,d\tau \\
&\le \frac{1}{\pi^2}\cdot\|f\|^2\cdot\int_{(-\infty,T)} \frac{4}{V^2(t-T)^4}\,d\tau \\
&= \frac{1}{\pi^2}\cdot\|f\|^2\cdot\frac{4}{3V^2(t-T)^3},
\end{aligned}$$

and for $t < -S$ a similar argument gives

$$\begin{aligned}
[\iota_V[f](t)]^2 &= \left[\frac{1}{\pi}\int_{(-T,T)} f(\tau)\mathscr{F}_V(t-\tau)\,d\tau\right]^2 \\
&\le \frac{1}{\pi^2}\cdot\int_{(-T,T)} [f(\tau)]^2\,d\tau \cdot \int_{(-T,T)} [\mathscr{F}_V(t-\tau)]^2\,d\tau \\
&\le \frac{1}{\pi^2}\cdot\|f\|^2\cdot\int_{(-T,\infty)} \frac{4}{V^2(t-T)^4}\,d\tau \\
&= \frac{1}{\pi^2}\cdot\|f\|^2\cdot\frac{4}{3V^2(t+T)^3}.
\end{aligned}$$

For each $\epsilon > 0$ it follows that

$$\begin{aligned}
\|f - \iota_V[f]\|^2 &= \int_{(-\infty,-S)} [\iota_V[f](t)]^2\,dt + \int_{(-S,S)} [f(t) - \iota_V[f](t)]^2\,dt \\
&\quad + \int_{(S,\infty)} [\iota_V[f](t)]^2\,dt \\
&\le \frac{2}{\pi^2}\cdot\|f\|^2\cdot\frac{4}{9V^2(S-T)^2} + 2\epsilon^2\cdot 2S
\end{aligned}$$

when V is sufficiently large. Since ϵ is arbitrarily chosen, it follows that $\iota_V[f]$ converges to f in the Euclidean space $\mathscr{C}_E$. Therefore, provided $f \in \mathscr{C}_E$, the sequence $\{\iota_V[f]\}_{V\in\mathbb{R}^+}$ converges to f in L^2. Despite the convergence of the sequence of averages of the partial integrals, we can show that in L^2 the original sequence of partial integrals provides a better approximation. Indeed, for every $f \in L^2$, we will show that the best possible approximation to f in L^2 by a spectral density on the interval $[0, V]$ is given by the partial integral $I_V[f] \in \mathscr{C}_E$ defined by

$$I_V[f](t) = \int_{(0,V)} [A(u)\cos ut + B(u)\sin ut]\,du,$$

where

$$A(u) = \frac{1}{\pi}\int_{\mathbb{R}} f(\tau)\cos u\tau \, d\tau \quad \text{and} \quad B(u) = \frac{1}{\pi}\int_{\mathbb{R}} f(\tau)\sin u\tau \, d\tau$$

for each $u \in \mathbb{R}$ are the respective Fourier cosine and sine transforms for f.

Lemma 9.11. *Let $f \in L^2$, and let*

$$P_V[\alpha,\beta](t) = \int_{(0,V)} [\alpha(u)\cos ut + \beta(u)\sin ut] \, du$$

be a partial integral with continuously differentiable spectral densities $\alpha : [0,V] \to \mathbb{R}$ and $\beta : [0,V] \to \mathbb{R}$. Let $E_V[f,\alpha,\beta] = ||f - P_V[\alpha,\beta]||^2$ denote the mean square error for the approximation of f by $P_V[\alpha,\beta]$ in L^2. The minimum value of $E_V[f,\alpha,\beta]$ is attained when $\alpha(u) = A(u)$ and $\beta(u) = B(u)$ for almost all $u \in [0,V]$, where $A(u)$ and $B(u)$ are the Fourier transforms of f. In this case $P_V[\alpha,\beta] = I_V[f]$, and the mean square error is

$$E_V[A,B] = ||f||^2 - \pi \int_{(0,V)} \left[A(u)^2 + B(u)^2\right] du.$$

Proof: We begin with the important observation that $\cos ut$ and $\sin ut$ are not elements of L^2. This means that the calculation of the various error formulae is more complicated than was the case with the corresponding formulae for Fourier series. To begin we note that

$$\int_{(0,V)} \alpha(u)\cos ut \, du \;=\; \alpha(V)\cdot\frac{\sin Vt}{t} - \int_{(0,V)} \alpha'(u)\cdot\frac{\sin ut}{t}\, du \tag{9.23}$$

and

$$\int_{(0,V)} \beta(u)\sin ut \, du \;=\; \beta(V)\cdot\frac{1-\cos Vt}{t} - \int_{(0,V)} \beta'(u)\cdot\frac{1-\cos ut}{t}\, du. \tag{9.24}$$

This rearrangement allows us to compute various integrals in the error expression

$$E_V[f,\alpha,\beta] = ||f - P_V[\alpha,\beta]||^2.$$

To make the computation as transparent as possible, we will begin by computing the various relevant integrals. We will make repeated use of the standard integral

$$\int_{\mathbb{R}} \frac{\sin^2 wt}{t^2}\, dt = \pi|w| \tag{9.25}$$

for $w \in \mathbb{R}$. In the first instance we have

$$\int_{\mathbb{R}} [f(t)]^2 \, dt = ||f||^2 \tag{9.26}$$

and

$$
\begin{aligned}
&\int_{\mathbb{R}} f(t)\left\{\int_{(0,V)} [\alpha(u)\cos ut + \beta(u)\sin ut]\, du\right\} dt \\
&= \int_{(0,V)} \left[\alpha(u)\left(\int_{\mathbb{R}} f(t)\cos ut\, dt\right) + \beta(u)\left(\int_{\mathbb{R}} f(t)\sin ut\, dt\right)\right] du \\
&= \pi \int_{(0,V)} [\alpha(u)A(u) + \beta(u)B(u)]\, du. \qquad (9.27)
\end{aligned}
$$

In the second instance we turn our attention to the term

$$\int_{\mathbb{R}} \left\{\int_{(0,V)} [\alpha(u)\cos ut + \beta(u)\sin ut]\, du\right\}^2 dt,$$

where we use the alternative expressions (9.23) and (9.24) for the inner integrals. We have

$$
\begin{aligned}
&\int_{\mathbb{R}} \left\{\int_{(0,V)} \alpha(u)\cos ut\, du\right\}^2 dt \\
&= \int_{\mathbb{R}} \left\{\alpha(V)\cdot\frac{\sin Vt}{t} - \int_{(0,V)} \alpha'(u)\cdot\frac{\sin ut}{t}\, du\right\}^2 dt \\
&= \alpha(V)^2 \int_{\mathbb{R}} \frac{\sin^2 Vt}{t^2}\, dt - 2\alpha(V)\int_{(0,V)} \alpha'(u)\left(\int_{\mathbb{R}} \frac{\sin Vt \sin ut}{t^2}\, dt\right) du \\
&\qquad + \int_{(0,V)}\int_{(0,V)} \alpha'(u)\alpha'(v)\left(\int_{\mathbb{R}} \frac{\sin ut \sin vt}{t^2}\, dt\right) du dv \\
&= \pi V\alpha(V)^2 - 2\pi\alpha(V)\int_{(0,V)} u\alpha'(u)\, du + \pi\int_{(0,V)} \alpha'(u)\left[\int_{(0,u)} v\alpha'(v)\, dv\right] du \\
&\qquad + \pi\int_{(0,V)} \alpha'(v)\left[\int_{(0,v)} u\alpha'(u)\, du\right] dv \\
&= \pi V\alpha(V)^2 - 2\pi\left[V\alpha(V)^2 - \alpha(V)\int_{(0,V)} \alpha(u)\, du\right] \\
&\qquad + 2\pi\left[\frac{1}{2}V\alpha(V)^2 - \alpha(V)\int_{(0,V)} \alpha(u)\, du + \frac{1}{2}\int_{(0,V)} \alpha(u)^2\, du\right] \\
&= \pi\int_{(0,V)} \alpha(u)^2\, du, \qquad (9.28)
\end{aligned}
$$

where we have used (9.25) and standard trigonometric formulae to show that

$$\int_{\mathbb{R}} \frac{\sin ut \sin vt}{t^2}\, dt = \begin{cases} \pi u & \text{when } u < v, \\ \pi v & \text{when } v < u. \end{cases}$$

In a similar fashion we can use (9.25) and standard trigonometric formulae to show that

$$\int_{\mathbb{R}} \frac{(1-\cos ut)(1-\cos vt)}{t^2}\, dt = \begin{cases} \pi u & \text{when } u < v, \\ \pi v & \text{when } v < u, \end{cases}$$

from which an argument similar to that used for the previous integral gives

$$\begin{aligned}
&\int_{\mathbb{R}} \left\{ \int_{(0,V)} \beta(u) \sin ut \, du \right\}^2 dt \\
&= \int_{\mathbb{R}} \left\{ \beta(V) \cdot \frac{1-\cos Vt}{t} - \int_{(0,V)} \beta'(u) \cdot \frac{1-\cos ut}{t} \, du \right\}^2 dt \\
&= \beta(V)^2 \int_{\mathbb{R}} \frac{(1-\cos Vt)^2}{t^2} \, dt \\
&\qquad - 2\beta(V) \int_{(0,V)} \beta'(u) \left(\int_{\mathbb{R}} \frac{(1-\cos Vt)(1-\cos ut)}{t^2} \, dt \right) du \\
&\qquad\qquad + \int_{(0,V)} \int_{(0,V)} \beta'(u)\beta'(v) \left(\int_{\mathbb{R}} \frac{(1-\cos ut)(1-\cos vt)}{t^2} \, dt \right) du dv \\
&= \pi V \beta(V)^2 - 2\pi \beta(V) \int_{(0,V)} u\beta'(u) \, du + \pi \int_{(0,V)} \beta'(u) \left[\int_{(0,u)} v\beta'(v) \, dv \right] du \\
&\qquad\qquad + \pi \int_{(0,V)} \beta'(v) \left[\int_{(0,v)} u\beta'(u) \, du \right] dv \\
&= \pi V \beta(V)^2 - 2\pi \left[V\beta(V)^2 - \beta(V) \int_{(0,V)} \beta(u) \, du \right] \\
&\qquad\qquad + 2\pi \left[\frac{1}{2} V\beta(V)^2 - \beta(V) \int_{(0,V)} \beta(u) \, du + \frac{1}{2} \int_{(0,V)} \beta(u)^2 \, du \right] \\
&= \pi \int_{(0,V)} \beta(u)^2 \, du.
\end{aligned} \tag{9.29}$$

It remains to show that

$$\int_{\mathbb{R}} \left\{ \int_{(0,V)} \alpha(u) \cos ut \, du \right\} \left\{ \int_{(0,V)} \beta(u) \sin ut \, du \right\} dt = 0. \tag{9.30}$$

Once again it is necessary to use the alternative expressions (9.23) and (9.24) for the inner integrals, but we leave this as an exercise for the reader. By adding together appropriate multiples of the integrals (9.26), (9.27), (9.28), (9.29), and (9.30) we obtain

$$\begin{aligned}
E_V[f,\alpha,\beta] &= \|f\|^2 - 2\pi \int_{(0,V)} [\alpha(u)A(u) + \beta(u)B(u)] \, du \\
&\qquad\qquad + \pi \int_{(0,V)} \left[\alpha(u)^2 + \beta(u)^2 \right] du \\
&= \|f\|^2 + \pi \int_{(0,V)} \{ [\alpha(u) - A(u)]^2 + [\beta(u) - B(u)]^2 \} \, du \\
&\qquad\qquad - \pi \int_{(0,V)} \left[A(u)^2 + B(u)^2 \right] du,
\end{aligned}$$

which is minimized by choosing $\alpha(u) = A(u)$ and $\beta(u) = B(u)$ for all $u \in (0, V)$. It is necessary to know that $A(u)$ and $B(u)$ are continuously differentiable functions. This follows from the continuity of both functions and the observation that $A'(u) = -uB(u)$ and $B'(u) = uA(u)$. The minimum mean square error is given by

$$E_V[A,B] = \|f\|^2 - \pi \int_{(0,V)} \left[A(u)^2 + B(u)^2\right] du \tag{9.31}$$

when $P_V[f,\alpha,\beta] = I_V[f]$. This completes the proof. □

Corollary 9.4. *For each $f \in L^2$ we have*

$$\pi \int_{(0,\infty)} [A(u)^2 + B(u)^2]\, du \le \|f\|^2. \tag{9.32}$$

This is Bessel's inequality.

When $f \in \mathscr{C}_E$ we have $\|f - I_V[f]\| \le \|f - \iota_V[f]\|$, and since $\|f - \iota_V[f)]\| \to 0$ as $V \to \infty$, it follows that $\|f - I_V[f]\| \to 0$ as $V \to \infty$. Thus, for all $f \in \mathscr{C}_E$, the sequence $\{I_V[f]\}$ of partial Fourier integrals converges in L^2 to f. Therefore, we can write $f = I[f]$ in L^2, where

$$I[f](t) = \int_{(0,\infty)} [A(u)\cos ut + B(u)\sin ut]\, du$$

when $f \in \mathscr{C}_E$, but we must remember that this representation does not imply pointwise convergence. Since $\|f - I_V[f]\| \to 0$ as $V \to \infty$, equation (9.31) now shows us that

$$\|f\|^2 = \pi \int_{(0,V)} \left[A(u)^2 + B(u)^2)\right] du$$

for all $f \in \mathscr{C}_E$. In general, when $f \in L^2$, we know that there is a Cauchy sequence $\{f_n\} \subseteq \mathscr{C}_E$ with $\|f_n - f\| \to 0$ as $n \to \infty$. Choose $\epsilon > 0$ and choose $k = k(\epsilon)$ so that

$$\|f_k - f\| < \frac{\epsilon}{2}.$$

Since $f_k \in \mathscr{C}_E$ we can find $V = V(k)$ so that

$$\|\iota_V[f_k] - f_k\| < \frac{\epsilon}{2}.$$

Therefore, $\|\iota_V[f_k] - f\| < \epsilon$. However, Lemma 9.11 tells us that the best spectral representation on the interval $[0, V]$ of f in L^2 is given by $I_V[f]$. Hence

$$\|I_V[f] - f\| < \epsilon.$$

Since $\epsilon > 0$ is arbitrary, it follows that $\|I_V[f] - f\| \to 0$ as $V \to \infty$. Therefore, we write $f = I[f]$ in L^2, where

$$I[f] = \int_{(0,V)} [A(u)\cos ut + B(u)\sin ut]\, du \tag{9.33}$$

provided we remember, once again, that this representation does not imply pointwise convergence. From (9.31) it follows that

$$\|f\|^2 = \pi \int_{(0,V)} \left[A(u)^2 + B(u)^2\right] du \tag{9.34}$$

for all $f \in L^2$. This equation is Parseval's identity, and it tells us that the square of the magnitude of the function f is, except for a scale factor, the integral of the squares of the Fourier spectral densities. If

$$I[f](t) = \int_{(0,\infty)} [A(u)\cos ut + B(u)\sin ut]\, du$$

and

$$I[g](t) = \int_{(0,\infty)} [\alpha(u)\cos ut + \beta(u)\sin ut]\, du,$$

we can use Parseval's identity to show that

$$\begin{aligned} \langle f, g \rangle &= \frac{1}{4}\left[\|f+g\|^2 - \|f-g\|^2\right] \\ &= \frac{\pi}{4}\int_{(0,\infty)} \left\{[A(u)+\alpha(u)]^2 + [B(u)+\beta(u)]^2\right\} du \\ &\qquad - \pi\int_{(0,\infty)} \left\{[A(u)-\alpha(u)]^2 + [B(u)-\beta(u)]^2\right\} du \\ &= \pi\int_{(0,\infty)} [A(u)\alpha(u) + B(u)\beta(u)]\, du. \end{aligned}$$

Thus, except for a scale factor, the inner product of f and g is the integral of the products of the corresponding Fourier spectral densities. We can describe L^2 as the linear space of all Fourier integrals

$$I[f] = \int_{(0,\infty)} [A(u)\cos ut + B(u)\sin ut]\, du$$

for which

$$\int_{(0,\infty)} \left[A(u)^2 + B(u)^2\right] du < \infty.$$

Although we will not pursue this issue rigorously, it is important to note that the Fourier representation is symmetric in the sense that the complete set of functions $f \in L^2(\mathbb{R})$ is generated by the complete set of densities $A, B \in L^2((0,\infty))$.

9.12 ▪ The Hilbert Space $H_0^1(\mathbb{R})$

The space $H_0^1 = H_0^1(\mathbb{R})$ is the Hilbert space of real-valued continuous functions f with generalized derivative $f' \in L^2(\mathbb{R})$. The space $\mathscr{C}_0^1 = \mathscr{C}_0^1(\mathbb{R})$ is the space of all real-valued functions with compact support which are continuous and have a continuous first derivative on $\mathbb{R}$. If we define a norm $\|\cdot\|_1 : \mathscr{C}_0^1 \to \mathbb{R}$ by the formula

$$\|f\|_1^2 = \int_{\mathbb{R}} \left[[f(t)]^2 + [f'(t)]^2\right] dt = \|f\|^2 + \|f'\|^2 < \infty$$

for all $f \in \mathscr{C}_0^1$, then the important properties of a norm are all satisfied. Because the norm also satisfies the property

$$\|f+g\|_1^2 + \|f-g\|_1^2 = 2\left[\|f\|_1^2 + \|g\|_1^2\right]$$

there is a well-defined inner product $\langle \cdot,\cdot \rangle_1 : \mathscr{C}_0^1 \times \mathscr{C}_0^1 \to \mathbb{R}$ given by the formula

$$\begin{aligned}\langle f,g \rangle_1 &= \frac{1}{4}\left[\|f+g\|_1^2 - \|f-g\|_1^2\right]\\ &= \frac{1}{4}\int_{[-\pi,\pi]} [[f(t)+g(t)]^2 + [f'(t)+g'(t)]^2 + [f(t)-g(t)]^2\\ &\quad + [f'(t)-g'(t)]^2]\,dt\\ &= \int_{[-\pi,\pi]} \left[f(t)g(t)+f'(t)g'(t)\right] dt\\ &= \langle f,g \rangle + \langle f',g' \rangle.\end{aligned}$$

The important properties of an inner product are all satisfied. With these definitions of norm and inner product the space $\mathscr{C}_0^1$ becomes a Euclidean space $(\mathscr{C}_0^1)_E = (\mathscr{C}_0^1)_E(\mathbb{R})$. We will show that the Euclidean space $(\mathscr{C}_0^1)_E$ can be extended to form a Hilbert space H_0^1.

9.12.1 ▪ The generalized derivative

We revise the concept of a generalized derivative. If $f, g \in L^2(\mathbb{R})$ and

$$\langle f,\varphi' \rangle = \int_{\mathbb{R}} f(t)\varphi'(t)\,dt = (-1)\int_{\mathbb{R}} g(t)\varphi(t)\,dt = (-1)\langle g,\varphi \rangle$$

for all test functions $\varphi \in (\mathscr{C}_0^1)_E$, then we say that g is the generalized derivative of f, and we write $g = f'$. Note that if we also have $f \in (\mathscr{C}_0^1)_E$, then integration by parts shows us that

$$\begin{aligned}\langle f,\varphi' \rangle = \int_{\mathbb{R}} f(t)\varphi'(t)\,dt &= f(t)\varphi(t)\Big|_{-\infty}^{\infty} - \int_{\mathbb{R}} f'(t)\varphi(t)\,dt\\ &= (-1)\int_{\mathbb{R}} f'(t)\varphi(t)\,dt = (-1)\langle f',\varphi \rangle,\end{aligned}$$

and hence the generalized derivative extends our original concept of differentiation. Using reasoning similar to that used in Subsection 9.5.1, we deduce that the function $G : \mathbb{R} \to \mathbb{R}$ given by

$$G(t) = \begin{cases} (-1)\displaystyle\int_{(t,0)} g(\tau)\,d\tau & \text{for } -\pi \le t < 0,\\ \displaystyle\int_{(0,t)} g(\tau)\,d\tau & \text{for } 0 \le t \le \pi \end{cases}$$

is well defined with $G \in \mathscr{C}(\mathbb{R})$. Indeed, we find that G is uniformly continuous on $\mathbb{R}$. We note that the primitive function G is differentiable almost everywhere with $G'(t) = g(t)$ for almost all $t \in \mathbb{R}$, and we apply our previous arguments to show that $f(t) = G(t) + c$ for some $c \in \mathbb{R}$ and almost all $t \in \mathbb{R}$.

9.12.2 ▪ Defining the elements of $H_0^1(\mathbb{R})$

Let $\{f_n\} \subseteq (\mathscr{C}_0^1)_E$ be a Cauchy sequence with $\|f_m - f_n\|_1 \to 0$ as $m,n \to \infty$. Since $\|f_m - f_n\|_1^2 = \|f_m - f_n\|^2 + \|f_m' - f_n'\|^2$, it follows that $\{f_n\}$ and $\{f_n'\}$ are Cauchy sequences in L^2, and hence there exist functions $f, g \in L^2$ such that $f_n \to f$ and $f_n' \to g$. Since

$$\langle f, \varphi' \rangle = \lim_{n\to\infty} \langle f_n, \varphi' \rangle = (-1) \lim_{n\to\infty} \langle f_n', \varphi \rangle = (-1)\langle g, \varphi \rangle$$

for all $\varphi \in (\mathscr{C}_0^1)_E$, it follows that $g = f'$. From the previous subsection we know that $f(t) = G(t) + c$ for almost all $t \in \mathbb{R}$. Thus we may suppose, without loss of generality, that $f \in \mathscr{C}$. Indeed, we note that f is uniformly continuous on $[-\pi, \pi]$. Therefore, we can describe $H_0^1([-\pi, \pi])$ as the linear space of all functions $f \in \mathscr{C}$ such that f is differentiable almost everywhere with derivative $f' \in L^2$.

9.12.3 ▪ The completeness of $H_0^1(\mathbb{R})$

Let $\{f_n\} \subseteq H_0^1(\mathbb{R})$ be a Cauchy sequence with $\|f_m - f_n\|_1 \to 0$ as $m, n \to \infty$. Since $\|f_m - f_n\|_1^2 = \|f_m - f_n\|^2 + \|f_m' - f_n'\|^2$, it follows that $\{f_n\}$ and $\{f_n'\}$ are Cauchy sequences in L^2, and hence there exist functions $f, g \in L^2$ such that $f_n \to f$ and $f_n' \to g$. Since

$$\langle f, \varphi' \rangle = \lim_{n\to\infty} \langle f_n, \varphi' \rangle = (-1) \lim_{n\to\infty} \langle f_n', \varphi \rangle = (-1) \lim_{n\to\infty} \langle g, \varphi \rangle$$

for all $\varphi \in (\mathscr{C}_0^1)_E$, it follows that $g = f'$. Our earlier arguments can be used to show that $f \in H_0^1(\mathbb{R})$ and that $\|f_n - f\|_1 \to 0$ as $n \to \infty$. It follows that $H_0^1(\mathbb{R})$ is complete and hence is a Hilbert space. Note that $H_0^1(\mathbb{R})$ is an elementary example of a Sobolev space.

9.13 ▪ Fourier Integrals in $H_0^1(\mathbb{R})$

Let $f \in H_0^1(\mathbb{R})$. Since $f, f' \in L^2$, we can represent both functions by their respective Fourier integrals:

$$f(t) = \int_{(0,\infty)} [A(u)\cos ut + B(u)\sin ut]\, du$$

and

$$f'(t) = \int_{(0,\infty)} [A^\dagger(u)\cos ut + B^\dagger(u)\sin ut]\, du.$$

If we assume, to begin, that $f \in \mathscr{C}_0(\mathbb{R})$, then

$$\begin{aligned} A^\dagger(u) &= \frac{1}{\pi}\int_{\mathbb{R}} f'(t)\cos ut\, dt \\ &= \frac{1}{\pi}\left[f(t)\cos ut\Big|_{-\infty}^{\infty} + \int_{\mathbb{R}} f(t)u\,\sin ut\, dt \right] \\ &= u\,B(u) \end{aligned}$$

and

$$\begin{aligned} B^\dagger(u) &= \frac{1}{\pi}\int_{\mathbb{R}} f'(t)\sin ut\, dt \\ &= \frac{1}{\pi}\left[f(t)\sin ut\Big|_{-\infty}^{\infty} - \int_{\mathbb{R}} f(t)u\,\cos ut\, dt \right] \\ &= (-1)\,u\,A(u). \end{aligned}$$

These relationships can be generalized to allow all $f \in H_0^1(\mathbb{R})$ by taking appropriate limits. Hence Parseval's identity becomes

$$||f||_1^2 = ||f||^2 + ||f'||^2 = \pi \int_{(0,\infty)} (1+u^2)\left[A(u)^2 + B(u)^2\right] du < \infty.$$

In the case where $f, g \in H_0^1(\mathbb{R})$ a similar argument shows us that if

$$f(t) = \int_{(0,\infty)} [A(u)\cos ut + B(u)\sin ut]\, du,$$

and

$$g(t) = \int_{(0,\infty)} [\alpha(u)\cos ut + \beta(u)\sin ut]\, du,$$

then

$$\langle f, g \rangle_1 = \langle f, g \rangle + \langle f', g' \rangle = \pi \int_{(0,\infty)} (1+u^2)[A(u)\alpha(u) + B(u)\beta(u)]\, du < \infty.$$

In Fourier integral terminology $H_0^1(\mathbb{R})$ is the collection of all Fourier integrals

$$I[f](t) = \int_{(0,\infty)} [A(u)\cos ut + B(u)\sin ut]\, du$$

with real coefficients such that

$$\int_{(0,\infty)} (1+u^2)\left[A(u)^2 + B(u)^2\right] du < \infty.$$

9.14 ▪ The Complex Hilbert Space $L^2(\mathbb{R})$

It is possible to extend $L^2 = L^2(\mathbb{R})$ to a space of complex-valued square integrable functions on the real line $\mathbb{R}$. The extended space is a linear space over the field $\mathbb{C}$ of complex numbers. Because the extended complex space includes the original real space as a subspace, we will not use any distinguishing notation. Consider the space $\mathscr{C} = \mathscr{C}_0(\mathbb{R})$ of all complex-valued continuous functions with compact support on the real line $\mathbb{R}$. If we define a norm $||\cdot|| : \mathscr{C} \to \mathbb{R}$ by the formula

$$||f||^2 = \int_{\mathbb{R}} |f(t)|^2\, dt = \int_{\mathbb{R}} |p(t) + iq(t)|^2\, dt = \int_{\mathbb{R}} \left[\, [p(t)]^2 + [q(t)]^2 \right] dt < \infty$$

for all $f = p + iq \in \mathscr{C}$, where $p, q \in \mathscr{C}$ are real-valued functions, then

1. $||f|| \geq 0$ and $||f|| = 0$ if and only if $f = 0$,

2. $||f + g|| \leq ||f|| + ||g||$ (the triangle inequality), and

3. $||cf|| = |c|\, ||f||$ for all $c \in \mathbb{C}$,

and the important properties of a norm are all satisfied. The additional property

$$\begin{aligned}
\|f+g\|^2+\|f-g\|^2 &= \int_{\mathbb{R}} \left[\, |f(t)+g(t)|^2+|f(t)-g(t)|^2 \,\right] dt \\
&= \int_{\mathbb{R}} \left[\, [p(t)+r(t)]^2+[q(t)+s(t)]^2 \right. \\
&\qquad \left. +[p(t)-r(t)]^2+[q(t)-s(t)]^2 \,\right] dt \\
&= 2\int_{\mathbb{R}} \left[\, [p(t)]^2+[q(t)]^2+[r(t)]^2+[s(t)]^2 \,\right] dt \\
&= 2\int_{\mathbb{R}} \left[\, |f(t)|^2+|g(t)|^2 \,\right] dt \\
&= 2\left[\|f\|^2+\|g\|^2\right]
\end{aligned}$$

is also satisfied, and hence there is a well-defined inner product $\langle \cdot\,,\cdot \rangle : \mathscr{C}\times\mathscr{C} \to \mathbb{C}$ given by

$$\begin{aligned}
\langle f,g \rangle &= \frac{1}{4}\left[\|f+g\|^2-\|f-g\|^2\right]+\frac{i}{4}\left[\|f+ig\|^2-\|f-ig\|^2\right] \\
&= \frac{1}{4}\int_{\mathbb{R}} \left[[p(t)+r(t)]^2+[q(t)+s(t)]^2-[p(t)-r(t)]^2-[q(t)-s(t)]^2 \right. \\
&\qquad \left. +\,i[p(t)-s(t)]^2+i[q(t)+r(t)]^2-i[p(t)+s(t)]^2-i[q(t)-r(t)]^2\right] dt \\
&= \int_{\mathbb{R}} \left[\, p(t)r(t)+q(t)s(t)+i[q(t)r(t)-p(t)s(t)] \,\right] dt \\
&= \int_{\mathbb{R}} [p(t)+iq(t)][r(t)-is(t)]\, dt \\
&= \int_{\mathbb{R}} f(t)\overline{g(t)}\, dt,
\end{aligned}$$

where $f = p+iq, g = r+is \in \mathscr{C}$, and $p,q,r,s \in \mathscr{C}$ are real-valued functions. The important properties of an inner product in a complex space are

1. $\langle f+g,h \rangle = \langle f,h \rangle + \langle g,h \rangle$;
2. $\langle cf,g \rangle = c\,\langle f,g \rangle$ for all $c \in \mathbb{C}$;
3. $\langle f,g \rangle = \overline{\langle g,f \rangle}$; and
4. $\langle f,f \rangle = \|f\|^2$.

With these definitions of norm and inner product the space $\mathscr{C}$ becomes a complex Euclidean space $\mathscr{C}_E = \mathscr{C}_E(\mathbb{R})$. The complex Euclidean space $\mathscr{C}_E$ can be extended to form a complex Hilbert space $L^2(\mathbb{R})$ by the same method used for the corresponding real spaces.

9.15 ▪ Fourier Integrals in the Complex Space $L^2(\mathbb{R})$

By applying the formulae from the previous section and using the earlier Fourier integral results, it is easy to see that if $f = p+iq$, then the Fourier integrals of f are given by

$A(u) = A[p](u) + iA[q](u)$ and $B(u) = B[p](u) + iB[q](u)$ for each $u \in (0,\infty)$. Thus the norm of f is given by

$$\begin{aligned}
\|f\|^2 &= \|p\|^2 + \|q\|^2 \\
&= \pi \int_{(0,\infty)} \left[A[p](u)^2 + B[p](u)^2\right] du + \pi \int_{\mathbb{R}} \left[A[q](u)^2 + B[q](u)^2\right] du \\
&= \pi \int_{(0,\infty)} \left[|A(u)|^2 + |B(u)|^2\right] du.
\end{aligned}$$

If $g = r + is$ and the Fourier integrals are denoted by $\alpha(u) = A[r](u) + iA[s](u)$ and $\beta(u) = B[r](u) + iB[s](u)$ for each $u \in (0,\infty)$, then the inner product of f and g can be calculated from

$$\begin{aligned}
\langle f, g \rangle &= \langle p, r \rangle + \langle q, s \rangle + i\langle q, r \rangle - i\langle p, s \rangle \\
&= \pi \int_{(0,\infty)} [A[p](u)A[r](u) + B[p](u)B[r](u)]\, du \\
&\qquad + \pi \int_{(0,\infty)} [A[q](u)A[s](u) + B[q](u)B[s](u)]\, du \\
&\qquad\qquad + i\pi \int_{(0,\infty)} [A[q](u)A[r](u) + B[q](u)B[r](u)]\, du \\
&\qquad\qquad\qquad - i\pi \int_{(0,\infty)} [A[p](u)A[r](u) + B[p](u)B[r](u)]\, du \\
&= \pi \int_{(0,\infty)} \left[A(u)\overline{\alpha(u)} + B(u)\overline{\beta(u)}\,\right] du.
\end{aligned}$$

Bibliography

[1] M. Abbad and J.A. Filar, "Perturbation and stability theory for Markov control problems", *IEEE Trans. Auto. Contr.*, 37, pp. 1415–1420, 1992. (Cited on pp. 208, 243)

[2] M. Abbad and J.A. Filar, "Algorithms for singularly perturbed Markov control problems: A survey", in *Techniques in Discrete-Time Stochastic Control Systems*, C.T. Leondes (ed.), Control and Dynamic Systems, 73, Academic Press, New York, 1995. (Cited on p. 208)

[3] M. Abbad, J.A. Filar, and T.R. Bielecki, "Algorithms for singularly perturbed limiting average Markov control problems", *IEEE Trans. Auto. Contr.*, 37, pp. 1421–1425, 1992. (Cited on p. 243)

[4] W. Adams and P. Loustaunau, *An Introduction to Gröbner Bases*, Graduate Studies in Mathematics, 3, AMS, Providence, RI, 1994. (Cited on pp. 106, 108)

[5] A.R. Albrecht, P.G. Howlett, and C.E.M. Pearce, "Necessary and sufficient conditions for the inversion of linearly-perturbed bounded linear operators on Banach space using Laurent series", *J. Math. Anal. Appl.*, 383, pp. 95–110, 2011. (Cited on pp. 311, 312)

[6] E. Altman, K.E. Avrachenkov, and J.A. Filar, "Asymptotic linear programming and policy improvement for singularly perturbed Markov decision processes", *ZOR: Math. Meth. Oper. Res.*, 49, pp. 97–109, 1999. (Cited on pp. 149, 227, 228, 244)

[7] E. Altman and V.G. Gaitsgori, "Stability and singular perturbations in constrained Markov decision problems", *IEEE Trans. Auto. Control*, 38, pp. 971–975, 1993. (Cited on p. 244)

[8] K.E. Avrachenkov, *Analytic perturbation theory and its applications*, PhD Thesis, University of South Australia, 1999. (Cited on pp. 37, 132, 136, 208, 311, 312)

[9] K. Avrachenkov, R.S. Burachik, J.A. Filar, and V. Gaitsgory, "Constraint augmentation in pseudo-singularly perturbed linear programs", *Mathematical Programming, Ser. A*, 132, pp. 179–208, 2012. (Cited on p. 149)

[10] K. Avrachenkov, V. Ejov, and J.A. Filar, "On Newton's polygons, Grobner bases and series expansions of perturbed polynomial programs", *Banach Center Publications*, 71, pp. 29–38, 2006. (Cited on pp. 108, 150)

[11] K.E. Avrachenkov, J.A. Filar, and M. Haviv, "Singular perturbations of Markov chains and decision processes", in *Handbook of Markov Decision Processes: Methods and Applications*, E. Feinberg and A. Shwartz (eds.), Kluwer, Dordrecht, The Netherlands, 2002. (Cited on pp. 206, 208)

[12] K.E. Avrachenkov and M. Haviv, "Perturbation of null spaces with application to the eigenvalue problem and generalized inverses", *Lin. Alg. Appl.*, 369, pp. 1–25, 2003. (Cited on p. 75)

[13] K.E. Avrachenkov, M. Haviv, and P.G. Howlett, "Inversion of analytic matrix functions that are singular at the origin", *SIAM J. Matrix Anal. Appl.*, 22, pp. 1175–1189, 2001. (Cited on pp. 37, 311, 312)

[14] K.E. Avrachenkov and J.B. Lasserre, "Analytic perturbation of generalized inverses", *Lin. Alg. Appl.*, 438, pp. 1793–1813, 2013. (Cited on pp. 75, 207)

[15] K.E. Avrachenkov and J.B. Lasserre, "The fundamental matrix of singularly perturbed Markov chains", *Adv. Appl. Prob.*, 31, pp. 679–697, 1999. (Cited on p. 206)

[16] K. Avrachenkov, N. Litvak, and K.S. Pham, "Distribution of PageRank mass among principle components of the Web", in *Algorithms and Models for the Web-Graph*, Anthony Bonato and Fan R. K. Chung (eds.), Springer, Berlin, Heidelberg, pp. 16–28, 2007. (Cited on pp. 194, 195, 197, 202, 203, 205)

[17] K. Avrachenkov, N. Litvak, and K.S. Pham, "A singular perturbation approach for choosing the PageRank damping factor", *Internet Mathematics,* 5, pp. 47–69, 2009. (Cited on p. 208)

[18] H. Bart, *Meromorphic operator valued functions*, Thesis. Vrije Universiteit, Amsterdam, 1973 (Math. Center Tracts 44, Mathematical Center, Amsterdam, 1973). (Cited on pp. 37, 75)

[19] H. Bart, I. Gohberg, and M.A. Kaashoek, *Minimal Factorization of Matrix and Operator Functions*, Birkhäuser, Berlin, 1979. (Cited on p. 37)

[20] H. Bart, M.A. Kaashoek, and D.C. Lay, "Stability properties of finite meromorphic operator functions", *Nederl. Akad. Wetensch. Proc. Ser. A*, 77, pp. 217–259, 1974. (Cited on p. 75)

[21] H. Bart, M.A. Kaashoek, and D.C. Lay, "Relative inverses of meromorphic operator functions and associated holomorphic projection functions", *Math. Ann.*, 218, pp. 199–210, 1975. (Cited on p. 75)

[22] H. Baumgärtel, *Analytic Perturbation Theory for Matrices and Operators*, Birkhäuser, Basel, 1985. (Cited on pp. 4, 36, 75)

[23] A. Ben-Israel and T.N.E. Greville, *Generalized Inverses: Theory and Applications*, 2nd ed., Springer, New York, 2003. (Cited on p. 37)

[24] T.R. Bielecki and J.A. Filar, "Singularly perturbed Markov control problem: Limiting average cost", *Annals O.R.*, 28, pp. 153–168, 1991. (Cited on p. 243)

[25] T.R. Bielecki and L. Stettner, "Ergodic control of singularly perturbed Markov process in discrete time with general state and compact action spaces", *Appl. Math. Optimization*, 38, pp. 261–281, 1998. (Cited on pp. 207, 208, 243)

[26] D. Blackwell, "Discrete dynamic programming", *Ann. Math. Stat.*, 33, pp. 719–726, 1962. (Cited on pp. 205, 243)

[27] E. Bohl and P. Lancaster, "Perturbation of spectral inverses applied to a boundary layer phenomenon arising in chemical networks", *Lin. Alg. Appl.*, 180, pp. 35–59, 1993. (Cited on p. 75)

[28] J.F. Bonnans and A. Shapiro, "Optimization problems with perturbations: A guided tour", *SIAM Review*, 40, pp. 228–264, 1998. (Cited on p. 150)

[29] J.F. Bonnans and A. Shapiro, *Perturbation Analysis of Optimization Problems*, Springer Series in Operations Research and Financial Engineering, Springer, New York, 2000. (Cited on pp. 4, 150)

[30] V. S. Borkar, V. Ejov, J.A. Filar, and G.T. Nguyen, *Hamiltonian Cycle Problem and Markov Chains*, Springer, New York, 2012. (Cited on p. 244)

[31] S. Boyd and L. Vandenberghe, *Convex Optimization*, Cambridge University Press, Cambridge, UK, 2004. (Cited on p. 150)

[32] B. Buchberger, "Gröbner bases: A short introduction for systems theorists", in *Computer Aided Systems Theory - EUROCAST 2001*, LNCS, 2178, Springer, New York, pp. 1–19, 2001. (Cited on p. 108)

[33] S.L. Campbell, *Singular Systems of Differential Equations*, Research Notes in Mathematics, 40, Pitman, London, 1980. (Cited on p. 36)

[34] S.L. Campbell, *Singular Systems of Differential Equations II*, Research Notes in Mathematics, 61, Pitman, London, 1982. (Cited on p. 36)

[35] S.L. Campbell and C.D. Meyer, *Generalized Inverses of Linear Transformation*, Pitman, London, 1979. (Cited on p. 37)

[36] N. Castro-González, "Additive perturbation results for the Drazin inverse", *Lin. Alg. Appl.*, 397, pp. 279–297, 2005. (Cited on p. 75)

[37] N. Castro-González, E. Dopazo, and M.F. Martínez Serrano, "On the Drazin inverse of the sum of two operators and its application to operator matrices", *J. Math. Anal. Appl.*, 350, pp. 207–215, 2009. (Cited on p. 75)

[38] F. Chatelin, *Spectral Approximation of Linear Operators*, Academic Press, New York, 1983. (Cited on p. 75)

[39] F. Chatelin, *Eigenvalue of Matrices*, John Wiley & Sons, New York, 1993. (Cited on p. 75)

[40] M. Coderch, A.S. Willsky, S.S. Sastry, and D.A. Castanon, "Hierarchical aggregation of linear systems with multiple time scales", *IEEE Trans. Auto. Contr.*, 28, pp. 1029–1071, 1983. (Cited on p. 208)

[41] M. Coderch, A.S. Willsky, S.S. Sastry, and D.A. Castanon, "Hierarchical aggregation of singularly perturbed finite state Markov processes", *Stochastics*, 8, pp. 259–289, 1983. (Cited on p. 208)

[42] R.W. Cottle and C.E. Lemke (eds.), *Nonlinear Programming*, SIAM-AMS Proceedings, 9, AMS, Providence, RI, 1975. (Cited on p. 150)

[43] J.-M. Coulomb, J.A. Filar, and W. Szczechla, "Asymptotic Analysis of Perturbed Mathematical Programs", *J. Math. Anal. Appl.*, 251, pp. 132–156, 2000. (Cited on p. 150)

[44] R. Courant and D. Hilbert, *Methods of Mathematical Physics, Vol. 1*, Interscience Publishers, New York, 1953. (Cited on p. 312)

[45] P.J. Courtois, *Decomposability: Queueing and Computer System Applications*, Academic Press, New York, 1977. (Cited on pp. 207, 208)

[46] P.J. Courtois and G. Louchard, "Approximation of eigencharacteristics in nearly-completely decomposable stochastic systems", *Stoch. Process. Appl.*, 4, pp. 283–296, 1976. (Cited on pp. 207, 208)

[47] P.J. Courtois and P. Semel, "Bounds for the positive eigenvectors of non-negative matrices and their approximation by decomposition", *J. ACM*, 31, pp. 804–825, 1984. (Cited on pp. 207, 208)

[48] F. Delebecque, "A reduction process for perturbed Markov chain", *SIAM J. Appl. Math.*, 43, pp. 325–350, 1983. (Cited on p. 208)

[49] F. Delebecque and J.P. Quadrat, "Optimal control of Markov chains admitting strong and weak interactions", *Automatica*, 17, pp. 281–296, 1981. (Cited on pp. 207, 208, 243)

[50] C. Derman, *Finite State Markovian Decision Processes*, Academic Press, New York, 1970. (Cited on p. 243)

[51] N. Dunford and J. Schwartz, *Linear Operators, Part I: General Theory*, Wiley Classics, John Wiley and Sons, New York, 1988. (Cited on p. 312)

[52] B.C. Eaves and U.G. Rothblum, "A theory on extending algorithms for parametric problems", *Math. Oper. Res.*, 14, pp. 502–533, 1989. (Cited on p. 149)

[53] V. Ejov and J.A. Filar, "Gröbner bases in asymptotic analysis of perturbed polynomial programs", *Math. Meth. Oper. Res.*, 64, pp. 1–16, 2006. (Cited on pp. 108, 150)

[54] E.A. Feinberg, "Constrained discounted Markov decision processes and Hamiltonian cycles", *Math. Oper. Res.*, 25, pp. 130–140, 2000. (Cited on pp. 242, 244)

[55] A.V. Fiacco, *Introduction to Sensitivity and Stability Analysis in Nonlinear Programming*, Mathematics in Science and Engineering, 165, Academic Press, New York, 1983. (Cited on p. 149)

[56] A.V. Fiacco (ed.), *Mathematical Programming with Data Perturbations*, Marcel Dekker, New York, 1998. (Cited on p. 149)

[57] J.A. Filar, "Controlled Markov chains, graphs & Hamiltonicity", *Foundation and Trends in Stochastic Systems*, 1, pp. 77–162, 2006. (Cited on p. 244)

[58] J.A. Filar, E. Altman, and K.E. Avrachenkov, "An asymptotic simplex method for singularly perturbed linear programs", *Operations Research Letters*, 30, pp. 295–307, 2002. (Cited on pp. 115, 116, 149, 244)

[59] J.A. Filar, I.L. Hudson, T. Matthew, and B. Sinha, "Analytic perturbations and Systematic Bias on Statistical Modelling and Inference", *Institute of Mathematical Statistics (IMS) Collections Beyond Parametrics in Interdisciplinary Research: Festschrift in honour of Professor Pranab K. Sen. IMS Lecture Notes – Monograph Series*, 1, pp. 17–34, 2008. (Cited on pp. 36, 150)

[60] J.A. Filar and D. Krass, "Hamiltonian cycles and Markov chains", *Math. Oper. Res.*, 19, pp. 223–237, 1994. (Cited on p. 244)

[61] J.A. Filar and K. Vrieze, *Competitive Markov Decision Processes*, Springer, New York, 1997. (Cited on p. 243)

[62] V.G. Gaitsgori and A.A. Pervozvanskii, "Aggregation of states in a Markov chain with weak interactions", *Cybernetics*, 11, pp. 441–450, 1975. (Translation of Russian original in *Kibernetika*, 11, pp. 91–98, 1975.) (Cited on p. 207)

[63] V.G. Gaitsgory and A.A. Pervozvanskii, "Perturbation theory for mathematical programming problems", *JOTA*, pp. 389–410, 1986. (Cited on p. 149)

[64] T. Gal, *Postoptimal Analyses, Parametric Programming, and Related Topics*, 2nd ed., W. de Gruyter, Berlin, New York, 1995. (Cited on p. 149)

[65] T. Gal and H.J. Greenberg (eds.), *Advances in Sensitivity Analysis and Parametric Programming*, Kluwer Academic Publishers, Dordrecht, The Netherlands, 1997. (Cited on p. 149)

[66] I. Gohberg, S. Goldberg, and M.A. Kaashoek, *Classes of Linear Operators Vol. I*, Operator Theory: Advances and Applications, 49, Birkhäuser, Basel, 1990. (Cited on pp. 36, 311, 312)

[67] I. Gohberg, S. Goldberg, and M.A. Kaashoek, *Classes of Linear Operators Vol. II*, Operator Theory: Advances and Applications, 63, Birkhäuser, Basel, 1993. (Cited on pp. 36, 37)

[68] I. Gohberg, M.A. Kaashoek, and P. Lancaster, "General theory of regular matrix polynomials and band Toeplitz operators", *Integral Equations and Operator Theory*, 11, pp. 776–882, 1988. (Cited on p. 37)

[69] I. Gohberg, M. A. Kaashoek, and F. van Schagen, "On the local theory of regular analytic matrix functions", *Lin. Alg. Appl.*, 182, pp. 9–25, 1993. (Cited on pp. 36, 37, 311)

[70] I. Gohberg, P. Lancaster, and L. Rodman, *Matrix Polynomials*, Academic Press, New York, 1982. (Cited on pp. 36, 37)

[71] I. Gohberg, P. Lancaster, and L. Rodman, *Invariant Subspaces of Matrices with Applications*, 2nd ed., SIAM Classics in Applied Mathematics, 51, SIAM, Philadelphia, 2006. (Cited on pp. 36, 37, 311)

[72] I.C. Gohberg and E.I. Sigal, "An operator generalization of the logarithmic residue theorem and the theorem of Rouché", *Math. USSR Sbornik*, 13, pp. 603–625, 1971. (Cited on p. 37)

[73] R. Hassin and M. Haviv, "Mean passage times and nearly uncoupled Markov chains", *SIAM J. Disc. Math.*, 5, pp. 368–397, 1992. (Cited on pp. 206, 207, 208)

[74] M. Haviv and L. van der Heyden, "Perturbation bounds for the stationary probabilities of a finite Markov chain", *Adv. Appl. Prob.*, 16, pp. 804–818, 1984. (Cited on p. 208)

[75] M. Haviv and M.L. Puterman, "Bias optimality in controlled queueing systems", *J. Appl. Prob.*, 35, pp. 136–150, 1998. (Cited on p. 208)

[76] M. Haviv and Y. Ritov , "Series expansions for stochastic matrices", *Unpublished manuscript*, Department of Statistics, The Hebrew University, 1989. (Cited on pp. 205, 208)

[77] M. Haviv and Y. Ritov, "On series expansions and stochastic matrices", *SIAM J. Matrix Anal. Appl.*, 14, pp. 670–676, 1993. (Cited on pp. 205, 206, 207, 208)

[78] O. Hernandez-Lerma and J.B. Lasserre, *Discrete-Time Markov Control Processes: Basic Optimality Criteria*, Springer-Verlag, New York, 1996. (Cited on p. 243)

[79] E. Hewitt and K. Stromberg, *Real and Abstract Analysis*, Graduate Texts in Mathematics, 25, Springer-Verlag, New York, 1975. (Cited on p. 312)

[80] N.J. Higham, "A survey of componentwise perturbation theory in numerical linear algebra", in *Proceedings of Symposia in Applied Mathematics*, 48, W. Gautschi (ed.), AMS, Providence, RI, pp. 49–77, 1994. (Cited on p. 2)

[81] A. Hordijk, R. Dekker, and L.C.M. Kallenberg, "Sensitivity analysis in discounted Markovian decision problems", *OR Spectrum*, 7, pp. 143–151, 1985. (Cited on p. 244)

[82] A. Hordijk and L.C.M. Kallenberg, "Linear programming and Markov decision chains", *Management Science*, 25, pp. 352–362, 1979. (Cited on p. 242)

[83] R.A. Howard, *Dynamic Programming and Markov Processes*, MIT Press, Cambridge, MA, 1960. (Cited on p. 243)

[84] P.G. Howlett, "Input retrieval in finite dimentional linear systems", *J. Austral. Math. Soc. (Series B)*, 23, pp. 357–382, 1982. (Cited on pp. 36, 37, 311, 312)

[85] P.G. Howlett, A. Albrecht, and C. Pearce, "Laurent series for inversion of linearly perturbed bounded linear operators on Banach space", *J. Math. Anal. Appl.*, 366, pp. 112–123, 2010. (Cited on pp. 311, 312)

[86] P.G. Howlett and K. Avrachenkov, "Laurent series for the inversion of perturbed linear operators on Hilbert spaces", in *Optimization and Related Topics*, A. Rubinov and B. Glover (eds.), pp. 325–342, 2001. (Cited on pp. 37, 311)

[87] P.G. Howlett, K. Avrachenkov, C. Pearce, and V. Ejov, "Inversion of analytically perturbed linear operators that are singular at the origin", *J. Math. Anal. Appl.*, 353, pp. 68–84, 2009. (Cited on pp. 37, 311)

[88] P.G. Howlett, V. Ejov, and K. Avrachenkov, "Inversion of perturbed linear operators that are singular at the origin", in *Proceedings of 42nd IEEE Conference on Decision and Control*, Maui, Hawaii, pp. 5628–5631 (on CD), 2003. (Cited on p. 311)

[89] Y. Huang, "A canonical form for pencils of matrices with applications to asymptotic linear programs, *Lin. Alg. Appl.*, 234, pp. 97–123, 1996. (Cited on pp. 149, 150)

[90] Y. Huang and A.F. Veinott, Jr., "Markov branching decision chains with interest-rate-dependent rewards", *Probability in the Engineering and Information Sciences*, 9, pp. 99–121, 1995. (Cited on p. 244)

[91] J.J. Hunter, "Stationary distributions of perturbed Markov chains", *Lin. Alg. Appl.*, 82, pp. 201–214, 1986. (Cited on p. 208)

[92] J.J. Hunter, "The computation of stationary distributions of Markov chains through perturbations", *J. Appl. Math. Stoch. Anal.*, 4, pp. 29–46, 1991. (Cited on p. 208)

[93] J.J. Hunter, "A survey of generalized inverses and their use in applied probability", *Math. Chronicle*, 20, pp. 13–26, 1991. (Cited on p. 208)

[94] C.-P. Jeannerod, "On matrix perturbations with minimal leading Jordan structure", *Journal of Computational and Applied Mathematics*, 162, pp. 113–132, 2004. (Cited on p. 75)

[95] R.G. Jeroslow, "Asymptotic linear programming", *Oper. Res.*, 21, pp. 1128–1141, 1973. (Cited on pp. 149, 244)

[96] R.G. Jeroslow, "Linear programs dependent on a single parameter", *Disc. Math.*, 6, pp. 119–140, 1973. (Cited on pp. 149, 244)

[97] L.C.M. Kallenberg, *Linear Programming and Finite Markovian Control Problems*, Mathematical Centre Tracts, 148, Amsterdam, 1983. (Cited on pp. 242, 243)

[98] L.C.M. Kallenberg, "Survey of linear programming for standard and nonstandard Markovian control problems, Part I: Theory", *ZOR – Methods and Models in Operations Research*, 40, pp. 1–42, 1994. (Cited on p. 243)

[99] T. Kato, *Perturbation Theory for Linear Operators*, Springer-Verlag, Berlin, 1966. (Cited on pp. 1, 4, 36, 37, 74, 75, 208, 310, 311, 312)

[100] M.V. Keldysh, "On the characteristic values and characteristic functions of certain classes of non-self-adjoint equations", *Dokl. Akad. Nauk SSSR*, 77, pp. 11–14, 1951. (Cited on pp. 36, 311)

[101] J.G. Kemeny and J.L. Snell, *Finite Markov Chains*, Springer-Verlag, New York, 1976. (Cited on p. 205)

[102] J. Kevorkian and J.D. Cole, *Multiple Scale and Singular Perturbation Methods*, Springer, New York, 1996. (Cited on p. 4)

[103] M. Konstantinov, D. Gu, V. Mehrmann, and P. Petkov, *Perturbation Theory for Matrix Equations*, Elsevier, Amsterdam, 2003. (Cited on pp. 2, 4, 36)

[104] V.S. Korolyuk and A.F. Turbin, *Mathematical Foundations of the State Lumping of Large Systems*, Naukova Dumka, Kiev, 1978 (in Russian), translated by Kluwer Academic Publishers, Dordrecht, Boston, 1993. (Cited on pp. 36, 37, 75, 147, 206, 207)

[105] H.T. Kung and J.F. Traub, "All algebraic functions can be computed fast", *J. ACM*, 25, pp. 245–260, 1978. (Cited on pp. 107, 108)

[106] B.F. Lamond, "A generalized inverse method for asymptotic linear programming", *Math. Programming*, 43, pp. 71–86, 1989. (Cited on pp. 149, 150)

[107] B.F. Lamond, "An efficient basis update for asymptotic linear programming", *Lin. Alg. Appl.*, 184, pp. 83–102, 1993. (Cited on pp. 149, 150)

[108] P. Lancaster, "Inversion of lambda-matrices and application to the theory of linear vibrations", *Arch. Rational Mech. Anal.*, 6, pp. 105–114, 1960. (Cited on p. 36)

[109] P. Lancaster, *Lambda-Matrices and Vibrating Systems*, Pergamon Press, Oxford, New York, Paris, 1966. (Cited on pp. 36, 311)

[110] P. Lancaster and P. Psarrakos, "A Note on Weak and Strong Linearizations of Regular Matrix Polynomials", Numerical Analysis Report 470, Manchester Centre for Computational Mathematics, University of Manchester, 2005. (Cited on p. 37)

[111] C.E. Langenhop, "The Laurent expansion for a nearly singular matrix", *Lin. Alg. Appl.*, 4, pp. 329–340, 1971. (Cited on pp. 36, 37, 150)

[112] J.B. Lasserre, "A formula for singular perturbation of Markov chains", *J. Appl. Prob.*, 31, pp. 829–833, 1994. (Cited on pp. 206, 208)

[113] G. Latouche, "First passage times in nearly decomposable Markov chains", in *Numerical Solution of Markov Chains*, Pure Prob. Appl., 8, pp. 401–411, 1991. (Cited on p. 208)

[114] G. Latouche and G. Louchard, "Return times in nearly completely decomposable stochastic processes", *J. Appl. Prob.*, 15, pp. 251–267, 1978. (Cited on p. 208)

[115] V.B. Lidskii, "Perturbation theory of non-conjugate operators", *USSR Comput. Math. and Math. Phys.*, 1, pp. 73–85, 1965 (*Zh. Vychisl. Mat. i Mat. Fiz.*, 6, pp. 52–60, 1965). (Cited on p. 37)

[116] G. Louchard and G. Latouche, "Geometric bounds on iterative approximations for nearly completely decomposable Markov chains", *J. Appl. Prob.*, 27, pp. 521–529, 1990. (Cited on p. 208)

[117] D.G. Luenberger, *Optimization by Vector Space Methods*, Wiley, New York, 1979. (Cited on p. 312)

[118] D.G. Luenberger, *Linear and Nonlinear Programming*, 2nd ed., Addison-Wesley, Reading, MA, 1984. (Cited on p. 150)

[119] Y. Ma and A. Edelman, "Nongeneric perturbations of Jordan blocks", *Lin. Alg. Appl.*, 273, pp. 45–63, 1998. (Cited on p. 75)

[120] A.S. Markus, *Introduction to the Spectral Theory of Polynomial Operator Pencils*, Translations of Mathematical Monographs, AMS, Providence, RI, 1988. (Cited on p. 37)

[121] A.I. Markushevich, *Theory of Functions of a Complex Variable*, Chelsea Publishing Company, New York, 1977. (Cited on p. 147)

[122] B.L. Miller and A.F. Veinott, Jr., "Discrete dynamic programming with a small interest rate", *Ann. Math. Stat.*, 40, pp. 366–370, 1969. (Cited on p. 243)

[123] J. Moro, J.V. Burke, and M.L. Overton, "On the Lidskii–Vishik–Lyusternik perturbation theory for eigenvalues of matrices with arbitrary Jordan structure", *SIAM J. Matrix Anal. Appl.*, 18, pp. 793–817, 1997. (Cited on p. 75)

[124] I. Newton, "Methods of series and fluxions", in *The Mathematical Papers of Isaac Newton, Vol. III*, D.T. Whiteside (ed.), Cambridge University Press, Cambridge, UK, 1969. (Cited on p. 107)

[125] R.E. O'Malley, *Singular Perturbation Methods for Ordinary Differential Equations*, Springer, New York, 1991. (Cited on p. 4)

[126] A.A. Pervozvanski and V.G. Gaitsgori, *Theory of Suboptimal Decisions*, Kluwer Academic Publishers, Dordrecht, The Netherlands, 1988. (Translation from the Russian orginal: *Decomposition, aggregation and approximate optimization*, Nauka, Moscow, 1979.) (Cited on pp. 2, 3, 146, 149, 243)

[127] A.A. Pervozvanskii and I.N. Smirnov, "Stationary-state evaluation for a complex system with slowly varying couplings", *Cybernetics*, 10, pp. 603–611, 1974. (Translation of Russian original in *Kibernetika*, 10, pp. 45–51, 1974.) (Cited on p. 207)

[128] R.G. Phillips and P.V. Kokotovic, "A singular perturbation approach to modeling and control of Markov chains", *IEEE Trans. Auto. Contr.*, 26, pp. 1087–1094, 1981. (Cited on p. 208)

[129] V.A. Puiseux, "Recherches sur les fonctions algébriques", *J. Math.*, 15, pp. 365–480, 1850. (Cited on p. 107)

[130] M.L. Puterman, *Markov Decision Processes*, John Wiley & Sons, New York, 1994. (Cited on p. 243)

[131] J.P. Quadrat, "Optimal control of perturbed Markov chains: The multitime scale case", in *Singular Perturbations in Systems and Control*, M.D. Ardema (ed.), CISM Courses and Lectures, 280, Springer-Verlag, New York, 1983. (Cited on p. 208)

[132] F. Rellich, *Perturbation Theory of Eigenvalue Problems*, Gordon and Breach Science Publishers, New York, 1969. (Cited on p. 75)

[133] M. Ribaric and I. Vidav, "Analytic properties of the inverse $A^{-1}(z)$ of an analytic operator valued function $A(z)$", *Arch. Rational Mech. Anal.*, 32, pp. 298–310, 1969. (Cited on p. 36)

[134] J.R. Rohlicek and A.S. Willsky, "The reduction of Markov generators: An algorithm exposing the role of transient states", *J. ACM*, 35, pp. 675–696, 1988. (Cited on p. 207)

[135] M.K. Sain and J.L. Massey, "Invertibility of linear time invariant dynamical systems", *IEEE Trans. Auto. Contr.*, 14, pp. 141–149, 1969. (Cited on pp. 36, 37, 311, 312)

[136] S.V. Savchenko, "On the change in the spectral properties of a matrix under perturbations of sufficiently low rank", *Functional Analysis and Its Applications*, 38, pp. 69–71, 2004. (Cited on p. 75)

[137] P.J. Schweitzer, "Perturbation theory and finite Markov chains", *J. Appl. Prob.*, 5, pp. 401–413, 1968. (Cited on pp. 2, 207, 208)

[138] P.J. Schweitzer, *Perturbation Series Expansions of Nearly Completely-Decomposable Markov Chains*, Working paper 8122, Graduate School of Management, University of Rochester, August 1981. (Cited on pp. 206, 207, 208)

[139] P.J. Schweitzer, *The Laurent Expansion for a Nearly Singular Pencil*, Working Paper QM8413, Graduate School of Management, University of Rochester, 1984. (Cited on pp. 36, 37, 208)

[140] P.J. Schweitzer, "Perturbation series expansions for nearly completely-decomposable Markov chains", in *Teletrafic Analysis and Computer Performance Evaluation*, O.J. Boxma, J.W. Cohen, and H.C. Tijms (eds.), Elsevier Science Publishers B.V. (North-Holland), Amsterdam, pp. 319–328, 1986. (Cited on p. 207)

[141] P.J. Schweitzer and G.W. Stewart, "The Laurent expansion of pencils that are singular at the origin", *Lin. Alg. Appl.*, 183, pp. 237-254, 1993. (Cited on pp. 36, 37, 150, 205, 208, 311, 312)

[142] E. Seneta, "Sensitivity analysis, ergodicity coefficients, and rank-one updates for finite Markov chains", in *Numerical Solution of Markov Chains*, Workshop 1990, *Pure Prob. Appl.*, 8, pp. 121–129, 1991. (Cited on p. 208)

[143] N. Sidorov, B. Loginov, A. Sinitsyn, and M. Falaleev, *Lyapunov-Schmidt Methods in Nonlinear Analysis and Applications*, Springer, New York, 2002. (Cited on pp. 36, 37, 75)

[144] H.A. Simon and A. Ando, "Aggregation of variables in dynamic systems", *Econometrica*, 29, pp. 111–138, 1961. (Cited on p. 207)

[145] I. Singer, *Best Approximation in Normed Linear Spaces by Elements of Linear Subspaces*, Springer-Verlag, New York, 1970. (Cited on p. 312)

[146] G.W. Stewart, "On the perturbation of pseudo-inverses, projections and linear least squares problems", *SIAM Review*, 19, pp. 634–662, 1977. (Cited on p. 75)

[147] G.W. Stewart and J.-G. Sun, *Matrix Perturbation Theory*, Academic Press, San Diego, 1990. (Cited on pp. 2, 4, 36)

[148] G. Strang, *Linear Algebra and Its Applications*, 2nd ed., Academic Press, New York, 1980. (Cited on p. 37)

[149] F. Stummel, "Diskrete Konvergenz Linearer Operatoren II", *Math. Z.*, 120, pp. 231–264, 1971. (Cited on p. 311)

[150] W. Szczechla, S. Connell, J. Filar, and O. Vrieze, "On the Puiseux series expansion of the limit discount equation of stochastic games", *SIAM J. Contr. Opt.*, 35, pp. 860–875, 1997. (Cited on p. 150)

[151] M.M. Vainberg and V.A. Trenogin, *Theory of Branching of Solutions of Non-Linear Equations*, Noordhoff International Publishing, 1969. (Cited on pp. 36, 37, 75, 107)

[152] H. Vantilborgh, "Aggregation with an error of $O(\varepsilon^2)$", *J. ACM*, 32, pp. 161–190, 1985. (Cited on pp. 207, 208)

[153] A.B. Vasileva, V.F. Butuzov, and L.V. Kalachev, *The Boundary Function Method for Singular Perturbed Problems*, Studies in Applied and Numerical Mathematics, SIAM, Philadelphia, 1987. (Cited on p. 4)

[154] A.F. Veinott, Jr., "Discrete dynamic programming with sensitive discount optimality criteria", *Ann. Math. Stat.*, 40, pp. 1635–1660, 1969. (Cited on p. 243)

[155] A.F. Veinott, Jr., "Markov decision chains", in *Studies in Optimization*, G.B. Dantzig and B.C. Eaves (eds.), pp. 124–159, 1974. (Cited on p. 243)

[156] F. Verhulst, *Methods and Applications of Singular Perturbations. Boundary Layers and Multiple Timescale Dynamics*, Springer, New York, 2005. (Cited on p. 4)

[157] M.I. Vishik and L.A. Lyusternik, "The solution of some perturbation problems in the case of matrices and self-adjoint and non-self-adjoint differential equations", *Uspechi Mat. Nauk*, 15, pp. 3–80, 1960. (Cited on pp. 36, 311)

[158] R.J. Walker, *Algebraic Curves*, Princeton University Press, Princeton, NJ, 1950. (Cited on p. 107)

[159] G. Wang, Y. Wei, and S. Qiao, *Generalized Inverses: Theory and Computations*, Science Press, 2004. (Cited on pp. 37, 75)

[160] H. Whitney, *Complex Analytic Varieties*, Addison-Wesley, Reading, MA, 1972. (Cited on p. 108)

[161] J. Wilkening, "An algorithm for computing Jordan chains and inverting analytic matrix functions", *Lin. Alg. Appl.*, 427, pp. 6–25, 2007. (Cited on p. 37)

[162] G.G. Yin and Q. Zhang, *Continuous-Time Markov Chains and Applications: A Singular Perturbation Approach*, Applications of Mathematics, 37, Springer-Verlag, New York, 1998. (Cited on pp. 4, 208)

[163] K. Yosida, *Functional Analysis*, 5th ed., Springer-Verlag, New York, 1978. (Cited on pp. 310, 312)

[164] F. Zhou, "A rank criterion for the order of a pole of a matrix function", *Lin. Alg. Appl.*, 362, pp. 287–292, 2003. (Cited on p. 37)

Index